Women and Depression

Throughout the world, rates of depression are greater among females than males, and this gender gap emerges during adolescence and persists throughout adulthood. Until recently, women's health has centered on the topic of reproductive health, because research focused almost exclusively on biological and anatomical differences distinguishing men and women. Social and behavioral research on gender differences in health now employs multiple disciplinary frameworks and methodologies, and researchers seek to understand the higher rates of specific diseases and disorders in women and men. Symptoms of depression and the diagnosis of depression are more prevalent in women, and research that focuses on biological, psychological, and sociopolitical explanations for this gender gap should now be brought together to better inform efforts at treatment and prevention. *Women and Depression* is a handbook that serves to move toward a more integrative approach to women's depression in particular and mental health for all.

Corey L. M. Keyes is a sociologist and social psychologist. He received his Ph.D. in sociology from the University of Wisconsin, Madison, and has been a member of the Emory University faculty since 1997, where he holds joint appointments in the Department of Sociology and the Department of Behavioral Sciences and Health Education of the Rollins School of Public Health. He is also an adjunct faculty member in the Department of Psychology. He is a leader in the new field of positive psychology and has published a new model of complete health along with initial measurements of optimal, complete health found in the U.S. adult population.

Sherryl H. Goodman is a professor in the Department of Psychology at Emory University, where she also has an appointment in the Department of Psychiatry and Behavioral Sciences. Her research and teaching interests focus on the fields of developmental psychopathology of the family. She is a Fellow of the American Psychological Association and of its Division of Clinical Psychology. She is an associate editor of the *Journal of Abnormal Psychology* and a former associate editor of the *Journal of Family Psychology*.

Women and Depression

A Handbook for the Social, Behavioral, and Biomedical Sciences

Edited by

COREY L. M. KEYES

Emory University

SHERRYL H. GOODMAN

Emory University

CAMBRIDGE
UNIVERSITY PRESS

CAMBRIDGE UNIVERSITY PRESS
Cambridge, New York, Melbourne, Madrid, Cape Town, Singapore, São Paulo

Cambridge University Press
40 West 20th Street, New York, NY 10011-4211, USA

www.cambridge.org
Information on this title: www.cambridge.org/9780521831574

First published 2006

Printed in the United States of America

A catalog record for this publication is available from the British Library.

Library of Congress Cataloging in Publication Data

Women and depression : a handbook for the social, behavioral, and biomedical sciences /
edited by Corey L. M. Keyes, Sherryl H. Goodman.
    p.   cm.
Includes bibliographical references and index.
ISBN-13: 978-0-521-53928-9 (pbk.)
ISBN-13: 978-0-521-83157-4 (hardback)
ISBN-10: 0-521-53928-5 (pbk.)
ISBN-10: 0-521-83157-1 (hardback)
1. Depression in women – Handbooks, manuals, etc.   2. Depression in women – Social
aspects – Handbooks, manuals, etc.   I. Keyes, Corey L. M.   II. Goodman, Sherryl H.
RC537.W658   2005
616.85′27′0082 – dc22   2005010711

ISBN-13   978-0-521-83157-4 hardback
ISBN-10   0-521-83157-1 hardback

ISBN-13   978-0-521-53928-9 paperback
ISBN-10   0-521-53928-5 paperback

This project was inspired by and is dedicated to the most important women in my life – Lisa Keyes, "Nana" Keyes, and Carrie Keyes.

– C. L. M. Keyes

This book is dedicated, with love, to my husband and son, Richard and Seth Snyder.

– S. H. Goodman

Contents

Contributors

Margaret Altemus
Weill Medical College
Cornell University
Ithaca, New York

Kristen G. Anderson
Department of Psychology
University of Kentucky
Lexington, Kentucky

Jeanne Brooks-Gunn
National Center for Children
  and Families
Columbia University
New York, New York

Rachel M. Ceballos
The Pennsylvania State University
University Park, Pennsylvania

Elizabeth J. Corwin
The Pennsylvania State University
University Park, Pennsylvania

Laura M. DeRose
National Center for Children
  and Families
Columbia University
New York, New York

Kirk W. Elifson
Georgia State University
Atlanta, Georgia

Joan S. Girgus
Princeton University
Princeton, New Jersey

Sherryl H. Goodman
Emory University
Atlanta, Georgia

Jean A. Hamilton
Department of Psychology
Duke University
Durham, North Carolina
Integral Science Institute
Durham, North Carolina

Christine Heim
Department of Psychiatry and
  Behavioral Sciences
Emory University
Atlanta, Georgia

Pamela Braboy Jackson
Department of Sociology
Indiana University
Bloomington, Indiana

Ronald C. Kessler
Department of Health
  Care Policy
Harvard Medical School
Cambridge, Massachusetts

Corey L. M. Keyes
Emory University
Atlanta, Georgia

Laura Cousino Klein
The Pennsylvania State
   University
University Park, Pennsylvania

Ania Korszun
Bart's and The London Queen
   Mary's School of Medicine
   and Dentistry
London, United Kingdom

Mary Clare Lennon
Department of Sociomedical
   Sciences
Mailman School of
   Public Health
Columbia University
New York, New York

Jeanne Marecek
Swarthmore College
Swarthmore, Pennsylvania

Tamar Mendelson
University of California,
   San Francisco
San Francisco, California

Stephanie Mullins-Sweatt
Department of Psychology
University of Kentucky
Lexington, Kentucky

Ricardo F. Muñoz
University of California,
   San Francisco
San Francisco, California

Charles B. Nemeroff
Department of Psychiatry and
   Behavioral Sciences
Emory University
Atlanta, Georgia

D. Jeffrey Newport
Department of Psychiatry and
   Behavioral Sciences
Emory University
Atlanta, Georgia

Susan Nolen-Hoeksema
University of Michigan
Ann Arbor, Michigan

Brenda W. J. H. Penninx
Sticht Center on Aging
Wake Forest University
Wake Forest, North Carolina
Department of Psychiatry
Vrije Universiteit
Amsterdam, The Netherlands

Kristin M. Penza
Department of Psychiatry and
   Behavioral Sciences
Emory University
Atlanta, Georgia

Kim Ragan
Department of Psychiatry and
   Behavioral Sciences
Emory University
Atlanta, Georgia

Nancy Felipe Russo
Department of Psychology
Arizona State University
Tempe, Arizona

Shekhar Saxena
Department of Mental Health and
   Substance Dependence
World Health Organization
Geneva, Switzerland

Pratap Sharan
Department of Mental Health and
   Substance Dependence
World Health Organization
Geneva, Switzerland

Wendy Somerset
Department of Psychiatry and
    Behavioral Sciences
Emory University
Atlanta, Georgia

Claire E. Sterk
Emory University
Atlanta, Georgia

Zachary N. Stowe
Department of Psychiatry and
    Behavioral Sciences and
    Department of Gynecology
    and Obstetrics
Emory University
Atlanta, Georgia

Katherine P. Theall
Emory University
Atlanta, Georgia
Tulane University
New Orleans, Louisiana

Natalie Tolejko
Department of Psychology
University of Colorado
Boulder, Colorado

Erin Tully
Emory University
Atlanta, Georgia

Lauren M. Weinstock
Department of Psychology

University of Colorado
Boulder, Colorado

Mark A. Whisman
Department of Psychology
University of Colorado
Boulder, Colorado

Thomas A. Widiger
Department of Psychology
University of Kentucky
Lexington, Kentucky

Kay Wilhelm
University of New South Wales
Faculty of Medicine
School of Psychiatry
St Vincent's Hospital
Sydney, Australia

David R. Williams
Department of Sociology and
    Epidemiology and Institute for
    Social Research
University of Michigan
Ann Arbor, Michigan

A. Jordan Wright
National Center for Children
    and Families
Columbia University
New York, New York

Elizabeth A. Young
Mental Health Research Institute
University of Michigan
Ann Arbor, Michigan

Foreword

– Rosalynn Carter, Director of the Mental Health Program of the Carter Center and former First Lady of the United States of America

For more than 30 years I have been involved in efforts to improve the lives of those with mental illnesses. During that time, unprecedented knowledge has been gained in understanding mental health. The past decade was a particularly exciting time for the field because we achieved greater recognition of the fact that most mental illnesses are biologically based, just as other physical illnesses. We know that, although family and social conditions interact in important ways with biological functioning, mental illnesses are not the result of weak will or misguided parenting, and we have learned that it is best to use approaches for studying and understanding mental illnesses that integrate each of these constructs.

Unfortunately, although much progress has been made in recent years, depression remains one of the most common and disabling mental illnesses. According to the *World Health Report 2001*, depression was the fourth leading cause of disability for all diseases in 2000 as measured by disability-adjusted life years, or DALYs, and if current trends continue, it is estimated it will become the second leading cause of disability in 2020.

The World Health Organization also reports that the illness is more common in women than in men. These statistics are distressing. Although the gap in depression rates between women and men has been narrowing, it is tragic that so many people continue to suffer unnecessarily. *Women and Depression* is a truly valuable book because, unlike many of its predecessors, it integrates the latest information from a wide variety of international experts and fields including sociology, psychology, psychiatry, and public health to address all of the issues concerning depression. The handbook provides a comprehensive overview of the most current theories and examines how we can best use this collective knowledge to prevent and treat depression in women.

By examining the interaction among the worlds of medicine and social science, *Women and Depression* will pave the way for greater progress toward understanding, treating, and preventing depression. I believe that the more we learn about the ways biological factors interact with psychological and social conditions, the closer we come to one day being able to live in a world in which depression is no longer the chronic and debilitating disease it is today. It is my hope that the handbook will serve not only as an instructive primer on the current state of knowledge concerning women and depression, but also will provide a model for continued international and interdisciplinary research in all areas of mental illnesses.

Preface

Women's health research has historically centered on the topic of reproductive health. Until recently, research focused almost exclusively on biological and anatomical systems and their differences distinguishing men and women. Today, social and behavioral research on gender differences in health employs multiple disciplinary frameworks and multiple methodologies. Moreover, researchers are now seeking to better understand the causes and mechanisms that explain the higher rates of physical diseases and mental disorders in women and men.

Symptoms of depression and the diagnosis of depression are more prevalent in women. The evidence is now overwhelming that nondepressed individuals function better and are more productive than depressed individuals. Moreover, depression is prevalent, is often comorbid with other psychiatric disorders, recurs throughout the lifespan, is costly to treat, and generates substantial indirect costs to society in terms of lost productivity. Depression, then, has serious consequences whether it affects men or women. However, the fact that it is two to three times more likely to happen to women has become historically troubling because a greater percentage of women today than ever before are participating in the paid labor force in addition to their more traditional roles of raising children and tending to their families. When depression strikes women, it is disabling a central lynchpin in the structure of society.

Why does depression strike in the first place? Research has now identified biological, psychological, and sociopolitical explanations for this gender gap, but researchers and the journals in which they publish on the etiology of depression tend to be segregated into disciplines: psychiatry (biological), clinical psychology (cognitive or interpersonal), and sociology (sociopolitical). Science is reductive, which is at once its strength and a weakness. Reductive science reveals increasingly precise findings, but does so at the cost of breaking the phenomenon into pieces or parts. When it comes to the scientific study of a phenomenon such as depression, reductive science has broken human life down into multiple pieces: biological (genetic, neurohormonal), psychological (emotional, cognitive, behavioral), and sociological (cultural, social conditions, organizations). We are left with a science of pieces of life. Can this Humpty Dumpty be put back together again?

It is our belief that these literatures must be brought together to better inform efforts at treatment and prevention of depression as well as other mental disorders. The aims of this handbook are threefold: (1) to increase the reader's understanding of the social, psychological, and biomedical exposures that increase women's vulnerability to depression; (2) to thoroughly review etiological theories and findings of the social, psychological, and biomedical exposures that increase women's risk for depression and which serve to guide the design and implementation of primary, secondary, and policy and legislative interventions; and (3) to synthesize risk research to critically examine treatment, prevention, and social policy approaches to reducing depression in women.

Unlike other books and edited volumes on the topic of depression or mental illness and women, this handbook is multidisciplinary – including sociology, public health, psychology, and psychiatry. The primary aim of the handbook is to provide a comprehensive viewpoint on the primary question, Why are rates of depression higher in women than in men? so that we may begin to better answer the call to reducing rates of depression and preventing the onset of depression. Although *multi*disciplinary, neither the research field nor this handbook presents an *inter*disciplinary perspective on women and depression. A first step toward a future interdisciplinary field will be training graduate students with handbooks such as this one that attempt to provide each discipline's leading models, theories, and studies. When students and researchers know as much about another discipline's view of depression as their own, then perhaps a new paradigm of research that is guided by interdisciplinary questions will emerge en masse.

The distinguishing characteristics of this handbook should be its appeal to a broad spectrum of readers. With the rapid growth in the recognition, prevention, and treatment advances in women's health, this handbook should appeal to health care providers (e.g., nurses and physicians), sociologists, psychologists, health policy experts, women's studies scholars, and public health practitioners and researchers, as well as students in these fields and cognate areas. Importantly, this handbook can be readily adopted for upper-level undergraduate as well as graduate courses, and it can serve as a supplementary text for more specialized courses on mental health and illness.

– Corey Keyes and Sherryl Goodman, Atlanta, Georgia

PART I

NOSOLOGY, MEASUREMENT, AND THE EPIDEMIOLOGY OF WOMEN AND DEPRESSION

1

Depression

From Nosology to Global Burden

Kay Wilhelm

Depression is now identified as a major health issue, being a major cause of both psychological and physical morbidity. It is predicted to be second only to ischaemic heart disease in terms of total burden of disease burden to 2020 (Murray & Lopez, 1996). However, the definition of depression has undergone a number of changes over time, reflecting clinical and research fashions, evolving theories concerning psychological and biological causations of disease and the wider social context in which the problem is perceived. The changing fashions have affected how depression is viewed and how the construct is measured. This chapter will explore the evolving concepts involved in the nosology of depression, including a consideration of the implications for both genders.

MELANCHOLIA AND DEPRESSION

Melancholia has had much wider currency than *depression* through most of recorded history (Akiskal, 2000; Rush, 1986). The term referred to the presence of black bile associated with coldness, blackness, and dryness, so that melancholic individuals possessed a temperament that was essentially depressive. In a summary of Greco-Roman concepts, Akiskal (2000) noted the description of Soranus of Ephesus, in the second century BC, who characterised those with melancholia as exhibiting "mental anguish and distress, dejection, silence, animosity towards members of their household, sometimes a desire to live, and at other times a longing for death; suspicion that a plot is being hatched against him, weeping without reason, meaningless muttering, and occasional joviality."

Rush (1986) noted that Burton, in his *Anatomy of Melancholy* (1621), first differentiated grief from melancholy and recognised suicide as

I acknowledge Heather Niven and Lucy Wedgwood for editing and background research, supported by NHMRC Program Grant 222708.

a manifestation of melancholy. Kraepelin (1921) distinguished manic-depressive insanity from schizophrenia, and then Bleuler (1930) first used the term *affective disorders* to include psychoneurotic depressive reactions and involutional melancholia. Today, the characteristic features of melancholia described many centuries ago are still recognisable in current descriptions of the construct, signifying a severe and disabling form of depression with a high risk of suicide. Much of the nosological discussion of depression in the past has revolved around nonmelancholic forms of depression. Areas of particular interest have included the definition and quantification of the nature of depression in different populations and applications and how the concept differs from normal human experience.

THE CONTEXT FOR CLASSIFYING DEPRESSION

Stress, distress, and disease originally all meant much the same thing (Rees, 1976) and implied a relationship between lack of emotional well-being and external precipitants. Rees (1976) noted that in the fifteenth century, *distress* was first shortened to *stress*, whereas *disease* meant discomfort or lack of well-being. The term *depression* was first used in the 19th century by cardiologists to describe the effects of a state of lowered (cardiac) function. Since then the term depression has been used more widely than melancholia. It covers a broader range of experience than that conveyed by melancholia, but in the process, it has become less meaningful and harder to define.

There are a number of meanings for the word depression, ranging from scientific to economic to psychological (*Oxford English Dictionary*, 1989). Most commonly, in general parlance, depression is used to describe a normal human emotion. It may also be used to convey an affect (the external manifestation of mood), a predicament (a state of being or condition that is unpleasant, trying, or dangerous), a symptom (a complaint reported), a disease (a constellation of symptoms and signs implying an underlying pathological process), or an illness (a manifestation of disease in the social context).

The purpose of classifying depression relates to the context. In a clinical setting, a diagnosis is linked to clinical goals such as making a prognosis about the likely outcome and management planning. The important diagnostic distinctions from depression in this setting are grief (where there can be similar symptoms and signs, but a clear precipitant leading to the grief and no associated loss of self-esteem) and anxiety disorders (which have their own characteristic sets of diagnostic features and often occur prior to or concurrent with depressive episodes).

In a research setting, the concept of *caseness* is used to determine population trends in morbidity, risk factors for epidemiological studies, and health resource planning. The objective is to determine a threshold for inclusion of subjects who are *cases*, irrespective of whether the subject considers him or herself to be a case, from *noncases*. The definition relates to the severity and subjective impact of the disorder by determining the impact on indicators such as daily psychological and physical functioning and days absent from normal life roles. There are gender implications as women are overrepresented in clinical studies as they engage in more help-seeking than men. This highlights the importance of using data from general population studies to ascertain rates. The reasons for men not seeking help include fears of their own vulnerability and denial concerns about stigma, all related to male sex-role conditioning (Moller-Leimkuhler, 2002). All of these factors can lead to inflation of women's rates relative to men's.

DEBATE ABOUT THE NATURE OF DEPRESSION

Apart from the ongoing debate about what how depression is conceptualised (in terms of reproducible symptom clusters, duration of episodes, and the boundaries with other diagnostic categories), there is a more fundamental question about the nature of diagnostic categories. The question of whether the construct of depression is dimensional or categorical is important in determining how depression is conceptualised (whether on a continuum with normal experience or something that is qualitatively different) and how it is best measured.

Lewis (1938) proposed a dimensional or spectrum model, with depression ranging from mild (being normal mood state/adjustment reaction, then anxiety state) to severe (endogenous/melancholic depression and finally psychotic depression). Others, following on from great European clinicians such as Kraepelin and Bleuler, had taken a categorical approach. The models of depression pre–*Diagnostic and Statistical Manual*, third edition (*DSM-III*; American Psychiatric Association, 1980) and in the *Manual of the International Standard Classification of Diseases, Injuries and Causes of Death*, ninth edition (ICD 9; World Health Organisation [WHO], 1977) were predicated on the assumption of two dichotomous groups. One type, variably labelled as *endogenous, autonomous,* or *psychotic* depression, was not considered to be the result of a psychosocial precipitant, but to have a biological causation requiring biological treatments. This endogenous form of depression was seen as synonymous with the early accounts of melancholia and viewed as qualitatively different from the other type, variously named *exogenous, reactive, neurotic,* or *characterological* depression. Kiloh, Andrews, Neilson, and Bianchi (1971) summarised the position, "Psychotic or endogenous depression is a condition . . . with an imputed genetic or biochemical basis, whilst so-called neurotic depression is a diffuse entity encompassing some of the ways in which the patient utilises his defence mechanisms to cope with his own neuroticism and concurrent environmental stress."

One criticism of the binary model was that the endogenous/melancholic type was much more clearly identified as a coherent construct than the neurotic/reactive type. The latter tended to be a default category with features more closely identified with anxiety disorders than depression – the "diffuse entity" that Kiloh et al. (1971) alluded to. Another problem was that the terms *psychotic* and *endogenous* often came to be used interchangeably, with many British writers using the term *psychotic* to denote severity of depression (Carney, Roth, & Garside, 1965) rather than the presence of specific melancholic or psychotic features.

For the endogenous depressions, a further distinction (Leonhard, Korff, & Schulz, 1962) was made between bipolar disorder (subjects having a history of manic and depressive episodes) and monopolar disorder (where there is a history of only manic *or* depressive episodes), based on family history studies. Subsequently, the concept of *monopolar depression* has been retained in the current term *unipolar depression*, whereas purely manic episodes have been subsumed under the category of bipolar disorder (American Psychiatric Association, 1980; Boyd & Weissman, 1981; Spitzer, Endicott, & Robins, 1978; World Health Organisation, 1977). There are some subtle differences in phenomenology between unipolar and bipolar depression (Mitchell, Parker, Gladstone, Wilhelm, & Austin, 2001) but they are not sufficiently characteristic to render this a useful concept.

There are generally smaller gender differences in rates and experience of more biological depressions. There are minimal or absent gender differences in rates of bipolar disorder and melancholic depressions, with differences more in nonmelancholic types of major depression and more minor depressions (Jenkins, 1985), where comorbid anxiety plays a significant role (Breslau, Chilcoat, Peterson, & Schlultz, 2000). Thus, men and women seem to have similar expression of established episodes of major depression (in terms of type of symptoms and severity); the differences are more in comorbidities and coping styles (Nolen-Hoeksema, 1987, 2000; Wilhelm, Roy, Mitchell, Brownhill, & Parker, 2002).

THE CHANGING FORTUNES OF THE CONCEPT
OF NEUROTIC DEPRESSION

In the ninth edition of the ICD system, ICD-9 (World Health Organisation, 1977), neurotic depression is defined as a "neurotic disorder characterised by disproportionate depression which has usually recognizably ensued on a distressing experience.... there is often preoccupation with the psychic trauma which preceded the illness." A further category, adjustment reaction, covers "mild or transient disorders lasting longer than acute stress reactions ... often relatively circumscribed or situation-specific, generally reversible"; these may be brief, including grief reactions, or prolonged, lasting up to a few months. The term *adjustment reaction* implies an understandable reaction to a specific stressor, whereas neurotic depression stipulates a level of depression disproportionate to the presumed stressor. There is also a further category *depressive disorder, not elsewhere classified* for "states of depression, usually of moderate but occasionally of marked intensity, which have no specifically manic-depressive or other psychotic features, and which do not appear to be associated with stressful events or other features specified under neurotic depression."

There were some implicit assumptions in contrasting *neurotic* or *reactive* depressions to *endogenous* or *psychotic* depressions in terms of presumed aetiological factors. The assumption was that neurotic or reactive depressions had an underlying psychological causation (i.e., psychosocial stressors or conflicts), whereas endogenous depression arose more spontaneously and was linked to psychotic depression, two terms that came to be used synonymously (Andreasen, 1982). From the mid-1970s leading up to the publication of the third edition of *DSM* (American Psychiatric Association, 1980), these concepts were rigorously investigated.

Neurotic or reactive depression was characterised partly by the lack of features characterising endogenous depression and was associated with reactivity of mood to environmental stimuli, initial insomnia, self-pity, or doubt rather than guilt and anxiety symptoms (World Health Organisation, 1977). It was defined more by what supposedly caused it rather than by a standard set of symptoms and signs. The concept of neurosis (embedded in that of neurotic depression) implied anxiety driven by conflict of motivations and had origins in psychoanalytic thought.

Neurotic conditions were reflected in diagnostic classifications until the mid-1970s, when Akiskal, Bitarm, Puzantian, Rosenthal, and Walker (1978) criticised the concept of neurosis as covering a multiplicity of conditions, including nonpsychotic depression, mild depression, coexisting neurotic symptoms, reactive/psychogenic depression, and characterological depression. Akiskal's views were important in the construction of new constructs for the third edition of the *Diagnostic and Statistical Manual* (American Psychiatric Association, 1980). These were termed major depression, adjustment disorder, and dysthymia.

The endogenous type of depression has to a large extent been replaced by the construct of major depression. The category of major depression has a set of features that are intended to increase interrater reliability (Kendell & Jablensky, 2003). These developments have been thought to have changed the focus "from a clinically-based biopsychosocial model to a research-based medical model" (Wilson, 1993). The *DSM-III* categories were intended to be descriptive rather than presuming any specific aetiology and have continued into *DSM-IV* with only minor modifications (American Psychiatric Association, 1994a). The 10th edition of the ICD (World Health Organisation, 1992) has now derived similar constructs based on similar premises. This classification system used *depressive episode*

(DE) for a syndrome very similar to major depression and notes that it includes the diagnoses of depressive reaction, psychogenic depression, and reactive depression. The ICD-10 system grades severity on number of symptoms, so that the category of mild DE involves a lower threshold for symptoms to the DE category, with moderate DE having more, while severe DE has superimposed extra symptoms, with or without psychotic symptoms. The concept of neurosis is still present in the category of neurotic, stress-related and somatoform disorders rather than for depression.

The increased interest and confidence in description of symptoms and signs without placing them in a context suggesting aetiology has come at a price. It represents a "significant narrowing of psychiatry's clinical gaze" (Wilson, 1993) at the expense of description of the psychological inner life of depressed people and (arguably) ongoing critical evaluation of the concepts.

Some Americans, notably Winokur (1985) and Wolpe (1986), have argued that the concept of neurotic depression is useful, albeit flawed, and have recommended revisiting the concept – in the light of current thinking. Winokur (1997) conceptualised the depressions as endogenous/psychotic and "those that occur in the context of marked emotional instability," a looser category (p. 105). Wolpe (1986) saw neurotic depressions in terms of their presumed relationship to anxiety states. He proposed a number of purer subtypes, in which depression was hypothesised to be a consequence of various types of anxiety states. These included severe conditioned anxiety (based on erroneous cognitions), anxiety-based interpersonal inadequacy, and an overreaction to bereavement. Wakefield (1998) criticised the drive for a set of atheoretical *DSM* categories as a failure to differentiate between the actual disorder the disorder and its symptoms. He noted that the importance of the role that anxiety plays in the onset of neurotic disorders and the impossibility of doing this without making references of the causation of under lying psychic conflicts.

There are gender implications in that women tend to rate higher on scales reflecting neuroticism and anxiety and have higher rates of anxiety disorders (Breslau et al., 2000). The greatest gender differences in rates of depression are within the potential childbearing age bands (Kessler, McGonagle, & Swartz, 1993; Wilhelm, Mitchell, Slade, Brownhill, & Andrews, 2003), when women also have higher rates of anxiety disorders that have been thought to be contributory (Breslau et al., 2000). This effect could be mediated as an effect of gonadal hormones on limbic system hyperactivity and may carry a biological advantage at times when women needed to be alert (Parker & Brotchie, 2004).

METHODS OF RATING DEPRESSION

The method of measuring depression will also reflect whether it is seen as dimensional (in which caseness is determined by a minimum score on a dimensional scale) or categorical (i.e., certain characteristics are necessary for inclusion in the category).

In screening populations, an instrument is applied to a wide range of patients to detect those who may benefit from closer assessment, including identifying whether individuals are likely to be cases. Screening tools may include self-report measures, such as the Center for Epidemiological Studies-Depression Scale (CES-D) (Radloff, 1977), the K-10 (Kessler et al., 2002), and the Beck Depression Inventory (BDI) (Beck et al., 1961); clinician-rated tools, such as the PRIME-MD (Spitzer et al., 1994); or case finding instruments (used by clinicians or lay interviewers, of which the most widely used is the Composite International Diagnostic Interview (CIDI) (World Health Organisation, 1996). Upon identification of subjects at increased risk, assessment tools are used to assist in making a diagnosis. These may be clinician rated, such as the CORE (Parker et al., 1994) for melancholic depression, or self-report instruments, such as the BDI (Beck et al., 1961). Finally, outcome measures are used to quantify the effect of interventions and progress of patients. Self-report measures such as the BDI, and clinician-rated instruments such as the Hamilton Depression Rating Scale (HDRS) (Hamilton, 1960) and the Montgomery–Asberg Rating Scale (Montgomery & Asberg, 1979) are commonly used.

In general, women tend to score slightly higher on these types of self-report measures. This seems to reflect differences in perception of distress as well as their tendency to report a wider emotional range in both directions (Wilhelm, Parker, & Dewhurst, 1998). Only more recently has there been much emphasis on positive affect and on whether positive affective states, such as happiness and contentment, are the polar opposites of negative affective states. This approach is reflected in measures of positive and negative emotion (e.g., the Positive and Negative Affect Schedule [PANAS; Watson, Clark, & Tellegen, 1988]) and positive and negative attitudes to life (e.g., the Life Orientation Test [LOT; Scheier & Carver, 1985]).

DEPRESSION CATEGORIES IN CURRENT DIAGNOSTIC SYSTEMS

The current diagnostic systems in clinical and research use are the International Classification of Diseases (ICD), originating in Europe under the auspices of the WHO, and the *Diagnostic and Statistical Manual*, currently in its fourth edition, originating in the United States. The prevailing definition of a depressive episode equates with the construct of major depression based on operational criteria (symptoms and signs), with cut-offs for inclusion in a diagnostic category and minimal reference to presumed aetiology. The DSM system relies on a multiaxial classification to provide other important contextual information, with the diagnosis of psychiatric disorder (i.e., symptom diagnosis) located on the first axis, with four further axes describing personality, any concurrent medical conditions, predisposing life events, and level of functioning in the preceding year.

DIAGNOSTIC SYSTEMS USING OPERATIONAL CRITERIA

The development of the Present State Examination (PSE; Wing, Cooper, & Sartorius, 1974) in England included an index of definition for sub-threshold, threshold, and definite cases, which fitted with the then-current *International Classification of Disease* (World Health Organisation, 1977) categories. The Research Diagnostic Criteria (RDC; Spitzer, Endicott, & Robins, 1978), which were developed from an earlier diagnostic system (Feichner, Robins, & Guze, 1972), similarly relied on strict operationalised criteria intended to increase interrater and test-retest reliability of diagnosis for research purposes. In the RDC, primary depression was divided into major (unipolar) depression and bipolar depression (subjects who had also experienced manic episodes). Two new categories, minor depression and intermittent depressive disorder, were also included. These categories were intended to afford a broad coverage of depressive experience and to encompass the endogenous and neurotic depression categories that had been discarded.

As the RDC system was intended for use in both clinical and nonclinical situations, allocation to RDC categories also requires the imposition of functional impairment criteria (i.e., seeking professional help, taking medication for the episode, or a subjective judgement of a significant impact on life of the episode) studies The *DSM-III* criteria were intended for clinical use rather than primarily for research, with the term *disorder* intended to imply that the episode is clinically significant. The reader is referred to a paper comparing *DSM-III* and RDC criteria if more specific information is required (Williams & Spitzer, 1982).

In the 1950s, the advent of antidepressant medications brought a new focus to the discussion of depression types. The biogenic amine dysfunction theory of depression (Akiskal & McKinney, 1973) was advanced in response to the greater appreciation of the potential role for neurotransmitters, stating that depression was maintained by dysfunction of the noradrenergic transmitter system in the central nervous system. Not long after, Beck published the first book on his cognitive treatment of depression (Beck, Rush, Shaw & Emery, 1979), which was part of the vanguard for a plethora of structured psychological approaches for depression. Although it was cognitive therapy, the theory stated that depression was precipitated and maintained by negative thinking patterns, there was a presumption of a final common pathway at a cellular level, such as that suggested by Akiskal.

With the advent of the structured definitions in the later DSM and ICD systems, *clinical depression* has become synonymous with *major depression*, a depressive episode with functional impairment warranting treatment. Akiskal et al. (1978) had argued against the inclusion of the subjective experience of conflict and other personal experiences that were difficult to validate and replicate. Akiskal et al. (1980) had raised the notion of char-acterological depression and dysthymia as alternatives to more chronic depressions. Besides major depression, the *DSM* affective disorders included dysthymia and *adjustment disorder with depressed mood*. The con-struct of adjustment disorder is different from other affective disorder categories in its being the only one in which there is a link between the presumed precipitant and the disorder. Despite the reservations implicit in the two current diagnostic systems concerning neurotic depression, a recent UK general population study elected to use the Clinical Interview Schedule-R (Lewis, Pelosi, Araya & Dunn, 1992), a structured case-finding instrument developed from the PSE (Wing et al., 1974) to consider neurotic disorders (which included *depressive episode* and *mixed anxiety and depressive episode*). This illustrates the issue that studies in primary care and general population settings will generate more cases that are subthreshold or have mixed symptoms of anxiety and depression and the need to convey more information about these episodes than is apparent in the major depression category. By contrast, in secondary and tertiary settings, there are likely to be different issues, as studies will generate more depressive episodes with melancholic and psychotic features, greater severity, chronicity, and more extensive medical and psychological comorbidities (Wilhelm et al., 2002).

Thus the field has both evolved and gone full circle. Both *DSM* and ICD-10 major depression categories are based on a number of objective criteria rather than any putative causative factors. Both systems describe mild, moderate, and severe forms, fitting with the earlier unitary model, but then also describe a biological form. This then begs the question of the status of major depression, which has become the most widely used term to categorise depressive episodes. Paykel's review (1987) of the current classification systems concluded that they "serve to operationalize rather than solve the problem of classifying depression (p. 69)." Current diagnosis and treatment is based on the assumption that major depression is a single entity that varies mainly in its severity and duration, with one type of treatment being effective for all patients. Kendell and Jablensky (2003) have noted that the improvement in reliability in diagnosis "has shifted attention to the more fundamental issue of the validity" of diagnosis (p. 4). The emergence of papers on minor depression, subthreshold depression, and subsyndromal depression implies that there are a number of depression types that are not being captured in the current diagnostic rubric.

Thus, although the major depression category appears to be relatively homogeneous, it appears to consist of a number of potential subtypes (Parker, 2003), including those with anxious and irritable depressions, those who are very withdrawn, and those with atypical features (Parker, 2003; Parker et al., 2002). Specifically, an Australian group has argued for the need to differentiate between melancholic depression and nonmelan-cholic depression. Within the category of melancholic depression, which generally appears as severe major depression, the subtypes of psychotic depression and most episodes of bipolar depression are subsumed. The second category, nonmelancholic depression, is seen to encompass a number of subtypes that reflect underlying personality styles (Parker et al., 2002). Generally, these styles are all lumped together as mild or moderate major depression. Similarly for dysthymia, the construct is not as homogeneous as it first appears. It may reflect those who have a longstanding clinical depression, a characterological disorder, a sustained period of demoralisation or learned helplessness, or a partially treated major depression.

Andreason and Winokur (1979) had previously suggested that depression could be further subdivided into familial pure depressive disease, sporadic depressive disease, and depressive spectrum disease, with depressive spectrum disease being "a type of major depression characterized by families in which male relatives are alcoholic and females are depressed" (Cadoret et al., 1996). They continued to propose similar ideas to the Australian group: "Depressions, though similar clinically, are of heterogeneous etiologies," which can be separated into endogenous/psychotic groups and a group of depression spectrum diseases that "occur in the context of marked emotional instability" and are separabl on the basis of personality, clinical, follow-up, familial and treatment variables (Winokur, 1985, 1997, p. 105). Winokur's (1997) depressions caused by marked emotional instability represent another attempt to characterise these nonmelancholic depressions.

An international group (Szadoczky, Rozsa, Patten, Araro, & Furedi, 2003) weighed into this debate by using a grade of membership model to produce six subtypes of depression: severe bipolar depression with early onset (which is essentially melancholic); nonmelancholic somatisation, with late onset; nonmelancholic nonsevere bipolar depression (with male preponderance); depression secondary to anxiety (with female preponderance); melancholic depression with suicidal ideation; and melancholic depression with panic attacks (with female preponderance).

A recent review (Haslam, 2003) of the arguments for categorical and dimensional models of mental disorder noted that psychiatrists (i.e., medically trained researchers and clinicians) have favoured categorical approaches to classification, whereas most psychologists and those trained with a psychodynamic framework have generally favoured a dimensional model. He stated that taxometric studies of mood disorders have considered three issues – the continuity versus discontinuity of depression in general, of depressive subtypes, and of temperamental vulnerability. He concluded that depression cannot be regarded as a single entity and concluded that a categorical model works best for melancholic depression, where there is an identifiable disorder, whereas a dimensional model works best for neurotic depression, where depression types are less discrete and tend to "shade smoothly into everyday unhappiness (p. 698)."

This conclusion is similar to those of Parker's and Winokur's groups, but is arrived at by different means.

Thus, at the current state of knowledge, it may be best to view depression as depressions, involving both categorical and dimensional approaches. Here melancholic depression is a categorical diagnosis, whereas nonmelancholic depressions may be seen as dimensional categories, reflecting affective instability related to responses to a variety of psychosocial and biological precipitants.

SUICIDE AND DEPRESSION

Depression and suicide are related concepts but are not synonymous, as many people have episodes of depression and may or may not have suicide ideation, attempts, or completed suicide. Women generally have higher rates of deliberate self-harm, with a number of studies demonstrating a male/female ratio of 1:1.3 to 1:3. Although men have higher rates of completed suicide (male/female ratio of 4:1) at all ages, both sexes have had an increase in suicide rates in developed countries (Hawton, Fagg, Simkin, 1997; National Institute of Mental Health [NIMH], 2003; Weissman et al., 1999; Winokur, 1985, 1997). The rates vary with age, with among the highest rates being for young white men (aged 15–19 and 20–24 years) (Hawton et al., 1997; Hawton, 2005; Hawton & James, 2005).

The 16-site WHO Multicentre Study on the epidemiology of parasuicide, the first major registration study of deliberate self-harm (Bille-Brahe et al., 1997), revealed a large variation in the rates across sites from 2.6 to 542 per 100,000 population. Females at all sites were found to have higher rates of self-harm than males, with comparison data of the female-to-male ratio ranging from 1.36:1 to 1.57:1. In all countries, the highest rates for males are in the 25–34 year range and for females the 15–24 year age range.

Thus, the consistent gender differences in rates of deliberate self-harm and completed suicide are different from the reported trends for depression. Rates of completed suicide are related to depression type (higher in those with melancholic and psychotic depression), comorbidity (higher in those with chronic medical illness, drug and alcohol abuse and dependence, presence of personality disorders), and a number of social issues including poor social support. A full review of this area is beyond the scope of this chapter.

CROSS-CULTURAL PERSPECTIVES

The literature from which this chapter is drawn has been from a predominantly Western perspective. However, there is now a much greater awareness of cross-cultural perspectives (Bhugra & Matrogianni, 2004) and needs (Murray & Lopez, 1996). For example, a recent review of the prevalence of depression in the Chinese, the largest population on earth (Bhugra & Matrogianni, 2004; Parker, Gladstone, & Chee, 2001), concluded that depression appeared to be less evident and was more likely to be expressed somatically than in the West. This was due to a combination of factors: lower reporting rates due to stigma and greater acceptance of distress, differences in representation of mind–body connections, greater use of neurasthenia as a concept, and (until recently) lack of detection and identification of cases. Prior to the advent of operationalised diagnostic systems and more standardised case-finding instruments, cross-cultural studies tended to report culture-specific syndromes but these instruments have enabled cross-national studies (Cheng, 2001).

Depression is now seen as an important disorder in the global context. The greater uniformity in case-finding techniques has an emphasis on disability and economic burden rather than simply on clinical issues (Murray & Lopez, 1996). Cheng's (2001) review of the methodological issues involved in cross-national studies concludes that cultural variation is more in the presenting features than the "nature and frequency of the underlying neuropsychiatric impairments and disorders (p. 103)," and that the evidence-based approaches reported in one part of the world can be applied to benefit those in other parts of the world. Depression is described as a "highly prevalent disorder" that can "serve as a paradigm in the discussion of the impact of globalisation in the prevalence of mental disorders, idioms of distress and pathways to care" (Bhugra & Matrogianni, 2004, p. 13; Murray & Lopez, 1996). Additionally, reviewing the concepts from the cross-cultural perspective adds depth to the appreciation of how the concepts of depression had developed in the West (Murray & Lopez, 1996).

IMPLICATIONS FOR MEN AND WOMEN

Epidemiological Implications

The material presented in this chapter is building a case that the term *depression* covers a wide range of experience. It is not a simple homogeneous concept but a series of constructs subsumed under a single word. Although the gender differences are often taken as a given, it is not surprising that on closer examination, the situation is not quite so straightforward. In terms of diagnostic subtypes of affective disorders, the rates for bipolar disorder are roughly equal, although women are more likely to have a rapid cycling course. For melancholic and psychotic subtypes of major depression, there are no differences in rates, age of onset, or course.

The differences are greatest (i.e., in the range of male/female ratio of 2:1) for mild to moderate severity of major depression, minor depression, and dysthymia. Differences in rates are greatest around the ages of 18–55 years, the years of potential childbearing for women. When considering gender differences in the longer-term course of depression, the National Comorbidity Shedy (NCS) data (Kessler, 2000; Kessler et al., 1993) showed no gender differences in the probability of being chronically depressed or having an acute relapse in the year following an acute episode, but found that women had higher rates of later recurrence.

There is evidence of cross-cultural variation (Weissman et al., 1996), with lifetime rates of major depression ranging from 1.5% in Taiwan to 19.0% in Beirut (Lebanon), mean ages of onset from 25.6 years in the United States to 34.9 years in Florence (Italy), and female to male ratios ranging from 1.6:1 in Taiwan and Beirut to 3.1:1 in Germany. This raises the possibility that the gender differences in current prevalence rates are not universal, but vary with factors such as depression type, how depression is conceptualised, age, setting, and the time period under observation. The differences in rates of depression are affected by the changing fashions in diagnosis, which instruments are selected, and the time course.

Women had previously been considered to have higher rates of some neuroses; indeed, the concept of hysteria was predicated around references to female anatomy. The considerable literature around the higher rates of major depression and dysthymia, as well as anxiety disorders (Breslau, Chilcoat, Peterson & Schlultz, 2000; Weissman & Klerman, 1977), has shed some light on the area. There are few differences in the actual experience of established episodes of depression (Wilhelm et al., 2002), and the female preponderance is related to increased onsets of nonmelancholic depressions that emerge in adolescence and continue into adulthood (Kessler, 2000). This may be related to increased rates of preceding anxiety disorders (Breslau et al., 2000) as well as some factors related to female physiology and help-seeking (Nolen-Hoeksema, 1987; Weissman & Klerman, 1977). Alternatively, a series of factors may decrease the rates in men, such as masking by alcohol and substance abuse, increased rates of suicide, shorter life span, and decreased acknowledgement and help-seeking. Men may have modes of expression of depression that are not identified by the generally accepted diagnostic criteria (Brownhill, Wilhelm, Barclay, & Parker, 2002; Mitchell, Parker, Gladstone, Wilhelm, & Austin, 2003; Moller-Leimkuhler, 2002). Women have higher rates of physical and mental morbidity and help-seeking and higher rates of anxiety disorders and suicide attempts. There has also been added concern that depression may be seen as a "women's problem," a concept that can become self-reinforcing. Presentations in men may be overlooked or downplayed (Blacker & Clare, 1987; Brownhill, Wilhelm, Barelay, & Parker, 2002). However, if health is associated with strength and control (said to be manly virtues), then loss of health may mean loss of masculinity, with attendant anxiety, sadness, and fear (Moynihan, 1998). With their emphasis on reliability and validity, the issues of the effects of sex role and culture are often left unaddressed by the current diagnostic systems but deserve more emphasis in future editions. There are specific gender issues for both men and women that need to be considered, as well as those for different cultures. This has a positive side in that the variation between individuals in the social and cultural realms broadens our perspective and understanding of the concept of depressions.

Economic Implications

In the Global Burden of Disease study (Murray & Lopez, 1996), more than 80% of the healthy years lost occur between the ages of 15 and 44 years, with women losing twice as many years of healthy life to depression as men. In terms of years lost to disability (YLD) only, depression was the leading cause of burden for both men and women (6.2% and 9.8% of total YLD respectively). The report has been described as a "watershed for psychiatry" (Scott & Dickey, 2003, p. 92) as it provides an effective means of communicating the global impact of depression, while also allowing cross-national comparisons and the evaluation of the economic impact of depressions and their treatments.

More recently, methods have been developed for evaluating the burden of disease (Andrews, Sanderson, & Beard, 1998) that can be applied in general populations and across cultures. One of the best known is the use of disability adjusted life year used in the Global Burden of Disease study (Murray & Lopez, 1996), which found that depression is the second leading cause of healthy life years lost in the developed world, and fourth leading cause in the developing world, causing 5 to 6% of total global burden globally. In these studies, depression generally refers to major depression (as defined by DSM-IV and ICD-10), the most widely used and accepted concept of depression when assessing economic and social impact.

Later chapters in this book will deal with the areas of economic burden in more depth. To set the stage, a recent review found that self-reported anxiety or depression is the most important cause of workplace absenteeism in the United Kingdom and has a significant effect on workplace productivity in the United States, as well as affecting self-esteem and social networks in the workplace (Knapp, 2003). Further, a recent study of the cost of depression for adults (from 15 years) in the United Kingdom in 2000 (Thomas & Morris, 2003) calculated direct clinical costs, consumption of prescribed drugs, and indirect costs from morbidity and mortality data. They estimated the cost of depression to the nation as over £9 billion, "despite the availability of effective treatment (p. 518)." They noted that 72% of the cases were female, and 20% of cases occurred in the 35–44 year age band, with women losing over twice as many days from depression-related work absence as men. However, they also noted that depression-related death rates were 2.5 to 4.5 times higher for men in every age band.

This chapter has presented the historical background to some of the notions about depressions. Both the concepts themselves and the social context in which they are placed are constantly evolving. It is critical that researchers continue to appraise the definitions of depression types and the purposes for such definitions and to evaluate the risk factors for both men and women.

CONCLUSION

The classification of depressions reflects current understanding of neuro-biological mechanisms, current trends in psychological mechanisms, and current research needs with clinical, economic, and health planning imperatives. It is important to review the historical roots of our nosology and to observe the consistent themes, as well as the themes that may change according to fashion. The word *depressions* is used advisedly to cover a number of types, of which a categorical type (i.e., melancholia) and a number of dimensional types (various personality characteristics amplified by a number of stressors) may exist. These personality characteristics may be biological (i.e., genetic, familial, resulting from illness or injury) or psychosocial (as part of earlier or current environment) and are likely to be multifactorial.

The focus on gender differences provides a useful framework for examining various psychosocial issues related to the appearance of depression. Notwithstanding, there is evidence that when men and women have equal social opportunities (such as work, social support, and education), gender differences in rates of depression are diminished or disappear (Egeland & Hostetter, 1983; Jenkins, 1985; Weissman & Klerman, 1977; Wilhelm & Parker, 1989, 1994). Thus, it is important to continue to question observed gender differences in depression and to evaluate the risk factors of depression for both genders, while continuing to appraise the accepted definitions and conceptualisations of depression.

References

Akiskal, H. (2000). Mood disorders: Introduction and overview. In B. J. Sadock & V. A. Sadock (Eds.), *Comprehensive textbook of psychiatry*. 7th ed. (pp. 1284–1298). Philadelphia: Lippincott Williams & Wilkins.

Akiskal, H., Bitarm, A., Puzantian, V., Rosenthal, T., & Walker, P. (1978). The nosological status of neurotic depression. *Archives of General Psychiatry, 35*, 756–766.

Akiskal, H., & McKinney, W. (1973). Depressive disorders: Toward a unified hypothesis. *Science, 182*, 20.

Akiskal, H., Rosenthal, T., Haykal, R., Lemmi, H., Rosenthal, R., & Scott-Strauss, A. (1980). Characterological depressions. Clinical and sleep EEG findings separating 'subaffective dysthymias' from 'character spectrum disorders.' *Archives of General Psychiatry, 37*, 777–783.

American Psychiatric Association. (1980). *Diagnostic and statistical manual of mental disorders* (3rd ed.). Washington, DC: American Psychiatric Association.

American Psychiatric Association (1987). (3rd ed. revised) *Diagnostic and statistical manual of mental disorders*, Washington DC.

American Psychiatric Association. (1994a). *Diagnostic and statistical manual of mental disorders* (4th ed.). Washington, DC: American Psychiatric Association.

Andreasen, N. *Concepts, diagnosis and classification* (a *Handbook of Affective Disorders*, Paykel ESCED). New York: Churchill Livingstone.

Andreasen, N., & Winokur, G. (1979). Newer experimental methods of classifying depression. *Archives of General Psychiatry, 36*, 447–452.

Andrews, G., Sanderson, K., & Beard, J. (1998). Burden of disease: Methods of calculating disability from mental disorder. *British Journal of Psychiatry, 173*, 123–131.

Beck, A., Rush, A., Shaw, B., & Emery, G. (1979). *Cognitive therapy of depression*. New York: Guilford.

Beck, A., Ward, C., Mendelsohn, M., Mock, M., Erbaugh, J., & Beck, J. (1961). An inventory for measuring depression. *Archives of General Psychiatry, 4*, 561–571.

Bhugra, D., & Matrogianni, A. (2004). Globalisation and mental disorders: Overview with relation to depression. *British Journal of Psychiatry, 184*, 10–20.

Bille-Brahe, U., Kerkhof, A., De Leo, D., Schmidtke, A., Crepet, P., Lonnqvist, J., et al. (1997). A repetition-prediction study of European parasuicide populations: A summary of the first report from part II of the WHO/EURO Multicentre Study on Parasuicide in co-operation with the EC concerted action on attempted suicide. *Acta Psychiatrica Scandinavica, 95*, 81–86.

Blacker, C., & Clare, A. (1987). Depressive disorder in primary care. *British Journal of Psychiatry, 150*, 737–751.

Bleuler, E. (1930). *Lehrbuch der Psychiatrie* (5th ed.). Berlin. Verlag von Julious Springer.

Boyd, J., & Weissman, M. (1981). Epidemiology of affective disorders. *Archives of General Psychiatry, 38*, 1039–1048.

Breslau, N., Chilcoat, H., Peterson, E., & Schlultz, L. (2000). Gender differences in major depression: The role of anxiety. In E. Frank (Ed.), *Gender and its effects on psychopathology* (pp. 131–151). Washington, DC: American Psychopathological Association.

Brownhill, S., Wilhelm, K., Barclay, L., & Parker, G. (2002). Detecting depression in men: A matter of guesswork. *International Journal of Mens' Health, 1*, 259–380.

Cadoret, R., Winokur, G., Langbehn, D., Troughton, E., Yates, W., & Stewart, M. (1996). Depression Spectrum Disease, I: The role of gene–environment interaction. *American Journal of Psychiatry, 153*, 892–899.

Carney, M., Roth, M., & Garside, R. (1965). The diagnosis of depressive syndromes and the prediction of ECT response. *British Journal of Psychiatry, 111*, 659–674.

Cheng, A. (2001). Case definition and culture: Are people all the same? *British Journal of Psychiatry, 179*, 1–3.

Egeland, J. A., & Hostetter, A. M. (1983). Amish Study: I. Affective disorders among the Amish, 1976–1980. *American Journal of Psychiatry, 140*, 56–61.

Feichner, J., Robins, E., & Guze, S. (1972). Diagnostic criteria for use in psychiatric research. *Archives of General Psychiatry, 26*, 57–63.

Hamilton, M. (1960). A rating scale for depression. *Journal of Neurology, Neurosurgery, and Psychiatry, 23*, 56–62.

Haslam, N. (2003). Categorial versus dimensional models of mental disorder: The taxometric evidence. *Australian and New Zealand Journal of Psychiatry, 37,* 696–704.

Hawton, K., Fagg, J., & Simkin, S. (1997). Trends in deliberate self-harm in Oxford, 1985–1995. Implications for clinical services and the prevention of suicide. *British Journal of Psychiatry, 171,* 556–560.

Hawton, K., James, A. (2005). Suicide and deliberate self harm in young people. *British Medical Journal, 330,* 891–894.

Jenkins, R. (1985). Sex differences in minor psychiatric morbidity: A survey of a homogeneous population. *Social Science & Medicine, 20,* 887–899.

Kendell, R., & Jablensky, A. (2003). Distinguishing between the validity and utility of psychiatric diagnoses. *American Journal of Psychiatry, 160,* 4–12.

Kessler, R. (2000). Gender differences in major depression. In E. Frank (Ed.), *Gender and its effects on psychopathology* (pp. 61–85). Washington, DC: American Psychopathological Association.

Kessler, R., Andrews, G., Colpe, L., Hiripi, E., Mroczek, D., Normand, S., et al. (2002). Short screening scales to monitor population prevalences and trends in nonspecific psychological distress. *Psychological Medicine, 32,* 959–976.

Kessler, R., McGonagle, K., & Swartz, M. (1993). Sex and depression in the National Comorbidity Survey I: Lifetime prevalence, chronicity and recurrence. *Journal of Affective Disorders, 29,* 85–96.

Kiloh, L., Andrews, G., Neilson, M., & Bianchi, G. (1971). The relationship of the syndromes called endogenous and neurotic depression. *British Journal of Psychiatry, 121,* 183–196.

Klerman, G. (1999). Overview of affective disorders. In I. Kaplan, A. Freedman, & B. Sadock (Eds.), *Comprehensive textbook of psychiatry*. Baltimore: Williams and Wilkins.

Knapp, M. (2003). Hidden costs of mental illness. *British Journal of Psychiatry, 183,* 477–478.

Kraepelin, E. (1921). *Manic-depressive insanity and paranoia.* Edinburgh: Livingstone.

Leonhard, K., Korff, I., & Schulz, H. (1962). Temperament in families with monopolar and bipolar phasic psychoses. *Psychiatric Neurology, 143,* 416–434.

Lewis, A. (1938). States of depression: Their clinical and aetiological differentiation. *British Journal of Psychiatry, 2,* 875–878.

Lewis, G., Pelosi, A., Araya, R., & Dunn, G. (1992). Measuring psychiatric disorder in the community: A standardized assessment for use by lay interviewers. *Psychological Medicine, 22,* 465–486.

Mitchell, P., Parker, G., Gladstone, G., Wilhelm, K., & Austin, M.-P. (2003). Severity of stressful life events in first and subsequent episodes of depression: The relevance of depressive subtype. *Journal of Affective Disorders, 73,* 245–252.

Mitchell, P. B., Wilhelm, K., Parker, G. B., Austin, M.-P., Rutgers, P., & Mahli, G. (2001). The clinical features of bipolar depression: A comparison with matched major depressive disorder patients. *Journal of Clinical Psychiatry, 62,* 212–216.

Moller-Leimkuhler, A. (2002). Barriers to help-seeking by men: A review of sociocultural and clinical literature with particular reference to depression. *Journal of Affective Disorders, 71,* 1–9.

Montgomery, S., & Asberg, M. (1979). A new depression scale designed to be sensitive to change. *British Journal of Psychiatry, 134,* 382–389.

Moynihan, C. (1998). Theories in health care and research: Theories of masculinity. *British Medical Journal, 317,* 1072–1075.

Murray, C., & Lopez, A. (Eds.) (1996). *The global burden of disease, injuries and risk factors in 1990 and projected to 2020.* Cambridge, MA: Harvard University Press.

Nolen-Hoeksema, S. (1987). Sex differences in unipolar depression: Evidence and theory. *Psychological Bulletin, 101,* 259–282.

Nolen-Hoeksema, S. (2000). The role of rumination in depressive disorders and mixed anxiety/depressive symptoms. *Journal of Abnormal Psychology, 109,* 504–511.

Parker, G. (2003). Modern diagnostic concepts of the affective disorders. *Acta Psychiatrica Scandinavica, 418* (Suppl), 24–28.

Parker, G., & Brotchie, H. (2004) From diathesis to dimorphism: The biology of gender differences in depression. *Journal of Nervous & Mental Disease, 192,* 210–216.

Parker, G., Gladstone, G., & Chee, K. (2001). Depression in the planet's largest ethnic group: The Chinese. *American Journal of Psychiatry, 158,* 857–864.

Parker, G., Hadzi-Pavlovic, D., Wilhelm, K., Hickie, I., Brodaty, H., Boyce, P., et al. (1994). Defining melancholia: Properties of a refined sign-based system. *British Journal of Psychiatry, 164,* 316–326.

Parker, G., Roy, K., Mitchell, P., Wilhelm, K., Malhi, G., & Hadzi-Pavlovic, D. (2002). Atypical depression: A reappraisal. *American Journal of Psychiatry, 159,* 1470–1479.

Paykel, E. (1987). Melancholia. *Journal of Psychopharmacology, 1,* 67–70.

Radloff, L. (1977). The CES-D scale: A new self-report depression scale for research in the general population. *Applied Psychological Measurement, 1,* 385–401.

Rees, W. L. (1976). Stress, distress and disease. The presidential address at the annual meeting of the Royal College of Psychiatrists, held in London, July 9, 1975. *British Journal of Psychiatry, 128,* 3–18.

Rush, A. (1986). Diagnosis of affective disorders. In A. Rush & K. Z. Altshuler (Eds.), *Depression: Basic mechanisms, diagnosis, and treatment* (pp. 1–32). New York: Guilford.

Scheier, M., & Carver, C. (1985). Optimism, coping, and health: Assessment and implications of generalized outcome expectancies. *Health Psychology, 4,* 219–247.

Scott, J., & Dickey, B. (2003). Global burden of depression: the intersection of culture and medicine. *British Journal of Psychiatry, 183,* 92–94.

Spitzer, R., Endicott, J., & Robins, E. (1978). Research diagnostic criteria: Rationale and reliability. *Archives of General Psychiatry, 35,* 773–832.

Spitzer, R., Williams, J., Kroenke, K., Linzer, M., Verloin deGruy, F., Hahn, S., et al. (1994). Utility of a new procedure for diagnosing mental disorders in primary care: The PRIME-MD 1000 study. *Journal of the American Medical Association, 272,* 1749–1756.

Szadoczky, E., Rozsa, S., Patten, S., Araro, M., & Furedi, J. (2003). Lifetime patterns of depressive symptoms in the community and among primary care attenders: An application of grade of membership analysis. *Journal of Affective Disorders, 77,* 31–39.

Thomas, C. M., & Morris, S. (2003). Cost of depression among adults in England in 2000. *British Journal of Psychiatry, 183,* 514–519.

Trumble W. R., Brown, L., Stevenson, A., Siefing, J. (Ed.). *Shorter Oxford English Dictionary* (2002). 5th ed. Oxford: Oxford University Press.

Wakefield, J. C. (1998). Meaning and melancholia: Why the DSM-IV cannot (entirely) ignore the patient's intentional system. In J. W. Barron (Ed.), *Making diagnosis meaningful: Enhancing evaluation and treatment of psychological disorders* (pp. 29–72). Washington, DC: American Psychological Association.

Watson, D., Clark, L., & Tellegen, A. (1988). Development and validation of brief measures of positive and negative affect: The PANAS scales. *Journal of Personality and Social Psychology, 54*, 1063–1070.

Weissman, M., Bland, R., Canino, G., Greenwald, S., Hwu, H., Joyce, P., et al. (1999). Prevalence of suicide ideation and suicide attempts in nine countries. *Psychological Medicine, 29*, 9–17.

Weissman, M., & Klerman, G. (1977). Sex differences and the epidemiology of depression. *Archives of General Psychiatry, 34*, 98–111.

Weissman, M. M., Bland, R. C., Canino, G. J., Faravelli, C., Greenwald, S., Hwu, H. G., et al. (1996). Cross-national epidemiology of major depression and bipolar disorder. *Journal of the American Medical Association, 276*, 293–299.

Wilhelm, K., Mitchell, P., Slade, T., Brownhill, S., & Andrews, G. (2003). Prevalence and correlates of DSM-IV major depression in an Australian national survey. *Journal of Affective Disorders, 75*, 155–162.

Wilhelm, K., & Parker, G. (1989). Is sex necessarily a risk factor to depression? *Psychological Medicine, 19*, 401–413.

Wilhelm, K., & Parker, G. (1994). Sex differences in depression: Fact or artifact? *Psychological Medicine, 24*, 97–111.

Wilhelm, K., Parker G., & Dewhurst J. (1998). Examining for sex differences in the impact of anticipated and actual life events. *Journal of Affective Disorders, 48*, 37–45.

Wilhelm, K., Roy, K., Mitchell, P., Brownhill, S., & Parker, G. (2002). Gender differences in depression risk and coping factors in a clinical sample. *Acta Psychiatrica Scandinavica, 106*, 45–53.

Williams, J. B., & Spitzer, R. L. (1982). Research diagnostic criteria and DSM-III: An annotated comparison. *Archives of General Psychiatry, 39*, 1283–1289.

Wilson, M. (1993). DSM-III and the transformation of American psychiatry: A history. *American Journal of Psychiatry, 150*, 399–410.

Wing, J., Cooper, J. E., & Sartorius, N. (1974). *The description and classification of psychiatric symptoms: An instruction manual for the PSE and CATEGO system.* London: Cambridge University Press.

Winokur, G. (1985). The validity of neurotic-reactive depression: New data and reappraisal. *Archives of General Psychiatry, 42*, 1116–1122.

Winokur, G. (1997). All roads lead to depression: Clinically homogeneous, etiologically heterogeneous. *Journal of Affective Disorders, 45*, 97–108.

Wolpe, J. (1986). The positive diagnosis of neurotic depression as an etiological category. *Comprehensive Psychiatry, 27*, 449–460.

World Health Organisation. (1977). *Manual of the International Standard Classification of diseases, injuries and causes of death* (9th ed.). Geneva: Author.

World Health Organisation. (1992). *The ICD-10 classification of mental and behavioral disorders: Clinical descriptions and diagnostic guidelines.* Geneva: Author.

World Health Organisation. (1996). *Composite International Diagnostic Interview, Version 2.0.* Geneva: Author.

2

The Epidemiology of Depression among Women

Ronald C. Kessler

Nonbipolar major depression is estimated by the World Health Organization (WHO) Global Burden of Disease (GBD) Study to be the leading cause of disease-related disability among women in the world today (Murray & Lopez, 1996). This conclusion is based on epidemiological data that have documented a high prevalence of major depression among women around the world in conjunction with estimates of disease burden based on expert judgements. This chapter reviews the epidemiological research that underlies the WHO GBD estimates as well as more recent epidemiological research on women and depression. The discussion begins with a review of basic descriptive epidemiological patterns and then turns to epidemiological evidence about the cause of the higher prevalence of depression among women than men.

PREVALENCE

The fact that women have a higher prevalence of depression than do men is one of the most widely documented findings in psychiatric epidemiology, having been found throughout the world using a variety of diagnostic schemes and interview methods (Andrade et al., 2003; Nolen-Hoeksema, 1987). The prevalence of major depression among women in community epidemiological studies is typically estimated to be between one and a half and three times that of men, although there is enormous variation in the estimated total population prevalence of major depression across

Preparation of this paper was partially supported by grants R01 MH41135, R01 MH46376, R01 MH49098, R37 MH42714, K05 MH00507, and U01-MH60220 from the U.S. National Institute of Mental Health with supplemental support from the National Institute of Drug Abuse, the WT Grant Foundations (Grant 90135190), the Robert Wood Johnson Foundation (Grant 044708), and the John W. Alden Trust. Portions of this paper previously appeared in Kessler (2003) and are reproduced here with permission of the publisher.

these studies. Lifetime prevalence estimates of major depressive episode range between 6 and 17%. Twelve-month prevalence estimates range between 1 and 10%. Current prevalence estimates range between less than 1 and 6%.

In addition to studying major depression, quite a few epidemiological studies have examined gender differences in other types of depression. Chronic minor depression and dysthymia have been studied in some community surveys (e.g., Angst & Merikangas, 1997). These studies have consistently found a female to male (F:M) prevalence ratio of approximately 2:1. Lifetime prevalence estimates of dysthymia are typically in the range of 6–8% when diagnostic hierarchy rules are not made and 2–3% when they are (Kessler et al., 1994). Smaller numbers of epidemiological studies have examined gender differences in minor depression (Kessler, Zhao, Blazer, & Swartz, 1997) or brief recurrent depression (Angst & Merikangas, 1997) and consistently have found higher prevalence among women than men. It is also noteworthy that certain types of depression are gender specific, including postpartum depression, perimenopausal depression, and premenstrual dysphoric disorder. Little epidemiological evidence exists about the extent to which these disorders account for the higher prevalence of more general types of depression among women than men.

A final type of clinical depression that is very important to consider is bipolar depression. No meaningful gender difference exists in the lifetime prevalence of bipolar disorder either in epidemiological surveys (Kessler et al., 1998) or in clinical studies (Goodwin & Jamison, 1990). This means that the proportionally higher lifetime prevalence of major depressive episode among women than men increases when we exclude depressive episodes that are part of a bipolar disorder. Similarly, among men and women with bipolar disorder, the proportion of lifetime episodes that are depressive versus manic is significantly higher among women than men (Arnold, 2003). This is important because it means that a disorder exists that does not differ by gender either in lifetime prevalence or in general treatment response (Burt & Rasgon, 2004) that leads to more depressive episodes in women than men. As discussed later in the chapter, speculation exists that bipolar disorder might not be the only disorder of this type. Between 10 and 20% of all major depressive episodes among women in the general population occur as part of a bipolar disorder.

In addition to clinical depression, a great many community surveys have studied gender differences in mean levels of depressed mood. These studies consistently find a significantly higher mean level of depressed mood among female than male children (Hankin & Abramson, 1999), adolescents (Wichstrom, 1999), and adults (Lennon, 1987). These differences are not due entirely to differences in clinical depression, as daily diary studies have shown that women are also significantly more likely than men to have depressed mood in the nonclinical range (Almeida & Kessler, 1998).

PATTERNS OF ONSET AND PERSISTENCE

Male and female children appear not to differ in clinical depression even though they do differ in depressed mood. It is not until the age range 11–14 when a gender difference emerges in clinical depression (Angold, Cossello, & Worthman, 1998). This raises obvious questions about the role of sex hormones in the high prevalence of depression among women, especially as many women report changes in depressed mood associated with other experiences that cause changes in levels of sex hormones, such as menopause (Hunter, Battersby, & Whitehead, 1986), use of oral contraceptives (Cullberg, 1972), and use of hormone replacement therapy (Zweifel & O'Brien, 1997). However, systematic epidemiological studies consistently fail to find that rates of major depression are associated with any of these experiences (Bosworth et al., 2001; Yonkers, Bradshaw, & Halbreich, 2000). In addition, the limited research that has studied the effects of pregnancy on onset and recurrence of major depression has consistently failed to find significant discrepancies from the rates found in nonpregnant controls (Gotlib, Whiffen, Mount, Milne, & Cordy, 1989). Furthermore, the emergence of the gender difference in depression is not constant across all races and ethnic groups in the United States (Hayward, Gotlib, Schraedley, & Litt, 1999), further suggesting that more is at work than pure hormonal effects. The only other related situation in which rates of major depression increase substantially is during the postpartum period (Gotlib et al., 1989; Wisner, Peindl, & Hanusa, 1993). Even here, though, the cases are atypical in the sense that a much higher proportion of postpartum first onsets than other first onsets occur to women with a strong family history of depression (Sichel, 2000).

A question exists whether lifetime F:M depression prevalence ratios are the same regardless of the age of respondents. This question is motivated by the possibility suggested in recent research that the prevalence of depression in women and men might have been getting more similar in recent years (Weissman, Bland, Joyce, & Newman, 1993). Only one direct examination of this possibility across the full life course has been reported (Kessler et al., 1994). This study made use of retrospective age-of-onset reports in a nationally representative U.S. cross-sectional survey and compared reports across subsamples that differed in age at interview. Cohort curves generated in this way showed that the gender difference in depression emerged in the age range 11–15 and was consistently higher at later ages through the end of midlife in successively younger cohorts. The age range of the sample extended only to 55, making it impossible to study trends among the elderly. This similarity across cohorts is particularly striking in light of the fact that the same retrospective data suggest that there has been a roughly fivefold increase in the lifetime prevalence of depression in the United States over the four decades between the early 1950s and early 1990s.

Some theories about the reasons for gender differences in depression emphasize the importance of differential persistence. For example, sex-role theories suggest that the chronic stresses associated with traditional female roles lead to a higher prevalence of depression among women than men (Mirowsky & Ross, 1989). Rumination theory suggests that women are more likely than men to dwell on problems and, because of this, to let transient symptoms of dysphoria grow into clinically significant episodes of depression (Nolen-Hoeksema, 1990). Both of these perspectives imply that the higher point prevalence of clinically significant depression among women than among men is due, at least in part, to a higher chronicity.

Epidemiological data can be used to shed light on this prediction. Some retrospective epidemiological studies show that women do, in fact, report a more chronic course of depression than men. However, methodological studies show fairly convincingly that this is due to a differential recall bias (Ernst & Angst, 1992). In surveys in which special procedures are used to stimulate lifetime recall through active memory search, no gender difference in chronicity recurrence risk of depression is found (Kessler et al., 1994). The same studies find no evidence of gender differences either in speed of episode recovery or in chronicity of depression. These epidemiological results are consistent with a recent report from the prospective National Institute of Mental Health (NIMH) Collaborative Program on the Psychobiology of Depression, which found no gender difference in the course of depression (Mueller et al., 1999; Simpson, Nee, & Endicott, 1997). Again, though, it is important to be clear that these results exclude people with bipolar disorders, where women have more recurrent depressive episodes than men (Robb, Young, Cooke, & Joffe, 1998).

Epidemiological research on daily depression obtained from mood diaries also sheds light on the issue of gender differences in the persistence of depression. Almeida and Kessler (1998) documented in such a study that women are significantly more likely than men to have days characterized by depressed mood. However, analysis of intertemporal consistency showed that this difference was due entirely to women having a significantly higher probability than men of becoming depressed the next day after a day when they were not depressed. There was no gender difference, in comparison, in the probability of remaining depressed the next day after a day when they were depressed. This pattern is inconsistent with the view that rumination leads to an expansion of minor dysphoric episodes into episodes of clinical depression among women compared to men. Instead, the data suggest that women are either more likely than men to have endogenous short depressive episodes, to have more stress-provoked short depressive episodes than men, or both.

MEASUREMENT ERROR

Before accepting the results about gender differences in prevalence, onset, and persistence of depression at face value, it is important to note that concerns have been raised that these apparent gender differences might be due, at least in part, to self-report bias. The argument is that women may be more willing than men to admit their depression to an interviewer (Young, Fogg, Scheftner, & Keller, 1990). However, the available evidence is inconsistent with this hypothesis. A higher prevalence of current depression among women than men is found not only in studies that rely on self-report, but also in those that use informant reports (Kendler, Davis, & Kessler, 1997). In addition, a number of methodological studies using standard psychometric methods to assess potential biasing factors such as social desirability, expressivity, lying, and yea-saying or nay-saying have found no evidence that the gender difference in self-reported psychological distress is due to these biasing factors (e.g., Gove & Geerken, 1977). Finally, symptom-level assessments are inconsistent with the response-bias argument in that response bias should make women more likely than men to admit ever having a period of being sad, blue, or depressed lasting 2 weeks or longer, but should not affect reports of the less stigmatizing symptoms associated with depressed mood, such as sleep disturbance, eating disturbance, and lack of energy. The opposite is true, though, both in clinical (Young et al., 1990) and community (Kessler et al., 1994) samples.

Another methodological possibility is that men are as likely to be depressed as women, but that men are more likely than women to have other core symptoms than dysphoria or anhedonia. This possibility is akin to the situation with bipolar disorder, in which women and men are equally likely to have a lifetime history of the disorder, but the presentation of women (more depressive than manic episodes, more mixed episodes, more rapid cycling) is different from that of men. The most widely discussed possibility of this sort concerns irritability as a functional equivalent of dysphoria or anhedonia (Pasquini, Picardi, Biondi, Gaetano, & Morosini, 2004). This equivalence is recognized in current DSM-IV and ICD-10 diagnostic systems for depression among children, in which it is said that irritability often is the core symptom due to the fact that children lack insight into their conditions. It is conceivable, though, that a similar broadening of the diagnostic criteria for adults would lead to a higher increase in the number of men than women who meet criteria for major depression. A complication here, of course, is that irritability is also a core symptom of bipolar disorder, requiring great care in differential diagnosis of irritable depressive episodes versus hypo-manic episodes (Benazzi, 2004). An intriguing result relevant to this possibility is that several antidepressant medications have been found to be effective in treating pathological aggression (Fava, 1997). Nonetheless, although several clinical reports exist on the possibility of a specifically male subtype of depression characterized more by irritability–hostility–aggression than dysphoria–anhedonia (Fava, 1998; Moller-Leimkuhler, Bottlender, Strauss, & Rutz, 2004; Pasquini et al., 2004), no epidemiological studies have as yet examined the implications for studying gender differences in depression of broadening the criteria to include this wider set of core symptoms.

HISTORICAL TRENDS

Retrospective analyses of age of onset reports in epidemiological surveys suggest that the prevalence of major depression has increased dramatically over the past few decades among both women and men (Andrade et al., 2003; Cross-National Collaborative Group, 1992). Neither the population distribution of sex hormones nor the gene pool changes this quickly, which means that the influence of changing environmental conditions must account for these dramatic increases, if they are true. Before considering what such environmental conditions might be, though, it is important to recognize that controversy exists about the meaning of the cohort effect because the evidence for this effect comes almost entirely from cross-sectional surveys that use retrospective age-of-onset reports. Recall failure might increase with age and, if so, would create the false impression that the prevalence of depression had increased in recent cohorts. In addition, as depression is currently perceived as more stigmatizing in older than younger generations, reluctance to admit depression might also increase with age as a result of a cohort effect. Consistent with these possibilities, a simulation study that assumed a fairly modest level of increasing recall failure with age was able to reproduce the intercohort variation in reported lifetime prevalence of major depression quite well (Giuffra & Risch, 1994). Another possibility is that selective mortality or other forms of sample censoring due to depression might increase with age.

At least some indirect evidence is consistent with the possibility that age-related recall bias leads to an exaggeration of the cohort effect (Simon & VonKorff, 1995). However, other evidence based on several long-term prospective epidemiological surveys suggests that the cohort effect might represent a real temporal increase (Hagnell, Lanke, Rorsman, & Ojesjo, 1982; Kessler & McRae, 1981; Srole & Fisher, 1980). Additional data supporting the claim that the cohort effect is real come from specifications in cross-sectional epidemiological data. Perhaps the most intriguing of these is the finding that the cohort effect for major depression is much more pronounced for secondary than primary disorders (Kessler et al., 1996). Such a finding is unlikely to be caused by recall bias. Indeed, in light of the fact that comorbid depression is generally more severe than pure depression and the plausible assumption that severe disorders are less likely to go unreported than nonsevere disorders, we might expect the opposite pattern to have occurred if recall error explained the cohort effect. Another finding that we would not expect if recall error was the primary reason for the observed intercohort differences in the survey data is that the concentration of the cohort effect in secondary depression has been due to increases in the prevalences of other primary disorders such as anxiety and drug addiction, not to increases in the transition probabilities from these primary disorders to secondary depression.

EPIDEMIOLOGICAL EVIDENCE ABOUT
PSYCHOSOCIAL DETERMINANTS

It has been known for more than three decades that women report higher levels of depressed mood than men in community surveys and that this gender difference is stronger for married people than for the unmarried (Gove, 1972). This specification is the main empirical basis for the sex-role theory of female depression (Barnett, Biener, & Baruch, 1987). The basic claim of this theory is that women are more depressed than men because of the higher levels of stress and lower levels of fulfillment in female versus male sex roles. The specification by marital status, according to this account, is due to the fact that married women are more strongly exposed to traditional sex-role experiences than single women. However, the empirical data are inconsistent with this line of thinking. Epidemiological data show that the gender difference in first onset of major depression is the same among the married, never married, and previously married. Two other processes lead to a stronger gender difference in depression among married than unmarried people. First, depression affects marital stability differently for women than men. Second, although no gender difference is seen in the chronicity or recurrence of major depression, the environmental experiences that are associated with chronicity and recurrence are different for women and men. For example, financial pressures are more depressogenic for men than women whereas family problems are more depressogenic for women than men (Kessler & McLeod, 1984). Together, these two processes create a stronger association between gender and depression among married than unmarried people.

Most epidemiological research on risk factors for depression has focused on the predictors of episode onset of major depression without distinguishing between first episodes and recurrences. A number of consistently significant risk factors have been found, including family history, childhood adversity, various aspects of personality, social isolation, and exposure to stressful life experiences (Kendler, Kuhn, & Prescott, 2004; Klein, Durbin, Shankman, & Santiago, 2002; Monroe & Hadjiyannakis, 2002; Sanathara, Gardner, Prescott, & Kendler, 2003). It is important to note that the results regarding family history have generally shown little specificity; that is, family histories of anxiety or alcoholism or other psychiatric disorders have often been just as important as a family history of mood disorders in predicting depression (e.g., Merikangas, Risch, & Weissman, 1994). The same has been true for the effects of childhood adversity, as the intercorrelations among the many different types of adversity that predict depression are so strong that it is impossible to pinpoint any individual type of adversity as being especially important (Mullen, Martin, Anderson, Romans, & Herbison, 1996). In comparison, some specificity has been found in the effects of stressful life experiences. Stressors involving loss are strongly related to risk of depression, whereas stressors involving danger are more strongly related to risk of anxiety, and stressors involving a combination of danger and loss are most strongly related to risk of mixed anxiety-depression (Brown, Harris, & Eales, 1993).

Despite the considerable research carried out on risk factors for episode onset of depression, failure to distinguish first episodes from recurrences makes it impossible to draw firm conclusions about the importance of these risk factors. A good example of this confusion can be found in the work of Nazroo, Edwards, and Brown (1997) on the differential effects of stress on episode onset of major depression among women and men. Nazroo and colleagues argued that the gender difference in risk of past-year episodes of major depression is due in large part to a greater vulnerability to the effects of stressful life events and difficulties among women than men. However, the authors failed to distinguish first onsets from recurrences in the analysis of past-year episode onset and so confounded three separable processes: the significant gender difference in risk of first onset in the subsample of respondents who had never been depressed prior to the past year; the (presumably, based on previous research) insignificant gender difference in recurrence risk in the subsample of respondents who were asymptomatic at the beginning of the past year, but who had a subsequent recurrence of major depression; and the significant gender difference in the proportion of respondents who have an elevated risk of episode recurrence due to the fact that they have a history of depression. Exposure to stressful life experiences is higher among people with a history of depression than among those without such a history (Kessler & Magee, 1993). History of major depression, in turn, is positively related to risk of a future episode. Given that women are more likely than men to have a history of depression, these differences in exposure and conditional risk will create the impression that the higher rate of past-year episode onset among women than men is due to a combination of differential exposure and differential reactivity to stress when, in fact, it might be largely due to a higher proportion of women than men having a history of depression.

EPIDEMIOLOGICAL EVIDENCE ABOUT BIOLOGICAL DETERMINANTS

The emergence of the gender difference in depression during puberty was analyzed in a prospective epidemiological study by Angold, Costello, Erkanli, & Worthman (1999) that followed cohorts of prepubescent boys and girls through puberty using direct measures of sex hormones from blood samples. An early report from this study found that the increase in depression among girls relative to boys occurred sharply at mid-puberty (Tanner stage III) and that change in body morphology was more important than increase in age in predicting the gender difference in depression (Angold et al., 1998). A subsequent report showed that statistical control for changes in sex hormones eliminated the effects of body morphology (Angold et al., 1999). This led Angold and his colleagues to argue that the onset of the gender difference in depression at mid-puberty is due to biology rather than to societal reactions to physical maturation.

This conclusion might be premature. An intriguing specification found in earlier research is that a gender difference in low self-esteem emerges in the seventh grade when students are situated in a school system that has a seventh through ninth-grade middle school, but not until the ninth grade when students hail from a system that has a K through eighth-grade primary school with a 4-year high school (Simmons & Blyth, 1987). A related specification is that girls who mature early physically experience more psychological distress than their on-time or late-maturing age-mates. This elevated distress is exacerbated by having mixed-sex rather than same-sex friends (Ge, Conger, & Elder, 1996), suggesting that environmental stresses related to sex roles potentiate the effects of sex hormones. In addition, a large Scandinavian study found that age-related changes in gender socialization (e.g., increased concerns with weight and relationships with boys) partly explained the increase in depressed mood among girls, but not boys, at mid-puberty in Norway (Wichstrom, 1999).

Searching for a biological basis of the gender difference in depression at the genetic level is also possible. Twin research clearly shows that a strong and equivalent heritability of major depression exists among both adolescent girls and boys (Eaves et al., 1997) and adult women and men (Kendler & Prescott, 1999) and suggests that the genetic and environmental factors that predispose to depressive symptoms do not differ by gender (Agrawal, Jacobson, Gardner, Prescott, & Kendler, 2004). However, some evidence indicates that genetic factors influence increased vulnerability to depressogenic life events among postpubertal girls compared to prepubertal girls and both pre- and postpubertal boys (Silberg et al., 1999). The extent to which sex hormones and societal responses to the physical maturity of girls and boys play a part in these specifications remains unexamined up to now. The ability to study complex specifications of this sort in a subtle way is hampered by the inability of the behavioral genetic models used in twin research to identify parameter estimates in models with unrestricted interactions. This means that much more subtle analyses will become possible if candidate genes are found that influence risk of major depression.

THE ROLE OF COMORBIDITY

Breslau (1995) proposed that the gender difference in depression is partly due to prior anxiety. She supported this claim in the analysis of a community survey in which the odds-ratio (OR) of gender-predicting major depression substantially attenuates when controls are introduced for the retrospectively reported prior existence of anxiety. A similar result was subsequently reported by Wilhelm, Parker, and Hadzi-Pavlovic (1997). This finding is indirectly consistent with the result, reported above, that the cohort effect for major depression is largely confined to secondary depression. However, this finding is limited in that it focuses on a predictor that is characteristic of women (i.e., anxiety) while ignoring other comparable predictors that are more characteristic of men (e.g., substance abuse, conduct disorder). The limitations of this approach were illustrated in an analysis of data from a large national survey (Kessler, 2000), which began by replicating Breslau's and then went on to show that the F:M OR increased when another model was estimated that controlled for prior substance-use disorders and conduct disorder (more often found among men than women) rather than for prior anxiety disorders. And when a model was estimated that controlled simultaneously for prior anxiety disorders, substance use disorders, and conduct disorder, the F:M OR was found to be exactly what it was in the model that had no controls. Hettema, Prescott, & Kendler (2003) subsequently replicated this set of results. These studies show that history of prior psychiatric disorders does, in fact, mediate the observed gender difference in onset risk of major depression, but through pathways that differ for women and men. These differences well might need to be taken into consideration as our understanding of gender differences in depression becomes more complete.

CONCLUSIONS AND FUTURE DIRECTIONS

The available evidence makes it clear that depression is a problem of enormous importance among women. Given that depression often begins early in life and that the gender difference emerges with puberty, the focus of prevention has so far been on children and adolescents (Dryfoos, 1990; Hamburg, 1992). Although a great deal of risk factor research has been carried out in an effort to inform these interventions, the confounding of information about prior history with information about first onset described earlier in the paper hampers the development of evidence-based preventive interventions. More focused risk factor research carried out in conjunction with intervention planning and implementation is needed to overcome these limitations.

This future risk factor research must be based on a clearer appreciation than in the past that interactions are likely to exist between biological and social determinants. Descriptive epidemiological information is still lacking on basic patterns of these joint effects. We do not yet know, for example, whether the effects of sex hormones are most pronounced among early-maturing girls. Nor do we know whether the effects of body morphology are attenuated in subsamples of girls who are not in contact with older boys (e.g., in all-girls schools) or in societies with social norms that protect young girls from the stressful social expectations associated with physical maturity in modern western societies. The NIMH has initiated the first national survey of adolescent mental health to begin collecting data of this sort in the United States, and the World Health Organization (WHO) is coordinating parallel surveys in a number of other countries, including less developed countries in which the roles of women are very different from those in the United States. In these surveys, biological data on sex hormones and survey data on social context and mental disorders are being collected from nationally representative samples of adolescents in the age range 12–17.

In addition, the WHO is currently carrying out a series of psychiatric epidemiological surveys among adults in 28 countries designed to help answer questions about interactions between biological influences and cultural influences on female depression. These surveys, known as the WHO World Mental Health (WMH) Surveys, will provide information on cross-national variation in the relationship between gender and depression throughout the world. Early WMH results have shown that the gender difference in depression exists throughout all regions of the world (Demyttenaere et al., 2004). More detailed analyses are currently underway to investigate specifications involving effects of cross-national differences in exposure to stressor events, access to coping resources, the social position of women (e.g., rates of labor force participation, access to birth control, economic and legal rights), age at menarche (which varies substantially across countries), and interactions between age at menarche and social factors in predicting time–space variation in the relationship between gender and depression.

References

Agrawal, A., Jacobson, K. C., Gardner, C. O., Prescott, C. A., & Kendler, K. S. (2004). A population based twin study of sex differences in depressive symptoms. *Twin Research, 7,* 176–181.

Almeida, D. M., & Kessler, R. C. (1998). Everyday stressors and gender differences in daily distress. *Journal of Personality and Social Psychology, 75,* 670–680.

American Psychiatric Association. *Diagnostic and Statistical Manual of Mental Disorders, (DSM-IV),* 4th ed. Washington, DC: American Psychiatric Association; 1994.

Andrade, L., Caraveo-Anduaga, J. J., Berglund, P., Bijl, R. V., De Graaf, R., Vollebergh, W., et al. (2003). The epidemiology of major depressive episodes: Results from the International Consortium of Psychiatric Epidemiology (ICPE) Surveys. *International Journal of Methods in Psychiatric Research, 12,* 3–21.

Angold, A., Costello, E. J., Erkanli, A., & Worthman, C. W. (1999). Pubertal changes in hormone levels and depression in girls. *Psychological Medicine, 29,* 1043–1053.

Angold, A., Costello, E. J., & Worthman, C. W. (1998). Puberty and depression: The roles of age, pubertal status, and pubertal timing. *Psychological Medicine, 28,* 51–61.

Angst, J., & Merikangas, K. (1997). The depressive spectrum: Diagnostic classification and course. *Journal of Affective Disorders, 45,* 31–39.

Arnold, L. M. (2003). Gender differences in bipolar disorder. *Psychiatric Clinics of North America, 26,* 595–620.

Barnett, R. C., Biene, G. K., & Baruc, G. K. (1987). *Gender and stress.* New York: The Free Press.

Benazzi, F. (2004). Intra-episode hypomanic symptoms during major depression and their correlates. *Psychiatry and Clinical Neurosciences, 58,* 289–294.

Blazer, D. G., Kessler, R. C., McGonagle, K. A., & Swartz, M. S. (1994). The prevalence and distribution of major depression in a national community sample: The National Comorbidity Survey. *American Journal of Psychiatry, 151,* 979–986.

Bosworth, H. B., Bastian, L. A., Kuchibhatla, M. N., Steffens, D. C., McBride, C. M., Skinner, C. S., et al. (2001). Depressive symptoms, menopausal status, and climacteric symptoms in women at midlife. *Psychosomatic Medicine, 63,* 603–608.

Breslau, N. (1995). Sex differences in depression: A role for preexisting anxiety. *Psychiatry Research, 58,* 1–12.

Brown, G. W., Harris, T. O., & Eales, M. J. (1993). Aetiology of anxiety and depressive disorders in an inner-city population: 2. Comorbidity and adversity. *Psychological Medicine, 23,* 155–165.

Burt, V. K., & Rasgon, N. (2004). Special considerations in treating bipolar disorder in women. *Bipolar Disorders, 6,* 2–13.

Cross-National Collaborative Group. (1992). The changing rate of major depression: Cross-national comparisons. *Journal of the American Medical Association, 268,* 3098–3105.

Cullberg, J. (1972). Mood changes and menstrual symptoms with different gestagen/estrogen combinations. A double blind comparison with placebo. *Acta Psychiatrica Scandinavica, Supplementum, 236,* 1–86.

Demyttenaere, K., Bruffaerts, R., Posada-Villa, J., Gasquet, I., Kovess, V., Lepine, J. P., et al. (2004). Prevalence, severity, and unmet need for treatment of mental disorders in the World Health Organization World Mental Health Surveys. *Journal of the American Medical Association, 291,* 2581–2590.

Dryfoos, J. G. (1990). *Adolescents at risk: Prevalence and prevention.* New York: Oxford University Press.

Eaves, L. J., Silberg, J. L., Meyer, J. M., Maes, H. H., Simonoff, E., Pickles, A., et al. (1997). Genetics and developmental psychopathology: II. The main effects of genes and environment on behavioral problems in the Virginia Twin Study of Adolescent Behavioral Development. *Journal of Child Psychology and Psychiatry, and Allied Disciplines, 38,* 965–980.

Ernst, C., & Angst, J. (1992). The Zurich study XII. Sex difference in depression. Evidence from longitudinal epidemiological data. *European Archives of Psychiatry and Clinical Neuroscience, 241,* 222–230.

Fava, M. (1997). Psychopharmacologic treatment of pathologic aggression. *Psychiatric Clinics of North America, 20,* 427–451.

Fava, M. (1998). Depression with anger attacks. *Journal of Clinical Psychiatry, 59* Suppl 18, 18–22.

Ge, X., Conger, R. D., & Elder, G. H., Jr. (1996). Coming to age too early: Pubertal influences on girls' vulnerability to psychological distress. *Child Development, 67,* 3386–3400.

Giuffra, L. A., & Risch, N. (1994). Diminished recall and the cohort effect of major depression: A simulation study. *Psychological Medicine, 24,* 375–383.

Goodwin, F. K., & Jamison, K. J. (1990). *Manic-depressive illness.* New York: Oxford University Press.

Gotlib, I. H., Whiffen, V. E., Mount, J. H., Milne K., & Cordy N. I. (1989). Prevalence rates and demographic characteristics associated with depression in pregnancy and the postpartum. *Journal of Consulting and Clinical Psychology, 57,* 269–274.

Gove, W. R. (1972). The relationship between sex roles, marital status, and mental illness. *Social Forces, 51,* 34–44.

Gove, W. R., & Geerken, M. R. (1977). Response bias in surveys of mental health: An empirical investigation. *American Journal of Sociology, 82,* 1289–1317.

Hagnell, O., Lanke, J., Rorsman, B., & Ojesjo, L. (1982). Are we entering an age of melancholy? Depressive illness in a prospective epidemiologic study over 25 years: The Lunby Study, Sweden. *Psychological Medicine, 12,* 279–289.

Hamburg, D. A. (1992). *Today's children: Creating a future for a generation in crisis.* New York: Times Books.

Hankin, B. L., & Abramson, L. Y. (1999). Development of gender differences in depression: Description and possible explanations. *Annals of Medicine, 31,* 372–379.

Hayward, C., Gotlib, I. H., Schraedley, P. K., & Litt, I. F. (1999). Ethnic differences in the association between pubertal status and symptoms of depression in adolescent girls. *Journal of Adolescent Health, 25,* 143–149.

Hettema, J. M., Prescott, C. A., & Kendler, K. S. (2003). The effects of anxiety, substance use and conduct disorders on risk of major depressive disorder. *Psychological Medicine, 33,* 1423–1432.

Hunter, M., Battersby, R., & Whitehead, M. (1986). Relationships between psychological symptoms, somatic complaints and menopausal status. *Maturitas, 8,* 217–228.

Kendler, K. S., Davis, C. G., & Kessler, R. C. (1997). The familial aggregation of common psychiatric and substance abuse disorders in the National Comorbidity Survey: A family history study. *British Journal of Psychiatry, 170,* 541–548.

Kendler, K. S., Kuhn, J., & Prescott, C. A. (2004). The interrelationship of neuroticism, sex, and stressful life events in the prediction of episodes of major depression. *American Journal of Psychiatry, 161,* 631–636.

Kendler, K. S., & Prescott, C. A. (1999). A population-based twin study of lifetime major depression in men and women. *Archives of General Psychiatry, 56,* 39–44.

Kessler, R. C. (2000). Gender differences in major depression: Epidemiologic find-ings. In E. Frank (Ed.), *Gender and its effect on psychopathology* (pp. 61–84). Washington, DC: American Psychiatric Press.

Kessler, R. C., & Magee, W. J. (1993). Childhood adversities and adult depression: Basic patterns of association in a US National Survey. *Psychological Medicine, 23,* 679–690.

Kessler, R. C., McGonagle, K. A., Nelson, C. B., Hughes, M., Swartz, M. S., & Blazer, D. G. (1994). Sex and depression in the National Comorbidity Survey II: Cohort effects. *Journal of Affective Disorders, 30,* 15–26.

Kessler, R. C., & McLeod, J. D. (1984). Sex differences in vulnerability to undesirable life events. *American Sociological Review, 49,* 620–631.

Kessler, R. C., & McRae, J. A., Jr. (1981). Trends in the relationship between sex and psychological distress: 1957–1976. *American Sociological Review, 46,* 443–452.

Kessler, R. C., Nelson, C. B., McGonagle, K. A., Liu, J., Swartz, M. S., & Blazer, D. G. (1996). Comorbidity of DSM-III-R major depressive disorder in the general population: Results from the US National Comorbidity Survey. *British Journal of Psychiatry, 168,* 17–30.

Kessler, R. C., Wittchen, H.-U., Abelson, J. M., McGonagle, K. A., Schwarz, N., Kendler, K. S., et al. (1998). Methodological studies of the Composite Interna-tional Diagnostic Interview (CIDI) in the U.S. National Comorbidity Survey. *International Journal of Methods in Psychiatric Research, 7,* 33–55.

Kessler, R. C., Zhao, S., Blazer, D. G., & Swartz, M. S. (1997). Prevalence, correlates and course of minor depression and major depression in the NCS. *Journal of Affective Disorders, 45,* 19–30.

Klein, D. N., Durbin, C. E., Shankman, S. A., & Santiago, N.J. (2002). Depression and personality. In I. H. Gotlib & C. L. Hammen (Eds.), *Handbook of depression.* (pp. 115–140). New York: Guilford.

Lennon, M. C. (1987). Sex differences in distress: the impact of gender and work roles. *Journal of Health and Social Behavior, 28,* 290–305.

Merikangas, K. R., Risch, N.J., & Weissman, M. M. (1994). Comorbidity and co-transmission of alcoholism, anxiety and depression. *Psychological Medicine, 24,* 69–80.

Mirowsky, J., & Ross, C. E. (1989). *Social causes of psychological distress.* New York: Aldine De Gruyter.

Moller-Leimkuhler, A. M., Bottlender, R., Strauss, A., & Rutz, W. (2004). Is there evidence for a male depressive syndrome in inpatients with major depression? *Journal of Affective Disorders, 80,* 87–93.

Monroe, S. M., & Hadjiyannakis, K. (2002). The social environment and depression: focusing on severe life stress. In I. H. Gotlib & C. L. Hammen (Eds.), *Handbook of depression* (pp. 314–340). New York: Guilford.

Mueller, T. I., Leon, A. C., Keller, M. B., Solomon, D. A., Endicott, J., Coryell, W., et al. (1999). Recurrence after recovery from major depressive disorder during 15 years of observational follow-up. *American Journal of Psychiatry, 156,* 1000–1006.

Mullen, P. E., Martin, J. L., Anderson, J. C., Romans, S. E., & Herbison, G. P. (1996). The long-term impact of the physical, emotional, and sexual abuse of children: A community study. *Child Abuse & Neglect, 20,* 7–21.

Nazroo, J. Y., Edwards, A. C., & Brown, G. W. (1997). Gender differences in the onset of depression following a shared life event: A study of couples. *Psychological Medicine, 27*, 9–19.

Nolen-Hoeksema, S. (1987). Sex differences in unipolar depression: Evidence and theory. *Psychological Bulletin, 101*, 259–282.

Nolen-Hoeksema, S. (1990). Sex differences in depression. Palo Alto, CA: Stanford University Press.

Pasquini, M., Picardi, A., Biondi, M., Gaetano, P., & Morosini, P. (2004). Relevance of anger and irritability in outpatients with major depressive disorder. *Psychopathology, 37*, 155–160.

Robb, J. C., Young, L. T., Cooke, R. G., & Joffe, R. T. (1998). Gender differences in patients with bipolar disorder influence outcome in the medical outcomes survey (SF-20) subscale scores. *Journal of Affective Disorders, 49*, 189–193.

Sanathara, V. A., Gardner, C. O., Prescott C. A., & Kendler, K. S. (2003). Interpersonal dependence and major depression: Aetiological inter-relationship and gender differences. *Psychological Medicine, 33*, 927–931.

Sichel, D. (2000). Postpartum psychiatric disorders. In M. Steiner, K. A. Yonkers, & E. Eriksson (Eds.), *Mood disorders in Women* (pp. 313–328). London: Martin Dunitz.

Silberg, J., Pickles, A., Rutter, M., Hewitt, J., Simonoff, E., Maes, H., et al. (1999). The influence of genetic factors and life stress on depression among adolescent girls. *Archives of General Psychiatry, 56*, 225–232.

Simmons, R. G., & Blyth, D. A. (1987). Moving into adolescence: The impact of pubertal change and school context. New York: Aldine de Gruyter.

Simon, G. E., & VonKorff, M. (1995). Recall of psychiatric history in cross-sectional surveys: Implications for epidemiologic research. *Epidemiological Reviews, 17*, 221–227.

Simpson, H. B., Nee, J. C., & Endicott, J. (1997). First-episode major depression: Few sex differences in course. *Archives of General Psychiatry, 54*, 633–639.

Srole, L., & Fisher, A. K. (1980). The Midtown Manhattan Longitudinal Study vs 'the Mental Paradise Lost' doctrine. A controversy joined. *Archives of General Psychiatry, 37*, 209–221.

Weissman, M. M., Bland, R., Joyce, P. R., & Newman, S. (1993). Sex differences in rates of depression: Cross-national perspectives [Special Issue: Toward a new psychobiology of depression in women]. *Journal of Affective Disorders, 29*, 77–84.

Wichstrom, L. (1999). The emergence of gender difference in depressed mood during adolescence: the role of intensified gender socialization. *Developmental Psychology, 35*, 232–245.

Wilhelm, K., Parker, G., & Hadzi-Pavlovic, D. (1997). Fifteen years on: Evolving ideas in researching sex differences in depression. *Psychological Medicine, 27*, 875–883.

Wisner K. L., Peindl, K., & Hanusa, B. H. (1993). Relationship of psychiatric illness to childbearing status: A hospital-based epidemiologic study. *Journal of Affective Disorders, 28*, 39–50.

World Health Organization. *The ICD-10 Classification of Mental and Behavioral Disorders: Diagnostic Criteria for Research*. Geneva, Switzerland: World Health Organization; 1993.

Yonkers, K. A., Bradshaw, K. D., & Halbreich, U. (2000). Oestrogens, progestins and mood. In M. Steiner, K. A. Yonkers, & E. Eriksson (Eds.), *Mood disorders in women* (pp. 207–232). London: Martin Dunitz.

Young, M. A., Fogg, L. F., Scheftner, W. A., Keller, M. B., & Fawcett, J. A. (1990). Sex differences in the lifetime prevalence of depression. *Journal of Affective Disorders, 18*, 187–192.

Zweifel, J. E., & O'Brien, W. H. (1997). A meta-analysis of the effect of hormone replacement therapy upon depressed mood. *Psychoneuroendocrinology, 22*, 189–212.

PART II

BIOLOGICAL, DEVELOPMENTAL, AND AGING MODELS OF RISK

3

The Biological Underpinnings of Depression

Ania Korszun, Margaret Altemus,
and Elizabeth A. Young

INTRODUCTION

Stress and Depression

Depressive disorders are widely regarded as stress-related conditions. Although genetic vulnerability is critical to the development of depression, in the absence of environmental stressors, the incidence of depressive disorders is very low (Kendler et al., 1995), and in approximately 75% of cases of depression there is a precipitating life event (Brown & Harris, 1978; Frank, Anderson, Reynolds, Ritenour, & Kupfer, 1994). Living organisms survive by maintaining a complex dynamic equilibrium or homeostasis that is constantly challenged by intrinsic or extrinsic stressors. These stressors set in motion responses aimed at preserving homeostasis, including activation of a wide variety of neurotransmitters and neuromodulators. The hypothalamic pituitary adrenal (HPA) axis is the body's main stress hormonal system. Corticotropin releasing hormone (CRH) is the principal central effector of the stress response (Chrousos & Gold, 1992). CRH triggers the release of adrenocorticotropic hormone (ACTH) from the anterior pituitary corticotrope, which, in turn, triggers the release of adrenal glucocorticoids. The stress response is terminated by glucocorticoid feedback at brain and pituitary sites.

Depression has been conceptualized as maladaptive, exaggerated responses to stress. Abnormalities of the HPA axis, as manifested by hypercortisolemia and disruption of the circadian rhythm of cortisol secretion, are well established phenomena in depression (Carroll, Curtis, & Mendels, 1976; Sachar et al., 1973). The assumption that the pathophysiology of depression involves exaggerated responses to stress is supported by evidence that CRH is activated in these patients (Altemus et al., 1992; Bhagwagar, Hafizi, & Cowen, 2003; Butler & Nemeroff, 1990; Charney, Woods, Goodman, & Heninger, 1987; Raadsheer, Hoogendijk,

Stam, Tilders, & Swaab, 1994; Southwick, Krystal, & Morgan, 1993; Young, Altemus, Parkison, & Shastry, 2001a).

Sex Differences in Depression

As well as their association with stress, another striking feature of mood disorders is the greater prevalence of these conditions in women. Several lines of evidence indicate that sex steroid hormones play a role in the increased vulnerability of women to anxiety disorders and depression. Women have a higher incidence of unipolar depression (Weissman & Olfson, 1995), which arises at puberty. The immediate postpartum period, in particular, is a time of greatly increased risk for new onset or recurrence of mood disorders (Altshuler, Hendrick, & Cohen, 1998; Dean, Williams, & Brockington, 1989). In some women, recurrent depressive symptoms occur only during the premenstrual period and in others, chronic depression is often exacerbated premenstrually (Rubinow & Roy-Byrne, 1984). Reports indicate that estrogen may be an effective treatment for postpartum (Gregoire, Kumar, Everitt, Henderson, & Studd, 1996) and perimenopausal depression (Zweifel & O'Brien, 1997).

HPA Axis Abnormalities in Depression

Overactivity of the HPA axis, as manifested by an increase in cortisol secretion, is a well-established phenomenon in depression (Carroll et al., 1976; Sachar et al., 1973). The original studies of Sachar and colleagues (1973) demonstrated increased cortisol secretory activity in depressed patients as measured by mean plasma cortisol concentration, the number of cortisol secretory episodes, and the number of minutes of active secretion. Later studies have continued to validate this hypercortisolemia in depression (Carroll et al., 1976; Halbreich, Asnis, Schindledecker, Zurnoff, & Nathan, 1985; Pfohl, Sherman, Schlecte, & Stone, 1985; Rubin, Poland, Lesser, Winston, & Blodgett, 1987). As many as two thirds of endogenously depressed patients fail to suppress cortisol, or show an early escape of cortisol, following overnight administration of 1 mg of dexamethasone, using a cortisol cutoff of 5 μg/dl to define *escape* (Carroll et al., 1981). Although nonsuppression of cortisol to dexamethasone is strongly associated with endogenous depression, this finding is less robust in outpatients with depression. Although both hypercortisolemia and feedback abnormalities to dexamethasone are present in depressed patients, they do not necessarily occur in the same individuals (Carroll et al., 1981; Halbreich et al., 1985). Abnormal glucocorticoid fast feedback (Young, Haskett, Watson, & Akil, 1991) and a blunted ACTH response to oCRF have also been reported in depressed patients (Gold et al., 1986; Holsboer, Bardeleden, Gerken, Stalla, & Muller, 1984; Young et al., 1990). The blunted response to oCRF appears to be dependent upon increased baseline cortisol, since blockade of cortisol production with metyrapone normalizes the ACTH response (von Bardeleben, Stalla, Mueller, & Holsboer, 1988; Young, Akil, Haskett, & Watson, 1995). It was expected that the increased cortisol would be accompanied by an increased level of ACTH in plasma, but this expectation has been difficult to validate. Some studies (Linkowski et al., 1985; Pfohl et al., 1985; Young, Carlson, & Brown, 2001b) have been able to demonstrate small differences between normal controls and depressed subjects in their mean 24-hour plasma ACTH levels. The demonstration of enhanced sensitivity to ACTH 1–24 in depressed patients suggests that increased ACTH secretion is not necessarily the cause of increased cortisol secretion (Amsterdam, Winokur, Abelman, Lucki, & Richels, 1983). However, other studies using very low threshold doses of ACTH 1–24 have not been able to demonstrate an increased sensitivity to ACTH in depressed patients (Krishnan, Ritchie, Saunders, Nemeroff, & Carroll, 1990), which suggests that increased cortisol secretion is secondary to increased ACTH secretion. Our 24-hour studies of ACTH and cortisol secretion demonstrated that subjects with increased mean cortisol also demonstrated increased mean ACTH, supporting a central origin of the HPA axis overactivity (Young, Carlson et al., 2001b). Finally, our studies with metyrapone in major depression support the presence of increased central nervous system (CNS) drive, at least in the evening (Young et al., 1994; Young, Lopez, Murphy-Weinberg, Watson, & Akil, 1997). It appears likely that there is increased corticotropin-releasing factor (CRF)/ACTH secretion, which is then probably amplified at the adrenal level leading to increased cortisol. These changes in cortisol secretion are commonly considered to be state changes and resolve when the depression resolves. However, almost all studies examining the HPA axis in major depression in euthymic subjects have examined patients on tricyclic antidepressants, which exert direct effects on the HPA axis. Two recent studies of ours, as well as a recent report by Bhagwagar and colleagues, have found that saliva cortisol is increased in subjects with *lifetime* major depression in subjects who had no current mood symptoms (Bhagwagar et al., 2003; Young, Aggen, Prescott, & Kendler, 2000; Young & Breslau, 2004). The overall picture in depression is of increased production of activational elements of the HPA axis together with a reduction in feedback inhibition.

Chronic antidepressant treatment inhibits stress responsive systems at multiple sites, through reductions in CRH and tyrosine hydroxylase activity, enhanced glucocorticoid receptor activity (Barden, Reul, & Holsboer, 1995; Brady, Whitfield, Fox, Gold, & Herkenham, 1991), and down regulation of arousal producing ß-adrenergic receptors (Heninger & Charney, 1987). Chronic treatment with antidepressant agents also reduces both behavioral and endocrine responses to stress (Barden et al., 1995, Murua & Molina, 1992; Reul, Stec, Soder, & Holsboer, 1993). Consequently, antidepressant actions may be through stress systems, as well as classical neurotransmitter systems that affect neurobiological systems involved in the pathophysiology of depression.

SEX DIFFERENCES IN HPA AXIS REGULATION IN DEPRESSION

Morning and Evening Cortisol Hypersecretion

We studied baseline cortisol secretion in the morning in 16 depressed patients and 16 age- and sex-matched control patients and found predictably increased cortisol secretion in the group as a whole (Young et al., 1991). However, there were also clear sex differences: male patients and their matched controls demonstrated the same plasma cortisol concentration, whereas female depressed patients demonstrated a significantly higher mean plasma cortisol concentration (11.3 ± 0.9 µg/dl) than that of the matched control group (8.1 ± 0.95 µg/dl; significant by a two-tailed t-test, $p = 0.033$). Removal of glucocorticoid negative feedback by metyrapone demonstrated increased central drive in depressed patients in the evening (Young et al., 1994). The response to metyrapone also showed sex differences, in that only the female depressed patients manifested rebound β-LPH/β-end secretion in comparison to their matched controls (ANOVA, $F = 8.8$, $df = 1$, $p = .01$), whereas the males did not.

Dexamethasone Nonsuppression and Menopause

We also examined the effect of loss of gonadal steroids at menopause on HPA axis regulation in depressed women (Young et al., 1993). We conducted studies using a protocol examining baseline and postdexamethasone secretion of β-LPH/β-end and cortisol over the course of the day (8 AM–4 PM). These were carried out on 51 depressed women, 36 of whom were premenopausal and 15 postmenopausal. The premenopausal women demonstrated a significantly lower incidence of pituitary (β-LPH/β-end) nonsuppression (44%; $n = 36$) than the postmenopausal women (nonsuppressor $= 81\%$; $n = 15$). To determine which of a number of potential variables were associated with β-LPH/β-end nonsuppression in women, a stepwise regression analysis was used. Independent variables included age, menopausal status, baseline β-LPH/β-end and cortisol, severity of depression (Hamilton Depression rating scores), and the number of previous episodes of depression. The dependent variable was β-LPH/β-end nonsuppression. We found that age had a significant effect on pituitary nonsuppression, but when age and menopausal status were compared, menopausal status showed a stronger correlation and combined with cortisol gave a correlation coefficient of 0.817. This suggests that menopausal status, in conjunction with cortisol hypersecretion, is a critical variable in the development of HPA dysregulation, as manifested by resistance to dexamethasone, and accounts for 65% of the variance.

In summary, depressed women demonstrate greater HPA axis activation than depressed men. Furthermore, this activation is linked to insensitivity to dexamethasone and thus appears to reflect the development of glucocorticoid receptor down-regulation following a period of hypercortisolemia. Menopause is not associated with increases in plasma cortisol concentrations in depressed women, but it is associated with an increase in dexamethasone resistance.

EFFECTS OF GONADOSTEROIDS ON THE HPA AXIS

Animal Studies

Studies in rodents support the existence of sex differences in several of the elements of the HPA axis. Female rats appear to have a more robust HPA axis response to stress than do male rats, and there is evidence that estrogen is at least partly responsible for this sexual dimorphism. For example, compared with male rats, female rats have a faster onset of corticosterone secretion after stress and a faster rate of rise of corticosterone (Jones, Brush, & Neame, 1972). The increased corticosterone response is accompanied by a greatly increased ACTH response to stress in female rodents (Young, 1996). Furthermore, corticosteroid binding globulin (CBG) is positively regulated by estrogen and thus higher in female rats; however, estrogen and progesterone have been demonstrated to affect the HPA axis independent of the effects of CBG (Young, 1996). In addition, chronic estrogen treatment of ovariectomized female rats enhances their corticosterone response to stress and slows their recovery from stress (Burgess & Handa 1992). Studies by Viau and Meaney (1991) demonstrate a greater ACTH and corticosterone stress response in acute estradiol treated rats compared with ovariectomized female rats, or with estradiol plus progesterone treated female rats, after short-term (24 hr) but not long-term (48 hr) estradiol treatment. However, our studies (Young, Altemus, Parkison, & Shastry, 2001) found that two centrally active estradiol antagonists, tamoxifen and CI-628, led to a greater stress response in intact female rats. Also, estradiol replacement of ovariectomized rats led to decreased stress responses. Similar inhibitory effects of estradiol on stress response have been found in sheep and humans (Komesaroff, Esler, Clarke, Fullerton, & Funder, 1998; Komesaroff, Esler, & Sudhir, 1999). Studies showing inhibitory effects of estradiol on the stress response have used lower doses of estradiol than those showing activation of the HPA axis.

Work by Keller-Wood and colleagues (1988) in pregnant ewes and ewes given progesterone infusions demonstrate that progesterone can diminish the effectiveness of cortisol feedback on stress responsiveness in vivo.

In addition, progesterone demonstrates antiglucocorticoid effects on feedback in intact rats in vivo and in vitro (Duncan & Duncan, 1979; Svec, 1988). Progesterone binds to the glucocorticoid receptor; although it does so with a faster binding time than glucocorticoid itself, progesterone binding is to a different site on the receptor than glucocorticoid binding (Svec, 1988). Progesterone can also increase the rate of dissociation of glucocorticoids from the glucocorticoid receptor (Rousseau, Baxter, & Tomkins, 1972). In addition, binding studies with expressed human mineralocorticoid receptor (MR) have demonstrated an affinity of progesterone for MR receptor in a range similar to that of dexamethasone (Arriza et al., 1987). Furthermore, there was an increase in MR binding following progesterone treatment of female rats (Carey, Deterd, de Koning, Helmerhorst, & Dekloet, 1995). Finally, female rats have a greater number of glucocorticoid receptors in the hippocampus than male rats (Turner & Weaver, 1985), and progesterone modulates immunoreactive glucocorticoid receptor (GR) distribution in the hippocampus of rats (Ahima, Lawson, Osei, & Harlan, 1992). It should be noted that binding studies do not distinguish agonist effects from antagonist effects; therefore even increases in number could result from antagonist effects at GR.

Human Studies

Until recently, the lack of a reliable stress test limited studies on sex differences in stress response in humans. In the Trier Social Stress Test, subjects undergo a mock job interview in front of a panel of interviewers who are instructed not to provide any verbal or nonverbal feedback; it is a reliable and robust stressor in normal subjects (Kirschbaum, Pirke, & Hellhammer, 1995). It has now been shown that oral contraceptives decrease the free cortisol response to a social stressor in women (Kirschbaum et al., 1995), whereas the treatment of normal men with estradiol for 48 hours results in an enhanced ACTH and cortisol response to a social stressor (Kirschbaum et al., 1996). These data of estrogen treatment in men are consistent with results of studies in rats (Burgess & Handa, 1992; Viau & Meaney, 1991). However, results from studies of oral contraceptives are harder to interpret because they are synthetic steroids given at constant doses for a prolonged period of time and may differ from endogenous steroids in their effects. Direct comparison of the ACTH response to this social stressor in men and women has demonstrated a smaller ACTH response in women but a similar cortisol response in both sexes (Kirschbaum, Kudielka, Gaab, Schommer, & Hellhammer, 1999; Young, Abelson, & Cameron, 2004). These data are in agreement with studies demonstrating that estradiol decreases stress responsiveness in women (Komesaroff et al., 1999)

With respect to the influence of changes in ovarian hormones across the menstrual cycle in women, recent studies by Altemus and colleagues

(1997) have found increased resistance to dexamethasone suppression during the luteal phase of the menstrual cycle, compared to during the follicular phase, a change that may again be related to either increased estradiol or progesterone during the luteal phase. In addition, ACTH, vasopressin, and cortisol responses to stress are enhanced in the luteal phase compared to the follicular phase of the menstrual cycle (Altemus et al., 1997; Galliven, et al., 1997), suggesting that decreases in glucocorticoid receptors may explain the decreased response to dexamethasone. In a design that allowed investigators to distinguish the effects of progesterone from those of estrogen, Roca and colleagues (1998a, 1998b, 2003) studied control women first treated with Lupron, a gonadotrophin-releasing hormone (GnRH) agonist, which causes suppression of both estrogen and progesterone secretion, and then given sequential replacement of the two hormones. They examined the response to exercise stress as well as to dexamethasone feedback and found that the exercise stress response was increased. Response to dexamethasone feedback was decreased during the progesterone add-back phase but not during the estrogen add-back phase. Again, these data suggest that progesterone acts as a glucocorticoid antagonist. Thus, data from human studies suggest that ovarian steroids may modulate stress responses in opposite directions so that estradiol decreases stress response and progesterone decreases sensitivity to negative feedback.

GENETIC VULNERABILITY TO DEPRESSION

While stress is clearly a precipitating factor in the onset of depression, not everyone who experiences stress develops a depressive episode, suggesting an underlying vulnerability (Kendler, 1998). It has long been known that depression clusters in families and extensive evidence has now accumulated that depression is a complex disorder resulting from the interaction of genetic and environmental factors (Jones, Kent, & Croddock, 2002). Twin studies of unipolar depression have shown variable degrees of heritability depending on the type of sample and the definitions of depression used. In a recent meta-analysis of twin data, Sullivan (Sullivan, Neale, & Kerdler, 2000) estimated the heritability of depression to be 37% with a significant contribution of environmental events specific to individuals but only a small contribution from shared familial environmental effects. However, in studies of subjects recruited from psychiatric treatment centers, rather than community samples, the heritability was much higher (60–70%) (Kendler et al., 1995; McGuffin et al., 1996), indicating a possibly greater heritability of more severe forms of depression.

Twin studies have examined the genetic risk factors for major depression separately in men and women with conflicting findings. Two studies have suggested that there is equal heritability for depression in the two sexes (Kendler, Pedersen, et al., 1995; McGuffin, Katz, Watkins, & Rutherford,

1996). However, in studies in which a broader definition of depression was used, genetic factors accounted for a significantly greater proportion of the liability to develop depression in women compared with men (Bierut et al., 1999; Lyons et al., 1998). Although the genes that influence risk for depression in the two sexes are correlated, they are probably not entirely the same and it is possible that genes conferring risk for depression differ in men and women (Kendler, Gardner, Neale, & Prescott, 2001). Kendler goes on to suggest that environmental factors may bring out distinct genetic variation in the two sexes or, alternatively, that biological factors, such as hormonal variations during the menstrual cycle and pregnancy, could elicit discrete genetic variation in women and men. Genetic influences may also operate through temperamental dimensions such as neuroticism and through the effect of different developmental influences such as the onset of puberty marking an increased vulnerability to depression resulting from negative life events in girls (Rutter, 2002).

NEUROANATOMICAL BASIS OF DEPRESSION

Postmortem studies of depressed patients have often involved patients who die by suicide, thus conflating depression with suicide, which only occurs in a small minority of patients and may involve comorbid disorders, particularly alcohol and substance abuse. Furthermore, because of the nature of postmortem studies, a priori hypotheses are usually tested rather than new hypotheses about functional systems and their relationship to depression. Postmortem data have confirmed low serotonin in the cerebrospinal fluid of suicide victims, as well as down-regulation of 5HT1a terminal receptors in cortical regions (Asberg, Traskman, & Thoren, 1976; Mann & Arango, 1990). They have also demonstrated increased CRH and vasopressin in the hypothalamus and down-regulation of CRH receptor binding in the frontal cortex (Raadsheer et al., 1994). Structural magnetic resonance imaging studies have found smaller hippocampi, often taken to indicate hypercortisolism associated with depression (Sheline, Wang, Gado, Csernansky, & Vannier; Sheline, Gado, & Kraemer, 2003). Functional imaging gives us another opportunity to understand the networks involved in depression. However, abnormalities found on functional imaging may not necessarily identify the anatomical lesion of depression but may also demonstrate compensatory coping mechanisms.

The majority of functional imaging studies have identified abnormalities in the prefrontal cortex, particularly dorsolateral and ventrolateral areas, as the primary finding in depression (Baxter et al., 1989; Buchsbaum et al., 1986; Drevets et al., 1992; George, Ketter, & Post, 1994; Ketter, George, Kimbrell, & Post, 1996; Mayberg, 1994; Mayberg, Lewis, Regenold, & Wagner, 1994). Studies of emotion in normal subjects find these same key areas activated in response to sad mood (George et al., 1995). Other areas identified in some studies include limbic areas such as amygdala, temporal cortex, and insula but there is less consensus for these findings. Mayberg has outlined a functional circuit in depression involving brainstem nuclei, striatum, and thalamus (Mayberg et al., 2000). Treatment data suggest that pharmacological therapies target the brainstem, limbic, and subcortical areas in addition to frontal cortical areas, whereas psychotherapies target the frontal cortical abnormalities (Mayberg et al., 2000, 2002). The issue of sex differences has not yet been addressed in either postmortem or in vivo imaging studies.

SEROTONIN AND DEPRESSION

A fundamental hypothesis of the etiology of depressive disorders is that these disorders may be due to a relative deficiency of serotonin. Major depression is accompanied by down-regulation of 5HT1A receptors and mRNA, in brain and lymphocytes, as seen in postmortem studies (Arango et al., 2001; López, Vazquez, Chalmers, Akil, & Watson, 1997) or in vivo imaging (Drevets et al., 1999; 2000) as well as increased 5HT2a binding (Arango et al., 1990; Hrdina, Demeter, Vu, Sotonyi, & Palkovits, 1993; Yates et al., 1990). Furthermore, these two serotonin receptors are targets for antidepressant action, particularly tricyclic antidepressants. Rodent studies have shown that chronic antidepressant administration results in functional up-regulation of the postsynaptic 5-HT1a receptor in the hippocampus (Blier & de Montigny, 1994). Some studies have also reported a modest *increase* in 5-HT1a receptor number in the hippocampus following antidepressant administration to rodents (Klimek, Zak-Knapik, & Mackowiak, 1994; Welner, DeMontigny, Desroches, Desjardines, & Suranyi-Cadotte, 1989). Studies have reported decreases in 5-HT2a binding in the prefrontal cortex after chronic antidepressant administration (Peroutka & Shyder, 1980). These findings have led some investigators to propose that postsynaptic 5-HT1a and 5-HT2a receptors have functionally opposing effects (Schreiber & De Vry, 1993), that a disturbed balance of these receptors may be contributing to the pathophysiology of depression (Berendsen, 1995), and that restoration of this balance is necessary for effective antidepressant action (Borsini, 1994).

Both stress and glucocorticoids modulate serotonin transmission. Acute stress levels of glucocorticoids increase serotonin turnover and increase the responsiveness of hippocampal neurons to 5HT1A receptor stimulation (McEwen, 1995; Meijer & de Kloet, 1998). When elevated levels of glucocorticoids persist, such as following chronic social stress, down-regulation of hippocampal 5HT1A receptors occur while 5HT2 receptors in the cerebral cortex are up-regulated (McEwen, 1995). In addition, 5HT2C receptors are increased following corticosterone adrenectomy and normalize following corticosterone replacement (Meijer & de Kloet, 1998). Animal studies demonstrate that chronic treatment with high doses of glucocorticoids lead to decreased serotonin receptor mediated responses, similar to the picture observed in depressed patients, although the exact mechanism of this hypofunctional serotonin state is unclear (Meijer & de Kloet, 1998). This hypofunctional serotonin state may have further consequences for glucocorticoid secretion, since serotonin appears to be an important regulator of glucocorticoid feedback. Antidepressants that increase serotonin cause increases in glucocorticoid receptor number and can reverse the increased glucocorticoid secretion seen in depressed humans and in transgenic mice who have been genetically altered to demonstrate reduced glucocorticoid receptors and increased glucocorticoid secretion (partial GR knockout) (Barden et al, 1995). Lesions of the serotonergic input to the hippocampus, an important site in inhibiting glucocorticoid secretion, do produce decreased glucocorticoid receptor expression and increased glucocorticoid secretion (Seckl & Fink, 1991).

SEX DIFFERENCES IN SEROTONIN SYSTEMS

Gonadal steroids appear to modulate mood, at least in part, through effects on serotonergic systems. Overall, the literature suggests that estrogen enhances the efficiency of serotonergic neurotransmission. Basic science studies indicate that there are clear sex differences in brain serotonin systems, some of which may depend upon estrogen and others upon testosterone. The serotonin content and uptake in multiple areas of the forebrain, hypothalamus, and limbic system are higher in females than males (Borisova, Proshlyakova, Sapronova, & Ugrumov, 1996; Carlsson & Carlsson, 1988; Haleem, Kennett, & Curzon, 1990). 5HT1C and 5HT2 receptor binding in the dentate gyrus or CA4 regions of the hippocampus is similar in males and females in rats; however, the 5HT1A receptor binding in the CA1 region of the hippocampus is higher in female rats and ovariectomy had no effect on this sex difference (Mendelson & McEwen, 1991). Stress has been shown to cause greater increases in serotonin in female rats in multiple areas of the brain (Heinsbroek et al., 1990). More recent studies have found that estradiol increased serotonin transporter binding in female rat brains (McQueen, Wilson, & Fink, 1997), as well as stimulating an increase in 5HT2A binding sites in the limbic cortex (Fink, Sumner, Rosie, Grace, & Quinn, 1996). Limited evidence suggests that like estrogen, progesterone may up-regulate 5HT2 receptors (Biegon, Reches, Snyder, & McEwen, 1983) and increase serotonin content (Pecins-Thompson, Brown, Kohama, & Bethea, 1996). However, no consistent effects of progesterone administration on serotonin function have been identified. There also is evidence that testosterone has opposing effects to estrogen on serotonergic activity. Reductions in androgenic steroid have been associated with enhancement of central serotonergic activity (Bonson, Johnson, Fiorella, Robin, & Winter,

1994; Fishette, Biegon, & McEwen, 1984; Matsuda, Nakano, Kanda, & Iwata, 1991). In addition, administration of testosterone has been associated with reductions in central serotonergic activity (Martinez-Conde, Leret, & Diaz, 1985; Mendelson & McEwen, 1990).

Studies of sex differences in serotonin systems in humans are more limited. Sex differences in the prolactin and cortisol responses to serotonin agonists have been reported, with women showing greater response to the serotonergic challenges (Gelfin, Lerer, Lesch, Gorfine, & Allolio, 1995; Lerer et al., 1996; Monteleone, Catapano, Tortorella, & Maj, 1997; Ryan et al., 1992). However, estrogen regulates prolactin synthesis as well as cortisol secretion, so greater responses in women do not necessarily indicate serotonin receptor differences. One study examining 5HT2 receptors on platelets in children found a suggestion of increased binding in teenage girls after the age of 14, but the study was clearly limited by a small sample size of postpubertal adolescents (Biegon & Greuner, 1992). Incubation of human platelets with sex steroids had no direct effects on serotonin uptake (Ehrenkranz, 1976). Depressed women have a higher density than male depressed patients of 5HT2 receptors on platelets (Hrdina, Bakish, Chudzik, Ravindran, & Lapierre, 1995). No sex differences have been found in serotonin metabolites in cerebrospinal fluid of normal subjects (Leckman et al., 1994; Yoshino, 1982). One postmortem study examining serotonin binding in the human brain found no sex differences (Marcusson, Oreland, & Winblad, 1984), whereas another reported increased 5HT2 binding in frontal cortex in women (Arato, Frecska, Tekes, & MacCrimmon, 1991). One positron emission tomography imaging study found a sex differences in 5HT2A in brain with men showing more 5HT receptors than women. (Biver et al., 1996). It should be noted that many of the human studies were conducted before an understanding of multiple serotonin receptors existed, so many of the compounds used to detect serotonin receptors were nonspecific. It is likely that there are sex differences in serotonin systems in humans and the differences may be larger in depressed women compared to depressed men than the differences seen in normal subjects. Of note, the two illnesses shown to have a specific response to serotonergic antidepressants, premenstrual syndrome (Eriksson, Hedberg, Andersch, & Sundblad, 1995) and obsessive-compulsive disorder (Greist, Jefferson, Kobak, Katzelnick, & Serlin, 1995), appear to be particularly sensitive to changes in gonadal steroids.

CONCLUSIONS

The fact that women have much greater and repetitive fluxes in reproductive hormones over the lifespan may enhance the potential for dysregulation of a wide variety of brain neurochemical systems. In addition, as noted above, organizational differences between male and female brains result from exposure to high levels of gonadal steroids during the pre- and perinatal periods. The interactions of these organizational effects in females with cyclical gonadal steroid hormone changes following puberty, then followed by menopause when there is loss of these same steroids, suggests that stress responsiveness and susceptibility to stress-related disorders could vary substantially over the lifetime of women. There is certainly evidence that women's increased vulnerability to depression arises at puberty, when gonadal steroids could further enhance HPA axis responsiveness (Kessler, McGonagle, Swartz, Blazer, & Nelson, 1993). Additionally, the evidence linking stress and glucocorticoids to hippocampal damage and subsequent memory problems (Issa, Rowe, & Meaney, 1990), and the important role that gonadal steroids may play in protection from these effects in premenopausal women, implies that further research is needed into the interaction of stress, menopause, and memory impairment.

References

Ahima, R. S., Lawson, A. N. L., Osei, S. Y. S., & Harlan, R. E. (1992). Sexual dimorphism in regulation of type II corticosteroid receptor immunoreactivity in the rat hippocampus. *Endocrinology, 131*, 1409–1416.

Altemus, M., Pigott, T., Kalogeras, K. T., Demitrack M, Dubbert B., Murphy D. L., et al. (1992). Abnormalities in the regulation of vasopressin and corticotropin releasing factor secretion in obsessive-compulsive disorder. *Archives of General Psychiatry, 49*, 9–20.

Altemus, M., Redwine, L., Yung-Mei, L., Yoshikawa, T., Yehuda, R., Detera-Wadleigh, S., et al. (1997). Reduced sensitivity to glucocorticoid feedback and reduced glucocorticoid receptor mRna expression in the luteal phase of the menstrual cycle. *Neurosychopharmacology, 17*, 100–109.

Altshuler, L. L., Hendrick, V., & Cohen, L. S. (1998). Course of mood and anxiety disorders during pregnancy and the postpartum period. *Journal of Clinical Psychiatry, 59* (Suppl 2), 29–33.

Amsterdam, J. C., Winokur, A., Abelman, E., Lucki, I., & Richels, K. (1983). Co-syntropin (ACTH a$^{1-24}$) stimulation test in depressed patients and healthy subjects. *American Journal of Psychiatry, 140*, 907–909.

Arango, V., Ernsberge, r P., Marzuk, P. M., Chen, J. S., Tierney, H., Stanley, M., et al. (1990). Autoradiographic demonstration of increased serotonin 5-HT^2 and b-adrenergic receptor binding sites in the brain of suicide victims. *Archives of General Psychiatry, 47*, 1038–1047.

Arango, V., Underwood, M. D., Boldrini, M., Tamir, H., Kassir, S. A., Hsiung, S., et al. (2001). Serotonin 1A receptors, serotonin transporter binding and serotonin transporter mRNA expression in the brainstem of depressed suicide victims. *Neuropsychopharmacology, 25*, 892–903.

Arato, M., Frecska, E., Tekes, K., & MacCrimmon, D. J. (1991). Serotonergic interhemispheric asymmetry: gender difference in the orbital cortex. *Acta Psychiatrica Scandinavica, 84*, 110–111.

Arriza, J. L., Weinberger, C., Cerelli, G., Glaser, T. M., Handelin, B. L., Housman, D. E., et al. (1987). Cloning of human mineralocorticoid receptor complementary DNA: Structural and functional kinship with the glucocorticoid receptor. *Science*, *237*, 268–275.

Asberg, M., Traskman, L., & Thoren, P. (1976) 5-HIAA in the cerebrospinal fluid. A biochemical suicide predictor? *Archives of General Psychiatry*, *33*, 1193–1197.

Barden, N., Reul, J. M. H. M., & Holsboer, F. (1995). Do antidepressants stabilize mood through actions on the hypothalamic-pituitary-adrenocortical system? *Trends in Neuroscience*, *18*, 6–10.

Baxter, L. R., Jr., Schwartz, J. M., Phelps, M. E., Mazziotta, J. C., Guze, B. H., Selin, C. E., et al. (1989). Reduction of prefrontal cortex glucose metabolism common to three types of depression. *Archives General Psychiatry*, *46*, 243–250.

Berendsen, H. H. (1995). Interactions between 5-hydroxytryptamine receptor subtypes: is a disturbed receptor balance contributing to the symptomatology of depression in humans? *Pharmacology Therapeutics*, *66*, 17–37.

Bhagwagar, Z., Hafizi, S., & Cowen, P. J. (2003). Increase in concentration of waking salivary cortisol in recovered patients with depression. *American Journal of Psychiatry*, *160*, 1890–1891.

Biegon, A., & Greuner, N. (1992). Age-related changes in serotonin 5HT2 receptors on human blood platelets. *Psychopharmacology*, *108*, 210–212.

Biegon, A., Reches, A., Snyder, L., McEwen, B. S. (1983). Serotonergic and noradrenergic receptors in the rat brain: modulation by chronic exposure to ovarian hormones. *Life Sciences*, *32*, 2015–21.

Bierut, L. J., Heath, A. C., Bucholz, K. K., Dinwiddie, S. H., Madden, P. A. F., Statham, D. J., et al. (1999). Major depressive disorder in a community-based twin sample: Are there different genetic and environmental contributions for men and women? *Archives General Psychiatry*, *56*, 557–563.

Biver, F., Lotstra, F., Monclus, M., Wikler, D., Damhaut, P., Mendlewicz, J., et al. (1996). Sex difference in 5HT2 receptor in the living human brain. *Neuroscience Letters*, *204*, 25–28.

Blier, P., & de Montigny, C. (1994). Current advances and trends in the treatment of depression. *Trends in Pharmacological Sciences*, *15*, 220–226.

Bonson, K. R., Johnson, R. G., Fiorella, D., Rabin, R. A., & Winter, J. C. (1994). Serotonergic control of androgen-induced dominance. *Pharmacology, Biochemistry & Behavior*, *49*, 313–322.

Borisova, N. A., Proshlyakova, E. V., Sapronova, A. Y., & Ugrumov, M. V. (1996). Androgen-dependent sex differences in the hypothalamic serotoninergic system. *European Journal of Endocrinology*, *134*, 232–235.

Borsini, F. (1994). Balance between cortical 5-HT1A and 5-HT2 receptor function: hypothesis for a faster antidepressant action. *Pharmacological Research*, *30*, 1–11.

Brady, L., Whitfield, H. J., Fox, R. J., Gold, P. W., & Herkenham, M. (1991). Long-term antidepressant administration alters corticotropin releasing hormone, tyrosine hydroxylase and mineralocorticoid receptor gene expression in rat brain: Therapeutic implications. *Journal of Clinical Investigations*, *87*, 831–837.

Brown, G. W., & Harris, T. (1978). *Social origins of depression: A study of psychiatric disorder in women*. New York: The Free Press.

Buchsbaum, M. S., Wu, J., DeLisi, L. E., Holcomb, H., Kessler, R., Johnson, J., et al. (1986). Frontal cortex and basal ganglia metabolic rates assessed by positron emission tomography with FDG in affective illness. *Journal Affective Disorders, 10*, 137–152.

Burgess, L. H., & Handa, R. J. (1992). Chronic estrogen-induced alterations in adrenocorticotropin and corticosterone secretion, and glucocorticoid receptor-mediated functions in female rats. *Endocrinology, 131*, 1261–1269.

Butler, P. D., & Nemeroff, C. B. (1990). Corticotropin releasing factor as a possible cause of comorbidity in anxiety and depressive disorders. In J. D. Maser & C.R. Cloninger (Eds.), *Comorbidity of mood and anxiety disorders.* Washington, DC: American Psychiatric Press.

Carey, M. P., Deterd, C. H., de Koning, J., Helmerhorst, & DeKloet, E. R. (1995). The influence of ovarian steroids on hypothalamic-pituitary-adrenal regulation in the femal rat. *Journal of Endocrinology, 144*, 311–332.

Carlsson, M., & Carlsson, A. (1988). A regional study in sex differences in rat brain serotonin. *Progress in Neuro-Psychopharmacology & Biological Psychiatry, 12*, 53–61.

Carroll, B. J., Curtis, G. C., & Mendels, J. (1976). Neuroendocrine regulation in depression I. Limbic system-adrenocortical dysfunction. *Archives of General Psychiatry, 33*, 1039–1044.

Carroll, B. J., Feinberg, M., Greden, J. F., Tarika, J., Albala, A. A., Haskett, R. F., et al. (1981). A specific laboratory test for the diagnosis of melancholia. *Archives of General Psychiatry, 38*, 15–22.

Charney, D., Woods, S., Goodman, W., & Heninger G. (1987). Neurobiological mechanisms of panic anxiety: Biochemical and behavioral correlates of yohimbine-induced panic attacks. *American Journal of Psychiatry, 144*, 1030–1036.

Chrousos, G. P., & Gold, P. W. (1992). The concepts of stress and stress system disorders: Overview of physical and behavioral homeostasis. *Journal of the American Medical Association, 267*, 1244–1252.

Dean, C., Williams, R. J., & Brockington, I. F. (1989). Is puerperal psychosis the same as bipolar manic-depressive disorder? A family study. *Psychological Medicine, 19*, 637–647.

Drevets, W. C., Frank, E., Price, J. C., Kupfer, D. J., Greer, P. J., & Mathis, C. (2000). Serotonin type-1A receptor imaging in depression. *Nuclear Medicine and Biology, 27*, 499–507.

Drevets, W. C., Frank, E., Price, J. C., Kupfer, D. J., Holt, D., Greer, P. J., et al. (1999). PET imaging of serotonin 1A receptor binding in depression. *Biological Psychiatry, 46*, 1375–1387.

Drevets, W. C., Videen, T. O., Price, J. L., Preskorn, S. H., Carmichael, S. T., & Raichle, M. E. (1992). A functional anatomical study of unipolar depression. *Journal of Neuroscience, 12*, 3628–3641.

Duncan, M. R., & Duncan, G. R. (1979). An in vivo study of the action of antiglucocorticoids on thymus weight ratio, antibody titre and the adrenal-pituitary-hypothalamus axis. *Journal of Steroid Biochemistry, 10*, 245–259.

Ehrenkranz, J. R. (1976). Effects of sex steroids on serotonin uptake in blood platelets. *Acta Endocrinologica, 83*, 420–428.

Eriksson, E., Hedberg, M. A., Andersch, B., & Sundblad, C. (1995). The serotonin reuptake inhibitor paroxetine is superior to the noradrenaline reuptake inhibitor maprotiline in the treatment of premenstrual syndrome. *Neuropsychopharmacology*, *12*, 167–176.

Fink, G., Sumner, B. E., Rosie, R., Grace, O., & Quinn J. P. (1996). Estrogen control of central neurotransmission: effect on mood, mental state, and memory. *Cellular & Molecular Neurobiology*, *16*, 325–344.

Fishette, C. T., Biegon, A., & McEwen, B. S. (1984). Sex steroid modulation of the serotonin behavioral syndrome. *Life Sciences*, *35*, 1197–1206.

Frank, E., Anderson, B., Reynolds, C., Ritenour, A., & Kupfer, D. J. (1994). Life events and the research diagnostic criteria endogenous subtype: A confirmation of the distinction using the Bedford College methods. *Archives of General Psychiatry*, *51*, 519–524.

Galliven, E. A., Singh, A., Michelson, D., Bina, S., Gold, P. W., & Deuster, P. A. (1997). Hormonal and metabolic responses to exercise across time of day and menstrual cycle phase. *Journal of Applied Physiology*, *6*, 1822–1831.

Gelfin, Y., Lerer, B., Lesch, K. P., Gorfine, M., & Allolio, B. (1995). Complex effects of age and gender on hypothermic, adrenocorticotrophic hormone and cortisol responses to ipsapirone challenge in normal subjects. *Psychopharmacology*, *120*, 356–364.

George, M. S., Ketter, T. A., Parekh, P. I., Horwitz, B., Herscovitch, P., & Post, R. M. (1995). Brain activity during transient sadness and happiness in healthy women. *American Journal of Psychiatry*, *152*, 341–351.

George, M. S., Ketter, T. A., & Post, R. M. (1994). Prefrontal cortex dysfunction in clinical depression. *Depression*, *2*, 59–72.

Gold, P. W., Loriaux, D. L., Roy, A., Kling, M. A., Calabrese, J. R., Kellner, C. H., et al. (1986). Response to corticotropin-releasing hormone in the hypercortisolism of depression and Cushing's disease. *New England Journal of Medicine*, *314*, 1329–1335.

Gregoire, A., Kumar, R., Everitt, B., Henderson, A. F., & Studd, J. W. (1996). Transdermal oestrogen for treatment of severe postnatal depression. *Lancet*, *347*, 930–933.

Greist, J. H., Jefferson, J. W., Kobak, K. A., Katzelnick D. J., & Serlin R. C. (1995). Efficacy and tolerability of serotonin transport inhibitors in obsessive-compulsive disorder. *Archives of General Psychiatry*, *52*, 53–60.

Halbreich, U., Asnis, G. M., Schindledecker, R., Zurnoff, B., & Nathan, R. S. (1985). Cortisol secretion in endogenous depression I. Basal plasma levels. *Archives of General Psychiatry*, *42*, 909–914.

Haleem, D. J., Kennett, G. A., & Curzon, G. (1990). Hippocampal 5-hydroxytryptamine synthesis is greater in female rats than in males and more decreased by the 5-HT1A agonist 8-OH-DPAT. *Journal of Neural Transmission*, *79*, 93–101.

Heinsbroek, R. P., van Haaren, F., Feenstra, M. G., van Galen, H., Boer, G., & van de Pool, N. E. (1990). Sex differences in the effects of inescapable footshock on central catecholaminergic and serotonergic activity. *Pharmacology, Biochemistry & Behavior*, *37*, 539–550.

Heninger, G. R., & Charney, D. S. (1987). Mechanism of action of antidepressant treatment: Implications for the etiology and treatment of depressive disorders. In H.Y. Meltzer (Ed.), *Psychopharmacology: The third generation of progress* (pp. 535–544). New York: Raven Press.

Holsboer, F., Bardeleden, U., Gerken, A., Stalla, G., & Muller, O. (1984). Blunted corticotropin and normal cortisol response to human corticotropin-releasing factor in depression. *New England Journal of Medicine, 311,* 1127.

Hrdina, P. D., Bakish, D., Chudzik, J., Ravindran, A., & Lapierre, Y. D. (1995). Serotonergic markers in platelets of patients with major depression: upregulation of 5-HT2 receptors. *Journal of Psychiatry & Neuroscience, 20,* 11–19.

Hrdina, P. D., Demeter, E., Vu, T. B., Sotonyi, P., & Palkovits, M. (1993). 5-HT uptake sites and 5-HT^2 receptors in brain of antidepressant-free suicide victims/depressives: Increase in 5-HT$_2$ sites in cortex and amygdala. *Brain Research, 614,* 37–44.

Issa, A. M., Rowe, W., & Meaney, M. J. (1990). Hypothalamic-pituitary-adrenal activity in aged, cognitively impaired and cognitively unimpaired rats. *Journal of Neuroscience, 10,* 3247–3254.

Jones, I., Kent, L., & Craddock, N. (2002). Genetics of affective disorders. In P. McGuffin, M. J. Owen, & Gottesman II (Eds.), *Psychiatric genetics & genomics* (pp. 211–245). Oxford: Oxford University Press.

Jones, M. T., Brush, F. R., & Neame, R. L. B. (1972). Characteristics of fast feedback control of corticotrophin release by corticosteroids. *Journal of Endocrinology, 55,* 489.

Keller-Wood, M., Silbiger, J., & Wood, C. E. (1988). Progesterone attenuates the inhibition of adrenocorticotropin responses by cortisol in nonpregnant ewes. *Endocrinology, 123,* 647–651.

Kendler, K. S. (1998). Major depression and the environment: A psychiatric genetic perspective. *Pharmacopsychiat, 31,* 5–9.

Kendler, K. S., Gardner, C. O., Neale, M. C., & Prescott, C. A. (2001). Genetic risk factors for major depression in men and women: similar or different heritabilities and same or partly distinct genes? *Psychological Medicine, 31,* 605–616.

Kendler, K. S., Kessler, R. C., Walters, E. E., MacLean, C., Neale, M. C., Heath, A. C., et al. (1995). Stressful life events, genetic liability and onset of an episode of major depression in women. *American Journal of Psychiatry, 152,* 833–842.

Kendler, K. S., Pedersen, N. L., Neale, M. C., & Mathe, A. A. (1995). A pilot Swedish twin study of affective disorders including hospital- and population-ascertained subsamples: Results of model fitting. *Behavioral Genetics, 25,* 217–232.

Kessler, R. C., McGonagle, K. A., Swartz, M., Blazer, D. G., & Nelson, C. B. (1993). Sex and depression in the National Comorbidity Survey I: Lifetime prevalence, chronicity and recurrence. *Journal of Affective Disorders, 29,* 85–96.

Ketter, T. A., George, M. S., Kimbrell, T. A., & Post R. M. (1996). Functional brain imaging, limbic function, and affective disorders. *Neuroscientist, 2,* 55–65.

Kirschbaum, C., Kudielka, B. M., Gaab, J., Schommer, N. C., & Hellhammer, D. H. (1999). Impact of gender, menstrual cycle phase and oral contraceptives on the hypothalamic-pituitary-adrenal axis. *Psychosomatic Medicine, 64,* 154–162.

Kirschbaum, C., Pirke, K-M., & Hellhammer, D. H. (1995). Preliminary evidence for reduced cortisol responsivity to psychological stress in women using oral contraceptive medication. *Psychoneuroendocrinology, 20,* 509–514.

Kirschbaum, C., Schommer, N., Federenko, I., Gaab, J., Neumann, O., Oellers, M., et al. (1996). Short-term estradiol treatment enhances pituitary-adrenal axis and sympathetic responses to psychosocial stress in healthy young men. *Journal of Clinical Endocrinology and Metabolism, 81,* 3639–3643.

Klimek, V., Zak-Knapik, J., & Mackowiak, M. (1994). Effects of repeated treatment with fluoxetine and citalopram, 5-HT uptake inhibitors, on 5-HT1A and 5-HT2 receptors in the rat brain. *Journal Psychiatry and Neuroscience, 19,* 63–67.

Komesaroff, P. A., Esler, M., Clarke, I. J., Fullerton, M. J., & Funder, J. W. (1998). Effects of estrogen and estrous cycle on glucocorticoid and catecholamine responses to stress in sheep. *American Journal of Physiology, 275,* E671–678.

Komesaroff, P. A., Esler, M. D., & Sudhir, K. (1999). Estrogen supplementation attenuates glucocorticoid and catecholamine responses to mental stress in perimenopausal women. *Journal of Clinical Endocrinology and Metabolism, 84,* 606–610.

Krishnan, K. R. R., Ritchie, J. C., Saunders, W. B., Nemeroff, C. B., & Carroll, B. J. (1990). Adrenocortical sensitivity to low-dose ACTH administration in depressed patients. *Biological Psychiatry, 27,* 930–933.

Leckman, J. F., Goodwin, W. K., North, W. G., Cappell, P. B., Price, L. H., Pauls, D. L., et al. (1994). The role of central oxytocin in obsessive-compulsive disorder and related normal behavior. *Psychoneuroendocrinology, 19,* 723–749.

Lerer, B., Gillon, D., Lichtenberg, P., Gorfine, M., Gelfin, Y., & Shapira, B. (1996). Interrelationship of age, depression, and central serotonergic function: evidence from fenfluramine challenge studies. *International Psychogeriatrics, 8,* 83–102.

Linkowski, P., Mendelwicz, J., LeClercq, R., Brasseur, M., Hubain, P., Goldstein, J., Copinschi, G., van Cauter, E. (1985). The 24-hour profile of ACTH and cortisol in major depressive illness. *Journal of Clinical Endocrinology and Metabolism, 61,* 429–438.

López, J. F., Vázquez, D. M., Chalmers, D. T., Akil; H., & Watson, S. J. (1997). Regulation of 5-HT receptors and the Hypothalamic-Pituitary-Adrenal axis: Implications for the neurobiology of suicide. *Annals of New York Academy of Science, 836,* 106–134.

Lyons, M. J., Eisen, S. A., Goldberg, J., True, W., Lin, N., Meyer, J. M., et al. (1998). A registry-based twin study of depression in men. *Archives of General Psychiatry, 55,* 468–472.

Mann J. J., Arango V., & Underwood M. D. (1990). Serotonin and suicidal behavior. *Annals of New York Academy Sciences, 600,* 476–484.

Marcusson, J., Oreland, L., & Winblad, B. (1984). Effect of age on human brain serotonin (S-1) binding sites. *Journal of Neurochemistry, 43,* 1699–1705.

Martinez-Conde, E., Leret, M. L., & Diaz, S. (1985). The influence of testosterone in the brain of the male rat on levels of serotonin (5-HT) and 5-hydroxyindoleacetic acid (5-HIAA). *Comparative Biochemistry and Physiology, 80,* 411–414.

Matsuda, T., Nakano, Y., Kanda, T., & Iwata, H. (1991). Gonadal hormones affect the hypothermia induced by serotonin1A (5HT1A) receptor activation. *Life Sciences, 48,* 1627–1632.

Mayberg, H., Brannan, S., Mahurin, R. K., Jerebak, P. A., Brickman, J. S., Tekell, J. L., et al. (1997). Cingulate function in depression: A potential predictor of treatment response. *NeuroReport, 8,* 1057–1061.

Mayberg, H. S. (1994). Frontal lobe dysfunction in secondary depression. *Journal of Neuropsychiatry and Clinical Neurosciences, 6,* 428–442.

Mayberg, H. S., Brannan, S. K., Tekell, J. L., Silva, J. A., Mahurin, R. K., McGinnis, S., et al. (2000). Regional metabolic effects of fluoxetine in major depression: Serial changes and relationship to clinical response. *Biological Psychiatry, 48,* 830–843.

Mayberg, H. S., Lewis, P. J., Regenold, W., & Wagner, H. N, Jr. (1994). Paralimbic Hypoperfusion in Unipolar Depression. *Journal of Nuclear Medicine, 35,* 929–934.

Mayberg, H. S., Silva, J. A., Brannan, S. K., Tekell, J. L., Mahurin, R. K., McGinnis, S., et al. (2002). The functional neuroanatomy of the placebo effect. *American Journal of Psychiatry, 159,* 728–737.

McEwen, B. S. (1995). Adrenal steroid action on brain: Dissecting the fine line between protection and damage. In M. J. Friedman, D.S. Charney, & A. Y. Deutch (Eds.), *Neurobiological and clinical consequences of stress: From normal adaptation to PTSD.* Philadelphia: Lippincott-Raven.

McGuffin, P., Katz, R., Watkins, S., & Rutherford, J. (1996). A hospital-based twin register of the heritability of DSM-IV unipolar depression. *Archives General Psychiatry, 53,* 129–136.

McQueen, J. K., Wilson, H., & Fink, G. (1997). Estradiol-17 beta increases serotonin transporter (SERT) mRNA levels and the density of SERT-binding sites in female rat brain. *Brain Research. Molecular Brain Research, 45,* 13–23.

Meijer, O. C., & de Kloet, R. (1998). Corticosterone and serotonergic neurotransmission in the hippocampus: Functional implications of central corticosteroid receptor diversity. *Critical Reviews in Neurobiology, 12,* 1–20.

Mendelson, S. D., & McEwen, B. A. (1990). Testosterone increases the concentration of (3H)8-hydroxy-2-(di-n-propylamino)tetralin binding at 5-HT1A receptors in the medial preoptic nucleus of the castrated male rat. *European Journal of Pharmacology, 181,* 329–331.

Mendelson, S. D., & McEwen, B. S. (1991). Autoradiographic analyses of the effects of restraint-induced stress on 5-HT1A, 5-HT1C and 5-HT2 receptors in the dorsal hippocampus of male and female rats. *Neuroendocrinology, 54,* 454–461.

Monteleone, P., Catapano, F., Tortorella, A., & Maj, M. (1997). Cortisol response to d-fenfluramine in patients with obsessive-compulsive disorder and in healthy subjects: evidence for a gender-related effect. *Neuropsychobiology, 36,* 8–12.

Murua, V. S., & Molina, V. A. (1992). Effects of chronic variable stress and antidepressant drugs on behavioral inactivity during an uncontrollable stress: interaction between both treatments. *Behavioral Neural Biology, 57,* 87–89.

Pecins-Thompson, M., Brown, N. A., Kohama, S. G., & Bethea, C. L. (1996). Ovarian steroid regulation of tryptophan hydroxylase mRNA expression in rhesus macaques. *Journal of Neuroscience, 16,* 7021–7029.

Peroutka, S. J., & Snyder, S, H. (1980). Regulation of serotonin2 (5-HT2) receptors labeled with [3H]spiroperidol by chronic treatment with the antidepressant amitriptyline. *Journal of Pharmacology and Experimental Therapeutics, 215,* 582–587.

Pfohl, B., Sherman, B., Schlecte, J., & Stone, R. (1985). Pituitary/adrenal axis rhythm disturbances in psychiatric patients. *Archives General Psychiatry, 42,* 897–903.

Raadsheer, F. C., Hoogendijk, W. J., Stam, F. C., Tilders, F. J., & Swaab, D. F. (1994). Increased numbers of corticotropin-releasing hormone expressing neurons in the hypothalamic paraventricular nucleus of depressed patients. *Neuroendocrinology, 60,* 436–444.

Reul, J. M., Stec, I., Soder, M., & Holsboer, F. (1993). Chronic treatment of rats with the antidepressant amitriptyline attenuates the activity of the hypothalamic-pituitary adrenocortical system. *Endocrinology, 133,* 312–320.

Roca, C. A., Altemus, M., Galliven, E., Schmidt, P. J., Deuster, P., Gold, P., et al. (1998a). Effect of reproductive hormones on the hypothalamic-pituitary-adrenal axis response to stress. *Biological Psychiatry, 43*, 46S.

Roca, C. A., Schmidt, P. J., Altemus, M., Dananceau, M., & Rubinow, D. (1998b, June 21–23). Effects of reproductive steroids on the Hypothalamic-pituitary-adrenal axis response to low dose dexamethasone. *Abstract at Neuroendocrine Workshop on Stress*, New Orleans.

Roca, C. A., Schmidt, P. J., Altemus, M., Deuster, P., Danaceau, M. A., Putnam, K., et al. (2003). Differential menstrual cycle regulation of hypothalamic-pituitary-adrenal axis in women with premenstrual syndrome and controls. *Journal Clinical Endocrinology and Metabolism, 88*, 3057–3063.

Rousseau, G. G., Baxter, J. D., & Tomkins, G. M. (1972). Glucocorticoid receptors: relations between steroid binding and biological effects. *Molecular Biology, 67*, 99–115.

Rubin, R. T., Poland, R. E., Lesser, I. M., Winston, R. A., & Blodgett, N. (1987). Neuroendocrine aspects of primary endogenous depression I. Cortisol secretory dynamics in patients and matched controls. *Archives General Psychiatry, 44*, 328–336.

Rubinow, D. R., & Roy-Byrne, P. P. (1984). Premenstrual syndromes: overviews from a methodologic perspective. *American Journal of Psychiatry, 141*, 163–172.

Rutter, M. (2002). The interplay of nature, nurture, and developmental influences. *Archives General Psychiatry, 59*, 996–1000.

Ryan, N., Birmaher, B., Perel, J. M., Dahl, R. E., Meyer, V., Al-Shabbout, M., et al. (1992). Neuroencocrine response to L-5-hydroxytryptophan challenge in prepubertal major depression. *Archives of General Psychiatry, 49*, 843–851.

Sachar, E. J., Hellman, L., Roffwarg, H. P., Halpern, F. S., Fukush, D. K., & Gallagher, T. F. (1973). Disrupted 24 hour patterns of cortisol secretion in psychotic depressives. *Archives General Psychiatry, 28*, 19–24.

Schreiber, R., & De Vry, J. (1993). Neuronal circuits involved in the anxiolytic effects of the 5-HT1A receptor agonists 8-OH-DPAT ipsapirone and buspirone in the rat. *European Journal of Pharmacology, 249*, 341–351.

Seckl, J. R., & Fink, G. (1992). Use of in situ hybridization to investigate the regulation of hippocampal corticosteroid receptors by monoamines. *Journal of Steroid Biochemistry and Molecular Biology, 40*, 685–688.

Sheline, Y. I., Gado, M. H., & Kraemer, H. C. (2003). Untreated depression and hippocampal volume loss. *American Journal of Psychiatry, 160*, 1516–1518.

Sheline, Y. I., Wang, P. W., Gado, M. H., Csernansky, J. G., & Vannier, M. W. (1996). Hippocampal atrophy in recurrent major depression. *Proceedings of the National Academy of Sciences, USA, 93*, 3908–3913

Southwick, S., Krystal, J., & Morgan, C. (1993). Abnormal noradrenergic function in posttraumatic stress disorder. *Archives General Psychiatry, 50*, 266–274.

Sullivan, P. F., Neale, M. C., & Kendler, K. S. (2000). Genetic epidemiology of major depression: review and meta-analysis. *American Journal of Psychiatry, 157*, 1552–1562.

Svec, F. (1988). Differences in the interaction of RU 486 and ketoconazole with the second binding site of the glucocorticoid receptor. *Endocrinology, 123*, 1902–1906.

Turner, B. B., & Weaver, D. A. (1985). Sexual dimorphism of glucocorticoid binding in rat brain. *Brain Research, 343,* 16–23.

Viau, V., & Meaney, M. J. (1991). Variations in the hypothalamic-pituitary-adrenal response to stress during the estrous cycle in the rat. *Endocrinology, 129,* 2503–2511.

von Bardeleben, U., Stalla, G. K., Mueller, O. A., & Holsboer, F. (1988). Blunting of ACTH response to CRH in depressed patients is avoided by metyrapone pretreatment. *Biological Psychiatry, 24,* 782–786.

Weissman, M. M., & Olfson, M. (1995). Depression in women: implications for health care research. *Science, 269,* 799–801.

Welner, S. A., De Montigny, C., Desroches, J., Desjardins, P., & Suranyi-Cadotte, B. E. (1989). Autoradiographic quantification of serotonin 1A receptors in rat brain following antidepressant drug treatment. *Synapse, 4,* 347–352.

Yates, M., Leake, A., Candy, J. M., Fairbairn, A. F., McKeith, I.G, & Ferrier, I. N. (1990). 5HT2 receptor changes in major depression. *Biological Psychiatry, 27,* 489–496.

Yoshino, K. (1982). Concentrations of monoamines and monoamine metabolites in cerebrospinal fluid determined by high-performance liquid chromatography with electrochemical detection. *Brain & Nerve, 34,* 1099–1106.

Young, E. A. (1996). Sex differences in response to exogenous corticosterone. *Molecular Psychiatry, 1,* 313–319.

Young, E. A., Abelson, J. L., & Cameron, O. G. (2004). Effect of comorbid anxiety disorders on the HPA axis response to a social stressor in major depression. 56:113–120. *Biological Psychiatry,* in press.

Young, E. A., Aggen, S. H., Prescott, C. A., & Kendler, K. S. (2000). Similarity in saliva cortisol measures in monozygotic twins and the influence of past major depression. *Biological Psychiatry, 48,* 70–74.

Young, E. A., Akil, H., Haskett, R. F., & Watson, S. J. (1985). Evidence Against Changes In Corticotroph CRF Receptors In Depressed Patients. *Biological Psychiatry, 37,* 355–363.

Young, E. A., Altemus, M., Parkison, V, & Shastry, S. (2001). Effects of estrogen antagonists and agonists on the ACTH response to restraint stress. *Neuropsychopharmacology, 25,* 881–891.

Young, E. A., & Breslau, N. (2004). Cortisol and catecholamines in posttraumatic stress disorder: A community study. *Archives General Psychiatry, 61,* 394–401

Young, E. A., Carlson, N. E., & Brown, M. B. (2001). 24 Hour ACTH and Cortisol Pulsatility in Depressed Women, *Neuropsychpoharmacology, 25,* 267–276.

Young, E. A., Haskett, R. F., Grunhaus, L., Pande, A., Weinberg, V. M., Watson, S. J., et al. (1994). Increased circadian activation of the hypothalamic pituitary adrenal axis in depressed patients in the evening. *Archives General Psychiatry, 51,* 701–707.

Young, E. A, Haskett, R. F., Watson, S. J. & Akil, H. (1991). Loss of glucocorticoid fast feedback in depression. *Archives of General Psychiatry, 48,* 693–699.

Young, E. A., Kotun, J., Haskett, R. F., Grunhaus, L., Greden, J. F., Watson, S. J., et al. (1993). Dissociation between pituitary and adrenal suppression to dexamethasone in depression. *Archives General Psychiatry, 50,* 395–403.

Young, E. A., Lopez, J. F., Murphy-Weinberg, V., Watson, S. J., & Akil, H. (1997). Normal pituitary response to metyrapone in the morning in depressed patients: Implications for circadian regulation of CRH secretion, *Biological Psychiatry*, *41*, 1149–1155.

Young, E. A., & Vazquez, D. (1996). Hypercortisolemia, hippocampal glucocorticoid receptors and fast feedback. *Molecular Psychiatry*, *1*, 149–159.

Young, E. A., Watson, S. J., Kotun, J., Haskett, R. F., Grunhaus, L., Murphy-Weinberg, V., et al. (1990). Response to low dose oCRH in endogenous depression: role of cortisol feedback. *Archives General Psychiatry*, *47*, 449–457.

Zweifel, J. E., & O'Brien, W. H. (1997). A meta-analysis of the effect of hormone replacement therapy upon depressed mood. *Psychoneuroendocrinology*, *22*, 189–212.

4

Depressive Disorders in Women

From Menarche to beyond the Menopause

Wendy Somerset, D. Jeffrey Newport, Kim Ragan,
and Zachary N. Stowe

INTRODUCTION

In the United States, over 30 million people experience clinical depression each year (Kessler et al., 2003), with the majority of these patients being female. The rate of depression in women is typically twice that of men, with several studies reporting variability in the lifetime ratios in different countries, for example ratios ranging from 1.6 in Beirut and Taiwan to 3.1 in West Germany (Weissman et al., 1996). The identification and treatment of depression in women has garnered increasing attention over the past decade, particularly with respect to the impact of reproductive life events on mood disorders. The National Institutes of Health has issued several announcements requesting applications to investigate this understudied area.

A major impetus for this increased research focus is that distribution of major depression across the female reproductive life cycle is variable. Women are at greatest risk for the first episode of major depression during the childbearing years (Angold, Costello, & Worthman, 1998; Bebbington et al., 1998; Weissman, 1996). The overlap between the symptoms of depression and many complaints considered by clinicians to be the normal sequelae of reproductive life events, such as menstruation, pregnancy, postpartum, and the transition to menopause, presents challenges to the accurate diagnosis as well as calls to question the validity of applying the same diagnostic criteria to women during these life events (Stowe & Newport, 1998). The importance of recognizing and treating depression in women of childbearing age is underscored by the potential impact such illnesses have on interpersonal relationships such as marriage and problematic relationships with and interactions with their children (Hammen, 2003). Although they are important, the details of these studies extend beyond the scope of this review.

Recent literature consistently reports a higher incidence of depression in women and numerous hypotheses have been proposed to explain this finding. This chapter provides an overview of depressive illnesses at various stages in the female reproductive life cycle: menarche, the premenstrual phase of the menstrual cycle, pregnancy, the postpartum period, the perimenopause, and postmenopause.

DEPRESSIVE DISORDERS SPECIFIC TO WOMEN

Why women are at increased risk for depression compared to men remains a widely debated topic. Theoretical proposals have noted the higher incidence of predisposing factors, greater contact with clinicians and help-seeking behaviors, and increased variation and potential aberrations in neuroendocrine axes in women relative to men. The answer is likely multifactorial, including a variety of psychological, social, and biological underpinnings.

Psychosocial factors, such as childhood trauma, personality traits, and the demands of myriad social roles undoubtedly impact a woman's vulnerability to depression. Evidence clearly demonstrates that early life trauma is associated with an increased incidence of adulthood depression (Kendler et al., 1995). Consequently, the effects of developmental trauma may in part explain the higher prevalence of depressive disorders among women, who are more likely than men to have experienced childhood sexual abuse and other forms of childhood trauma (Kessler, Sonnega, Bromet, Hughes, & Nelson, 1995). Other investigators report that certain personality factors, such as interpersonal sensitivity, seen more often in women contribute to the higher incidence of depression (Boyce, Parker, Barnett, Cooney, & Smith, 1991). In addition, women more often exhibit a passive, ruminative style of coping that is more likely to be associated with depression than the active, distraction-oriented style more prevalent in men (Nolen-Hoeksema, Larson, & Grayson, 1999). The multiple, conflicting, and changing social roles of women (i.e., concurrently serving as employee, homemaker, parent, and caretaker of aging parents) have been considered as contributors to the higher rates of depression in women (Seeman, 1997). Nevertheless, such *psychosocial* variables are not present in all women and certainly have not been adequately delineated in all investigations of gender differences. A popular line of investigation and speculation has involved the interface between these personality and psychosocial issues with neuroendocrine vulnerability in increasing the risk for depression in women.

It has been hypothesized that the cyclical alterations in sex steroids (estrogens, progestins) during childbearing years may contribute to the higher rate of mood and anxiety disorders in women (Seeman, 1997). It is certainly a common clinical complaint – "something is wrong with my hormones" – that is often reinforced by support groups and clinicians alike. However, there is sparse empirical data to support hormonal abnormalities based on peripheral measures in previous investigations. Mood disturbances experienced by women are unlikely to be solely attributable to alterations in gonadal steroids; however, they may indirectly impact mood by altering limbic neuroactivation of the hypothalamic-pituitary-adrenal (HPA) axis (Wisner & Stowe, 1997). Furthermore, the peaks and troughs of estrogen secretion may modulate glucocorticoid effects at different time points in the menstrual cycle (Seeman, 1997). Other evidence indicating that gonadal steroids modulate serotonin activity is derived from preclinical data. For example, estrogen has numerous proserotonergic activities including (1) activation of tryptophan hydroxylase (Adams et al., 1973; Price, Thornton, & Mueller, 1967); (2) inhibition of monoamine oxidase activity (Robinson & Nies, 1980); (3) decreased expression of serotonin-1 receptors (Biegon & McEwen, 1982); and (4) up-regulation of serotonin-2 receptors (Biegon & McEwen, 1982). In contrast, progesterone demonstrates antiserotonergic activity such as increased serotonin uptake and turnover (Hackmann, Wirz-Justice, & Lichtsteiner, 1973; Ladisich, 1977) and modulates GABA neurotransmission in a fashion similar to benzodiazepines (Smith, Waterhouse, & Woodward, 1987). Additional investigations have begun to focus on the role(s) of metabolites and gain insight into central markers via neuroimaging techniques of these sex steroids (Epperson, Wisner, & Yamamoto, 1999). Although these data suggest that gonadal steroids may exert modulatory influences over neuroendocrine axes and neurotransmitter systems, there is sparse clinical utility to standard laboratory measures. Additional evidence supporting a role(s) for sex hormone abnormalities is derived indirectly from data on medications that modulate such hormones. For example, medications that interfere with menstrual functioning via alteration in gonadotropin releasing hormone (GnRH) list side effects of depression and anxiety of $>10\%$ (*Physician's Desk Reference*, 2004).

Despite these data, the clinical utility of measuring sex steroid concentrations is limited at best, as there has been no confirmation of abnormalities in the concentrations of the sex steroid in psychiatric disorders in women. The potential clinical value is the appreciation of how various phases in a woman's reproductive life may influence risk for depression.

MENARCHE

Prior to puberty, the prevalence of depression is similar for males and females, if not even more common in prepubescent boys than girls (Cyranowski, Frank, Young, & Shear, 2000). By age 15, females are twice as likely as males to have had an episode of major depression, and this persists over the next four decades of life (Cyranowski et al., 2000). By young adulthood (ages 15 to 24), the lifetime prevalence of major depression in the United States is 20.6% for females but only 10.5% for males (Kessler & Walters, 1998).

It would be reductionistic to limit our concept of the impact of puberty on female psychological functioning to the simple recognition that the developing brain is newly exposed to higher and constantly fluctuating levels of various reproductive hormones (Seeman, 1997). Along with the morphological and biochemical changes of puberty, adolescents often experience significant social role transitions, including major shifts in school environment as well as relationships with parents, peers, and intimate partners (Cyranowski et al., 2000).

Female adolescence is clearly a time when biology and psychosocial changes intermingle, that is, morphological changes such as breast development and increased body fat have a psychosocial effect on self image and perception by others (Cyranowski et al., 2000). Thus, depressive vulnerability of women at this time is likely to be influenced by pubertal timing, age, self-image and morphological changes, and psychosocial support (Angold et al., 1998; Brooks-Gunn & Warren, 1989). For example, body shape changes at the time of puberty are generally welcomed by young men but are more likely to be experienced as distressful by young women, at least in industrialized cultures (Blyth, Hill, & Smyth, 1981).

Pubertal Changes and Stress Vulnerability

Girls going through pubertal changes are believed to be more susceptible to the untoward effects of stress than pre- or postpubertal girls (Caspi & Moffitt, 1991). The literature examining the impact of puberty on girls focuses on measures of "pubertal status" and "pubertal timing" (Angold et al., 1998; Kessler & Walters, 1998). Whereas pubertal status refers to the level of physical development, that is, Tanner stages, pubertal timing refers to the level of maturation of a given girl relative to her peers (Steiner, Dunn, & Born, 2003). One prospective study found that the prevalence of depression in females did not increase with chronological age, but with reaching Tanner III pubertal stage (Angold et al., 1998). It remains debated whether this finding is primarily a consequence of the psychosocial effects of changes in body morphology or alterations in circulating androgen and estrogen levels (Angold, Costello, Erkanli, & Worthman, 1999). Pubertal timing may also be an important factor, as young girls who develop secondary sexual characteristics earlier than their peers are more likely to experience depression during adolescence and are more likely to have difficulties with body image (Stattin & Magnusson, 1999). In fact, girls who mature either early or late have been found to experience higher rates of major depression than "on time" girls (30% vs. 22% and 34% vs. 22% respectively) (Lewinsohn, Rhode, & Seeley, 1998).

Cyranowski and colleagues theorize that pubertal maturation sensitizes females to the depressogenic impact of stressful life events (Cyranowski et al., 2000). More specifically, a subset of pubertal females with depression have been proposed to struggle with a hormonal drive for affiliation (possibly oxytocin-mediated) and difficulty transitioning to adolescence, combined with stressful life events (Cyranowski et al., 2000). Other risk factors for depression during adolescence include a lack of healthy parental attachment (Hammen et al., 1995), an anxious temperament (Allgood-Merten, Lewinsohn, & Hops, 1990), high level of self-consciousness (Allgood-Merten et al., 1990), and lack of good coping skills (Allgood-Merten et al., 1990).

Current data regarding the direct impact of pubertal hormones on mood are inconclusive (Steiner et al., 2003). At puberty, the HPA readjusts in response to fluctuation in gonadal hormones (Steiner et al., 2003). Some research suggests a correlation between negative affect in female adolescents and rapid increase in estradiol levels (Warren & Brooks-Gunn, 1989), higher levels of testosterone and cortisol (Susman, Dorn, & Chrousos, 1991), and lower levels of dehydroepiandrosterone sulfate (DHEAS) (Susman et al., 1991). As low serotonin levels have been demonstrated in adolescent depression, others have investigated the interplay of serotonin systems and gonadal hormones (Hughes, Petty, Sheikha, & Kramer, 1996). The hormonal milieu at menarche may impact the production of serotonin at the transcriptional level, thus altering central serotonin receptor function or distribution (Steiner et al., 2003). Though as with most forms of depression, the etiology is most likely multifactorial.

Identification of Pubertal Depression

Arguably, given the incidence of depression in peripubertal girls, pediatricians should be encouraged to assess adolescent girls with early development and girls at Tanner Stage III for depression. Moreover, information about the increased rate of depression with puberty and risk factors would be a valuable aspect of training for parents, middle school teachers, and guidance counselors. As puberty marks the onset of the marked difference in prevalence of depression based on gender, further research may elucidate other biological markers of mood disorders, risk factors, or preventive measures.

MENSTRUAL CYCLE AND MOOD

As early as 600 BC, Hippocrates noted that a subset of women experienced a cluster of symptoms, including suicidal thoughts and other psychiatric symptoms, prior to the monthly onset of menses (Eriksson, Andersch, Lo, Landen, & Sundblad, 2002). In 1931, Frank coined the term *premenstrual tension*, in reference to the cyclical recurrence of tension, anxiety, and suicidal ideation in 15 women during their premenstrual phase (Frank, 1931). Yet, it was not until 1987 that criteria for a premenstrual depressive disorder were established by the American Psychiatric Association (American Psychiatric Association, 1987). The diagnostic criteria and even the inclusion of a separate diagnosis for alterations in mood proximate to the menstrual cycle have undergone considerable debate. Current nosology and investigators typically divide alterations in mood over the course of the menstrual cycle into three categories: (1) premenstrual syndrome (PMS) or premenstrual tension syndrome – a common syndrome with a myriad of symptoms reported in the literature; (2) premenstrual dysphoric disorder (PMDD) – a more severe fluctuation in mood that requires prospective monthly charting for accurate diagnosis; and (3) premenstrual exacerbation (PME) – exacerbation of a preexisting psychiatric illness during the luteal phase.

Diagnosis: PMS, PMDD, PME

As many as 87% of women of child-bearing age experience premenstrual symptoms including mood fluctuations, irritability, anxiety, fatigue, joint or muscle pain, breast tenderness, or bloating (Johnson, McChesney, & Bean, 1988). Whereas PMS is principally a vasomotor disturbance with secondary dysphoria and irritability, PMDD is considered by many a true mood disorder. Women with PMDD experience a set of symptoms that overlaps significantly with the symptoms of major depression including social and occupational impairment (American Psychiatry Association, 1994). The primary difference is the duration of symptom criteria between PMDD and major depression. The symptoms of PMDD appear up to 10 days prior to the onset of menses and cease following the onset of menstrual bleeding (American Psychiatric Association, 1994). Furthermore, the diagnosis of PMDD requires daily documentation of symptoms both to confirm the presence and timing of symptoms and to distinguish PMDD from a premenstrual exacerbation of another preexisting psychiatric disorder (American Psychiatric Association, 1994). Approximately 3 to 6% of women fulfill criteria for PMDD, with the peak incidence in the mid-30s (Cohen et al., 2002; Johnson et al., 1988; Rivera-Tovar & Frank, 1990). There appears to be cultural variability in the clinical presentation of PMDD, with American women more likely to report affective symptoms while women from India and China are more likely to report a higher rate of somatic symptoms (Chang, Holroyd, & Chau, 1995; Chaturvedi & Chandra, 1991). In contrast to menstrual-cycle-related onset and cessation, PME of an undiagnosed psychiatric illness is a common problem. In a prospective diagnostic assessment, 38.1% of women responding to an advertisement for a premenstrual dysphoric disorder (PMDD) clinical treatment study were found to actually be suffering from undiagnosed major depression, bipolar disorder, or panic disorder (Bailey & Cohen, 1999). The incidence of premenstrual exacerbation (PME) has not been systematically investigated, though the clinician should be aware that some psychiatric disorders may present as premenstrual complaints. As a general rule, numerous medical conditions appear to worsen during the luteal phase relative to the follicular phase – asthma, epilepsy, migraine headache, depression, substance abuse, and sleep disorders. With respect to mood disorders, the potential for PME underscores the need to assess the presence of symptoms during the follicular phase to make an accurate diagnosis.

Risk Factors for PMDD

In a large community study, Cohen and colleagues (2002) examined premenstrual symptoms of 4,164 women from 36 to 44 years old and found (1) a strong association between PMDD and a previous history of depression, (2) less educated women were more likely to report significant PMDD symptoms, (3) women working outside the home were more likely to fulfill diagnostic criteria for PMDD, and (4) current cigarette smokers were approximately four times as likely to fulfill criteria for PMDD (Cohen et al., 2002). Numerous studies have reported that women with PMDD report more stressful life events and are more affected by stressors (Fontana & Badawy, 1997; Fontana & Palfai, 1994; Girdler, Pedersen, Stern, & Light, 1993). In particular, women with PMDD are more likely to have sexual abuse histories (Girdler et al., 1998).

Biological Theories of PMS and PMDD

Several studies have examined the underlying biology of premenstrual syndrome (PMS) and PMDD. One of the potential confounds in previous investigations has been the refinement of diagnostic criteria for PMDD, as such etiological investigations may have included a heterogeneous sample. The current hypotheses for PMDD (or PMS with significant mood symptoms) include (1) fluctuations in ovarian sex hormone secretion (Hammarback, Ekholm, & Backstrom, 1991); (2) central nervous system (CNS) sensitivity to ovarian steroids (Poromaa, Smith, & Gulinello, 2002); and (3) hyperandrogenicity (Eriksson et al., 2002).

Given the menstrual cycle phase-specific presentation of PMDD, it has been suggested that it is a direct consequence of fluctuations in reproductive steroids (Poromaa et al., 2002). In fact, women with PMDD often report their symptoms are alleviated by ovariectomy or by treatment with an ovulation-inhibiting gonadotropin-releasing hormone analog

(Hammarback et al., 1991; Muse, Cetel, Futterman, & Yen, 1984; Sundstrom, Nyberg, Bixo, Hammarback, & Backstrom, 1999). However, multiple studies comparing serum estradiol and progesterone concentrations and the change in these concentrations in women with PMDD versus healthy volunteers have failed to demonstrate any significant differences (Backstrom et al., 1983; Rubinow et al., 1988). Although circulating concentrations of estrogen and progesterone may not discriminate women with PMDD from controls, the issue of CNS sensitivity to hormonal fluctuations warrants attention.

Women with PMDD appear to be more sensitive to the effects of such hormonal fluctuations (Eriksson et al., 2002). This notion has been examined by studying the impact of administering exogenous estrogen/progesterone to (a) women with a history of PMDD who were symptom-free while treated with ovulation inhibitors (Leather, Stud, Watson, & Holland, 1999; Schmidt, Nieman, Danaceau, Adams, & Rubinow, 1998) and (b) menopausal women with a former history of PMDD (Eriksson et al., 2002). In both groups, the women with a history of PMDD were more likely to experience PMDD-like symptoms when exogenous female sex steroids were administered (Leather et al., 1999; Schmidt et al., 1998). In an elegant investigation, Schmidt and colleagues (1998) first injected leuprolide (a GnRH agonist) monthly for 3 months to women with premenstrual syndrome as well as to controls; after the 3-month treatment with either leuprolide or placebo, participants in this study received hormone replacement (transdermal 17 (beta) estradiol and progesterone vaginal suppositories). Women with a history of PMDD reported fewer symptoms during leuprolide treatment and more symptoms during the phase of hormone replacement, confirming that a subgroup of women are more susceptible to mood fluctuations in response to gonadal steroids.

The third hypothesis has focused on androgens. There are conflicting data regarding the association between hyperandrogenicity and PMDD. Higher serum testosterone levels in women with PMDD than controls have been described (Eriksson, Sundblad, Lisjo, Modigh, & Andersch, 1992); however, discordant reports exist (Bloch, Schmidt, Su, Tobin, Rubinow, 1998; Eriksson et al., 2002). In further support of this hypothesis, androgen antagonists have been reported to alleviate PMDD symptoms (Warren & Brooks-Gunn, 1989). Moreover, women with PMDD often have increased abdominal fat, leading to a higher waist-hip ratio that has been associated with hyperandrogenicity (Eriksson et al., 2002). Additional work in this area is warranted.

PMDD and Major Depression – Similarities and Differences

The overlap in the clinical symptoms of PMDD and major depression has prompted numerous investigations of biological markers that have been identified in major depression have been conducted in women with PMDD. One such study compared the effects of administering intravenous L-tryptophan, a dietary precursor to serotonin, to women with PMDD and control subjects (Bancroft, Cook, Davidson, Bennie, & Goodwin, 1991). This study reported blunted prolactin responses in both groups and an increased blunting of both growth hormone and cortisol secretion in women with PMDD in response to L-tryptophan (Bancroft et al., 1991). The literature is replete with studies of the HPA axis in major depression. In contrast to findings in major depression, several studies have failed to document altered HPA axis reactivity in women with PMDD (Rubinow et al., 1988; Steiner, Haskett, Carroll, Hays, & Rubin, 1984). Interestingly, women with premenstrual depression had decreased evening cortisol levels and a heightened response to exogenous administration of CRH compared to control subjects (Rabin et al., 1990). These findings are in contrast those typically observed in major depression but are consistent with a report of nondepressed adult female survivors of childhood abuse (Heim, Newport, Bonsall, Miller, & Nemeroff, 2001), an intriguing parallel given the association between childhood sexual abuse and the subsequent occurrence of PMDD. Perhaps the increased cortisol responses represent an adaptive response to the challenges of the premenstrual phase of the menstrual cycle (Odber, Cawood, & Bancroft, 1998). However, an evaluation of salivary cortisol levels of women with and without perimenstrual mood changes found salivary cortisol levels to be higher premenstrually in women with mild physical and emotional changes, whereas cortisol levels in women with more severe depressive symptoms were lower premenstrually (Odber et al., 1998). It is unclear if the more severe symptoms are associated with a down-regulation in the HPA axis, whereas milder symptoms do not produce such a down-regulation allowing the HPA axis to be reactive to the menstrual cycle. These biological data in women with PMDD have limited overlap with the data from major depression.

Additional investigations have demonstrated differences between PMDD and major depression. For example, loss of appetite and decreased sleep are common symptoms of major depression. In contrast, food craving (Bancroft, Cook, & Williamson, 1988) and increased sleep (Odber et al., 1998) are common symptoms in women with PMDD. Treatment response also differs; compared to women with major depression, those with PMDD preferentially respond to selective serotonin reuptake inhibitors (SSRIs) and often do so more quickly and at lower doses (Steiner et al., 1995). In fact, women with PMDD may respond to an SSRI in a few days, whereas treatment response for MDD requires 4 to 6 weeks on average (Steiner et al., 1995). Clinically efficacious antidepressants that do not alter serotonin uptake have failed to effectively treat women with PMDD. Similar to major depression treatment studies in community samples, a review of the PMDD treatment studies with the most rigorous enrollment criteria demonstrate a placebo response rate >40% (Steiner et al., 1995; Yonkers, Halbreich, Freeman, Brown, & Pearlstein, 1996)

DEPRESSION AND PREGNANCY

Historically, pregnancy has been perceived as a time of emotional well-being, but there is little evidence that pregnancy provides protection against psychiatric illness (Kendell, Wainwright, Hailey, & Shannon, 1976). Retrospective data suggests a lower rate of suicide during pregnancy (Marzuk et al., 1997). Otherwise, comparable rates of MDD have been reported in pregnant and nonpregnant women, with 10 to 14% of women experiencing a major depressive episode during pregnancy and 25% of pregnant women experiencing elevated levels of depressive symptomatology (Evans, Heron, Francomb, Oke, & Golding, 2001; Gotlib, Whiffen, Mount, Milne, & Cordy, 1989). Depression during pregnancy is easily overlooked, as there is a notable overlap between normal sequelae of pregnancy and symptoms of MDD (Kumar & Robson, 1984). Neurovegetative symptoms such as changes in appetite, body weight, sleep, libido, and energy may be manifestations of pregnancy rather than depressive symptomatology (Klein & Essex, 1995). Furthermore, medical disorders during pregnancy, such as anemia, gestational diabetes, and thyroid abnormalities, may also mimic or complicate the diagnosis of depression (Pedersen et al., 1993).

Recognizing and treating depression during pregnancy is important, as untreated depression may confer a risk to the mother-to-be and the fetus (Hedegaard, Henriksen, Sabroe, & Secher, 1993; Pagel, Smilkstein, Regen, & Montano, 1990; Steer, Scholl, Hediger, & Fischer, 1992; Zuckerman, Amaro, Bauchner, & Cabral, 1989). A depressed woman is at higher risk for poor nutrition, poor compliance with prenatal care, and exposure to tobacco, alcohol, or drugs, all of which may compromise the health of the developing baby (Hedegaard et al., 1993; Pagel et al., 1990; Zuckerman et al., 1989). Moreover, pregnant women with untreated depression have a higher risk for preterm deliveries and diminished fetal growth (Steer et al., 1992). Maternal depressive symptoms may even bear untoward neurodevelopmental consequences affecting the child's cognitive abilities and psychiatric vulnerability (Nulman et al., 2002). It is important to be familiar with the risk factors for maternal depression during pregnancy given such potential adverse effects (Henry, Beach, Stowe, & Newport, 2004).

Risk Factors for Depression during Pregnancy

Multiple risk factors for depression during pregnancy have been described. Risk factors for antenatal depression include prior history of depression (Gotlib et al., 1989), maternal youth (Gotlib et al., 1989), insufficient social support (O'Hara, 1986), marital discord (Kumar & Robson, 1984; O'Hara, 1986), lower socioeconomic status (Gotlib et al., 1989), poor relationship with woman's own mother (Murray, Cox, Chapman, & Jones, 1995), recent adverse life events (Martin, Brown, Goldberg, & Brockington, 1989), unwanted pregnancy or ambivalence regarding pregnancy (Kumar & Robson, 1984), occupational instability (Murray et al., 1995), and a greater number of children (Gotlib et al., 1989). In addition to these risk factors, closer scrutiny of the neuroendocrine alterations associated with pregnancy is warranted.

The physiological alterations of pregnancy directly parallel many of the biological indices of nongravid major depression: hypercortisolemic, cortisol nonsuppression on the dexamethasone suppression test (DST) blunted corticotropin releasing factor (CRF) stimulation test, and shortened rapid eye movement (REM) latency on sleep polysomnography. Though formal study of such indices in women with depression during pregnancy is absent, these similarities raise questions about the balance of neuroendocrine systems with respect to risk for developing depression. If such alterations unequivocally resulted in depression, then a higher incidence of depression during pregnancy would be expected. A systematic investigation of new onset major depression in pregnancy compared to those with recurrent depression may help to elucidate this apparent conundrum.

Treatment Issues for Depression during Pregnancy

The treatment of depression during pregnancy continues to be a complicated clinical dilemma. The majority of the literature has continued to focus on the use of antidepressants during pregnancy, though nonpharmacological interventions have demonstrated efficacy. The balancing of the risk–benefit of treatment has undergone extensive review (Newport & Stowe, 2003; Wisner et al., 2000) – arguably there are more review articles than systematic investigations. It suffices to summarize that treatment decisions remain on a case-by-case basis. One facet that largely complicates the treatment decision is the lack of data on the course of illness if previously treated depressed women choose to discontinue treatment when they become pregnant. A pair of prospective studies recently completed demonstrated a remarkably high rate of relapse (68–74%) during pregnancy for women discontinuing antidepressant therapy (Cohen, Altshuler, Stowe, & Faraone, 2004; Cohen, Altshuler, Stowe, Nonacs, et al., 2004). By comparison, an estimated 59% of patients (nongravid women and men) discontinuing antidepressant medication relapse when euthymia has persisted for <16 weeks, and 20% relapse within 8 weeks of medication discontinuation even when well for >16 weeks (Prien & Kupfer, 1986). This high rate of relapse during pregnancy may be indicative of the biological alterations noted above, though further research is warranted.

POSTPARTUM ONSET MOOD DISORDERS

Pregnancy and childbirth represents a major neuroendocrine and psychosocial stressor for women and the alterations encountered in the postpartum period rival any changes associated with other life events. Several investigations have demonstrated an increased rate of psychiatric hospitalization in the postpartum period relative to other times (Kendell, Chalmers, & Platz, 1987). Recognized throughout antiquity, only recently has postpartum-onset mental illness been officially included in psychiatric nosology (American Psychiatric Association, 1994). For mood disorders with symptomatology within the first 4 weeks after delivery, *Diagnostic and Statistical Manual of Mental Disorders* (*DSM-IV*) designates the postpartum onset modifier (American Psychiatric Association, 1994). The continuum of postpartum mood disturbance includes the relatively common postpartum blues, a major depressive episode termed postpartum depression, and the most severe – postpartum psychosis (PPS). Unfortunately, the majority of attention paid to postpartum onset mental illness often occurs following a tragic event with the child.

Postpartum Blues – Maternity Blues and Baby Blues

The blues occurs at such a high frequency (>60%) that it is considered a normal sequelae of childbirth. It is noteworthy that this purportedly normal condition appears to have many of the same risk factors seen for depression. Risk factors for postpartum blues include a personal or family history of depression, premenstrual dysphoria, recent stressful life events, depressive or anxiety symptoms during pregnancy, ambivalence about their role as mother, passive coping strategies, and excessive fear about labor (Cohen, Altshuler, Stowe, & Faraone, 2004; Cohen, Altshuler, Stowe, Nonacs, et al., 2004; Prien & Kupfer, 1986). Although clinical intervention is not typical, women experiencing the blues appear to be at greater risk for later depression. The role of the blues as a trigger or representing a brief episode of mood disorder in a vulnerable population warrants further attention.

Postpartum Depression

The rate of major depression in the postpartum period is highly variable with investigations citing 8–22% (Kumar & Robson, 1984; O'Hara, Zekoski, Phillips, & Wright, 1990). This range is largely dependent on the criteria or diagnostic instrument employed. There is evidence that teenage mothers and mothers living in poverty have a higher rate of PPD – 26% and 27%, respectively (Hobfoll, Ritter, Lavin, Hulsizer, & Cameron, 1995; Troutman & Cutrona, 1990). Overall, the incidence of depression triples the baseline rate in the first 5 weeks postpartum (Cox, Murray, & Chapman, 1993) and remains elevated through the first 6 months postpartum at which point the prevalence of depression declines to that seen in the general female population (Cox et al., 1993). The onset of PPD is typically within the first month postpartum. The symptoms of PPD closely resemble those seen in depression at other times (Suri & Burt, 1997), and like depression during pregnancy there is considerable overlap between depressive symptoms (alterations in sleep, libido decline, fatigue, appetite changes), the normal sequelae of childbirth, and the early postpartum period. This overlap can lead to an underdiagnosis of PPD and potential delays in appropriate treatment, as both women and clinicians may consider some symptoms as being normal. Several authors have noted the increased rate of obsessional symptoms in women with PPD, often focused on concerns about the baby's well-being (Suri & Burt, 1997). To aid in the diagnosis, rating scales like the Edinburgh Postnatal Depression Rating Scale and the Postpartum Depression Checklist, have been developed, validated, and translated into several different languages to help clinicians screen for depression in this population (Beck, 1995; Cox, Holden, & Sagovsky, 1987). These scales minimize the reliance on the presence of the overlapping symptoms, and rather emphasize the quality of the symptoms.

Risk Factors for PPD

The risk factors for PPD are similar to the risk factors for depression at other times of a woman's life. These risk factors include a history of depression (including past postpartum depression) (Campbell, Cohn, Flanagan, Popper, & Meyers, 1992; Playfair & Gowers, 1981; Watson, Elliott, Rugg, & Brough, 1984), anxious and depressive symptoms during pregnancy (O'Hara, 1986), family history of depression or postpartum depression (Playfair & Gowers, 1981), troubled relationship with partner (Campbell et al., 1992; O'Hara, 1986; Watson et al., 1984), limited social support (O'Hara, 1986), stressful life events during pregnancy (O'Hara, 1986), unplanned pregnancy (Campbell et al., 1992), and delivery complications or infants with difficult temperaments (Campbell et al., 1992). Additional risk factors that lend themselves to psychotherapeutic interventions include a pessimistic maternal attitude as well as ambivalence toward motherhood (Bernazzani, Saucier, David, & Borgeat, 1997) and pressures regarding the decision to breastfeed (Newport, Hostetter, Arnold, & Stowe, 2002). Anxiety and depressive symptoms during pregnancy appear to be one of the strongest predictors of depression in the postpartum period (Bernazzani et al., 1997). This raises significant questions regarding the actual timing of the onset of symptoms. Recent data from our group indicate that >20% of women with PPD actually have symptom onset during pregnancy (Stowe, Hostetter, & Newport, in press). PPD appears to be a highly recurrent illness, with a 50% chance of experiencing a subsequent episode of postpartum depression following another pregnancy (Garvey, Tuason, Lumry, & Hoffman, 1983). Early identification of maternal depression during the postpartum period is important given the impact of maternal depression on numerous facets of infant well-being – such issues are detailed in the companion chapters.

Biology of PPD

There remains a contention, largely supported by support groups, that PPD is a result of aberrations in sex hormones. However, systematic investigation has not confirmed any abnormalities in the sex steroid (Wisner & Stowe, 1997). Only a limited number of studies have identified some form of neuroendocrine aberrations and they lack replication. Increased plasma cortisol and decreased serum thyroid hormone (T3, T4) concentrations have been found in postpartum women with depressive symptomatology (Garvey et al., 1983). Other studies have investigated the interplay between postpartum autoimmune thyroiditis and postpartum depression, though no clear association has been described (Mallett et al., 1995; Pop et al., 1991). Smith and colleagues (1990) report that women who experienced an onset of depressed mood between their 38th week of pregnancy and postpartum day 2 had larger declines in plasma endorphin levels that those whose mood remained unchanged (Smith et al., 1990). Certainly, the myriad of biological alterations associated with childbirth lend themselves to etiological theories – further attention with diagnostic rigor (e.g., new onset postpartum vs. recurrent depression vs. symptom onset in pregnancy) is warranted. Without such studies the extant data remain equivocal.

Treatment of PPD

A recent pair of reviews detailed the psychosocial, psychotherapeutic, pharmacological, and nonpharmacological somatic interventions (light therapy, sleep deprivation) for PPD (Dennis, 2004; Dennis & Stewart, 2004). The author accurately concluded that the vast majority of investigations were far less than convincing and most lacked control groups. The treatment studies are further limited by a failure to distinguish between recurrent depression, depression continuing from pregnancy, and new onset postpartum. Despite these shortcomings, clinical practice is "open label" and most treatment studies demonstrated efficacy with open-label SSRI treatment, as well as interpersonal psychotherapy versus a wait list control group. The emerging clinical standard is the identification of high risk groups and preventive treatment planning. In women with a history of postpartum depression, cognitive behavioral therapy or interpersonal psychotherapy may help prevent relapse in some women (O'Hara, Stuart, Gorman, & Wenzel, 2000). A recent double-blind placebo controlled prevention study found that sertraline was highly effective in preventing recurrent PPD (Wisner et al., 2004).

The treatment of maternal depression during the postpartum period is often complicated by breastfeeding. These concerns have generated an enormous literature on the amount of antidepressants in human breast milk and serum concentrations in breast-fed infants exposed to antidepressants (Newport et al., 2002). Typically, nursing infant serum demonstrates either low or undetectable concentrations of antidepressants with adverse effects limited to gastrointestinal upset, fussiness, and irritability in the infants. It is noteworthy that antidepressants have more breast-feeding data (e.g., breast milk concentrations, nursing infant serum concentrations) than any other class of medications.

Postpartum psychosis

Postpartum psychosis represents a severe mental illness that is one of the few true psychiatric emergencies. PPS affects 0.2% of postpartum women, with the onset most commonly within the first week after delivery, but by definition within the first 4–6 weeks postpartum (Muller, 1985). Postpartum psychosis is a psychiatric emergency that requires aggressive treatment, always in an inpatient setting. The strongest risk factors for postpartum psychosis are a history of bipolar disorder or previous episode of postpartum psychosis (Marks, Wieck, Checkley, & Kumar, 1991; McNeil, 1987). While 80% of the women with postpartum psychosis have onset following the birth of their first child (McNeil, 1987), it is a highly recurrent illness for those women choosing to have another child (Davidson & Robertson, 1985). Follow-up data indicate that the majority of PPS cases are related to a mood disorder (e.g., bipolar depression, major depression). This disorder is characterized by intense mood lability, obsessive ruminations about the baby, hallucinations, sleep disturbances, and paranoia (Muller, 1985). Risk factors include a family history of bipolar disorder, primiparity, and arguably psychosocial stressors (Marks et al., 1991; McNeil, 1987). One study of 486 women admitted to an inpatient facility with postpartum psychosis found that more than one third had prior diagnoses of bipolar disorder as compared to less than 5% who carried a diagnosis of schizophrenia (Kendell et al., 1987).

Unfortunately, the treatment and impact data for PPS are extremely limited. Given the potential severity, women with a history of severe mood disorders or PPS should be observed closely during the postpartum period.

DEPRESSION AND MENOPAUSE

The impact of the transition to menopause on mood disorders has provided an interesting set of data that is often overlooked by clinicians. Several studies indicate higher rates of depression in the perimenopausal years relative to other periods in the reproductive life cycle, most notably among women with a history of a mood disorder (Schmidt, Roca, Bloch, & Rubinow, 1997; Weissman, 1996). For example, the 1996 Cross-National Study demonstrated an increase in the onset of depression in the 45–49 year age group of women (Weissman, 1996). Moreover, those in the 45–64 year age group have the highest rate of completed suicide among women (Anderson, Kochanek, & Murphy, 1997). The impact of the transition to menopausal status has received increased attention.

The transition to menopause (climacteric period) can be of varying duration and of varying etiology (natural, surgical, chemical, pharmacological). Several epidemiological studies have compared rates of depression in menopausal women to rates at other points in a woman's life and although no increase in depressive illness has been noted in postmenopausal women, more depressive symptoms have been reported in the climacteric period (Avis, Brambilla, McKinlay, & Vass, 1994; Ballinger, 1975; Bromberger et al., 2003; Bungay, Vessey, & McPherson, 1980). One study of 3,302 women reported the rate of persistent depressive symptoms to be 14.9 to 18.4% for perimenopausal women as compared to 8 to 12% for premenopausal women (Bromberger et al., 2003). Likewise, Tam and colleagues found significantly higher Beck Depression Inventory scores among perimenopausal women as compared to postmenopausal women (Tam, Stucky, & Hanson, 1999). Moreover, those postmenopausal women who were depressed or taking an antidepressant reported the onset of the depressive episode during the perimenopausal transition (Tam et al., 1999). Diagnosing depression in the perimenopausal woman can be challenging as there is overlap between depressive symptomatology and normal symptoms of menopause (Prior, 1998). For example, irritability, insomnia, libido changes, loss of energy, and concentration difficulties have been described as common symptoms of menopause (Prior, 1998).

Sorting out the clinical diagnosis requires additional questions regarding the symptoms that are specific to menopause (e.g., vaginal dryness) or unique to psychiatric illness (e.g., low self-esteem, thoughts of death and dying). The clinician should also be aware that both conditions can occur simultaneously and are not mutually exclusive.

Another confounding factor in looking at the relationship between depression and menopause is the association between a history of major depression and an early age of onset of menopause. One study found that undergoing a natural menopause at a younger (before age 47) than average age (50 to 55) was associated with a 2- to 3-fold increased likelihood of a

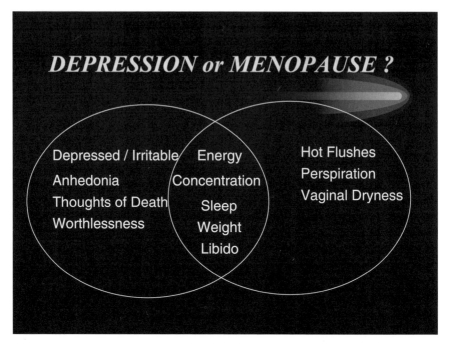

FIGURE 4.1.

history of depression (Harlow, Wise, Otto, Soares, & Cohen, 2003). However, this is confounded by a failure to control for nicotine, which increases the risk for earlier onset of menopause and is elevated in women with depression (Harlow et al., 2003).

Risk Factors for Menopausal Depression

The best predictor of perimenopausal depression is a past history of mood disorder; a past history significantly outweighs the impact of social circumstances or psychosocial stressors at this time (Schmidt et al., 1997). The presence and severity of vasomotor symptoms, history of PMS, lack of employment, and possibly African American race appear to increase the risk for depression at this time (Freeman et al., 2004).

Biology of Menopausal Depression

One of the leading theories to explain depression during menopause is the decline in gonadal hormones. Indirect support of this theory is derived from the finding that menopausal women with depressive symptoms are more likely to have experienced other purportedly gonadal hormone-related depressions such as PMDD, postpartum dysphoria, or depressive symptoms attributed to oral contraceptive use (Stewart, Boydell, Derzko, & Marshall, 1992), though convincing etiological data is lacking.

Treatment of Menopausal Depression

Although estrogen has consistently been shown to enhance cognitive function in women (Sherwin, 1997), the impact of estrogen replacement on mood has been less clear. Several studies report that hormone replacement therapy (HRT) is effective in treating depression in menopausal women (Carranza-Lira & Valentino-Figueroa, 1999; Zweifel & O'Brien, 1997); however, another investigation indicates that the women whose depression improves with estrogen replacement are the women who suffer from comorbid vasomotor symptoms (Hlatky, Boothroyd, Vittinghoff, Sharp, & Whooley, 2002). The decision of whether to treat perimenopausal and menopausal women with HRT specifically for depression is not well supported given the increasing data that HRT may carry additional risks. With respect to the treatment of hot flashes, there are now several double-blind placebo controlled studies demonstrating efficacy with antidepressants (venlafaxine, fluoxetine, paroxetine) in reducing hot flash severity (Barton et al., 2002; Loprinzi et al., 2002; North American Menopause Society, 2004; Yonkers, 2003).

CONCLUSION

Women endure multiple biopsychosocial changes during their lifetime. A major impediment to rendering definitive conclusions about the incidence of depression across these reproductive life events remains the complexity of the events and the overlap of what is deemed normal with symptom criteria. Arguably, there is little evidence that supports clinical measurement of gonadal steroids in women with depression, but rather a focus on other axes such as the HPA axis and thyroid function. To view the impact of the reproductive life cycle on the incidence of depression in women as purely biological is not supported by any avenue of research completed to date. However, as research techniques have advanced researchers have learned that reducing the hypothalamic-pituitary-gonadal (HPG) axis to estrogens and progestins was an oversimplification of a very complex system. Additional research involving the metabolites and other HPG products may help to delineate the role(s) of sex steroids in psychiatric illness.

The clinician is encouraged to consider the myriad of alterations associated with these life events – puberty, menses, pregnancy, postpartum, and menopause to provide a more comprehensive treatment plan. Such treatment planning includes the consideration of birth control pills and hormonal therapies as potentially mood altering and as able to influence the metabolism of medications. Similarly, the menstrual cycle, pregnancy, lactation, and menopause can affect the metabolism and distribution of standard pharmacotherapy agents. Although pregnancy and lactation complicate treatment, they do not preclude treatment. Further refinement of the risk–benefit assessment for the treatment of mental illness during pregnancy and lactation from systematic investigations will enhance treatment planning for these women. Presently, the reproductive safety data on antidepressants appears devoid of deleterious effects compared to the deleterious effects of untreated maternal mental illness on children. A thoughtful clinical interview detailing the temporal relationship of reproductive life events and exposure to exogenous hormone to alterations in mood should be the standard of care for women.

References

Adams, P. W., Rose, D. P., Foldard, J., Wynn, V., Seed, M., & Strong, R. (1973). Effect of pyridoxine hydrochloride (Vitamin B6) upon depression associated with oral contraception. *Lancet, 1*, 899–904.

Allgood-Merten, B., Lewinsohn, P., & Hops, H. (1990). Sex differences and adolescent depression. *Journal of Abnormal Psychology, 99*, 55–63.

American Psychiatric Association. (1987). *Diagnostic and statistical manual of mental disorders* (3rd ed., revised). Washington, DC: American Psychiatric Association.

American Psychiatric Association. (1994). *Diagnostic and statistical manual of mental disorders* (4th ed.). Washington, DC: American Psychiatric Association.

Anderson, R. N., Kochanek, K. D., & Murphy, S. L. (1997). *Report of final mortality statistics 1995*. Hyattsville, MD: National Center for Health Statistics.

Angold, A., Costello, E. J., Erkanli, A., & Worthman, C. M. (1999). Pubertal changes in hormone levels and depression in girls. *Psychological Medicine, 29*, 1043–1053.

Angold, A., Costello, E. J., & Worthman, C. M. (1998). Puberty and depression: The roles of age, pubertal status, and pubertal timing. *Psychological Medicine, 28*, 51–61.

Avis, N. E., Brambilla, D., McKinlay, S. M., & Vass, K. (1994). A longitudinal analysis of the assocation between menopause and depression: Results from the Massachusetts Women's Health Study. *Annals of Epidemiology, 4*, 214–220.

Backstrom, T., Sanders, D., Leask, R., Davidson, D., Warner, P., & Bancroft, J. (1983). Mood, sexuality, hormones, and the menstrual Cycle. II. Hormone levels and their relationship to the premenstrual syndrome. *Psychosomatic Medicine, 45*, 503–507.

Bailey, J. W., & Cohen, L. S. (1999). Prevalence of mood and anxiety disorders in women who seek treatment for premenstrual syndrome. *Journal of Womens Health & Gender Based Medicine, 8*, 1181–1184.

Ballinger, C. B. (1975). Psychiatric morbidity and the menopause: Screening of the general population sample. *British Medical Journal, 3*, 344–346.

Bancroft, J., Cook, A., Davidson, D., Bennie, J., & Goodwin, G. (1991). Blunting of neuroendocrine responses to infusion of L-tryptophan in women with perimenstrual mood change. *Psychological Medicine, 21*, 305–312.

Bancroft, J., Cook, A., & Williamson, L. (1988). Food craving, mood and the menstrual cycle. *Psychological Medicine, 19,* 855–860.

Barton, D., La, V. B., Loprinzi, C., Novotny, P., Wilwerding, M. B., & Sloan, J. (2002). Venlafaxine for the control of hot flashes: Results of a longitudinal continuation study. *Oncology Nursing Forum, 29,* 33–40.

Bebbington, P. E., Dunn, G., Jenkins, R., Lewis, G., Brugha, T. S., Farrell, M., et al. (1998). The influence of age and sex on the prevalence of depressive conditions: Report from the National Survey of Psychiatric Morbidity. *Psychological Medicine, 28,* 9–19.

Beck, C. T. (1995). Screening methods for postpartum depression. *Journal of Obstetric, Gynecologic, & Neonatal Nursing, 24,* 308–312.

Bernazzani, O., Saucier, J., David, H., & Borgeat, F. (1997). Psychosocial predictors of depressive symptomatology level in postpartum women. *Journal of Affective Disorders, 46,* 39–49.

Biegon, A., & McEwen, B. S. (1982). Modulation by estradiol of serotonin receptors in brain. *Journal of Neuroscience, 2,* 199–205.

Bloch, M., Schmidt, P. J., Su, T. P., Tobin, & M. B., Rubinow, D. R. (1998). Pituitary-adrenal hormones and testosterone across the menstrual cycle in women with premenstrual syndrome and controls. *Biological Psychiatry, 43,* 897–903.

Blyth, D. A., Hill, J. P., & Smyth, C. K. (1981). The influence of older adolescents on younger adolescents: Do grade-level arrangements make a difference in behaviors, attitudes, and experiences? *Journal of Adolescence, 1,* 85–110.

Boyce, P., Parker, G., Barnett, B., Cooney, M., & Smith, F. (1991). Personality as a vulnerability factor to depression. *British Journal of Psychiatry, 159,* 106–114.

Bromberger, N. J., Assman, S. F., Avis, N. E., Shocken, M., Kravitz, H. M., & Cordal, A. (2003). Persistent mood symptoms in a multiethnic community cohort of pre- and perimenopausal women. *American Journal of Epidemiology, 158,* 347–356.

Brooks-Gunn, J., & Warren, M. (1989). Biological and social contributions to negative affect in young adolescent girls. *Child Development, 60,* 251–264.

Bungay, G. T., Vessey, M. P., & McPherson, C. K. (1980). Study of Symptoms in the Middle Life with Special Reference to the Menopause. *British Medical Journal, 281,* 181–183.

Campbell, S. B., Cohn, J. F., Flanagan, C., Popper, S., & Meyers, T. (1992). Course and correlates of postpartum depression during the transition to parenthood. *Development and Psychopathology, 4,* 29–47.

Carranza-Lira, S., & Valentino-Figueroa, M. L. (1999). Estrogen therapy for depression in postmenopausal women. *International Journal of Gynaecology & Obstetrics, 65,* 35–38.

Caspi, A., & Moffitt, T. E. (1991). Individual differences are accentuated during periods of social change: The sample case of girls at puberty. *Journal of Personality and Social Psychology, 61,* 157–168.

Chang, A. M., Holroyd, E., & Chau, J. P. (1995). Premenstrual syndrome in employed Chinese women in Hong Kong. *Health Care for Women International, 16,* 551–561.

Chaturvedi, S. K., & Chandra, P. S. (1991). Sociocultural aspects of menstrual attitudes and premenstrual experiences in India. *Social Science & Medicine, 32,* 349–351.

Cohen, L. S., Altshuler, L. L., Stowe, Z. N., & Faraone, S. V. (2004). Reintroduction of antidepressant therapy across pregnancy in women who previously discontinued treatment. *Psychotherapy and Psychosomatics, 73,* 255–258.

Cohen, L. S., Altshuler, L. L., Stowe, Z. N., Nonacs, R. M., Suri, R., Newport, D. J., et al. (2004, May). *Relapse of depression during pregnancy following discontinuation of antidepressant treatment.* Poster presented at American Psychiatric Association Annual Meeting, New York.

Cohen, L. S., Soares, C., Otto, M., Sweeney, B. H., Liberman, R. F., & Harlow, B. F. (2002). Prevalence and predictors of premenstrual dysphoric disorder (PMDD) in older premenopausal women: The Harvard Study of Moods and Cycles. *Journal of Affective Disorders, 70,* 125–132.

Cox, J. L., Holden, J. M., & Sagovsky, R. (1987). Postnatal depression: Development of the 10-item Edinburgh Postnatal Depression Scale. *British Journal of Psychiatry, 150,* 782–786.

Cox, J. L., Murray, D., & Chapman, G. (1993). A controlled study of the onset, duration and prevalence of postnatal depression. *British Journal of Psychiatry, 163,* 27–31.

Cyranowski, J. M., Frank, E., Young, E., & Shear, M. K. (2000). Adolescent onset of the gender difference in lifetime rates of major depression: A theoretical model. *Archives of General Psychiatry, 57,* 21–27.

Davidson, J., & Robertson, E. (1985). A follow-up study of postpartum illness 1946–1978. *Acta Psychiatrica Scandinavica, 71,* 451–457.

Dennis, C. L. (2004). Treatment of postpartum depression, Part 2: A critical review of nonbiological interventions. *Journal of Clinical Psychiatry, 65,* 1252–1256.

Dennis, C. L., & Stewart, D. E. (2004). Treatment of postpartum depression, Part 1: A critical review of biological interventions. *Journal of Clinical Psychiatry, 65,* 1242–1251.

Epperson, C. N., Wisner, K. L., & Yamamoto, B. (1999). Gonadal steroids in the treatment of mood disorders. *Psychosomatic Medicine, 61,* 676–697.

Eriksson, E., Andersch, B., Ho, H. P., Landen, M., & Sundblad, C. (2002). Diagnosis and treatment of premenstrual dysphoria. *Journal of Clinical Psychiatry, 63*(suppl 7), 16–23.

Eriksson, E., Sundblad, C., Lisjo, P., Modigh, K., & Andersch, B. (1992). Serum levels of androgen are higher in women with premenstrual irritability and dysphoria than in controls. *Psychoneuroendocrinology, 17,* 195–204.

Evans, J., Heron, J., Francomb, H., Oke, S., & Golding, J. (2001). Cohort study of depressed mood during pregnancy and after childbirth. *British Medical Journal, 323,* 257–260.

Fontana, A. M., & Badawy, S. (1997). Perceptual and coping processes across the menstrual cycle: An investigation in a premenstrual syndrome clinic and a community sample. *Behavioral Medicine, 22,* 152–159.

Fontana, A. M., & Palfai, T. G. (1994). Psychosocial factors in premenstrual dysphoria: Stressors, appraisal, and coping processes. *Journal of Psychosomatic Research, 38,* 557–567.

Frank, R. T. (1931). The hormonal basis of premenstrual tension. *Archives of Neurology & Psychiatry, 26,* 1053–1057.

Freeman, E. W., Sammel, M. D., Lui, L., Gracia, C. R., Nelson, D. B., & Hollander, L. (2004). Hormones and menopausal status as predictors of depression in women in transition to menopause. *Archives of General Psychiatry, 61*, 62–70.

Garvey, M. J., Tuason, V. B., Lumry, A. E., & Hoffman, N. G. (1983). Occurrence of depression in the postpartum state. *Journal of Affective Disorders, 5*, 97–101.

Girdler, S. S., Pedersen, C. A., Stern, R. A., & Light, K. C. (1993). Menstrual cycle and premenstrual syndrome: Modifiers of cardiovascular and neuroendocrine responses to stress in premenstrual dysphoric disorder. *Psychiatry Research, 81*, 163–178.

Gotlib, I. H., Whiffen, V. E., Mount, J. H., Milne, K., & Cordy, N. I. (1989). Prevalence rates and demographic characteristics associated with depression in pregnancy and the postpartum. *Journal of Consulting & Clinical Psychology, 57*, 269–274.

Hackmann, E., Wirz-Justice, A., & Lichtsteiner, M. (1973). The uptake of dopamine and serotonin in rat brain during progesterone decline. *Psychopharmacologia, 32*, 183–191.

Hammarback, S., Ekholm, U., & Backstrom, T. (1991). Spontaneous anovulation causing disappearance of cyclical symptoms in women with premenstrual syndrome. *Acta Endocrinologica, 125*, 132–137.

Hammen, C. (2003). Social stress and women's risk for recurrent depression. *Archives of Women's Mental Health, 6*, 9–13.

Hammen, C., Burge, D., Daley, S., Davila, S. E., Paley, B., & Rudolph, K. D. (1995). Interpersonal attachment cognitions and prediction of symptomatic responses to interpersonal stress. *Journal of Abnormal Psychology, 104*, 436–443.

Harlow, B. L, Wise, L. A., Otto, M., Soares, C. M., & Cohen, L. S. (2003). Depression and its influence on reproductive endocrine and menstrual cycle markers associated with perimenopause: The Harvard Study of moods and cycles. *Archives of General Psychiatry, 60*, 29–36.

Hedegaard, M., Henriksen, T. B., Sabroe, S., & Secher, N. J. (1993). Psychological distress in pregnancy and preterm delivery. *British Medical Journal, 307*, 23–29.

Heim, C., Newport, D. J., Bonsall, R., Miller, A. H., & Nemeroff, C. B. (2001). Altered pituitary-adrenal axis responses to provocative challenge tests in adult survivors of childhood abuse. *American Journal of Psychiatry, 158*, 575–581.

Henry, A. L., Beach, A. J., Stowe, Z. N., & Newport, D. J. (2004). The fetus and maternal depression: Implications for antenatal treatment guidelines. *Clinical Obstetrics & Gynecology, 47*, 535–546.

Hlatky, M. A., Boothroyd, D., Vittinghoff, E., Sharp, P., & Whooley, M. A. (2002). Quality-of-life and depressive symptoms in postmenopausal women after receiving hormone therapy: Results from the Heart and Estrogen/Progestin Replacement Study (HERS) trial. *Journal of the American Medical Association, 287*, 591–597.

Hobfoll, S., Ritter, C., Lavin, J., Hulsizer, M. R., & Cameron, R. P. (1995). Depression prevalence and incidence among inner city pregnant and postpartum women. *Journal of Consulting and Clinical Psychology, 63*, 445–453.

Hughes, C. W., Petty, F., Sheikha, S., & Kramer, G. L. (1996). Whole-blood serotonin in children and adolescents with mood and behavior disorders. *Psychiatry Research, 65*, 79–95.

Johnson, S. R., McChesney, C., & Bean, J. A. (1988). Epidemiology of premenstrual symptoms in a non-clinical sample. I. Prevalence, natural history and help-seeking behavior. *Journal of Reproductive Medicine, 33,* 340–346.

Kendell, R. E., Chalmers, J. C., & Platz, C. (1987). Epidemiology of puerperal psychoses. *British Journal of Psychiatry, 150,* 662–673.

Kendell, R., Wainwright, S., Hailey, A., & Shannon, B. (1976). The influence of childbirth on psychiatric morbidity. *Psychological Medicine, 6,* 297–302.

Kendler, K., Kessler, R., Walters, E., MacLean, C., Neale, M. C., & Heath, A. C. (1995). Stressful life events, genetic liability, and onset of an episode of major depression in women. *American Journal of Psychiatry, 2,* 833–842.

Kessler, R., Berglund, P., Demler, O., Jin, R., Koretz, D., Merikangas, K. R., et al. (2003). The epidemiology of major depressive disorder: Results from the National Comorbidity Survey Replication (NCS-R). *Journal of the American Medical Association, 289,* 3095–3705.

Kessler, R. C., Sonnega, A., Bromet, E., Hughes, M., & Nelson, C. B. (1995). Post-traumatic stress disorder in the National Comorbidity Survey. *Archives of General Psychiatry, 52,* 1048–1060.

Kessler, R., & Walters, E. (1998). Epidemiology of DSM-III-R major depression and minor depression among adolescents and young adults in the National Comorbidity Survey. *Depression & Anxiety, 7,* 3–14.

Klein, M. H., & Essex, M. J. (1995). Pregnant or depressed? The effects of overlap between symptoms of depression and somatic complaints of pregnancy on rates of major depression in the second trimester. *Depression, 2,* 308–314.

Kumar, R., & Robson, K. M. (1984). A prospective study of emotional disorders in childbearing women. *British Journal of Psychiatry, 144,* 25–47.

Ladisich, W. (1977). Influence of progesterone on serotonin metabolism: A possible causal factor for mood changes. *Psychoneuroendocrinology, 2,* 257–266.

Leather, A. T., Stud, J. W., Watson, N. R., & Holland, E. F. (1999). The treatment of severe premenstrual syndrome with goserelin with and without "add-back" estrogen therapy: A placebo-controlled study. *Gynecological Endocrinology, 13,* 48–55.

Lewinsohn, P., Rhode, P., & Seeley, J. (1998). Major depressive disorder in older adolescents: Prevalence, risk factors, and clinical implications. *Clinical Psychiatry Review, 18,* 765–794.

Loprinzi, C. L., Sloan, J. A., Perez, E. A., Quella, S. K., Stella, P. J., Mailliard, J. A., et al. (2002). Phase III evaluation of fluoxetine for treatment of hot flashes. *Journal of Clinical Oncology, 20,* 1578–1583.

Mallett, P., Andrew, M., Hunter, C., Smith, J., Richard, C., Othman, S., et al. (1995). Cognitive function, thyroid status and postpartum depression. *Acta Psychiatrica Scandinavica, 91,* 243–246.

Marks, M. N., Wieck, A., Checkley, A., & Kumar, R. (1991). Life stress and postpartum psychosis: A preliminary report. *British Journal of Psychiatry, 158*(suppl 10), 45–49.

Martin, C. J., Brown, G. W., Goldberg, D. P., & Brockington, I. F. (1989). Psychosocial stress and puerperal depression. *Journal of Affective Disorders, 16,* 283–293.

Marzuk, P. M., Tardiff, K., Leon, A., Hirsch, C. S., Portera, L., Hartwell, N., et al. (1997). Lower risk of suicide during pregnancy. *American Journal of Psychiatry*, *154*, 122–123.

McNeil, T. F. (1987). A prospective study of postpartum psychoses in a high risk group: 2. Relationships to demographic and psychiatric history characteristics. *Acta Psychiatrica Scandinavica*, *75*, 35–43.

Muller, C. (1985). On the nosology of postpartum psychoses. *Psychopathology*, *18*, 181–184.

Murray, K., Cox, J. L., Chapman, G., & Jones, P. (1995). Childbirth: Life event or start of a long-term difficulty? Further data from the Stoke-on-Trent Controlled Study of Postnatal Depression. *British Journal of Psychiatry*, *166*, 595–600.

Muse, K., Cetel, N. S., Futterman, L. A., & Yen, S. C. (1984). The premenstrual syndrome. Effects of "medical ovariectomy." *New England Journal of Medicine*, *311*, 1345–1349.

Newport, D. J., Hostetter, A., Arnold, A. F., & Stowe, Z. N. (2002). The treatment of postpartum depression: Minimizing infant exposures. *Journal of Clinical Psychiatry*, *63*(suppl 7), 31–44.

Newport, D. J., & Stowe, Z. N. (2003). Clinical management of perinatal depression: Focus on paroxetine. *Psychopharmacology Bulletin*, *37*(suppl 1), 148–166.

Nolen-Hoeksema, S., Larson, J., & Grayson, C. (1999). Explaining the gender differences in depressive symptoms. *Journal of Personality and Social Psychology*, *77*, 1060–1072.

North American Menopause Society. (2004). Treatment of Menopause-associated Vasomotor Symptoms: Position Statement of The North American Menopause Society. *Menopause*, *11*, 11–33.

Nulman, I., Rovet, J., Stewart, D. E. Wolpin, J., Pace-Asciak, P., Shuhaiber, S., et al. (2002). Child development following exposure to tricyclic antidepressants or fluoxetine throughout fetal life: A prospective controlled study. *American Journal of Psychiatry*, *159*, 1889–1895.

Odber, J., Cawood, E. H., & Bancroft, J. (1998). Salivary cortisol in women with and without premenstrual mood changes. *Journal of Psychosomatic Research*, *45*, 557–568.

O'Hara, M. (1986). Social support, life events, and depression during pregnancy and the puerperium. *Archives of General Psychiatry*, *43*, 569–573.

O'Hara, M., Stuart, S., Gorman, L. L., & Wenzel, A. (2000). Efficacy of interpersonal psychotherapy for postpartum depression. *Archives of General Psychiatry*, *57*, 1039–1045.

O'Hara, M., Zekoski, E., Phillips, L., & Wright, E. J. (1990). Controlled prospective study of postpartum mood disorder: Comparison of childbearing and non-childbearing women. *Journal of Abnormal Psychology*, *99*, 3–15.

Pagel, M. D., Smilkstein, F., Regen, H., & Montano, D. (1990). Psychosocial influences on newborn outcome: A controlled prospective study. *Social Science of Medicine*, *30*, 597–604.

Pederson, C. A., Stern, R. A., Pate, J., Senger, M. A., Bowes, W. A., & Mason, G. A. (1993). Thyroid and adrenal measures during late pregnancy and the puerperium in women who have been major depressed or become dysphoric postpartum. *Journal of Affective Disorders*, *29*, 201–211.

Physician's Desk Reference. (58th ed.). (2004). Montvale, NJ: Thomson PDR.

Playfair, H. R., & Gowers, J. I. (1981). Depression following childbirth: A search for predictive signs. *Journal of the Royal College of General Practioners, 31,* 201–208.

Pop, V. J. M., DeRooy, H. A. M., Vader, H. L., van der Heide, D., van Son, M., Domproe, I. H., et al. (1991). Postpartum thyroid dysfunction and depression in an unselected population. *New England Journal of Medicine, 324,* 1815–1816.

Poromaa, I. S., Smith, S., & Gulinello, M. (2002). GABA receptors, progesterone, and premenstrual dysphoric disorder. *Archives of Women's Mental Health, 6,* 23–41.

Price, J. M., Thornton, M. J., & Mueller, L. M. (1967). Tryptophan metabolism in women using steroid hormones for ovulation control. *American Journal of Clinical Nursing, 20,* 452–456.

Prien, R. F., & Kupher, D. J. (1986). Continuation drug therapy for major depressive episodes: How long should it be maintained? *American Journal of Psychiatry, 143,* 18–23.

Prior, J. C. (1998). Perimenopause: The complex endocrinology of the menopausal transition. *Endocrine Reviews, 19,* 397–428.

Rabin, D. S., Schmidt, P. J., Campbell, G., Gold, P. W., Jensvold, M., Rubinow, D. R., et al. (1990). Hypothalamic-pituitary-adrenal function in patients with the premenstrual syndrome. *Journal of Clinical Endocrinology & Metabolism, 71,* 1158–1162.

Rivera-Tovar, A. D., & Frank, E. (1990). Late luteal phase dysphoric disorder in young women. *American Journal of Psychiatry, 147,* 1634–1636.

Robinson, D. S., & Nies, A. (1980). Demographic, biologic, and other variables affecting monoamine oxidase activity. *Schizophrenia Bulletin, 6,* 298–307.

Rubinow, D. R., Hoban, M. C., Grover, G. N., Galloway, G. S., Roy-Burne, P., Andersen, R., et al. (1988). Changes in plasma hormones across the menstrual cycle in patients with menstrually related mood disorder and in control subjects. *American Journal of Obstetrics and Gynecology, 158,* 5–11.

Schmidt, P. J., Nieman, I., Danaceau, M. A., Adams, L. F., & Rubinow, D. R. (1998). Differential behavioral effects of gonadal steroids in women with and in those without premenstrual syndrome. *New England Journal of Medicine, 338,* 209–216.

Schmidt, P. J., Roca, C. A., Bloch, M., & Rubinow, D. R. (1997). The perimenopause and affective disorders. *Seminars in Reproductive Endocrinology, 15,* 91–100.

Seeman, M. (1997). Psychopathology in women and men: Focus on female hormones. *American Journal of Psychiatry, 154,* 1641–1647.

Sherwin, B. B. (1997). Estrogen effects on cognition in menopausal women. *Neurology, 48,* s21–s26.

Smith, R., Cubis, J., Brinsmead, M., Lewin, T., Singh, B., Owens, P., et al. (1990). Ostetric experience and alterations in plasma cortisol, beta-endorphins, and corticotrophin releasing hormone during pregnancy and the puerperium. *Journal of Psychosomatic Research, 34,* 53–69.

Smith, S. S., Waterhouse, B. D., & Woodward, D. J. (1987). Sex steroid effects on extrahypothalamic CNS. II. Progesterone alone and in combination with estrogen, modulates cerebellar responses to amino acid transmitters. *Brain Research, 422,* 52–62.

Stattin, H., & Magnusson, D. (1999). *Paths through life – Vol. 2: Pubertal maturation in female development.* Hillsdale, NJ: Lawrence Erlbaum Associates.

Steer, R. A., Scholl, T. O., Hediger, M. L., & Fischer, R. L. (1992). Self-reported depression and negative pregnancy outcomes. *Journal of Clinical Epidemiology, 45*, 1093–1099.

Steiner, M., Dunn, E., & Born, L. (2003). Hormones and mood: From menarche to menopause and beyond. *Journal of Affective Disorders, 74*, 67–83.

Steiner, M., Haskett, R. F., Carroll, B. J., Hays, S. E., & Rubin, R. T. (1984). Circadian hormone secretory profiles in women with severe premenstrual tension syndrome. *British Journal of Obstetrics & Gynecology, 91*, 466–471.

Steiner, M., Steinberg, S., Stewart, D., Carter, D., Berger, C., Reid, R., et al. (1995). Fluoxetine in the treatment of premenstrual dysphoria. Canadian fluoxetine/premenstrual dysphoria collaborative study group. *New England Journal of Medicine, 332*, 1529–1534.

Stewart, D. E., Boydell, K., Derzko, C., & Marshall, V. (1992). Psychologic distress during menopausal years in women attending a menopause clinic. *International Journal of Psychiatry in Medicine, 22*, 213–220.

Stowe, Z. N., Hostetter, A., & Newport, D. J. (2005). The onset of postpartum depression: Implications for identification in obstetrical and primary care. *American Journal of Obstetrics and Gynecology, 192*, 522–526.

Stowe, Z. N., & Newport, D. J. (1998). Depression in women: Recognition and treatment. *Women's Health in Primary Care, 1*, 29–39.

Sundstrom, I., Nyberg, S., Bixo, M., Hammarback, S., & Backstrom, T. (1999). Treatment of premenstrual syndrome with gonadotropin-releasing hormone agonist in a low dose regimen. *Acta Obstetrica et Gynecologica Scandinavica, 78*, 891–899.

Suri, R., & Burt, V. (1997). The assessment and treatment of postpartum psychiatric disorders. *Journal of Practical Psychiatry & Behavioral Health, 3*, 67–77.

Susman, E. J., Dorn, L. D., & Chrousos, G. P. (1991). Negative affect and hormone levels in young adolescents: Concurrent and predictive perspectives. *Journal of Youth & Adolescence, 20*, 167–190.

Tam, L. W., Stucky, V., Hanson, R. E., & Parry, B. L. (1999). Prevalence of depression in menopause: A pilot study. *Archives of Womens Mental Health, 2*, 175–181.

Troutman, B., & Cutrona, C. (1990). Nonpsychotic postpartum depression among adolescent mothers. *Journal of Abnormal Psychology, 99*, 69–78.

Warren, M. P., & Brooks-Gunn, J. (1989). Mood and behavior at adolescence: Evidence for hormonal factors. *Journal of Clinical Endocrinology & Metabolism, 69*, 77–83.

Watson, J. P., Elliott, S. A., Rugg, A. J., & Brough, D. I. (1984). Psychiatric disorder in pregnancy and the first postnatal year. *British Journal of Psychiatry, 144*, 453–462.

Weissman, M. (1996, May). *Epidemiology of major depression in women.* Paper presented at American Psychiatric Assocation Annual Meeting, New York.

Weissman, M., Bland, R., Canino, G., Favavelli, C., Greenwald, S., Hwu, S. G., et al. (1996). Cross-national epidemiology of major depression and bipolar disorder. *Journal of the American Medical Association, 276*, 293–299.

Wisner, K. L., Perel, J. M., & Peindl, K. S., Hanusa, B. H., Piontek, C. M., & Findling, R. L. (2004). Prevention of postpartum depression: A pilot randomized clinical trial. *American Journal of Psychiatry, 161*, 1290–1292.

Wisner, K. L., & Stowe, Z. N. (1997). Psychobiology of postpartum mood disorders. *Seminars in Reproductive Endocrinology, 15*, 77–89.

Wisner, K. L., Zarin, D. A., Holmboe, E. S., Appelbaum, P. S., Gelenberg, A. J., Leonard, H. L., et al. (2000). Risk-benefit decision making for treatment of depression during pregnancy. *American Journal of Psychiatry, 157*, 1933–1940.

Yonkers, K. A. (2003). Paroxetine treatment of mood disorders in women: Premenstrual dysphoric disorder and hot flashes. *Psychopharmacology Bulletin, 37*(suppl 1), 135–147.

Yonkers, K. A., Halbreich, U., Freeman, E., Brown, C., & Pearlstein, T. (1996). Sertraline in the treatment of premenstrual dysphoric disorder. *Psychopharmacology Bulletin, 32*, 41–46.

Zuckerman, B., Amaro, H., Bauchner, H., & Cabral, H. (1989). Depressive symptoms during pregnancy: Relationship to poor health behaviors. *American Journal of Obstetrics and Gynecology, 160*, 1107–1111.

Zweifel, J. E., & O'Brien, W. H. (1997). A meta-analysis of the effect of hormone replacement therapy upon depressed mood. *Pschoneuroendocrinology, 22*, 189–212.

5

Does Puberty Account for the Gender Differential in Depression?

Laura M. DeRose, A. Jordan Wright,
and Jeanne Brooks-Gunn

The period of pubertal development coincides with a dramatic shift in the prevalence rates of depression. As depressive disorders rise in general during adolescence, puberty seems to introduce a divergence between the genders. Rates of depression in girls and boys are nearly indistinguishable up until this period of life (Nolen-Hoeksema & Girgus, 1994). During the transition from middle childhood to adolescence, girls begin to experience depression at a higher rate than boys. By mid-adolescence, the gender difference in both subclinical levels of depressive symptoms and diagnosable unipolar depression is at the rate of about 2:1 for girls to boys, which persists through adulthood (Nolen-Hoeksema, 2001).

The entry into adolescence is marked by the hormonal and physical changes of puberty; social changes in the family, peer group, and school environment; and concomitant individual changes in cognitive and socioemotional functioning. It is therefore important to consider models that examine the biological, psychological, and social components of adolescence that may contribute to depressive outcomes (Graber & Brooks-Gunn, 1996). The research to date has focused more on biological changes, especially the timing of the biological changes, than on psychological and social changes as a mechanism for depressive outcomes. The purpose of this chapter is to review several different sets of proposed models that may explain the gender differential in depression that emerges during adolescence.

A number of theories have been proffered to explain why this emergence of gender differences occurs coincidentally with pubertal development. With evidence of the impact of hormones, the degree of physical maturation, and pubertal timing on depression in girls via direct effect

Preparation of this chapter was supported in part by the National Institute of Mental Health, the National Institute of Child Health and Human Development, the National Institute of Child Health and Human Development Research Network on Child and Family Wellbeing, the Marx Family Foundation, and the William T. Grant Foundation.

models, several mechanisms have been proposed to link these pubertal phenomena and depression. Included among these models is the diathesis-stress model, in which pubertal development interacts with preexisting psychosocial vulnerability factors to elicit increased emotional stress in girls. The transitional stress model posits that the reproductive transition of puberty marks a period of development involving reorganization of biological and behavioral systems that may increase emotional arousal. A third set of theories concerns social challenges as being integral in the development of depression in girls, including girls facing more and different social challenges than boys and girls reacting differently to social challenges than boys.

The present chapter will begin with a description of the biological aspects of pubertal development, with attention to the gender differences in development. Next, methods of measuring puberty and the age of pubertal onset are addressed. Direct effects models linking aspects of pubertal development and depression will be presented and the research evaluating them discussed. Finally, the chapter will present the proposed mechanisms explaining the direct effects models, including empirical bases for the models and suggestions for future directions in research.

BIOLOGICAL ASPECTS OF PUBERTAL DEVELOPMENT

Pubertal development is a series of interrelated processes resulting in maturation and adult reproductive functioning. The physiological changes of puberty primarily involve the hypothalamic-pituitary-adrenal (HPA) axis and hypothalamic-pituitary-gonadal (HPG) axis. Pubertal development begins in middle childhood and takes 5 to 6 years for most adolescents to complete (Brooks-Gunn & Reiter, 1990; Petersen, 1987). A wide range of individual differences exists in the timing of onset of and rate of development through puberty. The following sections describe the physiological, physical, and central nervous system changes of pubertal development, with attention to gender differences in development. An explanation of how the different aspects of pubertal development are measured is also provided.

Physiological Changes of Puberty

Puberty is part of a continuum of events initiated at conception, mostly involving the HPG axis. The hypothalamic gonadotropin releasing hormone (GnRH) pulse generator, or *gonadostat*, is active prenatally and during early infancy, suppressed during childhood, and then reactivated at the onset of puberty (Fechner, 2003).[1] For puberty to begin, the brain's

---

[1] During early infancy, it may be the case that the gonadostat is not completely mature, meaning it is insensitive to the presence of gonadal sex steroids (Fechner, 2003). Maturation

sensitivity to the negative feedback of gonadal sex steroids (testosterone in males and estrogen in females) decreases, which then releases the HPA axis from inhibition. Puberty begins with the release of GnRH pulses, which activates pulsatile bursts of gonadotropins, luteinizing hormone (LH), and follicle stimulating hormone (FSH) from the pituitary gland. The LH and FSH pulses secreted in response to the GnRH occur first at night and then during the day. Increases in LH and FSH are some of the earliest measurable hormonal indications of pubertal development, and they have been found to rise progressively during puberty (Reiter & Grumbach, 1982). Episodic nocturnal bursts of low levels of LH are indicative of early pubertal stages (Grumbach & Styne, 1998). The gonads respond to LH and FSH by enlarging, maturing, and secreting increased amounts of gonadal sex steroids, androgens and estrogens.

Multiple gender differences in the mean levels and functions of hormone secretions are evident during the period of pubertal development. In females, the function of LH and FSH is to initiate follicular development in the ovaries, which stimulates them to produce estrogen. Estrogen-sensitive tissues, such as the breasts and uterus, then respond to the increase (Fechner, 2003). In males, increased LH stimulates the testes to secrete testosterone, resulting in an increase in testicular size, and FSH stimulates spermatogenesis. LH levels increase in both girls and boys at puberty, whereas FSH is higher in girls than boys during the prepubertal and pubertal years. Increased FSH levels stimulate the ovaries to produce estrogen. Although LH and FSH levels in both sexes are regulated by the negative feedback of the gonadal steroids and by the hormone inhibin, girls have a second control mechanism associated with their menstrual cycles, which is under positive feedback and is cyclic. When the estradiol level is high enough, it triggers an LH and, to a lesser extent, an FSH surge, each which lasts less than 2 days and stimulates ovulation. A corpus luteum forms from the ruptured follicle and begins to secrete progesterone. In the absence of pregnancy, the corpus luteum regresses and the progesterone and estrogen levels drop, triggering withdrawal bleeding and menstruation (Fechner, 2002).

Estrogen and testosterone levels also differ between the two sexes at puberty. Estradiol levels at puberty increase in females and then remain elevated during periods of each menstrual cycle. In males, estradiol levels increase until their growth spurt (at mid-puberty) and then decrease again. However, although males experience substantial increases in testosterone and androstenedione (a weaker androgen than testosterone) at puberty, there is only a slight rise in females. The sexes also differ in their levels of

---

of the gonadostat could be described as increased sensitivity to the negative feedback of gonadal sex steroids, which inhibits the GnRH pulse activator during childhood. The mechanism by which the GnRH pulse activator is "dis-inhibited" for puberty to commence is not clear.

TABLE 5.1. *The Five Pubertal Stages for Breast and Pubic Hair Growth in Girls.*

	Characteristic	
Stage	Pubic Hair Development	Breast Development
1.	No pubic hair.	No breast development.
2.	There is a small amount of long pubic hair chiefly along vaginal lips.	The first sign of breast development has appeared. This stage is sometimes referred to as the breast budding stage. Some palpable breast tissue under the nipple, the flat area of the nipple (areola) may be somewhat enlarged.
3.	Hair is darker, courser, and curlier and spreads sparsely over skin around vaginal lips.	The breast is more distinct although there is no separation between contours of the two breasts.
4.	Hair is adult in type, but area covered is smaller than in most adults. There is no pubic hair on the inside of the thighs.	The breast is further enlarged and there is greater contour distinction. The nipple including the areola forms a secondary mound on the breast.
5.	Hair is adult in type, distributed as an inverse triangle. There may be hair on the inside of the thighs.	Breast size may vary in the mature stage. The breast is fully developed. The contours are distinct and the areola has receded into the general contour of the breast.

Source: Table adapted and reproduced from W. A. Marshall and J. M. Tanner, Variations in the pattern of pubertal changes in girls, Archives of Disease in Childhood, *44* [1969] pp. 291–303. Copyright 1969 by BMJ Publishing Group.

dehydroepiandrosterone (DHEA) and DHEAS (sulfated form of DHEA), hormones that mark the beginning of adrenarche, the period of initial increases in adrenal androgen hormones, which occurs at 6 to 7 years of age in both sexes. Levels of DHEA and DHEAS are similar between the sexes until late puberty, when males begin to have higher levels than females. This difference persists into adulthood (Fechner, 2002).

Physical Changes of Puberty

In females, secondary sexual characteristic development is a result of estrogens from the ovaries. Breast budding is generally the first sexual characteristic to appear and is most commonly classified by Marshall and Tanner's (1969) five stages of development, as illustrated in Table 5.1. Breast development begins in girls living the United States between ages 8 and 13, with a mean age of 9.96 for White girls and a mean age of 8.87 for Black girls (Herman-Giddens et al., 1997). The process of developing mature breasts from breast budding takes approximately 4.5 years, regardless of whether girls enter puberty earlier or later than average (Brooks-Gunn & Reiter, 1990). Pubic hair development typically begins shortly after breast budding; however, approximately 20% of girls experience pubic hair development prior to breast budding.

Pubic hair development begins for girls in the United States between the ages of 8 and 13 years, with a mean age 10.5 years in White girls and 8.8 years for Black girls (Herman-Giddens et al., 1997). Table 5.1 illustrates the five stages of pubertal hair development in girls. Menarche is a late sign of pubertal development in girls and occurs following the peak in height velocity and during the rapid increase in weight and body fat (Tanner, 1978). The mean age of menarche in North America is 12.88 years for White girls and 12.16 years for Black girls (Herman-Giddens et al., 1997).

In males, secondary sexual characteristic development is a result of testosterone from the testes. The onset of testicular growth is the initial sign of pubertal development, which occurs on average between ages 11 and 11.5, but can begin as early as age 9.5 (Brooks-Gunn & Reiter, 1990). Similar to girls, the most common classification of testicular and pubic hair development in males is Tanner's five stages of development (Marshall & Tanner, 1969), as illustrated in Table 5.2. Pubic hair growth begins on average at about age 12; however, 41% of boys are in Tanner stage IV of testicular growth when initial pubic hair growth begins. The average length of time between initial genital growth and the development of mature genitalia in boys is 3 years (Brooks-Gunn & Reiter, 1990). Spermarche, or first ejaculation, usually occurs between 13 and 14 years of age. More noticeable physical changes in boys include voice changing and the development of facial hair, which occur predominantly in early adolescence (Brooks-Gunn & Reiter, 1990).

Compared to girls, not as many studies have been conducted with boys that compare timing of pubertal onset across ethnic groups. Based on data from the Third National Health and Nutrition Examination Survey (NHANES III), conducted between 1988 and 1994, Black boys had earlier median and mean ages for Tanner stages than the White and Mexican American boys (Sun et al., 2002). These findings parallel the findings that Black girls begin puberty earlier than White girls (Herman-Giddens et al., 1997).

The developmental course of physical changes during puberty for girls and boys is demonstrated in Figures 5.1 and 5.2, respectively. Boys typically begin pubertal development about a year later than girls. Gender differences are also evident in regard to the alterations in linear growth, body composition, and the regional distribution of body fat during puberty. The

TABLE 5.2. *The Five Pubertal Stages for Penile and Pubic Hair Growth in Boys.*

	Characteristic	
Stage	**Pubic Hair Development**	**Penile Development**
1.	There is no pubic hair, although there may be a fine velus over the pubes similar to that over other parts of the abdomen.	The infantile state that persists from birth until puberty begins. During this time the genitalia increase slightly in overall size but there is little change in general appearance.
2.	Sparse growth of lightly pigmented hair, which is usually straight or only slightly curled. This usually begins at either side of the base of the penis.	The scrotum has begun to enlarge, and there is some reddening and change in texture of the scrotal skin.
3.	The hair spreads over the pubic symphysis and is considerably darker and courser and usually more curled.	The penis has increased in length and there is smaller increase in breadth. There has been further growth of the scrotum.
4.	The hair is adult in character but covers an area considerably smaller than in most adults. There is no spread to the medial surface of the thighs.	The length and breadth of the penis have increased further and the glans has developed. The scrotum is further enlarged and the scrotal skin has become darker.
5.	The hair is distributed in an inverse triangle as in the female. It has spread to the medial surface of the thighs but not up the linea alba or elsewhere above the base of the triangle.	The genitalia are adult in size and shape. The appearance of the genitalia may satisfy criteria for one of these stages for a considerable time before the penis and scrotum are sufficiently developed to be classified as belonging to the next stage.

Source: Table adapted and reproduced from N.M. Morris and J.R. Udry, Validation of a self-administered instrument to assess stage of adolescent development, *Journal of Youth and Adolescence*, 9 [1980], pp. 275–276. Reprinted with kind permission of Springer Science and Business Media.

pubertal growth spurt begins about 2 years earlier for females compared to males, and it also occurs at an earlier stage in puberty in girls than it does it boys (Fechner, 2003). Girls average a peak height velocity of 9 cm/yr at Tanner stage II and a total height gain of 25 cm during pubertal growth (Marshall & Tanner, 1969). Boys attain a mean peak height velocity of 10.3 cm/yr during Tanner stage IV and gain 28 cm in height total (Marshall & Tanner, 1970). Increases and redistribution of body fat also occur in girls and boys during puberty. Prepubertally, lean body mass, bone mass, and

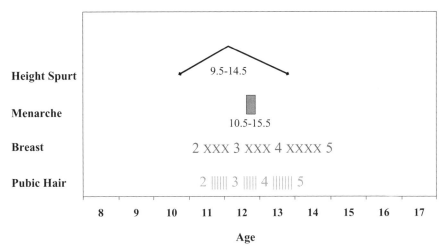

FIGURE 5.1. The developmental course of four pubertal processes for girls. *Source*: (From J. M. Tanner, *Growth at Adolescence*, Oxford: Blackwell Scientific [1962], p. 36. Copyright 1962 by Blackwell Scientific. Reprinted with permission.)

body fat are about equal in boys and girls. However, postpubertal boys have 1.5 times the lean body mass and bone mass of postpubertal girls, and postpubertal girls have twice as much body fat as postpubertal boys (Grumbach & Styne, 1998).

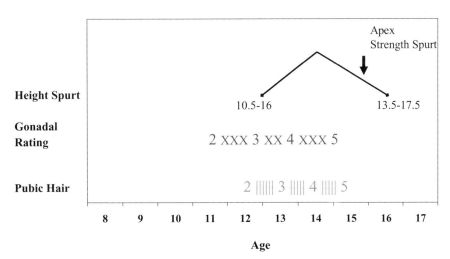

FIGURE 5.2. The developmental course of four pubertal processes for boys. (*Source*: From J. M. Tanner, *Growth at Adolescence*, Oxford: Blackwell Scientific [1962], p. 30. Copyright 1962 by Blackwell Scientific. Reprinted with permission.)

Central Nervous System and Tissue Sensitivity to Sex Steroids

Levels of circulating steroid hormones explain on average less than half the variance in morphological pubertal development and growth in girls and boys (Nottelmann et al., 1987). One reason for this is that most hormone measurements are collected at only one point in time. Some of the variance could also be accounted for by structural differences in the central nervous system (CNS) or in peripheral target tissues such as the breast. For example, breast development is partly controlled by the number and type of breast tissue receptors and other intracellular conditions (Layman, 1995). End-organ sensitivity in the CNS (that is, how reactive neural tissues are to hormones and neurotransmitters) may also play a role, although more research is needed in this area (Sanborn & Hayward, 2003).

Steroid hormones may affect sexual differentiation of the brain via influencing cell proliferation, cell migration, ontogenetic cell death, synaptogenesis, and neuroregulation (Casper, 1998; MacLusky & Naftolin, 1981; Phoenix, Goy, Gerall, & Young, 1959). As most sexual dimorphisms in brain morphology are established prenatally, changes in human brain anatomy have not been well studied during the pubertal development period (Giedd, Castellanos, Rajapakse, Vaituzis, & Rapoport, 1997). In one cross-sectional study that examined brain dimorphisms in children between 4 and 18 years of age, amygdala and hippocampal volume increased for both sexes but with the amygdala increasing significantly more in males than in females and hippocampal volume increasing more in females (Giedd et al., 1997). Research with rats that involves highly specific probes for estrogen action in the CNS has uncovered new classes of estrogen receptors in rat brains (Laflamme, Nappi, Drolet, Labrie, & Rivest, 1998; Mitchner, Garlick, & Ben-Jonathon, 1998; Osterlund, Kuiper, Custafsson, & Hurd, 1998). The nature of these estrogen receptors in humans and how they may relate to pubertal variation between individuals are not known.

MEASURING PUBERTY

Self- and Parent-Report Measures

The majority of studies that assess secondary sexual characteristic development include self- or parent-report ratings of Tanner's five stages of development (Marshall & Tanner, 1969). Although the most accurate Tanner ratings are those assessed by health professionals via visual inspection and sometimes palpitation of the breast, self- or parent-report ratings are much more feasible to obtain. Studies report correlations between parent and examiner ratings of Tanner stages ranging from .75 to .87 (Brooks-Gunn, Warren, Rosso, & Gargiulo, 1987; Dorn, Susman, Nottelmann,

Inoff-Germain, & Chrousos, 1990). Studies have also examined the validation of self-reported maturation based on Tanner drawings. Pearson correlation coefficients, when excluding testicular size, have been reported as .60 or above for the self- and physician reports (Morris & Udry, 1980). Dorn and colleagues (1990) reported correlations between self- and physician reports ranging between .77 and .91, which were slightly more accurate than the parent ratings in the same study. The Pubertal Development Scale (PDS) is another commonly used measure that includes questions about growth spurt, body hair (not specifically pubic hair) and skin change in boys and girls, facial hair growth and voice change in boys, and breast development and menarche in girls, rated on 4-point scales, from "no development" to "development already completed" (Petersen, Crockett, Richards, & Boxer, 1988). Correlations between physician Tanner ratings and self-reports of the PDS were between .61 and .67 (Brooks-Gunn et al., 1987).

When physician ratings are feasible, concerns have been raised regarding the validity of assessing breast development via visual inspection without palpation. By visual inspection only, it is sometimes difficult to distinguish fat tissue from real breast tissue, especially in overweight girls. In the Herman-Giddens and colleagues (1997) study of pubertal development in girls in pediatric settings, the data only included ratings of breast development by visual inspection. However, 39% of physicians assessed breast stage by palpation, as well as visual inspection. Only 4% of the girls with breast development by inspection had no breast tissue by palpation only 1.7% of girls who had no breast development by inspection had breast tissue by palpation (Kaplowitz, Slora, Wasserman, Pedlow, & Herman-Giddens, 2001). In addition, findings showed that the occasional misclassification of breast tissue was just as likely to occur in thin girls as in overweight girls. This follow-up evaluation of the pediatric setting study supports Tanner staging via inspection of breast development as a valid method of assessment, regardless of whether the girls are overweight.

In regard to measuring height, self-report data are fairly accurate, even for young adolescents. Correlations between self- and actual reports of weight and height range between .75 and .98 (Brooks-Gunn et al., 1987; Goodman, Hinden, & Khandelwal, 2000). It is unusual for researchers to ask parents to measure their children at home, so to our knowledge parental reports of height have not been validated.

Hormone Measures

Pubertal development can also be assessed via measuring hormonal biomarkers. As explained in the previous section, a major hormone involved in the regulation of puberty is GnRH. However, it is difficult to measure GnRH because it has a short half-life and is transported directly to the pituitary (Rockett, Lynch, & Buck, 2004). Therefore, pubertal development is most often measured through hormones regulated directly or indirectly by GnRH, including the gonadotropins (LH, FSH) and sex steroid hormones (testosterone and estrogen). Methods for measuring these hormones include blood draws, salivary collection, blood spots, and urinary collection.

Compared to blood draws, advantages of salivary and urinary assessments are that they are noninvasive, painless, and highly acceptable to most research participants. Drawbacks are that salivary and urinary studies provide limited estimates of total output by peripheral glands and preclude evaluation of central neuroendocrine regulation. Also, there are technical demands on salivary and urinary assays for steroids of low concentrations (e.g., estradiol) that may prevent their use in some developmental studies. Advantages of blood spots are that they are minimally invasive and require very little blood, they allow for multiple collections so that mean hormone concentrations can be measured, few factors compromise validity of blood spot samples for analysis, and assays for blood are more sensitive than the assays for urine or saliva (Worthman & Stallings, 1997).

Important Issues to Consider When Measuring Puberty

The appropriate measure(s) of puberty to use in a study must be "purpose-dependent" (Hayward, 2003). Indicators of puberty are correlated, but not equivalent, as each indicator captures a different aspect of the pubertal process (Brooks-Gunn & Warren, 1985; Graber, Petersen, & Brooks-Gunn, 1996). Each indicator involves limitation in how it is measured. For example, validity of Tanner self-ratings may vary by degree of body image disturbance (Hick & Katzman, 1999; Litt, 1999), and cross-sectional measurements of hormonal levels are difficult to interpret (Hayward, 2003). The ideal way to measure the pubertal process would be multiple indicators of puberty collected longitudinally.

It is also important to note that at different stages of puberty, the correlations between the various manifestations of puberty change dramatically (Angold, Worthman, & Costello, 2003). For example, as girls begin to cycle, age and Tanner stage may not be as correlated with FSH, LH, and estrogen, because their levels become primarily controlled by the menstrual cycle. Even though circulating hormone levels are the best available correlate of hormonal action in the CNS, self-report of breast development is a better measure of breast development than is circulating estrogen level (Angold et al., 2003). Therefore, researchers must consider the meaning of the pubertal indicator and the developmental range when designing studies involving puberty.

THE DECREASING AGE OF PUBERTAL ONSET

The data presented in the previous section on average age of breast and pubic hair development in girls and boys in the United States stem from study conducted by 225 clinicians in pediatric practices belonging to Pediatric Research in Office Settings (PROS), a practice-based research network that is part of the American Academy of Pediatrics (Herman-Giddens et al., 1997). Prior to this study, the study most often quoted as defining normative ages for puberty was the Marshall and Tanner (1969) longitudinal study of 192 White girls living in a children's home, assessed every 3 months from age 8 through age 18. Studies of pubertal onset comparable in scope to the British (Marshall and Tanner) study had not been conducted in the United States. The PROS study was cross-sectional, consisting of 17,000 girls between the ages of 3 and 12 whose breast and pubic hair development was assessed via practitioner ratings (Herman-Giddens et al., 1997). Findings from this study indicate that the mean age of pubertal onset for girls has decreased by as much as a year, compared to the Marshall and Tanner data. Furthermore, the study found that Black girls are beginning breast and pubic hair development about 6 months to a year earlier than White girls, although the reasons for the earlier age of secondary sexual characteristic onset for Black girls are not clear.

The National Health and Nutrition Examination Survey (third cycle) found similar results (Wu, Mendola, & Buck, 2002). Physician ratings of Tanner stages of breast and pubic hair development were available for 1,623 girls between ages 8 and 16 years. Mean age of onset of pubic hair and breast development was 9.5 and 9.5 years for Black girls, respectively, 10.3 and 9.8 years for Mexican American girls, and 10.5 and 10.3 years for White girls. These ethnic differences remained even after adjustment for current body mass index and several social and economic variables (Wu et al., 2002).

Changes in the age of menarche in the United States across the past few decades are not as clear. A large study of puberty in U.S. girls conducted between 1966 and 1970, the U.S. National Health Examination Survey, assessed age of menarche in girls between the ages of 12 and 17 (Harlan, Harlan, & Grillo, 1980). Data from this study indicated that the mean age of menarche was 12.8 years for White girls and 12.5 years for Black girls. The mean age of menarche for White girls in the PROS study, 12.88 years, was similar to the 1980 report, whereas the mean age for Black girls in the PROS study, 12.16 years, was slightly lower (Herman-Giddens et al., 1997). A study based in Chicago neighborhoods in the mid- to late 1990s, which included Black, White, and Latina girls from low, middle, and high socioeconomic status, reported younger ages of menarche than the previously mentioned studies (Obeidallah, Brennan, Brooks-Gunn, Kindlon, & Earls, 2000). Results showed that Latina girls reached menarche at a younger age (11.58 years) than Black girls (11.93 years), after controlling for socioeconomic factors. Adjusting for socioeconomic status, no significant differences were found between White girls (12.04 years) and Latinas or between White girls and Black girls. It should be noted that not all girls in this study had begun to menstruate, so the means are lower than they ultimately will be, although the ethnic differences should hold.

Although findings for boys indicate that there may be variations in puberty onset by race (Sun et al., 2002), no clear data indicate that boys are entering puberty earlier now than they were 20 or 30 years ago. Implications for potential effects of earlier pubertal timing on depression in girls will be reviewed later in the chapter.

Potential Explanations for the Decreasing Age of Pubertal Onset

Several hypotheses have been proposed that focus on environmental factors as an explanation for the earlier age of pubertal onset in girls. One such hypothesis is that exposure to environmental toxins may mimic estrogens in the body and thus stimulate pubertal development. Two epidemics of early puberty, one in Italy and the other in Puerto Rico, are suspected to have been caused by exposure to estrogens in food, specifically meat and poultry (Fara et al., 1979; Saenz de Rodriguez, Bongiovanni, & Conde de Borrego, 1985). However, no localized outbreaks of early breast development have been reported in the continental United States (Kaplowitz, 2004), and no published data provide evidence that an increased overall exposure to environmental estrogens leads to earlier puberty (Paretsch & Sippell, 2001). Another hypothesis is that manufactured compounds being released into the environment may interfere with the human endocrine system. Studies in this area are difficult because pubertal onset may occur many years after exposure to the chemical in question (Kaplowitz, 2004). Although a few studies have managed to collect data on exposure to environmental contaminants at the time of birth and relate it to growth and development (Michels Blanck et al., 2000; Colon, Caro, Bourdony, & Rosario, 2000; Gladen, Ragan, & Rogan, 2000; Krstevska-Komstantinova et al. 2001), the results do not provide enough compelling evidence that the overall trend for earlier puberty is linked with environmental contaminants. Hormone-containing hair products have been targeted as a reason for earlier pubertal onset (Zimmerman, Francis, & Poth, 1995), but this hypothesis requires further examination.

Other hypotheses regarding the earlier onset of puberty focus on more intrinsic factors. For example, perinatal factors, such as birth weight, have been found to play a role in subsequent pubertal development. In one study, girls who were smaller at birth but had a rapid catch-up period of growth between birth and age 6 were earlier maturers (Persson et al., 1999). Mechanisms for this association are not clear, but it is likely that the prenatal environment may influence subsequent timing of onset of development, given that sex hormones are active prenatally in organizing the brain for subsequent pubertal development and reproductive functioning (Fechner, 2002). A more widely discussed hypothesis is that higher body fat is associated with earlier maturation. In general, overweight girls tend to mature earlier than girls of normal weight, and thin girls tend to mature later. Numerous studies have indicated that in the past 30 years, there is an increasing prevalence of obesity in both sexes, at all ages, and in all racial and ethnic groups. Because obesity is widespread and prevalent throughout all parts of the United States, it makes sense to examine obesity as a link to earlier pubertal onset. Kaplowitz and colleagues (2001) examined the role of body mass index (BMI) in earlier pubertal onset using data from the PROS study. BMI standard deviation scores (z scores) were computed to compare each girl's BMI with what was normal for her age. A key finding of the study was that 6- to 9-year-old girls with early breast development had significantly higher BMI scores than the girls of the same age and race who were prepubertal. When the difference in BMI z scores between Black and White girls was controlled, Black girls still had an earlier onset of puberty than White girls. Furthermore, Black and White girls who had pubic hair but no breast development were also more overweight than prepubertal girls, even though hormonal regulation of pubic hair and breast development is quite different (Kaplowitz et al., 2001). The hormone leptin has been proposed as a mechanism linking body mass index with puberty (Clayton & Trueman, 2000). Leptin, a protein produced by fat cells, is involved in regulation of appetite and body composition. Evidence has accrued over the past 10 years that leptin also plays a role in the regulation of puberty (Barash et al., 1996). Leptin levels rise progressively during puberty in normal girls, beginning at age 7–8; their rise occurs before increases in LH and estradiol, which means that leptin could be a trigger for the production of puberty hormones (Ahmed et al., 1999; Blum et al., 1997; Garcia-Mayor et al., 1997).

A final hypothesis to be discussed is that psychosocial or environmental stress is associated with earlier puberty. In particular, stressful family situations have been linked to earlier onset of puberty (Graber, Brooks-Gunn, & Warren, 1995; Moffitt, Caspi, Belsky, & Silva, 1992; Surbey, 1990). Lower warmth in parent–child relationships has been associated with earlier age of menarche after controlling for the effect of maternal age at menarche and level of breast development (Graber et al., 1995). Father's absence in the childhood years has also been predictive of earlier maturation (Ellis & Garber, 2000; Surbey, 1990). In girls not living with their biological parents, the presence of a stepfather rather than the absence of a father was more strongly associated with earlier pubertal maturation (Ellis & Garber, 2000). A Polish study found that age of menarche in girls who experienced stressful family dysfunction was 0.4 years earlier than age of menarche in girls from families free of major trauma. Mechanisms for the relationship between family stress and early puberty are not clear. Estrogens may play a role, as there is an increasing body of evidence showing effects of stress on estrogen systems in adults (McEwen, 1994).

Potential Reasons for Ethnic Differences in Age of Pubertal Onset

A finding across three large-scale studies conducted in the 1980s and 1990s, the PROS study (Herman-Giddens et al., 1997), NHANES III (Wu et al., 2002), and the National Heart, Lung, and Blood Institute Growth and Health Study (Morrison et al., 1994), is that Black girls begin breast and pubic hair development about a year earlier than White girls and begin menses about half a year earlier. The reasons for the earlier age of secondary sexual characteristic onset for Black girls are not clear. Possible factors to consider include differences in diet and weight, environmental hazards or environmental estrogens, or differences in contextual stress and cultural attitudes between ethnic groups (Graber, 2003). Physiological differences that have been hypothesized to play a role in timing of pubertal onset include lower insulin sensitivity (Arslanian, Suprasongsin, & Janosky, 1997) and higher serum leptin levels (Wong et al., 1998) in Black children compared to White children. None of these hypotheses have been investigated extensively.

DEPRESSION IN ADOLESCENCE: TYPES AND PREVALENCE

In discussing depression during adolescence, it is important to distinguish between depressive affect, syndromes, and clinical disorders (Compas, Ey, & Grant, 1993; Petersen et al., 1993). Depressed affect or moods refer to states of sadness or unhappiness and are not assessed in connection with other symptoms or in terms of their duration. Depressive syndromes, or subclinical problems, are defined as constellations of symptoms that co-occur in a statistically consistent manner (Jensen, Brooks-Gunn, & Graber, 1999). Even though moderate levels of depressive symptoms may not meet criteria for a psychiatric disorder, they are associated with significant impairment in school and peer functioning in children (Nolen-Hoeksema, Girgus, & Seligman, 1992; Susman, Dorn, & Chrousos, 1991). A diagnosis of clinical depression is based on *DSM-IV* criteria (American Psychiatric Association, 1994) and is greater than depressive syndromes in severity and duration. A diagnosis of clinical depression must be derived from clinical interviews, whereas depressive affect and syndromes can be assessed via self- or parent-report scales. On the basis of point prevalence data, 15–40% of adolescents experience significant depressed mood, 5–6% exhibit depressive syndrome in the clinical range, and 1–3% of adolescents are diagnosed with a depressive disorder (Compas et al., 1993).

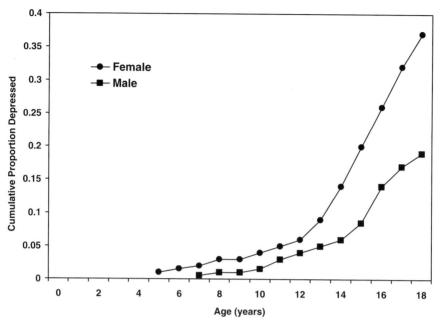

FIGURE 5.3. Probability of experiencing an episode of major depressive disorder as a function of age and gender. *Source*: Reprinted from *Clinical Psychology Review, 18,* Lewinsohn, P. M., Rohde, P., & Seeley, J. R., "Major depressive disorder in older adolescents: Prevalence, risk factors, and clinical implications," pp. 765–794, Copyright (1998), with permission from Elsevier.

As seen in Figure 5.3, it is now generally accepted that depression as a diagnosis (major depressive disorder) is twice as prevalent in adolescent and adult females compared to adolescent and adult males (American Psychiatric Association, 1994; Born & Steiner, 2001; Lewinsohn, Rhode, & Seeley, 1998). The divergence in rates of diagnosed depression seems to emerge around age 12 or 13, or around Tanner stage III of puberty in girls (Angold, Costello, & Worthman, 1998). Although results of studies of depressive *diagnoses* have been consistent in regard to the gender differential, findings from studies that use questionnaires and scale scores to assess depressive symptoms and depressed mood have been more mixed (Angold, Erkanli, Silberg, Eaves, & Costello, 2002). For example, a psychometric study of the Children's Depression Inventory (CDI), the most widely used questionnaire in this area (Kovacs, 1992), found that both 7- to 12-year-old and 13- to 17-year-old boys had higher mean scores than girls. The study did not find an increase in mean symptoms scores with age or an interaction between age and gender. Other studies that used the CDI or other scales found that girls had higher mean scores (Angold et al., 2002).

In general, any differences between boys and girls in mean scores on depression scales have not been as significant as differences between the sexes in diagnosed depression (Compas et al., 1997). However, results may differ when scale scores in the upper extreme are compared. In an examination of two large, longitudinal studies of twins and singletons aged 8–17, boys' overall mean scores fell over this age range whereas girls' mean scores fell from age 9 to age 11 then increased from age 12 to age 17. When the researchers considered a cut point with the upper 6% of scores of a 13-item self-report scale, the expected female-to-male 2:1 ratio emerged (Angold et al., 2002).

MODELS LINKING PUBERTAL PROCESSES AND DEPRESSION

The research that links puberty with depression involves two main categories of models – pubertal status and timing of puberty (Brooks-Gunn, Graber, & Paikoff, 1994; Buchanan, Eccles, & Becker, 1992; Graber, Brooks-Gunn, & Archibald, 2005; Graber, Brooks-Gunn, & Warren, in press). Pubertal status models refer to adolescents' degree of physical maturation and hormone levels. Models that examine hormone levels are considered direct effect models, and those that measure physical change secondary to hormone changes are considered indirect effect models. The theory behind status models is that girls may experience negative reactions or receive negative feedback from others about their development when they reach certain stages, or they may feel that certain behaviors are expected with increasing physical development. Pubertal timing models suggest that being either an early maturer or out-of-synch (earlier or later) with one's peers is what affects depressive outcomes. Other types of models suggest that it is not pubertal development per se, but factors that interact with the challenges of pubertal development that lead to more adjustment problems (Nolen-Hoeksema & Girgus, 1994). For example, risk factors for depression may be more common in girls than in boys before adolescence, but depression results when these factors interact with the challenges specific to early adolescence, such as pubertal development.

In general, any model describing the relationship between pubertal and social events in depressive symptomology should be mediated rather than direct, bidirectional rather than unidirectional, and interactive rather than additive (Brooks-Gunn et al., 1994), as Figure 5.4 illustrates. A framework with three potential mediational processes between the hormonal changes of puberty and short-term effects on affective states such as depression is illustrated. The first mediational pathway shows timing of secondary sexual characteristic development linking hormonal changes and affective states. The second mediational pathway highlights the effect of social experiences, including perceptions of puberty, on affective states; this effect stems partly from hormonal changes and is also a response to changes in

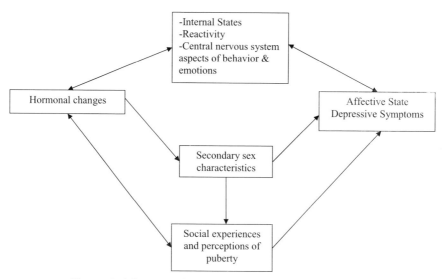

FIGURE 5.4. Theoretical framework model linking puberty with depressive symptoms in girls. *Source*: Figure is adapted from *Journal of Research on Adolescence, 4*, Brooks-Gunn, J., Graber, J. A., & Paikoff, R. L., "Studying links between hormones and negative affect: Models and measures," pp. 469–486, Copyright (1994), with permission from Blackwell Publishing.

physical development during puberty. The third mediational link refers to internal states, such as central nervous system changes that stem from the hormonal changes, and individual differences in arousal and physiological reactivity. Figure 5.4 illustrates that some of the pathways are bidirectional; for example, girls' social experiences and behaviors may have effects on the hormonal systems, which may affect the timing of pubertal development. The following sections review the empirical evidence associated with the different aspects of this theoretical framework.

Hormone Effects

The majority of studies linking pubertal hormone levels and depression in girls have examined how increases in testosterone and estradiol during the course of puberty are linked to depressive affect (e.g., see Buchanan et al., 1992). Findings reveal that effects vary across study, by gender, by hormone, and by outcome. In studies focused on hormone–behavior links in girls only, findings indicate that increases in estradiol, specifically during the most rapid period of increase during puberty, have been associated with negative affect (Brooks-Gunn & Warren, 1989; Warren & Brooks-Gunn, 1989). Estradiol levels increase dramatically during puberty, and they correlate strongly with many of the other hormone levels. In the first

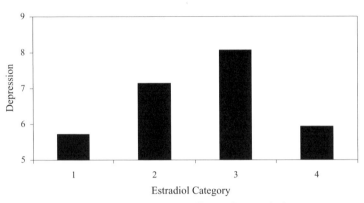

FIGURE 5.5. Significant quadratic effect of estradiol categories on depressive affect in girls. *Source*: Figure is adapted from *The Journal of Clinical Endocrinology and Metabolism, 69*, Warren, M. P., & Brooks-Gunn, J., "Mood and behavior at adolescence: Evidence for hormonal factors," pp. 77–83, Copyright 1989, The Endocrine Society.

set of analyses, ordinary least-squares regressions (OLS) were performed to compare findings with those of other researchers. Age and five hormones, FSH, LH, estradiol, testosterone, and DHEAS, were entered as possible predictors. None of the individual variables had significant beta weights in the linear analysis.

The second set of analyses tested both linear and nonlinear effects (the square of each hormone was entered to test for nonlinear associations). Findings indicated a nonlinear effect of estradiol for depressive affect and a negative linear effect of DHEAS for aggressive affect. Based on the significant nonlinear estradiol finding, girls were categorized into four hormonal stages based on the range of their estradiol levels; each range affects reproductive organs and functioning of the reproductive system differently. Stage I girls were considered prepubertal, stage II girls were experiencing beginning pubertal development, stage III girls were considered to be in mid- or late puberty, and stage IV girls were experiencing cyclic menstrual function. Estradiol levels were 0–25, 26–50, 51–74, and greater than 75 pg/mL, respectively, for each stage.

As seen in Figure 5.5, results indicated a significant quadratic effect for girls' depressive affect while controlling for age, with highest levels of depressive affect in the groups (stages II and III) that demonstrated initial increases in estradiol (Brooks-Gunn & Warren, 1989; Warren & Brooks-Gunn, 1989). In follow-up analysis, this hormone–affect association was found to persist over the course of 1 year (Paikoff, Brooks-Gunn, & Warren, 1991). The curvilinear nature of the hormone–affect association fits the premise that activational effects may be greatest when the endocrine system is being turned on. However, the magnitude of the hormone effect was small, accounting for only 4% of the variance in negative affect. Other studies have also indicated that when hormone–affect associations are found, they generally account for a small portion of variance in behavior (Buchanan et al., 1992). Social factors, as measured in this study by positive and negative life events, accounted for more variance than hormonal pubertal factors alone (8–18%), as did the interaction between negative life events and pubertal factors (9–15%). Thus, hormone effects on affect may be overshadowed by environmental events.

Angold and colleagues have conducted a number of studies on links between pubertal factors and depression. The first study assessed age, pubertal timing, and Tanner stage on the probability of depression in both boys and girls over four waves of data collection from the Great Smoky Mountains Study (Angold et al., 1998). Depression included three diagnoses: *DSM-IV* major depressive episode, dysthymia, and minor depressive disorder. Findings revealed that only after the transition to mid-puberty (Tanner stage III and above) were girls more likely than boys to be depressed. Timing of the pubertal transition did not affect depression, whether measured by onset of menarche or Tanner stage development. These findings imply that some aspect of puberty itself was related to depression, rather than the age at which the pubertal level was achieved. Further analyses considered HPG axis hormonal effects on depression in girls to disentangle the effects of the morphological changes of puberty and the hormonal changes underlying them (Angold, Costello, Erkanli, & Worthman, 1999). Results indicated that the effects of testosterone and estrogen in the model eliminated the apparent effect of Tanner stage. The odds ratio associated with Tanner stage was reduced from 2.9 to 1.0 by the addition of these hormones, meaning that before the hormones were added to the model, the likelihood of depression for girls at Tanner stage III and above was nearly three times as great as the likelihood of depression for girls below Tanner stage III. An odds ratio of 1 indicates that there is no difference in the likelihood of depression based on Tanner stage grouping. To examine the possibility that hormone thresholds may be present, the researchers divided the ranges of testosterone and estrogen into quintiles and plotted rates of depression for each quintile. The effect of testosterone was particularly marked and was manifested only above a certain threshold, above the 60th percentile of testosterone levels in this sample (corresponding to a level of 24.7 ng/dl). The relationship between testosterone and depression was nonlinear in that there is a sharp jump between the third and fourth quintiles, whereas the effects of estrogen on depression appeared to be reasonably linear. In this study, levels of FSH and LH had no effect on depression rates over and above those accounted for by testosterone and estrogen. The findings of this study parallel previous findings that negative affect is associated with higher levels of androgens and estrogens in adolescent girls and imply that hormones rather than physical status are the active ingredient in the effects of puberty on depression in girls (Angold et al., 1999).

More recently, Angold and colleagues hypothesized that the apparent effects of both testosterone and estrogen could represent an intracellular estrogenic effect (Angold et al., 2003), since when behavioral effects of testosterone in animals have been investigated at the level of the brain receptors involved, most have proven to occur via estrogen receptors following intracellular aromatization, or conversion, of testosterone to estrogen (Hutchison, Schumacher, Steimer, & Gahr, 1990; Rasmussen, Torres-Aleman, MacLusky, Naftolin, & Robbins, 1990). Using the same sample of girls from previously described studies, the researchers combined levels of testosterone and estrogen, instead of entering testosterone and estrogen separately in analyses. The rationale for this decision was that both may act on the same receptors in their effects on depression. The sum of the measured molarities of testosterone and estrogen was called sex steroid level (SSL). The distribution of SSL was divided into deciles, and the prevalence of depression was plotted in the deciles. The plot showed pronounced threshold effects of SSL groups based on three cut points; the difference between each group was significant, with the highest rate of depressive symptoms for the SSL group in the 80th percentile and above. Additional analyses showed that the effects of pubertal hormone status were not explained by either changes in levels of life events or by the interaction of life events and SSL status. The threshold effect of the SSL hormones is similar to the findings of Brooks-Gunn and Warren (1989), in which a significant quadratic effect was found for girls' depressive symptoms, with highest levels of depressive symptoms in the groups that demonstrated initial increases in hormones. Both studies underscore the need to consider linear and nonlinear models when examining effects of hormones on behaviors.

Pubertal Status Effects

Examining links between pubertal status and depression involves comparisons of outcomes among adolescents at different levels or stages of key external signs of pubertal development (e.g., breast growth, pubic hair, testicular changes). Pubertal staging is usually indexed by some measure of Tanner stages that ranges from no signs of development to completed development. Pubertal status is considered important because it signifies that the adolescent is more adult-like in appearance, which may result in different responses from family and peers in the adolescents' social world, as well as changes in how adolescents view themselves.

Because hormonal changes are the cause of the changes in physical growth and development, it is often difficult to disentangle hormonal and status effects on depression. For example, as described in the previous section, a study by Angold and colleagues (1998) found that only after reaching Tanner stage III were girls more likely than boys to experience higher rates of depressive disorder. However, subsequent analyses showed that effects of elevated estradiol and testosterone levels eliminated effects due to secondary sexual characteristics (Angold et al., 1999). This study suggests that when pubertal status effects on adjustment are found, they are likely be driven by hormonal changes.

For girls in particular, the larger body size that accompanies pubertal development, especially if there is increased adiposity, may play a role in the development of depressive symptoms. Increased weight is inconsistent with the Western cultural ideal of thinness. More advanced pubertal development has been associated with less satisfaction with weight and to perceptions of being overweight for girls but not for boys (Richards, Boxer, Petersen, & Albrecht, 1990). Weight-related negative body image, weight dissatisfaction, and weight concerns were associated with increased depressive symptoms in a sample of early adolescent girls, even when controlling for objective weight status (Rierdan & Koff, 1997). It is likely that girls more often experience increased body size negatively due to the media images in Western cultures that value the thin physique of a prepubertal body over the mature body for girls (Attie & Brooks-Gunn, 1989; Parker, Nichter, Nichter, Vuckovic, Sims, C., Ritenbaugh et al., 1995). The increase in weight that accompanies advancing pubertal status cannot be overlooked when examining links between puberty and depression.

Pubertal Timing Effects

A large body of literature involves studies examining links between pubertal timing and adjustment in adolescents, mainly because there is substantial variation among individuals regarding when puberty begins and how it progresses (Tanner, 1970). Girls typically exhibit the external manifestations of puberty about 1 or 2 years earlier than boys. Early-maturing girls can therefore develop 3 to 6 years ahead of boys as well as developing earlier than on-time or late-maturing girls. Variations in pubertal timing are most likely a combination of genetic differences and environmental factors such as nutrition, exercise, and health conditions.

Classifications of maturational timing may differ by study, even when the same pubertal status measure is used, such as the Tanner stages. Many studies group adolescents using population norms (Duke-Duncan, Ritter, Dornbusch, Gross, & Carlsmith, 1985) or classify adolescents according to sample distribution into early, on-time, and late developers. Researchers have typically considered the earliest 20% within the distribution for that sample as the early developers and the latest 20% as the late developers (Brooks-Gunn & Warren, 1985; Graber et al., 1996). Another method is to use sample values to indicate the deviation from the mean, such as defining more than one standard deviation above or below the group mean as the threshold for early or late pubertal maturation (Alsaker, 1992). Some researchers have used ratings of perceived pubertal timing, in which adolescents rate their timing as earlier, the same, or later than their peers (Dubas, Graber, & Petersen, 1991; Graber, Lewinsohn, Seeley, & Brooks-Gunn, 1997; Graber, Seeley, Brooks-Gunn, & Lewinsohn, 2004; Obeidallah, Brennan, Brooks-Gunn, & Earls, 2004). These varied methods result in cross-study variation in the maturational and chronological ages of adolescents classified in the same timing group. For the cases in which the same classification system is used, different samples, having different distributions, may exhibit various percentages of off-time and on-time girls (even if the same definition of timing is used). It is also important to consider that when assessing pubertal timing, studies often do not include a complete design, in that at any one grade level or age, a complete range of pubertal growth will not be seen, resulting in restricted ranges (Brooks-Gunn, Petersen, & Eichorn, 1985).

A few hypotheses have emerged to explain links between pubertal timing and psychological development. The off-time hypothesis is the most general one; it predicts that both earlier and later development in girls and boys compared to one's same-age, same-gender peers is a risk factor for problem behaviors (Caspi & Moffitt, 1991). A more specific hypothesis is the gendered deviation pattern of pubertal timing effects, in which early maturation is a risk factor for females and late maturation is a risk factor for males. This hypothesis is based on the developmental pattern that girls, on average, mature earlier than boys. Girls who mature earlier than their peers or boys who mature later than their peers are considered to be in the deviant categories (Brooks-Gunn et al., 1985; Petersen & Taylor, 1980).

The early maturation hypothesis is also referred to as the stage-termination hypothesis (Petersen & Taylor, 1980). This hypothesis posits that early maturation is a risk factor for adjustment problems among both females and males across a range of outcomes (Brooks-Gunn et al., 1985; Caspi & Moffitt, 1991; Ge, Conger, & Elder, 1996; Tschann, Adler, Irwin, Mill Stein, Turner, & Kegeles et al., 1994). Early maturation may be disadvantageous because early maturers experience social pressure to adopt more adult norms and engage in adult behaviors, even though they may not be socially, emotionally, or cognitively prepared (Brooks-Gunn et al., 1985; Caspi & Moffitt, 1991) for the new experiences. This hypothesis involves the notion of stage termination (Petersen & Taylor, 1980), which means that early maturation disrupts the normal course of development such that early maturers have less time and are less experienced to handle adult behaviors.

Early maturation has been repeatedly associated with more depressive symptoms in girls, compared to on-time or later maturing peers

(Brooks-Gunn et al., 1985; Ge et al., 1996; Graber et al., 1997, 2003; Hayward et al., 1997; Stattin & Magnusson, 1990). In general, early maturing girls report more negative emotions than on-time or later maturing peers (Hayward, Gotlib, Schraedley, & Litt, 1999). Earlier maturers tend to experience higher levels of psychological stress and are more vulnerable to the exacerbation of childhood psychological problems (Ge et al., 1996). In a longitudinal study investigating links between pubertal transition and depressive symptoms in rural White youth living in Iowa, girls began to experience more depression than boys in the eighth grade, and this difference persisted through mid- and late adolescence (Ge, Conger, & Elder, 2001a). Girls who experienced menarche at a younger age subsequently experienced a higher level of depressive symptoms than their on-time and late-maturing peers, at each annual assessment during the 6-year study. Additionally, the interaction between early menarche and recent life events predicted subsequent depressive symptoms for girls. Interestingly, the significant main effect of gender on depressive symptoms disappeared when pubertal transition, recent life events, and their interaction were included in models, suggesting that pubertal factors may explain a significant part of the observed gender differences in depressive symptoms during adolescence.

In one of the only studies that has examined long-term consequences of pubertal timing on psychopathology in young adults, young women who had been early maturers (based on self-reports of perceived timing relative to one's peers) continued to have higher lifetime prevalence rates of major depression, anxiety, disruptive behavior disorders, and hence any Axis I psychiatric disorder, as well as higher lifetime rates of attempted suicide in comparison to other women. The on-time and late maturers did not catch up in rates of disorder (at least by age 24), and differences in lifetime prevalence rates were maintained into adulthood (Graber et al., 2003). These results imply that early maturing girls are at a unique risk for persistent difficulty with depression and other disorders during adolescence and continuing into young adulthood.

On the other hand, in boys, findings on links between pubertal timing and depressive symptoms are more inconsistent. In the National Institutes Mental Health (NIMH) study of puberty and psychopathology, a higher rate of negative emotional tone has been found in late maturing boys in mid- and late adolescence as compared to their agemates (Nottelmann et al., 1987). However, studies have also found that early maturing boys experience more internalizing symptoms (Petersen & Crockett, 1985; Susman et al., 1985, 1991) or that both early and late maturing boys show more depressive tendencies (Alsaker, 1992). Graber and colleagues (1997) found that late maturing boys experienced more internalizing symptoms than their on-time peers, and that both early and late maturing boys showed significantly higher rates of depression than on-time maturing boys. In a longitudinal study with White boys living in a rural area, early maturing boys, compared to their on-time and late maturing peers, exhibited more internalized distress (Ge, Conger, & Elder, 2001b). Results from a large-scale study of Black boys also indicated that early maturers reported higher levels of internalizing symptoms (Ge, Brody, Conger, & Simons, in press). In sum, although there is a trend for early and late maturing boys to experience more internalizing and depressive symptoms than their on-time peers, the findings between pubertal timing and depression for boys are not as consistent as they are for girls.

Findings from studies that examine links between pubertal timing and adjustment across ethnic groups have also been inconsistent. One study found associations between early menarche and depressive symptoms in White girls but not Black or Hispanic girls (Hayward et al., 1999), whereas the results of another study including only Black children ($N = 639$) showed that early maturing Black girls had higher rates of depressive symptoms compared to their on-time peers (Ge et al., 2003). A study that examined pubertal timing effects across Hispanic, Black, and White children from economically diverse Chicago neighborhoods found that girls in each ethnic group who matured *off-time*, that is, earlier or later than their same-age, same-gender peers, experienced more clinical levels of depression or anxiety, with strongest effects found in White girls (Foster & Brooks-Gunn, manuscript under review). It is important for researchers to further address associations between pubertal change and adjustment in non-White girls and boys, as this group has been understudied. It is possible that girls of different ethnic groups may differ from White children in theoretically important ways, such as in their preparation for puberty or in the responses from others within their ethnic group to their pubertal changes and attitudes about it.

PROPOSED MECHANISMS FOR EXPLAINING PUBERTAL EFFECTS

Although tests of specific pathways that might be predicted from each of the preceding hypotheses have not been conducted, several pathways have been proposed. One such pathway is the individual diathesis-stress model. This model is based on the idea that it is not puberty per se that is associated with more depression in girls, but that puberty accentuates the effects of psychosocial factors that exist prior to the onset of puberty. The transitional stress model posits that pubertal development is a biological transition that may be linked with increases in emotional arousal or distress. Finally, contextual models, which focus on the role of the social factors and contexts that coincide with the period of pubertal development, will be discussed. We will address whether girls experience more social challenges than boys or have different social challenges than boys and if girls react differently than boys to social challenges.

Individual Diathesis-Stress Model

The individual diathesis-stress model posits that psychosocial vulnerability factors that exist prior to adolescence accentuate the probability of increases in emotional distress in interaction with pubertal development (Caspi & Moffitt, 1991; Dorn & Chrousos, 1997; Nolen-Hoeksema & Girgus, 1994; Susman, Dorn, & Schiefelbein, 2003). For example, this model was supported by a longitudinal study of the behavioral responses of adolescent girls to the onset of menarche (Caspi & Moffitt, 1991). Beginning at age 3, girls were assessed with a battery of psychological, medical, and sociological measures every 2 years, through age 15. Systematic interactions between premenarcheal personality and age of onset of menarche were examined. Results indicated that the early onset of menarche magnified and accentuated behavioral problems among girls who were predisposed to behavior problems earlier in childhood. The group that experienced the most adjustment difficulties throughout adolescence was the early maturing girls with a history of behavioral problems earlier in childhood. To further test this model, longitudinal studies are needed in which levels of adjustment, the risk factors for poor adjustment, and the pubertal challenges with which these risk factors might interact are tracked as children transition from childhood into adolescence (Nolen-Hoeksema & Girgus, 1994). Studies on puberty have rarely explored the full range of adolescence, including the transition from childhood and the transition into adulthood. As such, many studies miss the onset of puberty for a portion of their samples, which makes it challenging to validate this model.

Transitional Stress Model

The basis of the transitional stress model is that reproductive transitions are periods of development that involve reorganization of biological and behavioral systems (Susman, 1997, 1998). This reorganization may increase emotional arousal and the onset of psychiatric disorders (Dorn & Chrousos, 1997). In the case of depression, aroused physiological states may trigger increased moodiness, sudden mood changes, feelings of self-consciousness, or elevated intensity of moods, all of which, if interpreted negatively, could lead to adjustment problems. To test this proposed mechanism between pubertal hormones, pubertal timing, and depression (among other things), these issues were examined in a study of 100 adolescent girls (Graber, Brooks-Gunn, & Warren, in press). Two hormones were focused on in this study to identify mechanisms between hormones and depressive symptoms – estradiol and DHEAS. In this investigation, rapidly changing levels of estradiol had previously been found to be predictive of depressive symptom level (Warren & Brooks-Gunn, 1989). DHEAS taps adrenal androgen activity and has, along with other adrenal

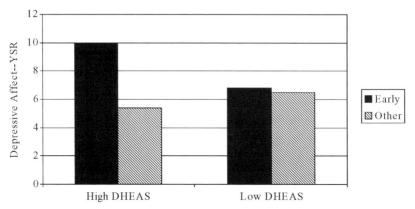

FIGURE 5.6. The interaction of high DHEAS levels (as an index of high adrenal activity) and early maturation on girl's depressive affect. *Source*: J. A. Graber, J. Brooks-Gunn, J., & A. B. Archibald, "Links between puberty and internalizing and exgternalizing behavior in girls: Moving from demonstrating effects to identifying pathways, in D. M. Stoff & E. J. Susman (Eds.), *Developmental psychobiology of aggression.* Copyright (2005), Cambridge University Press.

androgens, been associated with depressive symptoms (Angold et al., 1998; Susman, Nottelmann, Inoff-Germain, Dorn, & Chrousos, 1987). Whereas estradiol was associated with depressive symptoms, mediation was not found for arousal (tapped by mood changes and increased intensity of moods, including tension and self-consciousness), attention difficulties, or stressful life events. As such, no mechanism between estradiol and depressive symptoms could be identified.

Although arousal did not explain hormone effects on depression, the effect of pubertal timing was mediated by emotional arousal. Thus arousal offered a mechanism for the effect of pubertal timing on depression. However, the fact that it did not explain the hormonal effect as expected warranted further exploration. A possible interaction between hormonal links to depressive symptoms and early maturation in girls was examined; that is, whether early maturation was associated with hormonal arousal, a possible underlying factor in emotional arousal, was examined. Because a link between adrenal response (e.g., cortisol response) and stress has been noted, DHEAS levels were examined in association with early maturation. The interaction between hormonal arousal (the upper third of the distribution of DHEAS considered high hormonal arousal) and timing (early versus other) was predictive of depressive symptoms and had a trend toward predicting the emotional arousal construct. As indicated in Figure 5.6, girls who were early maturers and who had high hormonal arousal as tapped by higher levels of DHEAS had the highest reports of depressive symptoms; a similar pattern is seen for emotional arousal. That is, this subgroup of early maturing girls about age 12 with increased levels of DHEAS showed elevated emotional arousal and depressive symptoms.

Contextual Models

Another proposed pathway linking puberty and depression takes contextual interactions, including social support, into account. The biological and social changes of puberty may vary systematically with the family, peer, school, or neighborhood contexts in which they occur (Petersen & Taylor, 1980). Different contexts may amplify or attenuate the effects of puberty on depression. Factors that have been shown to amplify negative effects of pubertal timing include association with deviant peers, adverse parenting, living in dangerous neighborhoods, and negative life events in general (Brooks-Gunn & Warren, 1989; Ge, Conger, Lorenz, & Simons, 1994). Conversely, positive factors such as parental support and warmth have been found to buffer the potential stressful effects of the pubertal transition (Ge et al., 1994; Petersen, Sargiani, & Kennedy, 1991). There have been several models proposed to explain how life-events and social challenges relate to adjustment in adolescence (Graber & Brooks-Gunn, 1996). If contexts are systematically different for girls and boys during their transitions through puberty, this may contribute to the emergence of gender differences in rates of depression seen in adolescence.

Do Girls Have More Social Challenges Than Boys? Evidence suggests that although boys are exposed to more stressful life events than girls during middle childhood, girls likely face more stressful life events than boys in adolescence (Compas & Wagner, 1991; Ge et al., 1994; R. Larson & Ham, 1993). Cumulative change models posit that the experience of several simultaneous (or in close proximity) stressful events overtax the coping resources of adolescents, the combination of events significantly contributing to adolescent outcomes (Simmons & Blyth, 1987). Evidence from longitudinal studies suggests that adolescents experience more life events, both negative and positive, in early adolescence than in later adolescence, the number of events peaking around age 14 (Brooks-Gunn, 1991; Ge et al., 1994). These studies also showed associations between increased experiences of life events and depressive affect. Because girls tend to transition through puberty earlier than boys, the stressful events that co-occur with this transition are different for them, especially early maturing girls. For example, going through a school change at the same time as going through peak pubertal development, as early maturing girls are likely to do, has been identified as an experience that sets adolescents on course for poorer adjustment across adolescence (Petersen et al., 1991). Boys most often transition schools before the onset of puberty, making this cumulative stress situation particular to girls.

Additionally, adolescence marks a time when primary attachment systems shift from parental to peer and romantic relationships (Cyranowski, Frank, Young, & Shear, 2000; Gargiulo, Attie, Brooks-Gunn, & Warren, 1987). This transition is most often negotiated successfully by adolescents; however, this transition is likely to happen earlier for girls, who mature earlier, sometimes before they are cognitively ready to handle it, especially for early maturing girls. Adolescent females have been found to report significantly more negative interpersonal events than adolescent males, to perceive these events as more stressful (Wagner & Compas, 1990), and to be more vulnerable to stress in peer and family contexts (Greene & Larson, 1991). Thus, the fact that girls are navigating many social challenges in simultaneity with puberty, whereas boys most often experience the same social challenges during periods not concurrent with pubertal development, may contribute to the emergence of gender differences in depression during adolescence.

Do Girls Have Different Social Challenges Than Boys? The social challenges that are most salient to girls during this time of adolescence differ than the social challenges that are most salient to boys. Particularly during adolescence, negative mood states in girls are more linked with interpersonal domains, such as the family and peer group (Greene & R. Larson, 1991), whereas negative mood states in boys tend to be linked to activity based experiences such as competitive games (L. Larson & Asmussen, 1991). The added value placed on relationships by girls may lead to heightened vulnerability and stress when the relationships are perceived not to be going well. For example, girls have been found to be more aware of threat of conflict to friendships, a gender difference that intensifies with age (Laursen, 1996). Adolescent girls are also more likely than boys to be preoccupied with negative self-evaluations, including body image concerns (Graber et al., 1996). These concerns can make social interactions with peers especially important, as girls are sensitive to interpersonal cues related to these self-evaluations. Comparable concerns with negative self-evaluations in adolescent boys have not been found in the literature (Graber, 2003). In the family context, girls report more emotional autonomy from parents than do boys during adolescence (Steinberg & Silverberg, 1986). However, parents tend to employ higher levels of control in the absence of autonomy-granting behavior with girls than with boys (Pomerantz & Ruble, 1998). This discrepancy between girls' levels of perceived and actual autonomy may result in more tension and stress for girls during adolescence.

Because normatively girls mature earlier than boys, early maturing girls especially tend to befriend older peers, including boys, who are similar in physical development. These friendships, both because of associations with peers who are more cognitively developed than themselves and the fact that there is a change in social support, lead to social challenges unique to girls. These girls, especially early maturers, are less cognitively mature than the older peers whom they tend to befriend, and, as such, may not be ready to manage the challenges that arise in these older social situations. Social failures may lead to lowered self-esteem and depressive symptoms. The transition from the same-age friends who have served as social support to new, older friends may be perceived by early-maturing girls as decreased social support. Early-maturing girls, in comparison to on-time maturers, have consistently reported lower levels of social support from friends in mid-adolescence and young adulthood (Graber et al., 1997). This lower level of social support may serve as a mechanism to depressive symptoms. Thus girls not only face more social challenges at the time of puberty, but the challenges they do face are quite different than those that their male peers navigate.

Do Girls React Differently to Social Challenges Than Boys? In addition to facing more and different stressful life events and social challenges in adolescence, there is evidence that girls respond differently to social challenges than boys, most often reporting greater stress in relation to these events than boys (Compas & Wagner, 1991; Ge et al., 1994; Siddique & D'Arcy, 1984). For example, changes in uncontrollable events were associated with an increase in girls' but not boys' depressive symptoms (Ge et al., 1994). Similarly, girls reacted more negatively to stressful life events, in particular the transition to high school (Marcotte, Fortin, Potvin, & Papillon, 2002). Girls and boys may also react different to family conflict and stress. In a study by Crawford and colleagues (2001), there were no gender differences in the relation between parental distress and discord and internalizing symptoms in early adolescence; by midadolescence, parental discord was associated with internalizing symptoms in girls but not boys (Crawford, Cohen, Midlarsky, & Brook, 2001).

Girls have been found to experience higher levels of interpersonal stress than boys during adolescence, particularly in their friendships (Rudolph, 2002). Nolen-Hoeksema (1994) has suggested that adolescent boys and girls may use different methods to cope with stress, as well as different attributional styles. Specifically, she posits that girls are more likely than boys to attribute negative events to internal, stable, and global causes that are out of their control, an attribution style that has been linked to depressive symptoms (Abramson, Metalsky, & Alloy, 1989; Kaslow, Adamson, & Collins, 2000; Nolen-Hoeksema & Girgus, 1994). Additionally, girls are more likely to ruminate on the negative emotions that result from these hopeless attributions to negative life events.

Girls have been found to be more emotionally reactive in general to environmental stressors than boys (Ge et al., 1994). It has been suggested that this gender-typed characteristic difference becomes intensified, as individuals become more aware and attentive to these socialized differences, during adolescence (Alfieri, Ruble, & Higgins, 1996; Hill & Lynch, 1983). Specifically, Simmons and Blyth (1987) suggested that girls are more sensitive to problems with peer relations than boys.

Rudolph's (2002) review of the literature revealed that adolescent girls have greater investment in interpersonal success, with higher levels of distress associated with peer relationships, than boys. Additionally, girls are more likely to blame themselves for interpersonal problems than boys. Moreover, girls have higher interpersonal sensitivity and reactivity to interpersonal stress, which contribute to emotional distress. This may be because girls are socialized to be more affiliative, that is, preferring close emotional communication, intimacy, and responsiveness within interpersonal relationships, whereas boys are socialized to prefer independent activity and mastery (Cyranowski et al., 2000). One step further, however, is that emotional distress seems to cause stressful interpersonal situations, reciprocally maintaining the emotional distress. This cycle of stressful social situations and emotional distress seems to reinforce girls' greater emergence of depressive symptoms during adolescence.

Hormonal Sensitivity to Social Challenges. It is possible that girls may be more hormonally sensitive to social challenges or contextual stress compared to boys. For example, girls who live in more stressful family contexts tend to reach menarche at an earlier age. Father absence in the childhood years was associated with earlier maturation in a retrospective study of college-aged women (Surbey, 1990). Lower warmth in family relations (Graber et al., 1995), poorer quality of parent–child interactions (Ellis, McFadyen-Ketchum, Dodge, Pettit, & Bates, 1999), and maternal mood disorders (Ellis & Garber, 2000) are other factors that have been associated with earlier maturation in girls. Although the mechanisms linking family relationships to puberty development have not been identified, they presumably operate through the hormonal pathways controlling puberty, specifically the malleability of the estrogen system in girls' pubertal development.

CONCLUSIONS

As reviewed, multiple lines of evidence link particular aspects of pubertal development with increasing depressive symptoms in girls during adolescence. Most of the studies so far have examined only the short-term effects of pubertal development on depression (as an exception, see Graber et al., 1997). Studies that follow the same group of children through puberty and into adulthood would better elucidate the developmental pathways leading to depressive outcomes. Angold and Worthman (1993) have pointed out additional limitations to the research in this area. The child and adolescent epidemiologic studies that focus on depression have not included adequate measures of pubertal status, whereas the relevant endocrine studies have been limited by small sample sizes and scale measures of depression. Furthermore, many of the hormone studies include single hormones only and when multiple hormones were measured, the statistical analyses were compromised by small samples. Longitudinal studies that include multiple dimensions of pubertal development and social factors simultaneously, with representative population samples, would help elucidate how a combination of factors during adolescence may result in increased depression in females compared to males.

What seems clear, though, is that girls both are different physiologically and face quite different social challenges than boys when it comes to puberty. As such, girls may face more and more demanding challenges in early adolescence, and they may react quite differently to them. This interaction between biology and social mechanisms seems to constitute a complex mechanism leading to a greater prevalence of depression in girls beginning at puberty, a prevalence that persists throughout the lifespan, up until old age. More longitudinal, widespread research efforts assessing both the array of physiological factors and the many social factors at play will help pinpoint which are most important when considering girls' psychological adjustment.

References

Abramson, L. Y., Metalsky, G. I., & Alloy, L. B. (1989). Hopelessness depression: A theory-based subtype of depression. *Psychological Review*, *96*, 358–372.

Ahmed, M. L., Ong, K. K. L., Morrell, D. J., Cox. L. Deayee. Perry. L. et al. (1999). Longitudinal study of leptin concentrations during puberty: Sex differences and relationship to changes in body composition. *Journal of Clinical Endocrinology and Metabolism*, *84*, 899–905.

Alfieri, T., Ruble, D. N., & Higgins, E. T. (1996). Gender stereotypes during adolescence: Developmental changes and the transition to junior high school. *Developmental Psychology*, *32*, 1129–1137.

Alsaker, F. D. (1992). Pubertal timing, overweight, and psychological adjustment. *Journal of Early Adolescence*, *12*, 396–419.

American Psychiatric Association. (1994). *Diagnostic and statistical manual of mental disorders (4th ed.)*. Washington, DC: American Psychiatric Association.

Angold, A., Costello, E. J., Erkanli, A., & Worthman, C. W. (1999). Pubertal changes in hormone levels and depression in girls. *Psychological Medicine*, *29*, 1043–1053.

Angold, A., Costello, E. J., & Worthman, C. W. (1998). Puberty and depression: the roles of age, pubertal status and pubertal timing. *Psychological Medicine*, *28*, 51–61.

Angold, A., Erkanli, A., Silberg, J., Eaves, L., & Costello, E. J. (2002). Depression scale scores in 8–17-year-olds: effects of age and gender. *Journal of Child Psychology and Psychiatry*, *43*, 1052–1063.

Angold, A., Worthman, C. M., & Costello, E. J. (2003). Puberty and depression. In C. Hayward (Ed.), *Gender differences at puberty* (pp. 137–164). New York: Cambridge University Press.

Arslanian, S., Suprasongsin, C., & Janosky, J. (1997). Insulin secretion and sensitivity in black vs. white prepubertal healthy children. *Journal of Clinical Endocrinology and Metabolism, 82*, 1923–1927.

Attie, I., & Brooks-Gunn, J. (1989). Development of eating problems in adolescent girls: A longitudinal study. *Developmental Psychology, 25*, 70–79.

Barash, I. A., Cheung, C. C., Weigle, D. S., Ren, H., Kabigting, E. B., Kuijper, J., et al. (1996). Leptin is a metabolic signal to the reproductive system. *Endocrinology, 137*, 2144–3147.

Blum, W. F., Englaro, P., Hanitsch, S., Juul, A., Hektel, N. T., Muller, J., et al. (1997). Plasma leptin levels in healthy children and adolescents: Dependence on body mass index, body fat mass, gender, pubertal stage, and testosterone. *Journal of Clinical Endocrinology and Metabolism, 82*, 2904–2910.

Born, L., & Steiner, M. (2001). The relationship between menarche and depression in adolescence. *CNS Spectrums, 6*, 126–138.

Brooks-Gunn, J. (1991). How stressful is the transition to adolescence for girls? In M. E. Colton & Gore, S. (Eds.), *Adolescent stress: Causes and consequences* (pp. 131–149). New York: Aldine de Gruyter.

Brooks-Gunn, J., Graber, J., & Paikoff, R. L. (1994). Studying links between hormones and negative affect: Models and measures. *Journal of Research on Adolescence, 4*, 469–486.

Brooks-Gunn, J., Petersen, A. C., & Eichorn, D. (1985). The study of maturational timing effects in adolescence. *Journal of Youth and Adolescence, 14*, 149–161.

Brooks-Gunn, J., & Reiter, E. O. (1990). The role of pubertal processes. In S. S. Feldman & G. R. Elliott (Eds.), *At the threshold: The developing adolescent* (pp. 16–53). Cambridge, MA: Harvard University Press.

Brooks-Gunn, J., & Warren, M. P. (1985). Measuring physical status and timing in early adolescence: A developmental perspective. *Journal of Youth and Adolescence, 14*, 163–189.

Brooks-Gunn, J., & Warren, M. P. (1989). Biological and social contributions to negative affect in young adolescent girls. *Child Development, 60*, 40–55.

Brooks-Gunn, J., Warren, M. P., Rosso, J., & Gargiulo, J. (1987). Validity of self-report measures of girls' pubertal status. *Child Development, 58*, 829–841.

Buchanan, C. M., Eccles, J. S., & Becker, J. B. (1992). Are adolescents the victims of raging hormones: Evidence for activational effects of hormones on moods and behavior at adolescence. *Psychological Bulletin, 111*, 62–107.

Casper, R. (1998). Growing up female. In R. Casper (Ed.), *Women's health: Hormones, emotions, and behavior* (Vol. 18, pp. 1–14). Cambridge, UK: Cambridge University Press.

Caspi, A., & Moffitt, T. E. (1991). Individual differences are accentuated during periods of social change: The sample case of girls at puberty. *Journal of Personality and Social Psychology, 61*, 157–168.

Clayton, P. E., & Trueman, J. A. (2000). Leptin and puberty. *Archives of Disease in Childhood, 83*, 1–4.

Colon, I., Caro, D., Bourdony, C. J., & Rosario, O. (2000). Identification of phthalate esters in the serum of young Puerto Rican girls with premature breast development. *Environmental Health Perspectives, 108,* 895–900.

Compas, B., Ey, S., & Grant, K. E. (1993). Taxonomy, assessment, and diagnosis of depression during adolescence. *Psychological Bulletin, 114,* 323–344.

Compas, B., Oppendisano, G., Connor, J. K., Gerhardt, C. A., Hinden, B. R., Achenbach, T. M., et al. (1997). Gender differences in depressive symptoms in adolescence: comparison of national samples of clinically referred and nonreferred youths. *Journal of Consulting and Clinical Psychology, 65,* 617–626.

Compas, B., & Wagner, B. M. (1991). Psychosocial stress during adolescence: Intrapersonal and interpersonal processes. In M. E. Colton & S. Gore (Eds.), *Adolescent stress: Causes and consequences. Social institutions and social change* (pp. 67–85). New York: Aldine de Gruyter.

Crawford, T. N., Cohen, P., Midlarsky, E., & Brook, J. S. (2001). Internalizing symptoms in adolescents: Gender differences in vulnerability to parental distress and discord. *Journal of Research on Adolescence, 11,* 95–118.

Cyranowski, J. M., Frank, E., Young, E., & Shear, M. K. (2000). Adolescent onset of the gender difference in lifetime rates of major depression: A theoretical model. *Archives of General Psychiatry, 57,* 21–27.

Dorn, L. D., & Chrousos, G. P. (1997). The neurobiology of stress: Understanding regulation of affect during female biological transitions. *Seminars in Reproductive Endocrinology, 15,* 19–35.

Dorn, L. D., Susman, E. J., Nottelmann, E. D., Inoff-Germain, G., & Chrousos, G. P. (1990). Perceptions of puberty: Adolescent, parent, and health care personnel. *Developmental Psychology, 26,* 322–329.

Dubas, J. S., Graber, J. A., & Petersen, A. C. (1991). The effects of pubertal development on achievement during adolescence. *American Journal of Education, 99,* 444–460.

Duke-Duncan, P. M., Ritter, P., Dornbusch, S. M., Gross, R. T., & Carlsmith, J. M. (1985). The effects of pubertal timing on body image, school behavior, and deviance. *Journal of Early Adolescence, 14,* 227–235.

Ellis, B. J., & Garber, J. (2000). Psychosocial antecedents of variation in girls' pubertal timing: maternal depression, stepfather presence, and marital and family stress. *Child Development, 71,* 485–501.

Ellis, B. J., McFadyen-Ketchum, S., Dodge, K. A., Pettit, G. S., & Bates, J. E. (1999). Quality of early family relationships and individual differences in the timing of pubertal maturation in girls: A longitudinal test of an evolutionary model. *Journal of Personality and Social Psychology, 77,* 387–401.

Fara, G. M., Del Corvo, S., Bernuzzi, S., et al. (1979). Epidemic of breast enlargement in an Italian school. *Lancet, II,* 295–297.

Fechner, P. Y. (2002). Gender differences in puberty. *Journal of Adolescent Health, 30,* 44–48.

Fechner, P. Y. (2003). The biology of puberty: new developments in sex differences. In C. Hayward (Ed.), *Gender differences at puberty* (pp. 17–28). New York: Cambridge University Press.

Foster, H., & Brooks-Gunn, J. (manuscript under review). Pubertal timing effects by gender and ethnicity in the transition to adolescence.

Garcia-Mayor, R. V., Andrade, A., Rios, M., Lage, M., Dieguez, C. & Casanueva, F. F. (1997). Serum leptin levels in normal children: Relationship to age, gender, body mass index, pituitary-gonadal hormones, and pubertal stage. *Journal of Clinical Endocrinology and Metabolism, 82*, 2849–2855.

Gargiulo, J., Attie, I., Brooks-Gunn, J., & Warren, M. P. (1987). Girls' dating behavior as a function of social context and maturation. *Developmental Psychology, 23*, 730–737.

Ge, X., Brody, G. H., Conger, R. D., & Simons, R. L. (in press). Pubertal transition and African American children's internalizing and externalizing symptoms. *Journal of Youth and Adolescence*.

Ge, X., Conger, R. D., & Elder, G. H., Jr. (1996). Coming of age too early: Pubertal influences on girls' vulnerability to psychological distress. *Child Development, 67*, 3386–3400.

Ge, X., Conger, R. D., & Elder, G. H., Jr. (2001a). Pubertal transition, stressful life events, and the emergence of gender differences in adolescent depressive symptoms. *Developmental Psychology, 37*, 404–417.

Ge, X., Conger, R. D., & Elder, G. H., Jr. (2001b). The relationship between puberty and psychological distress in adolescent boys. *Journal of Research on Adolescence, 11*, 49–70.

Ge, X., Conger, R. D., Lorenz, F. O., & Simons, R. L. (1994). Parents' stressful life events and adolescent depressed mood. *Journal of Health and Social Behavior, 35*, 28–44.

Ge, X., Kim, I. J., Brody, G. H., Conger, R. D., Simons, R. L., Gibbons, F. X., et al. (2003). It's about timing and change: pubertal transition effects on symptoms of major depression among African American youths. *Developmental Psychology, 39*, 430–439.

Giedd, J. N., Castellanos, F. X., Rajapakse, J. C., Vaituzis, A. C., & Rapoport, J. L. (1997). Sexual dimorphism of the developing human brain. *Progress in Neuro-Psychopharmacology and Biological Psychiatry, 21*, 1185–1201.

Gladen, B. C., Ragan, N. B., & Rogan, W. J. (2000). Pubertal growth and development and prenatal and lactational exposure to polychlorinated biphenyls and dichlorodiphenyl dichloroethene. *Journal of Pediatrics, 136*, 190–196.

Goodman, E., Hinden, B. R., & Khandelwal, S. (2000). Accuracy of teen and parental reports of obesity and body mass index. *Pediatrics, 106*, 52–58.

Graber, J. A. (2003). Puberty in context. In C. Hayward (Ed.), *Gender differences at puberty* (pp. 307–325). New York: Cambridge University Press.

Graber, J. A., & Brooks-Gunn, J. (1996). Transitions and turning points: Navigating the passage from childhood through adolescence. *Developmental Psychology, 32*, 768–776.

Graber, J. A., Brooks-Gunn, J., & Archibald, A. B. (2005). Links between girls' puberty and externalizing and internalizing behaviors: Moving from demonstrating effects to identifying pathways. In D. M. Stoff & E. J. Susman (Eds.), *Developmental psychobiology of aggression*. New York: Cambridge University Press.

Graber, J. A., Brooks-Gunn, J., & Warren, M. P. (1995). The antecedents of menarcheal age: Heredity, family environment, and stressful life events. *Child Development, 66*, 346–359.

Graber, J. A., Brooks-Gunn, J., & Warren, M. P. (in press). Pubertal effects on adjustment in girls: Moving from demonstrating effects to identifying pathways. *Journal of Youth and Adolescence*.

Graber, J. A., Lewinsohn, P. M., Seeley, J. R., & Brooks-Gunn, J. (1997). Is psychopathology associated with the timing of pubertal development? *Journal of the American Academy of Adolescent Psychiatry, 36*, 1768–1776.

Graber, J. A., Petersen, A. C., & Brooks-Gunn, J. (1996). Pubertal processes: methods, measures, and models. In J. A. Graber, J. Brooks-Gunn & A. C. Petersen (Eds.), *Transitions through adolescence: Interpersonal domains and context* (pp. 23–53). Mahwah, NJ: Lawrence Erlbaum Associates.

Graber, J. A., Seeley, J. R., Brooks-Gunn, J., & Lewinsohn, P. M. (2004). Is pubertal timing associated with psychopathology in young adulthood? *Journal of the American Academy of Child and Adolescent Psychiatry, 43*, 718–726.

Greene, A. L., & Larson, R. W. (1991). Variation in stress reactivity during adolescence. In E. M. Cummings, A. L. Greene & K. H. Karraker (Eds.), *Life-span developmental psychology: Perspectives on stress and coping* (pp. 195–209). Hillsdale, NJ: Erlbaum.

Grumbach, M. M., & Styne, D. M. (1998). Puberty: Ontogeny, neuroendocrinology, physiology, and disorders. In J. D. Wilson, D. W. Foster & H. M. Kronenberg (Eds.), *Williams textbook of endocrinology* (pp. 1509–1625). Philadelphia: W. B. Saunders.

Harlan, W. R., Harlan, E. A., & Grillo, C. P. (1980). Secondary sex characteristics of girls 12–17 years of age: The U.S. Health Examination Survey. *Journal of Pediatrics, 96*, 1074–1078.

Hayward, C. (2003). Methodological concerns in puberty-related research. In C. Hayward (Ed.), *Gender differences at puberty* (pp. 1–16). New York: Cambridge University Press.

Hayward, C., Gotlib, I., Schraedley, P. K., & Litt, I. F. (1999). Ethnic differences in the association between pubertal status and symptoms of depression in adolescent girls. *Journal of Adolescent Health, 25*, 143–149.

Hayward, C., Killen, J. D., Wilson, D. M., Hammer, L. D., Litt, I. F., Kraemer, H. C., et al. (1997). Psychiatric risk associated with early puberty in adolescent girls. *Journal of the American Academy of Child and Adolescent Psychiatry, 36*, 255–262.

Herman-Giddens, M. E., Slora, E. J., Wasserman, R. C., Bourdony, C. J., Bhapkar, M. V., Koch, G. G., et al. (1997). Secondary sexual characteristics and menses in young girls seen in office practice: a study from the pediatric research in office settings network. *Pediatrics, 99*, 505–512.

Hick, K. M., & Katzman, D. K. (1999). Self-assessment of sexual maturation in adolescent females with anorexia nervosa. *Journal of Adolescent Health, 24*, 206–211.

Hill, J. P., & Lynch, M. E. (1983). The intensification of gender-related role expectations during early adolescence. In J. Brooks-Gunn & A. C. Petersen (Eds.), *Girls at puberty: Biological and psychological perspectives* (pp. 201–228). New York: Plenum Press.

Hutchison, J. B., Schumacher, M., Steimer, T., & Gahr, M. (1990). Are separable aromatase systems involved in hormonal regulation of the male brain? *Journal of Neurobiology, 21*, 743–759.

Jensen, P. S., Brooks-Gunn, J., & Graber, J. A. (1999). Dimensional scales and diagnostic categories: Constructing crosswalks for child psychopathology assessments. *Journal of the American Academy of Adolescent and Child Psychiatry, 38,* 118–120.

Kaplowitz, P. B. (2004). *Early puberty in girls.* New York: Random House.

Kaplowitz, P. B., Slora, E. J., Wasserman, R. C., Pedlow, S. E., & Herman-Giddens, M. E. (2001). Earlier onset of puberty in girls: Relation to increased body mass index and race. *Pediatrics, 108,* 347–353.

Kaslow, N.J., Adamson, L. B., & Collins, M. H. (2000). A developmental psychopathology perspective on the cognitive components of child and adolescent depression. In A. J. Sameroff, M. Lewis, & S. M. Miller (Eds.), *Handbook of developmental psychopathology* (Vol. 1, 2nd ed., pp. 491–510). New York: Plenum Press.

Kovacs, M. (1992). *Children's Depression Inventory Manual.* North Tonawanda, NY: Multi-Health Systems, Inc.

Krstevska-Komstantinova, M., Charlier, C., Craen, M., Du Caju, M., Heinrichs, C., de Beaufort, C., et al. (2001). Sexual precocity after immigration from developing countries to Belgium: Evidence of previous exposure to organochlorine pesticides. *Human Reproduction Update, 16,* 1020–1026.

Laflamme, N., Nappi, R. E., Drolet, G., Labrie, C., & Rivest, S. (1998). Expression and neuropeptidergic characterization of estrogen receptors (ER alpha and ER beta) throughout the rat brain: anatomical evidence of distinct roles of each subtype. *Journal of Neurobiology, 36,* 357–378.

Larson, L., & Asmussen, L. (1991). Anger, worry, and hurt in early adolescence: An enlarging world of negative emotions. In M. E. Colten & S. Gore (Eds.), *Adolescent stress: Causes and consequences* (pp. 21–41). New York: Aldine De Gruyter.

Larson, R., & Ham, M. (1993). Stress and "storm and stress" in early adolescence: The relationship of negative events with dysphoric affect. *Developmental Psychology, 29,* 130–140.

Laursen, B. (1996). Closeness and conflict in adolescent peer relationships: Interdependence with friends and romantic partners. In W. M. Bukowski, A. F. Newcomb, & W. W. Hartup (Eds.), *The company they keep: Friendship in childhood and adolescence* (pp. 186–210). New York: Cambridge University Press.

Layman, L. C. (1995). Molecular biology in reproductive endocrinology. *Current Opinion in Obstetrics and Gynecology, 7,* 328–339.

Lewinsohn, P. M., Rhode, P., & Seeley, J. R. (1998). Major depressive disorder in older adolescents: prevalence, risk factors and clinical implications. *Clinical Psychology Review,* 765–794.

Litt, I. F. (1999). Self-assessment of puberty: problems and potential (editorial). *Journal of Adolescent Health, 24,* 157.

MacLusky, N., & Naftolin, F. (1981). Sexual differentiation of the central nervous system. *Science, 211,* 1294–1302.

Marcotte, D., Fortin, L., Potvin, P., & Papillon, M. (2002). Gender differences in depressive symptoms during adolescence: Role of gender-typed characteristics, self-esteem, body image, stressful life events, and pubertal status. *Journal of Emotional and Behavioral Disorders, 10,* 29–42.

Marshall, W. A., & Tanner, J. M. (1969). Variations in the pattern of pubertal changes in girls. *Archives of Disease in Childhood, 44,* 291–303.

Marshall, W. A., & Tanner, J. M. (1970). Variations in the pattern of pubertal changes in boys. *Archives of Disease in Childhood, 45,* 13–23.

McEwen, B. S. (1994). How do sex and stress hormones affect nerve cells? *Annals of the New York Academy of Sciences, 743,* 1–18.

Michels Blanck, H., Marcus, M., Tolbert, P. E., Rubin, C., Henderson, A. K., Hertzberg, V. S. et al. (2000). Age at menarche and Tanner stage in girls exposed in utero and postnatally to polybrominated biphenyl. *Epidemiology, 11,* 641–647.

Mitchner, N. A., Garlick, C., & Ben-Jonathon, N. (1998). Cellular distribution and gene regulation of estrogen receptors alpha and beta in the rat pituitary gland. *Endocrinology, 139,* 3976–3983.

Moffitt, T. E., Caspi, A., Belsky, J., & Silva, P. A. (1992). Childhood experience and the onset of menarche: A test of a sociobiological model. *Child Development, 63,* 47–58.

Morris, N. M., & Udry, J. R. (1980). Validation of a self-administered instrument to assess stage of adolescent development. *Journal of Youth and Adolescence, 9,* 271–280.

Morrison, J. A., Barton, B. A., Biro, F., Sprecher, D. L., Falkner, F., & Obarzanek, E. (1994). Sexual maturation and obesity in 9- and 10-year old black and white girls: The National Heart, Lung, and Blood Institute Growth and Health Study. *Journal of Pediatrics, 124,* 889–895.

Nolen-Hoeksema, S. (2001). Gender differences in depression. *Current Directions in Psychological Science, 10,* 173–176.

Nolen-Hoeksema, S., & Girgus, J. S. (1994). The emergence of gender differences in depression during adolescence. *Psychological Bulletin, 115,* 424–443.

Nolen-Hoeksema, S., Girgus, J. S., & Seligman, M. E. P. (1992). Predictors and consequences of childhood depressive symptoms: A five-year longitudinal study. *Journal of Abnormal Psychology, 101,* 405–422.

Nottelmann, E. D., Susman, E. J., Inoff-Germain, G., Cutler, G. B., Loriaux, D. L., & Chrousos, G. P. (1987). Developmental processes in American early adolescence: Relations between adolescent adjustment problems and chronologic age, pubertal stage and puberty-related serum hormone levels. *Journal of Pediatrics, 110,* 473–480.

Obeidallah, D. A., Brennan, R. T., Brooks-Gunn, J., & Earls, F. (2004). Links between puberty timing, neighborhood contexts, and girls' violent behavior. *Journal of the American Academy of Child and Adolescent Psychiatry, 43,* 1460–1468.

Obeidallah, D. A., Brennan, R. T., Brooks-Gunn, J., Kindlon, D., & Earls, F. (2000). Socioeconomic status, race, and girls' pubertal maturation: Results from the Project on Human Development in Chicago Neighborhoods. *Journal of Research on Adolescence, 10,* 443–488.

Osterlund, M., Kuiper, G. G., Custafsson, J. A., & Hurd, Y. L. (1998). Differential distribution and regulation of estrogen receptor-alpha and -beta mRNA within the female rat brain. *Brain Research: Molecular Brain Research, 54,* 175–180.

Paikoff, R., Brooks-Gunn, J., & Warren, M. P. (1991). Predictive effects of hormonal change on affective expression in adolescent females over the course of one year. *Journal of Youth and Adolescence, 20,* 191–214.

Paretsch, C.-J., & Sippell, W. G. (2001). Pathogenesis and epidemiology of precocious puberty: Effects of exogenous oestrogens. *Human Reproduction Update, 7,* 292–302.

Parker, S., Nichter, M., Nichter, M., Vuckovic, N., Sims, C., Ritenbaugh, C. (1995). Body image and weight concerns among African American and white adolescent females: Differences that make a difference. *Human Organization, 54,* 103–114.

Persson, I., Ahlsson, F., Ewald, U., Tuvemo, T., Qingyuan, M., von Rosen, D., et al. (1999). Influence of perinatal factors on the onset of puberty in boys and girls: Implications for interpretation of link with risk of long term diseases. *American Journal of Epidemiology, 150,* 747–755.

Petersen, A. C. (1987). The nature of biological-psychosocial interactions: The sample case of early adolescence. In R. M. Lerner & T. T. Foch (Eds.), *Biological-psychosocial interactions in early adolescence: A life-span perspective* (pp. 35–61). Hillsdale, NJ: Erlbaum.

Petersen, A. C., Compas, B., Brooks-Gunn, J., Stemmler, M., Ey, S., & Grant, K. (1993). Depression in adolescence. *American Psychologist, 48,* 155–168.

Petersen, A. C., Crockett, L., Richards, M., & Boxer, A. (1988). A self-report measure of pubertal status: Reliability, validity, and initial norms. *Journal of Youth and Adolescence, 17,* 117–133.

Petersen, A. C., & Crockett, L. J. (1985). Pubertal timing and grade effects on adjustment. *Journal of Youth and Adolescence, 14,* 191–206.

Petersen, A. C., Sargiani, P. A., & Kennedy, R. E. (1991). Adolescent depression: Why more girls? *Journal of Youth and Adolescence, 20,* 247–271.

Petersen, A. C., & Taylor, B. (1980). The biological approach to adolescence: Biological change and psychological adaptation. In J. Adelson (Ed.), *Handbook of adolescent psychology* (pp. 117–155). New York: Wiley.

Phoenix, C. H., Goy, R. W., Gerall, A. A., & Young, W. C. (1959). Organizing action of prenatally administered testosterone propionate on the tissues mediating mating behavior in the female guinea pig. *Endocrinology, 65,* 369–382.

Pomerantz, E. M., & Ruble, D. N. (1998). The role of maternal control in the development of sex differences in child self-evaluative factors. *Child Development, 69,* 458–478.

Rasmussen, J. E., Torres-Aleman, I., MacLusky, N. J., Naftolin, F., & Robbins, R. J. (1990). The effects of estradiol on the growth patterns of estrogen receptor-positive hypothalamic cell lines. *Endocrinology, 126,* 235–240.

Reiter, E. O., & Grumbach, M. M. (1982). Neuroendocrine control mechanisms and the onset of puberty. *Annual Review of Physiology, 44,* 595–613.

Richards, M. H., Boxer, A. W., Petersen, A. C., & Albrecht, R. (1990). Relation of weight to body image in pubertal girls and boys from two communities, *Developmental Psychology, 26,* 313–321.

Rierdan, J., & Koff, E. (1997). Weight, weight-related aspects of body image, and depression in early adolescent girls. *Adolescence, 32,* 615–624.

Rockett, J. C., Lynch, C. D., & Buck, G. M. (2004). Biomarkers for assessing reproductive development and health: Part I-Pubertal Development. *Environmental Health Perspectives, 112,* 105–112.

Rudolph, K. D. (2002). Gender differences in emotional responses to interpersonal stress during adolescence. *Journal of Adolescent Health, 30,* 3–13.

Saenz de Rodriguez, C. A., Bongiovanni, A. M., & Conde de Borrego, L. (1985). An epidemic of precocious pubertal development in Puerto Rican children. *Journal of Pediatrics, 107,* 393–396.

Sanborn, K., & Hayward, C. (2003). Hormonal changes at puberty and the emergence of gender differences in internalizing disorders. In C. Hayward (Ed.), *Gender differences at puberty* (pp. 29–58). New York: Cambridge University Press.

Siddique, C. M., & D'Arcy, C. (1984). Adolescence, stress, and psychological well-being. *Journal of Youth and Adolescence, 13*, 459–473.

Simmons, R. G., & Blyth, D. A. (1987). *Moving into adolescence: The impact of pubertal change and school context.* New York: Aldine de Gruyter.

Stattin, H., & Magnusson, D. (1990). *Pubertal maturation in female development.* Hillsdale, NJ: Erlbaum.

Steinberg, L. D., & Silverberg, S. B. (1986). The vicissitudes of autonomy in early adolescence. *Child Development, 57*, 841–851.

Sun, S. S., Schubert, C. M., Cameron, W. C., Roche, A. F., Kulin, H. E., Lee, P. A., et al. (2002). National estimates of the timing of sexual maturation and racial differences among U.S. children. *Pediatrics, 110*, 911–919.

Surbey, M. K. (1990). Family composition, stress, and the timing of human menarche. In T. E. Ziegler & F. B. Bercovitch (Eds.), *Socioendocrinology of primate reproduction* (pp. 11–32). New York: John Wiley.

Susman, E. J. (1997). Modeling developmental complexity in adolescence: Hormones and behavior in context. *Journal of Research on Adolescence, 7*, 283–306.

Susman, E. J. (1998). Biobehavioural development: An integrative perspective. *International Journal of Behavioral Development, 22*, 671–679.

Susman, E. J., Dorn, L. D., & Chrousos, G. P. (1991). Negative affect and hormone levels in young adolescents: Concurrent and predictive perspectives. *Journal of Youth and Adolescence, 20*, 167–189.

Susman, E. J., Dorn, L. D., & Schiefelbein, V. (2003). Puberty, sexuality, and health. In M. Lerner, M. A. Easterbrooks, & J. Mistry (Eds.), *The comprehensive handbook of psychology* (Vol. 6, pp. 295–324). New York: Wiley.

Susman, E. J., Nottelmann, E. D., Inoff-Germain, G., Dorn, L. D., & Chrousos, G. P. (1987). Hormonal influences on aspects of psychological development during adolescence. *Journal of Adolescent Health Care, 8*, 492–504.

Susman, E. J., Nottelmann, E. D., Inoff-Germain, G., Dorn, L. D., Cutler, G. B., Loriaux, D. L., et al. (1985). The relation of relative hormonal levels and physical development and social-emotional behavior in young adolescents. *Journal of Youth and Adolescence, 14*, 245–264.

Tanner, J. M. (1970). Physical growth. In P. H. Mussen (Ed.), *Carmichael's manual of child psychology* (pp. 77–155). New York: Wiley.

Tanner, J. M. (1978). *Fetus into man: Physical growth from conception to maturity.* Cambridge, MA: Harvard University Press.

Tschann, J. M., Adler, N. E., Irwin, C. E. J., Millstein, S. G., Turner, R. A., Kegeles, S. M., et al. (1994). Initiation of substance use in early adolescence: The roles of pubertal timing and emotional distress. *Health Psychology, 13*, 326–333.

Wagner, B. M., & Compas, B. E. (1990). Gender, instrumentality, and expressivity: Moderators of the relation between stress and psychological symptoms during adolescence. *American Journal of Community Psychology, 18*, 383–406.

Warren, M. P., & Brooks-Gunn, J. (1989). Mood and behavior at adolescence: Evidence for hormonal factors. *Journal of Clinical Endocrinology and Metabolism, 69*, 77–83.

Wong, W. W., Nicolson, M., Stuff, J. E., Butte, N. F., Ellis, K. J., Hergenroeder, A. C., et al. (1998). Serum leptin concentrations in Caucasian and African-American girls. *Journal of Clinical Endocrinology and Metabolism, 83*, 3574–3577.

Worthman, C. M., & Stallings, J. F. (1997). Hormone measures in finger-prick blood spot samples: New field methods for reproductive endocrinology. *American Journal of Physical Anthropology, 104*, 1–21.

Wu, T., Mendola, P., & Buck, G. M. (2002). Ethnic differences in the presence of secondary sex characteristics and menarche among U.S. girls: The third National Health and Nutrition Examination Survey, 1988–1994. *Pediatrics, 110*, 752–757.

Zimmerman, P. A., Francis, G. L., & Poth, M. (1995). Hormone-containing cosmetics may cause signs of early sexual development. *Military Medicine, 160*, 628–630.

6

Women's Aging and Depression

Brenda W. J. H. Penninx

Population aging is a worldwide phenomenon. The oldest of the old, persons aged 85 years and older, comprise the fastest growing segment of the aged. Since the life expectancy has consistently been higher for women than for men, the elderly population is composed of more women than men. So, in absolute terms, aging is affecting women more than men. Among persons aged 65 years and older in the United States, almost 60% are women. The proportion of women in the older populations climbs with age to over 70% in those aged 85 years and older. This trend can be observed across the developed world, where women typically outlive men by 5 to 9 years. This chapter describes the extent of depression experienced in women and men in old age. Subsequently, important biological, physical, and psychosocial risk factors for depression in older women will be described. The chapter will finish with describing the physical consequences of late-life depression in women.

PREVALENCE OF DEPRESSION IN OLDER WOMEN AND MEN

Major depression (also called major depressive disorder) is diagnosed in the *Diagnostic and Statistical Manual* of the American Psychiatric Association (American Psychiatric Association, 1989) when a person exhibits five or more out of the following nine symptoms: depressed mood, lack of interest, feelings of worthlessness or inappropriate guilt, diminished ability to concentrate or make decisions, fatigue, psychomotor agitation or retardation, insomnia or hypersomnia, significant decrease or increase in weight or appetite, and recurrent thoughts of death or suicidal ideation. The identified symptoms should at least include one or both of the two core symptoms (depressed mood and lack of interest), should be present for at least 2 weeks, and should generally be severe enough to cause disruptions in a person's daily functioning. Overall, the symptoms of moderate to severe depression presented to the clinician are rather similar across older persons and persons in midlife. However, some subtle differences in symptom experience across age groups have been described. Melancholia (symptoms of noninteractiveness and psychomotor retardation or agitation) appears to be a little more frequent in later age than in younger age, with psychomotor disturbances being more obvious in older persons.

In a review of the international scientific literature, it was concluded that major depression is relatively rare among older community-dwelling persons, affecting about 1 to 2% (Beekman, Copeland, & Prince, 1999). In fact, major depression appears to be less frequent in late life than in middle life. In the recently conducted National Comorbidity Survey Replication Study among 9,090 U.S. adults, the prevalence of major depression was, for both men and women, highest among the persons in the 30 to 44 age range, and lowest among persons aged 60 years and older (Kessler et al., 2003). Even among the older adults, there appears to be a lower prevalence of major depression with older age. This is shown in Figure 6.1, using data from the Dutch Longitudinal Aging Study Amsterdam (LASA) in which more than 3,000 adults aged between 55 and 85 years underwent a psychiatric interview. This figure shows that even among older persons, the prevalence of major depressive disorder shows a slightly declining trend when age increases from 55 years to 85 years. Some have suggested that part of the declining prevalence of major depression with older age could be due to bias because those with a depression may die earlier and therefore not survive to old age or because older persons may underreport depression compared to younger persons. Nevertheless, overall, research findings do indicate that the prevalence of major depression in old age is at least not higher than in earlier age.

The prevalence rates of depression vary considerably depending on the older sample studied. Studies in clinical settings, such as hospitals or nursing homes, generally find much higher prevalences of major depression than studies in community settings. Of course, this is largely due to the higher disease and disability burden in these specific samples (see Risk Factors of Depression in Older Women section). In addition, it is important to realize that a large proportion of older persons with a major depressive disorder has had depressive episodes during earlier phases of their lives. Consequently, a personal history of depressive disorder is one of the strongest risk factors for a major depressive disorder in old age. Thus, major depressive disorders in old age most often represent recurring episodes of early-onset depressive disorders.

A female preponderance in depression has been very consistently found in both young and old persons. In general, depression prevalence rates are about twice as high in women than in men. As shown in Figure 6.1, a clear gender difference in the prevalence of major depressive disorders

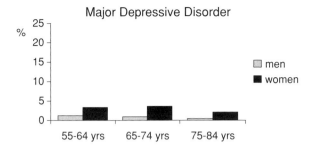

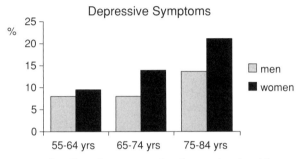

FIGURE 6.1. Prevalence rates for depression in older men and women in various older age groups. Results are from the Longitudinal Aging Study Amsterdam. Major Depressive Disorder is based on diagnostic DSM-criteria. Depressive symptoms is measured as a score ≥ 16 on the Center for Epidemiologic Studies Depression Scale.

also exists in the older population. Consistently for all older age groups, the prevalence of major depression is about twice as high for older women than for older men.

When compared to those of major depression, prevalence rates in older persons are quite different for milder symptoms of depression. These milder symptoms of depression are commonly assessed using depression symptoms questionnaires that ask for presence, intensity, or frequency of a series of symptoms. Some examples of symptom checklists commonly used in older populations are the Center for Epidemiologic Studies-Depression Scale and the Geriatric Depression Scale. These instruments are well-validated and have been proven to be valid and reliable instruments in older populations. It is possible to score relatively high on a symptom checklist without meeting diagnostic criteria for major depressive disorder. In fact, depressive symptoms checklists identify for the largest part persons with a significant high level of depressive symptoms but who do not fulfill the diagnostic threshold of major depressive disorder. This condition is often referred to as *depressed mood, significant depressive symptoms,* or minor depression in the medical literature.

When using a depressive symptom checklist it becomes obvious that a much higher proportion of the older population suffers from significant symptoms of depression. Prevalences of clinically significant levels of depressed mood in older community-based populations range from 12 to 20% (Beekman et al., 1999). Data from several community studies converge in suggesting that there is a curvilinear relationship between age and depression scores over the entire adult life span, with the highest scores among younger adults and those over age 75. Figure 6.1 illustrates the positive association between a significant level of depressive symptoms and increasing age in older participants of the Dutch LASA study. The elevation in scores among the very old has been shown not to be an artifact of greater endorsement of somatic symptoms. In line with findings for major depressive disorders, women clearly show higher rates of significant depressive symptoms than men. So, overall we can conclude that depression, especially in a milder form, is very common in older persons and does affect older women about twice as much as older men. The prevalence of significant feelings of depression, but not of major depressive disorders, does increase with increasing age.

RISK FACTORS OF DEPRESSION IN OLDER WOMEN

Most age-related changes, either biological, physical, or social, can be expected to constitute losses rather than gains. Several of these changes are potentially stressful to older persons. Chronic illness conditions, for instance, can introduce pain, disability, despair, and fear for treatment and pending death. Widowhood, the need for instrumental support, and the loss of personal relationships might cause feelings of loneliness, dependency, and helplessness. When adjustment to these burdening circumstances is inadequate, an older person might experience substantial psychological stress, which could result in depressive symptomatology. Although several of the changes that occur with aging put both older women and men at risk for the development of depressive symptomatology or even a major depressive disorder, older women may be more exposed to some of the important risk factors for depression. Because older women have a higher exposure than older men to some important risk factors for depression, in absolute terms, they are more at risk for depression in later life. The following sections provide a more detailed description of some of the most important risk factors for depression in older women.

Biological and Disease Risk Factors

For the most part, individual aging is associated with many adverse changes in human anatomy and physiology. Consequences of these changes are, for instance, losses in the sense of balance and movement, poorer hearing and vision, slower reactions, and weaker muscles. In later life, biological reserves are reduced, which can cause a weakening of one or another biological function essential to life. Consequently, conditions such as heart disease, cancer, respiratory infection, and kidney failure may arise. Furthermore, some acquired biological vulnerabilities may predispose an older individual for psychopathological disorders. Age-related physiological and brain structure changes have been suggested as plausible risk factors for late-onset depression because of parallels between these changes and pathological changes observed in depression, such as dysregulation of the monoamine system, increased monoamine oxidase activity, deficits in norepinephrine functioning, and down-regulation of serotonin receptors.

Much attention has been directed to vascular risk for late-life depression. Recent studies suggest that vascular lesions in selected regions of the brain may contribute to a unique variety of late-life depression (Kumar et al., 2002). The vascular depression impairments resemble impairments exhibited in frontal lobe syndromes. Magnetic resonance imaging of depressed patients has revealed structural abnormalities in areas related to the cortical-stratial-pallidal-thalamus-cortical pathway. Endocrine changes have also been associated with late-life depression. For instance, late-life depression has been associated with nonsuppression of cortisol (Davis et al., 1984). In addition, aging is associated with an increased responsiveness to corticotrophin-releasing factor (CRF), which is implicated in depression and is thought to mediate sleep and appetite disturbances, reduced libido, and psychomotor changes (Arborelius, Owens, Plotsky, & Nemeroff, 1999). Finally, low levels of sex steroid hormones, such as testosterone, estradiol, and dehydroepiandrosterone sulfate (DHEAS), which are especially apparent in older women, have been shown to be associated with late-life depression (Seidman, Araujo, Roose, & McKinlay, 2001; Yaffe et al., 1998, 2000).

The importance of physical health for the presence of late-life depression is undisputed. Chronic physical illnesses are consistently shown to be included among the strongest risk factors of depressed mood for both women and men in old age. For instance, depressive symptomatology appears to be higher in the presence of the following diseases: lung disease, arthritis, cancer, diabetes, stroke, and coronary heart disease and other cardiac illnesses (Ormel et al., 1997; Penninx et al., 1996). In addition to specific disease, depressive symptoms are also more prevalent in the presence of certain impairments, such us hearing and vision problems and cognitive impairment (Ormel et al., 1997). Table 6.1 lists the most important chronic conditions and impairments that put older women at risk for late-life depressed mood.

Several explanations can be given for the higher prevalence of depressive symptoms in the presence of chronic diseases and impairments.

TABLE 6.1. *Risk Factors for Depression in Older Women*

Chronic conditions and impairments
Cardiovascular disease
Stroke
Diabetes
Chronic obstructive pulmonary disease
Osteoarthritis
Cancer
Pain
Hearing impairment
Visual impairment
Cognitive impairment
Physical disability

Behavioral risk factors and individual characteristics
Low physical activity
High and low body mass index, weight loss
Heavy alcohol consumption
Older age
Low socioeconomic status (income, education)
Personality
Heredity

Social circumstances
Negative life-events
Reduced social contacts
Widowhood

Depression can occur as an outcome of certain somatic illnesses or medication, reflecting a biologically mediated process. For example, the structural and neurochemical changes involved in stroke and Parkinsonism can
lead to depression. Diseases and their specific symptoms also have several
adverse psychosocial and physical consequences. For example, loss of function, role, and independence; negative body image and sense of identity;
pain; and promoted sense of helplessness can be a reaction of being ill
and consequently cause increased feelings of depression. Some symptoms
of depression, especially the somatic symptoms such as the presence of a
low energy level and sleeping problems, may partly be a manifestation of
the disease. Among older persons, it is sometimes rather difficult to separate pure emotional consequences of a somatic condition from depression.
According to psychiatric criteria, a major depression is only present when
all the depressive symptoms are clearly not the consequence of one somatic
condition or the use of a certain pharmacological drug. So, this distinction
needs to be considered when evaluating the presence of a depression. However, it is very difficult to differentiate between the two, especially since

several symptoms, including the key symptoms of depressed mood and loss of interest, are not somatically oriented and will be present more often in certain persons but not in others. In addition, several clinicians and psychiatrists believe that even though various depressive symptoms may be caused by an underlying chronic condition, when a person has all the symptoms necessary for a depression diagnosis this person still may benefit from appropriate depression treatment. Several of the most important somatic disease risk factors of late-life depression, such as arthritis, pain, and cognitive and sensory impairments are more prevalent in older women than in older men, which indicates that in absolute terms these risk factors put more older women than older men at risk for late-life depression.

Both the biological age-related changes and the consequent development of chronic conditions and impairments have a large impact on the physical functioning of older persons. The extent of physical disability is an important factor in the development of psychological stress. People with acquired physical disabilities inevitably encounter additional losses, such as the loss of function, role, and body image, and may experience a greater dependency on others and a more negative view of themselves, their future, and their world, leading to more feelings of depression.

The number of older persons with difficulties in mobility and activities of daily living increases dramatically with increasing age. One of the key paradoxes in gerontology is that although women live longer than men, they live with more physical disability at older age. This is for a large part caused by the fact that the prevalence of nonlethal disabling chronic diseases is generally higher in older women than in older men. National estimates in the United States show that more women than men aged 65 years and older have arthritis (55.4 vs. 42.9%, respectively) and the prevalence of musculoskeletal joint symptoms such as pain and stiffness are typically higher in older women. Other prevalent nonlethal conditions in older women are hearing and vision impairment, osteoporosis (with resulting falls and fractures), and obesity. These conditions have a marked adverse impact on the physical functioning of older women. Of course, several fatal conditions such as heart disease, diabetes, and cancer are also rather common in older women and add to the high proportion of women who live with disability. Data from the U.S. National Health Interview Survey of Disability show that, of the women aged 65 years and older, 18.8% needs help with (instrumental) activities of daily living, relative to 10.9% of men. The gender difference is even more striking at the age of 85 years or older with 54.8% of the women and 36.9% of the men needing help with (instrumental) activities of daily living. Several studies among older persons have shown that symptoms of depression are more strongly predicted by the level of physical disability than by the number or specific type of chronic conditions. This indicates that the effect of chronic conditions on depressive symptomatology is partly mediated through the level of existing physical disability (Ormel et al., 1997).

Behavioral, Psychosocial, and Other Individual Risk Factors

In addition to the impact on biological and physical areas, other aspects of life such as the psychosocial and behavioral domains are affected by aging as well. Table 6.1 summarizes some of the most important psychosocial and behavioral risk factors of late-life depression in older women. Individual aging is accompanied by a series of social transitions, some entered into voluntarily and some imposed by circumstances. These transitions have consequences for the mental health of older women. In general, aging tends to be associated with relationship losses, because of retirement, widowhood, or death of age-peers such as siblings or friends. Increasing age also brings changes in relationship needs, for example as the result of increasing physical impairment. Older adults may become more dependent on others when they lose the ability to fulfill certain social or instrumental tasks themselves. The existing balance in their relationships may be disrupted, introducing strain and discomfort. It has been hypothesized that specific stressors, such as loss of partner or other intimates, are more normative in old age and more usual in that part of the life cycle than in younger age and therefore less disruptive. Older individuals have had the opportunity to learn how to cope with stressful circumstances and how to adjust their expectations to have fewer feelings of failure. On the basis of age and experience, older persons have developed more effective skills with which to manage stressful life events and to reduce emotional distress. Nevertheless, although their impact might be smaller than in younger age, social circumstances still have a significant impact on the presence of depressive symptoms in later life. For instance, the absence of a partner, fewer social relationships, and negative life events have consistently been associated with more feelings of depression in old age (Beekman et al., 1995; Kraaij, Arensman, & Spinhoven, 2002). The search for restitution secondary to the inevitable losses in late life is a major developmental task for aging individuals and a depression-like syndrome may appear, *a depletion syndrome*, when this task is not accomplished successfully. In line with the social disengagement theory (Lewinsohn, Rohde, Seeley, & Fischer, 1991), it appears that older persons who are less socially engaged are more depressed. This is illustrated in a study that showed that older persons who stopped driving had a greater risk of worsening depressive symptoms (Fonda, Wallace, & Herzog, 2001).

Personality also has been strongly linked to depression in older women and men. For instance, feelings of high neuroticism, low internal locus of control, and low extraversion have been associated with more depressive symptoms. The link between neuroticism and depression appears to be especially strong and therefore is one of the best indices of a person's underlying vulnerability to major depression (Roberts & Kendler, 1999). In addition to personal susceptibility, great interest has arisen in recent years in a search for genetic susceptibility to mood disorders across the life cycle. In community samples of elderly twins, genetic influences accounted for between 14 and 31% of the variance in reported depressive symptoms (Gatz, Pedersen, Plomin, Nesselroade, & McClearn, 1992; Jansson et al., 2004; McGue & Christensen, 1997). In a study in which they examined both older and middle-age twins, Johnson, McGue, Gaist, Vaupel, and Christensen (2002) found no age or sex difference in the magnitude of heritable influence. It appears that in old age, longstanding vulnerability factors such as personality and heredity are more important factors for major depressive disorder than for the milder type of depressive symptoms.

In the Prevalence of Depression in Older Women and Men section, we showed that the prevalence of significant depressive symptoms increases with increasing age. Nevertheless, the risk factor profile for minor depression over the life span is rather stable in terms of the domains represented. Risk factors in younger persons have also been found in old age (e.g., physical health problems, negative life events, lack of social support). However, the importance of risk factors relative to each other do vary across the life span, because the distribution of these risk factors is related to age. For instance, older persons report more social isolation and impoverishment and have more physical health problems and disability compared to younger persons. These relative increases in risk factors, especially in physical health problems, have been found to almost completely explain the increased prevalence of depression in the oldest-old (Blazer, Burchett, Service, & George, 1991). Thus, what seems to be an association between age and depressive symptoms appears to be largely the result of other risk factors. So, age per se is not a risk factor for depressive symptoms when other factors are taken into account.

The same is not true for the observed female preponderance in late-life depression. As previously mentioned, older women are about twice as likely to experience significant depression compared to older men. Part of the higher prevalence of depressive symptoms in older women might be due to the higher occurrence of physical and social circumstances, such as arthritis and pain-related conditions or widowhood. However, studies have indicated that, by adjusting for various differences in circumstances, it is not possible to completely explain the gender differentials in late-life depression and women continue to report more depression. The reason for the remaining gender differential is not well understood and is being debated. It has been suggested that there may be other causes such as differences in adolescent physical development, social role enactment, and gender-related patterns of life stress.

HEALTH CONSEQUENCES OF DEPRESSION IN OLDER WOMEN

Not only do medical comorbidity, functional impairment, and comorbid dementing disorders all adversely influence depressed mood, but depression in turn also adversely affects the outcome of the comorbid problems and may even cause the onset of comorbid problems. Overall, the adverse health consequences of late-life depression do appear to be rather consistent for older men and older women. The studies that have examined whether there is evidence of a gender-specific association of depression with major health outcomes are generally negative (Sevick, Rolih, & Pahor, 2000).

Most of the conducted scientific studies regarding the impact of depression on morbidity and mortality are in the area of cardiovascular disease. In reviewing these studies, it is clear the depression has a significant impact on the cardiovascular health of subjects (Wulsin & Singal, 2003). For instance, among persons with initially no cardiovascular disease, depression has been shown to increase the risk of developing coronary heart disease, heart failure, and stroke in studies with follow-up periods between 1 to 6 years. Researchers also found depression to increase mortality from cardiovascular, coronary, and cardiopulmonary conditions. The effect of depression on cardiovascular mortality was two times higher for major depressive disorder (threefold higher risk compared to nondepressed persons) than for depressive symptoms (1.5 times higher risk compared to nondepressed persons) (Penninx et al., 2001), which suggests that there is a dose–response association between the severity of depressive symptoms and the risk to die from cardiovascular disease. Depression has an impact on health outcomes not only in rather healthy older individuals, but also in those with cardiovascular disease. In one of the first studies illustrating this, older patients with depression following a myocardial infarction were much more likely to die in the first 4 months after the event compared to nondepressed patients with a myocardial infarction (26 vs. 7%) (Frasure-Smith, Lesperance, & Talajic, 1993). Since then, various studies have confirmed the negative effect that depression has on the prognosis of cardiovascular disease.

Depression has also been shown to affect functional status and disability over time. In one study, depression increased the risk for activities of daily living disability and mobility disability over 6 years by 67% and 73% respectively (Penninx, Leveille, Ferrucci, van Eijk, & Guralnik, 1999). Also when the physical function of older persons is objectively assessed through timed physical performance tests such as a walking and balance test, depressed persons showed significantly greater decline than nondepressed persons in performance over 4 years (Penninx et al., 1998). In older women, depression was also found to increase the risk of falls and the risk of nonvertebral fractures (Whooley et al., 1999).

Another health consequence of depression is in the cognitive domain. Severe depression has been shown to be a risk factor for the onset of Alzheimer's disease and for the development of (mild) cognitive impairment 3 years later (Comijs, Jonker, Beekman, & Deeg, 2001). Early depressive symptoms among subjects with minimal cognitive impairment may represent a preclinical sign and should be considered a risk factor for impending Alzheimer's disease or vascular dementia. Depression further complicates the course of Alzheimer's disease by increasing disability and physical aggression and leading to greater caregiver depression and burden.

Overall mortality is a significant adverse outcome resulting from late-life depression. In a review of 61 reports from 1997 to 2001, 72% of the reports demonstrate a positive association between depression and mortality in elderly people (Schulz, Drayer, & Rollman, 2002). Both severity and duration of depressive symptoms predict mortality in the elderly, with more severe and a longer duration of depressive symptoms being associated with the highest mortality risk. The association between depression and mortality holds in many of the published studies, despite the addition of potentially confounding variables in the analyses. For example, high levels of depressive symptoms independently predicted mortality when demographics, medical comorbidity, smoking, and body mass index were controlled for. Most conducted studies have focused on nonsuicidal mortality, so suicide cannot explain the observed link between late-life depression and mortality. However, there appears to be a definite link between suicide frequency and depression, which is rather consistent across the life cycle. In the United States, suicide frequency in the persons 65 years and older was 16.9 per 100,000 persons per year in 1998. The majority of the suicide cases were from older white males, who reach a suicide frequency of 62 per 100,000 persons (National Center for Health Statistics, 2001). Whereas the rate of completed suicides seems to increase with age, suicidal behaviors seem not to increase in age. There are approximately four attempts for each completed suicide in late life compared with 10 or more attempts per completed suicide earlier in life. Suicidal ideation is high among older adults, ranging from 5 to 10% of the population of older adults.

HOW CAN LATE-LIFE DEPRESSION LEAD TO ADVERSE HEALTH CONSEQUENCES?

First, the presence of depressive symptoms and especially major depressive disorder has physiological accompaniments, including altered autonomic balance and increased hypothalamic-pituitary-adrenal axis function, both of which are biologically plausible contributors to pathogenesis. When depression is induced in laboratory studies, persons exhibit larger sympathetic nervous-system-mediated cardiovascular responses.

Depressive symptoms appear to elevate resting heart rate and blood pressure, decrease heart rate variability, and increase ventricular arrhythmias and myocardial ischemia. In addition, depressed persons have been shown to exhibit hypersecretion of the adrenal steroid cortisol, adrenal hypertrophy, and an increased cortisol response to adrenocorticotrophic hormone. Depressed persons show increased sympathetic nervous system activation during everyday life, as documented by larger increases in daytime urinary epinephrine excretion and down-regulation of lymphocyte beta adrenergic receptors. There is also evidence that parasympathetic nervous system function is reduced in depressed persons. These direct physiologic processes may explain why depressed older persons are more vulnerable for subsequent health deterioration or for onset of some conditions such as cardiovascular disease.

Second, certain specific behavioral risk profiles in depressed persons may explain their higher risk for adverse health consequences. Behavioral risk factors appear to cluster in the same individuals. Increased smoking and alcohol consumption are well documented in depression and also among older depressed persons. Depressed persons not only smoke more often but they are found to be less likely to quit smoking and might inhale more deeply and smoke more of the cigarette than nondepressed smokers (Anda et al., 1990). In our own study of persons aged between 55 and 85 years, we found that depressed persons were 70% more likely than nondepressed persons to be smokers and, if they were smokers, they were smoking more cigarettes (van Gool et al., 2003). We did not find a difference in the link between depression and smoking for the older women and men. In addition, the food intake of depressed older persons may be less adequate and nutritious than that of nondepressed older persons. It has been shown that depressed persons have a higher 24-hour caloric intake than nondepressed persons. However, certain vitamin deficiencies, such as vitamin B12 and folate deficiencies, are more prevalent in depressed older persons (Penninx et al., 2000), which illustrates that certain depressed older persons may not get adequate nutrition.

Depressed older women and men also engage less in physical activities such as walking, gardening, and vigorous exercise activities such as sports as compared to their nondepressed peers (van Gool et al., 2003). This can partly explain why depressed older individuals are more at risk for adverse health outcomes, since a sedentary lifestyle is one of the most important risk factors for the onset of disability and mortality in old age. This is an important observation, because the level of physical activity is potentially modifiable through an exercise regimen. There are several clinical trials that illustrate that when depressed older persons are randomized to an exercise intervention, their depressed mood significantly improves compared to older persons who are not physically active (Penninx et al., 2002).

Third, related somatic symptoms of depression such as fatigue or sleeplessness may worsen older persons' health status and may therefore contribute to the adverse physical consequences of depression. This explanation is in line with findings that vital exhaustion (which has shared common characteristics with depression such as listlessness, loss of energy, irritability, and sleep problems) increases the risk of developing cardiovascular disease. Another mechanism linking depression to subsequent health deterioration may be via psychological mechanisms. Depressed mood may impede recovery processes by discouraging persons from obtaining adequate medical attention and rehabilitation and following treatment regimens. It has been described that depressed persons are generally less compliant to taking medications or following up on certain lifestyle regimens provided by health care professionals. Whether this is equally the case for younger versus older depressed persons and for depressed women and men has not been examined, but it has been described that older depressed persons experience multiple side-effects when taking antidepressants, which results in a low treatment compliance (Prabhakaran & Butler, 2002).

In sum, this chapter described that, although the prevalence of major depressive disorder in old age is rather low, significant depressive feelings are common in old age. Older women have about twice the prevalence of late-life depression than older men. The prevalence of significant depressive symptoms increases with increasing age. There are many risk factors for late-life depression, which cover biological, physical, and psychosocial domains. Although the importance of certain risk factors for depression in late life may not be very different from that in earlier life, the prevalence of certain risk factors in old age is much higher. For instance, (multiple) chronic conditions, disability, and loneliness because of a decreasing social network are all rather prevalent circumstances in old age. Therefore, especially in late life, clinically significant levels of depressive symptoms often appear to be a reactive depression because these symptoms are reactions to prevalent stressors encountered in later life such as worsened health status, economic stress, and a changing social context. For a psychiatric diagnosis of major depression in late life, longstanding personal or genetic vulnerability factors may play a relatively more important role and a depression diagnosis often represents recurring episodes of early-onset depressive disorders.

Although the risk factors for late-life depression may generally not be that different for older women and older men, older women are in relative terms more exposed to some of the most important risk factors of depression, which partly explain their higher prevalence of late-life depression. Late-life depression has several adverse health consequences ranging from an increased risk of mortality, decline in physical and cognitive function, and onset of cardiovascular disease. This subsequent health deterioration can be due to physiological changes induced by depression, the unhealthier behavioral risk profile of depressed older persons, the somatic accompaniments of depression, or the decreased motivation of depressed older persons for self-care.

References

American Psychiatric Association. (1989). _Diagnostic and statistical manual of mental disorders_ (3rd ed.), Washington, DC.

Anda, R. F., Williamson, D. F., Escobedo, L. G., Mast, E. E., Giovino, G. A., & Remington, P. L. (1990). Depression and the dynamics of smoking. A national perspective. _Journal of the American Medical Association, 264,_ 1541–1545.

Arborelius, L., Owens, M. J., Plotsky, P. M., & Nemeroff, C. B. (1999). The role of corticotropin-releasing factor in depression and anxiety disorders. _Journal of Endocrinology, 160,_ 1–12.

Beekman, A. T., Copeland, J. R., & Prince, M. J. (1999). Review of community prevalence of depression in later life. _British Journal of Psychiatry, 174,_ 307–311.

Beekman, A. T., Deeg, D. J., van Tilburg, T., Smit, J. H., Hooijer, C., & van Tilburg, W. (1995). Major and minor depression in later life: A study of prevalence and risk factors. _Journal of Affective Disorders, 36,_ 65–75.

Blazer, D., Burchett, B., Service, C., & George, L. K. (1991). The association of age and depression among the elderly: An epidemiologic exploration. _Journal of Gerontology Medical Sciences, 46,_ M210–M215.

Comijs, H. C., Jonker, C., Beekman, A. T., & Deeg, D. J. (2001). The association between depressive symptoms and cognitive decline in community-dwelling elderly persons. _International Journal of Geriatric Psychiatry, 16,_ 361–367.

Davis, K. L., Davis, B. M., Mathe, A. A., Mohs, R. C., Rothpearl, A. B., Levy, M. I., et al. (1984). Age and the dexamethasone suppression test in depression. _American Journal of Psychiatry, 141,_ 872–874.

Fonda, S. J., Wallace, R. B., & Herzog, A. R. (2001). Changes in driving patterns and worsening depressive symptoms among older adults. _Journal of Gerontology of Social Science, 56_ S343–S351.

Frasure-Smith, N., Lesperance, F., & Talajic, M. (1993). Depression following myocardial infarction. Impact on 6-month survival. _Journal of the American Medical Association, 270,_ 1819–1825.

Gatz, M., Pedersen, N. L., Plomin, R., Nesselroade, J. R., & McClearn, G. E. (1992). Importance of shared genes and shared environments for symptoms of depression in older adults. _Journal of Abnormal Psychology, 101,_ 701–708.

Jansson, M., Gatz, M., Berg, S., Johansson, B., Malmberg, B., McClearn, G. E., et al. (2004). Gender differences in heritability of depressive symptoms in the elderly. _Psychological Medicine, 34,_ 471–479.

Johnson, W., McGue, M., Gaist, D., Vaupel, J. W., & Christensen, K. (2002). Frequency and heritability of depression symptomatology in the second half of life: Evidence from Danish twins over 45. _Psychological Medicine, 32,_ 1175–1185.

Kessler, R. C., Berglund, P., Demler, O., Jin, R., Koretz, D., Merikangas, K. R., et al. (2003). The epidemiology of major depressive disorder: Results from the National

Comorbidity Survey Replication (NCS-R). *Journal of the American Medical Association, 289,* 3095–3105.

Kraaij, V., Arensman, E., & Spinhoven, P. (2002). Negative life events and depression in elderly persons: A meta-analysis. *Journal of Gerontology Social Sciences and Medicine, 57,* 87–94.

Kumar, A., Thomas, A., Lavretsky, H., Yue, K., Huda, A., Curran, J., et al. (2002). Frontal white matter biochemical abnormalities in late-life major depression detected with proton magnetic resonance spectroscopy. *American Journal of Psychiatry, 159,* 630–636.

Lewinsohn, P. M., Rohde, P., Seeley, J. R., & Fischer, S. A. (1991). Age and depression: Unique and shared effects. *Psychology and Aging, 6,* 247–260.

McGue, M., & Christensen, K. (1997). Genetic and environmental contributions to depression symptomatology: Evidence from Danish twins 75 years of age and older. *Journal of Abnormal Psychology, 106,* 439–448.

National Center for Health Statistics (2001). *Death rates for 72 selected causes by 5-year age groups, race and sex: United States, 1979–1998.* Washingon DC.

Ormel, J., Kempen, G. I., Penninx, B. W., Brilman, E. I., Beekman, A. T., & van Sonderen, E. (1997). Chronic medical conditions and mental health in older people: disability and psychosocial resources mediate specific mental health effects. *Psychological Medicine, 27,* 1065–1077.

Penninx, B. W., Beekman, A. T., Honig, A., Deeg, D. J., Schoevers, R. A., van Eijk, J. T., et al. (2001). Depression and cardiac mortality: Results from a community-based longitudinal study. *Archives of General Psychiatry, 58,* 221–227.

Penninx, B. W., Beekman, A. T., Ormel, J., Kriegsman, D. M., Boeke, A. J., van Eijk, J. T., et al. (1996). Psychological status among elderly people with chronic diseases: does type of disease play a part? *Journal of Psychosomatic Research, 40,* 521–534.

Penninx, B. W., Guralnik, J. M., Ferrucci, L., Fried, L. P., Allen, R. H., & Stabler, S. P. (2000). Vitamin B(12) deficiency and depression in physically disabled older women: Epidemiologic evidence from the Women's Health and Aging Study. *American Journal of Psychiatry, 157,* 715–721.

Penninx, B. W., Guralnik, J. M., Ferrucci, L., Simonsick, E. M., Deeg, D. J., & Wallace, R. B. (1998). Depressive symptoms and physical decline in community-dwelling older persons. *Journal of the American Medical Association, 279,* 1720–1726.

Penninx, B. W., Leveille, S., Ferrucci, L., van Eijk, J. T., & Guralnik, J. M. (1999). Exploring the effect of depression on physical disability: Longitudinal evidence from the established populations for epidemiologic studies of the elderly. *American Journal of Public Health, 89,* 1346–1352.

Penninx, B. W., Rejeski, W. J., Pandya, J., Miller, M. E., Di Bari, M., Applegate, W. B., et al. (2002). Exercise and depressive symptoms: A comparison of aerobic and resistance exercise effects on emotional and physical function in older persons with high and low depressive symptomatology. *Journal of Gerontology: Psychological Sciences, 57,* 124–132.

Prabhakaran, P., & Butler, R. (2002). What are older peoples' experiences of taking antidepressants? *Journal of Affective Disorders, 70,* 319–322.

Roberts, S. B., & Kendler, K. S. (1999). Neuroticism and self-esteem as indices of the vulnerability to major depression in women. *Psychological Medicine, 29,* 1101–1109.

Schulz, R., Drayer, R. A., & Rollman, B. L. (2002). Depression as a risk factor for non-suicide mortality in the elderly. *Biological Psychiatry, 52,* 205–225.

Seidman, S. N., Araujo, A. B., Roose, S. P., & McKinlay, J. B. (2001). Testosterone level, androgen receptor polymorphism, and depressive symptoms in middle-aged men. *Biological Psychiatry, 50,* 371–376.

Sevick, M. A., Rolih, C., & Pahor, M. (2000). Gender differences in morbidity and mortality related to depression: A review of the literature. *Aging (Milano.), 12,* 407–416.

van Gool, C. H., Kempen, G. I., Penninx, B. W., Deeg, D. J., Beekman, A. T., & van Eijk, J. T. (2003). Relationship between changes in depressive symptoms and unhealthy lifestyles in late middle aged and older persons: Results from the Longitudinal Aging Study Amsterdam. *Age and Aging, 32,* 81–87.

Whooley, M. A., Kip, K. E., Cauley, J. A., Ensrud, K. E., Nevitt, M. C., & Browner, W. S. (1999). Depression, falls, and risk of fracture in older women. Study of Osteoporotic Fractures Research Group. *Archives of Internal Medicine, 159,* 484–490.

Wulsin, L. R., & Singal, B. M. (2003). Do depressive symptoms increase the risk for the onset of coronary disease? A systematic quantitative review. *Psychosomatric Medicine, 65,* 201–210.

Yaffe, K., Ettinger, B., Pressman, A., Seeley, D., Whooley, M., Schaefer, C., et al. (1998). Neuropsychiatric function and dehydroepiandrosterone sulfate in elderly women: a prospective study. *Biological Psychiatry, 43,* 694–700.

Yaffe, K., Lui, L. Y., Grady, D., Cauley, J., Kramer, J., & Cummings, S. R. (2000). Cognitive decline in women in relation to non-protein-bound oestradiol concentrations. *Lancet, 356,* 708–712.

PART III

COGNITIVE, EMOTIONAL, AND INTERPERSONAL MODELS OF RISK

7

Cognition and Depression

Joan S. Girgus and Susan Nolen-Hoeksema

Cognitive models of depression have been popular over the past four decades (Abramson et al., 2002). Do these cognitive models help to explain why women are more prone to depression than men (cf. Nolen-Hoeksema, 2002)? In this chapter, we will address this question, examining each of four broad categories of cognitive variables that have been proposed as predisposing factors for depression. The first of these is the self-concept or the characteristic ways that people think about themselves. The second is interpersonal orientation or the characteristic ways that people think about their relationships with others. The third is cognitive style or the characteristic ways that people think about the things that happen to them and about what the future is likely to bring. The fourth is coping style or the characteristic ways that people deal with the stressful things that happen to them or with their depressed feelings.

Unfortunately, most of the research on the relationship between cognition and depression is concurrent in nature. That is, the measures of the cognitive variables and depression variables were administered at the same time. This makes it impossible to discern the causal direction of any relationship that is found, because it is as easy to imagine that depression affects the way people think as it is to imagine that the way people think affects how depressed they are (and, indeed, there is evidence for effects in both directions). Thus, we will focus on what longitudinal research there is to ascertain the extent to which cognitive factors are associated with increases in depressive symptoms and, in turn, with the gender difference in depression.

SELF-CONCEPT

There are three major categories of self-concept that have been explored in relation to depression. The first is body image, the way people think about how they look and how attractive they are. The second is self esteem, people's perceptions of their overall or global self-worth. The third is perceived mastery or control or self-efficacy, the extent to which people feel confident about their ability to accomplish things.

Body Image

Body image has been repeatedly hypothesized to be related to depression. Most research has focused on people's, and particularly on women's, overall satisfaction or dissatisfaction with their bodies. In addition, some of the research has looked at whether there is a connection between people's, and again particularly women's, estimates of how fat or big they think they are and depression. It is not clear whether estimates of satisfaction–dissatisfaction and estimates of size–shape are measuring the same or different things. It is possible that perception of fatness–thinness is the major factor contributing to overall body satisfaction or perceived physical attractiveness, especially for women. Direct tests of this are hard to find, but a few studies with women respondents have used multiple measures that ostensibly tap overall body satisfaction separately from perception of fatness–thinness. Unfortunately, these studies (Denniston, Roth, & Gilroy, 1992; Mable, Balance, & Galgan, 1986) found correlations between body image measures that range from .03 to .92, so a good deal more research is needed to understand the various components of body image as well as the relationship between these components and depression.

There is considerable evidence that women who are more dissatisfied with their bodies have higher levels of depressive symptoms than women who are more satisfied with their bodies (Davis & Katzman, 1997; Denniston et al., 1992; Hayaki, Friedman, & Brownell, 2002; Joiner, Wonderlich, Metalsky, & Schmidt, 1995; Koenig & Wasserman, 1995; Marsella, Shizuru, Brennan, & Kameoka, 1981; Sarwer, Wadden, & Foster, 1998; Strong, Williamson, Netemeyer, & Geer, 2000; Thompson & Psaltis, 1988; Walker, Timmerman, Kim, & Sterling, 2002). It is less clear that the same relationship between body dissatisfaction and depressive symptoms holds for men (Davis & Katzman, 1997; Marsella et al., 1981; Oates-Johnson & DeCourville, 1999; Strong et al., 2000).

Women are also consistently less satisfied with their bodies than men are (Barry, Grilo, & Masheb, 2002; Davis & Katzman, 1998; Furnham & Greaves, 1994; McCauley, Mintz, & Glenn, 1988; Oates-Johnson & DeCourville, 1999, but see Mable et al., 1986), not only in the United States but in other countries as well (Davis & Katzman, 1998), and this gender gap has been increasing over time (Feingold & Mazzella, 1998). Furthermore, women who perceive themselves as being overweight have higher levels of depressive symptoms than do women who perceive themselves as being of normal weight, but men who perceive themselves as overweight do not differ in their level of depressive symptoms from men who perceive themselves as being of normal weight (Quinn & Crocker, 1998). The connection to depression seems clearly to be based on how people perceive themselves, as actual body mass index does not correlate with depressive symptoms, even in women (Joiner et al., 1995; Sarwer et al., 1998).

Unfortunately, these cross-sectional studies cannot tell us whether women who are unhappy about their bodies or their weight become depressed or whether depression causes women to become dissatisfied with their bodies or weight. There is some evidence, however, that the relationship may be bilateral, with depression likely to lead to body dissatisfaction as well as body dissatisfaction likely to contribute to depression (Cohen-Tovee, 1993; Mori & Morey, 1991).

The research on body image and depression has not been designed in ways that tell us whether body dissatisfaction contributes to the gender difference in depression. Women are much less satisfied with their bodies then men are, however, and this, combined with the consistent correlation of body dissatisfaction with depressive symptoms (at least in women), certainly suggests that research explicitly designed to investigate whether body dissatisfaction plays a role in the gender difference in depression would be useful.

Self-Esteem or Perceived Competence

The largest body of research that relates the self-concept to depression involves the idea of self esteem. Although, in principle, people could be asked how much they like (or dislike) this or that aspect of the self, in practice, adults are generally asked to make judgments about their overall self-esteem or self-worth, and the most widely used scale is designed around this definition (Rosenberg, 1965, 1979). It is possible that global self-esteem is in fact a better predictor of depressive symptoms than the esteem a person has for any specific aspect of the self (Rosenberg, Schooler, Schoenbach, & Rosenberg, 1995).

The relationship between high levels of depressive symptoms and low self-esteem, as indexed by perception of general self-worth, is one of the strongest and most consistent in the depression literature (Berthiaume, David, Saucier, & Borgeat, 1998; Chou & Chi, 2001; Cotten, 1999; Culp & Beach, 1998; Katz, Joiner, & Kwon, 2002; Pakriev, Poutanen, & Salokangas, 2002; Ritter, Hobfoll, Lavin, Cameron, & Hulsizer, 2000; Scheier, Carver, & Bridges, 1994; Schieman & Turner, 2001; Smith & Betz, 2002; Stein & Nyamathi, 1999; Thomas & Vindhya, 2000; Turner & Butler, 2003; Turner, Lloyd, & Roszell, 1999; Zhang & Norvilitis, 2002). Similarly, low self-esteem is usually correlated with the occurrence of depressive episodes (Reinherz, Giaconia, Hauf, Wasserman, & Silverman, 1999; Roberts & Kendler, 1999; Turner et al., 1999; Wilhelm, Parker, Dewhurst-Savellis, & Asghari, 1999). The relationship between low self-esteem and depression holds both for women and men, for all ages from young adults through the elderly, and in countries other than the United States.

The correlation between general self-worth and depressive symptoms is frequently as high as the correlation between two administrations of either the self-esteem or the depression measure. But what this signifies is not clear. Longitudinal research would allow us to begin to answer the question of whether self-esteem bears a causal relationship with depression over and above prior depression, but there is relatively little evidence to this point. What there is suggests that self esteem does contribute to future depression over and above current depression (Fernandez, Mutran, & Reitzes, 1998; Kernis et al., 1998; Roberts, Shapiro, & Gamble, 1999; but see Ritter et al., 2000). This relationship between self-esteem and depression may reflect a straightforward effect; people with high self-esteem may be less likely to become depressed no matter what else is going on in their lives. But it is also possible that the relationship between self-esteem and depression operates only in the context of stressful events. That is, people who are experiencing stressful events may be more likely to become depressed if they have low self-esteem than if they have high self esteem, but people who are not experiencing stress in their lives may be unlikely to become depressed no matter what their levels of self-esteem.

A number of studies have examined whether self-esteem buffers or moderates the effect of stress on depression (Chou & Chi, 2001; Culp & Beach, 1998; Fernandez et al., 1998; Kernis et al., 1998; Ritter et al., 2000). In general (but not always, see Ritter et al., 2000), self-esteem does seem to buffer the effect of stress on depression: under low levels of stress, it does not matter very much whether people have high or low self-esteem, but, when stress levels are high, people with high self-esteem experience less depression than people with low self-esteem.

There is a popular belief that women have lower self-esteem than men, but most studies do not find any difference between the self-esteem of men and women (Culp & Beach, 1998; Smith & Betz, 2002; Stein & Nyamathi, 1999; Turner et al., 1999); those that do find a difference, however, find lower self-esteem levels among women (Katz et al., 2002; Zhang & Norvilitis, 2002). This latter possibility raises the question of whether self-esteem might mediate or account for all or part of the gender difference in depression. This has been little tested, and what results there are, are mixed (e.g., Katz et al., 2002; Turner et al., 1999). Given the existing data, it seems unlikely that self-esteem contributes directly to the gender difference in depression, although more research to this point would be useful.

Perceived Self-Efficacy, Perceived Mastery, and Perceived Control

Although perceived self-efficacy, mastery, and control are sometimes treated as separate constructs, they are conceptually very similar, and many investigators use the terms interchangeably (and measure them in the same way). We will take perceived self-efficacy, mastery, and control all to refer to people's beliefs in their ability to attain particular ends or to "produce desired effects by their actions" (Bandura, Pastorelli, Barbaranelli, & Caprara, 1999).

A substantial body of literature has explored whether there is a relationship between perceived self-efficacy, mastery, or control and depression in adults, often in the context of particular kinds of physical or other difficulties. Thus, there are data on the direct relationship between perceived self-efficacy, mastery, and control and depression and on the interaction between stress and these dimensions and depression.

The concurrent relationship between perceived self-efficacy, mastery, or control and depressive symptoms is strong and well-established (Badger, 2001; Bisconti & Bergeman, 1999; Bullers, 2000; Chou & Chi, 2001; Cotten, 1999; Danziger, Carlson, & Henly, 2001; Ennis, Hobfoll, & Schroder, 2000; Ginexi, Howe, & Caplan, 2000; Heilemann, Lee, & Kury, 2002; Hobfoll, Johnson, Ennis, & Jackson, 2003; Jang, Haley, Small, & Mortimer, 2002; Lachman & Weaver, 1998; Maciejewski, Prigerson, & Mazure, 2000; Macrodimitris & Endler, 2001; Maier & Lachman, 2000; Makaremi, 2000; Nolen-Hoeksema & Jackson, 2001; Regehr, Cadell, & Jansen, 1999; Scheier et al., 1994; Schieman & Turner, 2001; Schieman, Van Gundy, & Taylor, 2002; Smith & Betz, 2002; Turner et al., 1999). This relationship holds for all ages from young adults through the elderly, for people who have health problems or have been abused, for adults in other countries as well as in the United States, and for both women and men. There are also, however, a few studies in which no relationship has been found between standard measures of self-efficacy, mastery, or control and depressive symptoms (Hobfoll, Bansal, & Schurg, 2002; Osowiecki & Compas, 1999; Turner & Butler, 2003). Longitudinal results are less frequently reported, but those that do exist support the idea that perceived self-efficacy, mastery, or control predict increases in depressive symptoms over time even when prior levels of depressive symptoms are taken into account (Grant, Long, & Willms, 2002; Hobfoll et al., 2003; Lightsey, 1997; Lightsey & Christopher, 1997). It is not clear whether this effect holds for both men and women. Two of the studies (Grant et al., 2002; Hobfoll et al., 2003) report data on women alone, while the other two (Lightsey, 1997; Lightsey & Christopher, 1997) report data for men and women combined.

There is considerable evidence that perceived self-efficacy, mastery, and control can buffer stressful circumstances, both negative life events in general and specific circumstances such as economic strain and physical disability (Chou & Chi, 2001; Ennis et al., 2000; Jang et al., 2002; Lachman & Weaver, 1998; Lightsey, 1997; Lightsey & Christopher, 1997). When the level of stress is high, people with low levels of perceived self-efficacy, mastery, and control have considerably higher levels of depressive symptoms than people with high levels of perceived self-efficacy, mastery, and control. The difference in depressive symptoms between people with high and low levels of perceived self-efficacy, mastery, and control is much smaller when stress levels are low. With one exception (Ennis et al., 2000), these studies have asked about buffering effects in samples of men and women combined so we have no way of knowing if these effects hold equally for both sexes separately.

It is not clear whether there are gender differences in perceived self-efficacy, mastery, and control. In some studies, there is no difference between men and women in perceived self-efficacy, mastery, and control (Bisconti & Bergeman, 1999; Bullers, 2000; Danziger et al., 2001; Lightsey & Christopher, 1997; Maier & Lachman, 2000; Smith & Betz, 2002). In studies in which gender differences are found, however, men generally have higher levels of perceived self-efficacy, mastery, and control than women do (Nolen-Hoeksema & Jackson, 2001; Nolen-Hoeksema, Larson, & Grayson, 1999; Pearlin & Schooler, 1978; Turner et al., 1999), although occasionally the reverse is found (Makaremi, 2000). Thus, it is not clear whether perceived self-efficacy, mastery, and control are candidates for explanation of the gender difference in depression, although two studies have found that they partially mediate the gender difference in depression (Cotten, 1999; Nolen-Hoeksema et al., 1999) and more research along these lines would be helpful.

INTERPERSONAL ORIENTATION

One of the most consistent psychological differences between women and men is in interpersonal orientation (Feingold, 1994). In recent years, research has explored whether certain kinds of interpersonal orientation make people more vulnerable to depressive symptoms or episodes. Much of this work has been based on Beck's (1983) idea that some people are more sociotropic than others; that is, some people exhibit a heightened concern about what others think of them and a greater dependence on the approval of others. Sociotropy is sometimes conceptualized as a cognitive self-schema that influences how the individual determines self-worth and responds emotionally to particular types of stressors (Campbell & Kwon, 2001) and sometimes as a personality trait that causes individuals to depend on the love and attention of others for the maintenance of their self-esteem (Fairbrother & Moretti, 1998), an idea that has a long history in the self-esteem literature (e.g., James, 1890).

Sociotropy consistently correlates with measures of depressive symptoms and with diagnosed depressive episodes (Alford & Gerrity, 1995; Allen, Ames, Layton, Bennetts, & Kingston, 1997; Beck, Robbins, Taylor, & Baker, 2001; Beck, Taylor, & Robbins, 2003; Bieling, Beck, & Brown, 2000; Campbell & Kwon, 2001; Connor-Smith & Compas, 2002; Davila, 2001;

Fairbrother & Moretti, 1998; Mazure, Raghavan, Maciejewski, Jacobs, & Bruce, 2001; Moore & Blackburn, 1994, 1996; Nunn, Mathews, & Trower, 1997; Oates-Johnson & Decourville, 1999; Robins & Block, 1988; Sato & McCann, 1997, 1998; Steer & Clark, 1997). This correlation holds for both women and men and for all ages from young adults through the elderly.

Because sociotropy is conceptualized as a stable part of the self-concept or as a personality trait, it should predict increases in depressive symptoms and onset of depression episodes, but this is a hypothesis that has not really been tested. There is very little longitudinal research, and the few studies that do exist have inconsistent results (Fresco, Sampson, Craighead, & Koons, 2001; Robins, Hayes, Block, Kramer, & Villena, 1995).

There has, however, been a substantial amount of research on another aspect of the relationship of sociotropy to depression: the congruency hypothesis. Because sociotropy involves an excessive dependence on interpersonal relationships, sociotropic individuals should be more vulnerable to disruptions in those relationships. This means that stressors in the interpersonal realm (as opposed to some other realm) should interact with sociotropy to predict depression. The research that has tested this hypothesis has had somewhat mixed results: in most studies, sociotropy does interact with interpersonal stressors to predict depression (Bartelstone & Trull, 1995; Clark, Beck, & Brown, 1992; Fresco et al., 2001; Mazure & Maciejewski, 2003; Raghavan, Le, & Berenbaum, 2002; Robins et al., 1995; Rude & Burnham, 1993), but in some, it does not (Flett, Hewitt, Garshowitz, & Martin, 1997; Kwon & Whisman, 1998; Mazure, Bruce, Maciejewski, & Jacobs, 2000). (Sociotropy initially correlates with depression in all these studies; when the interaction is significant, sometimes sociotropy continues to correlate with depression and sometimes it does not.) The form of the significant interactions is as one would expect: when people do not experience interpersonal stressors, it does not matter whether they are sociotropic, they have low levels of depressive symptoms; when people do experience interpersonal stressors, however, those who are sociotropic have higher levels of depressive symptoms (or episode onset) than those who are not sociotropic. On the other hand, there is also some indication that sociotropy may interact with other kinds of negative life events (in addition to interpersonal ones) to predict depression (Fresco et al., 2001; Robins & Block, 1988; Robins et al., 1995). Thus, it is not clear that the congruency hypothesis (that depression will result only when the diathesis of sociotropy interacts with interpersonal stressors) is supported; rather, sociotropy may be a more general diathesis or risk factor for depression in the face of negative life events.

Because women generally have stronger interpersonal orientations (Feingold, 1994), one might expect that women are more sociotropic or dependent on relationships for their self-esteem than men are. This is probably the case (Campbell & Kwon, 2001; Sato & McCann, 1997), but it is far from well tested. If this is the case, then one might also expect that sociotropy mediates or partially mediates the gender difference in depression but, surprisingly, this hypothesis is essentially untested.

More recently, Helgeson and her colleagues (Helgeson & Fritz, 1998) proposed that unmitigated communion rather than a communal or interpersonal orientation makes an individual vulnerable to depression. Unmitigated communion, which bears considerable conceptual similarity to sociotropy, involves an excessive concern about relationships with others, which can lead individuals to silence their own wants and needs in favor of the needs of others in order to maintain a positive emotional tone in relationships. High scores on measures of unmitigated communion are also usually correlated with depression (Fritz, 2000; Fritz & Helgeson, 1998; Helgeson & Fritz, 1996; Katz et al., 2002); high scores on unmitigated communion also predict increases in depressive symptoms over time (Fritz, 2000).

Helgeson and others have suggested that such excessive concern about relationships is more characteristic of women than of men (Helgeson, 1994; Jack, 1991). In studies using measures to tap "placing others' need before your own" or "feeling too responsible for relationships," females have generally scored higher on such measures than males (Fritz & Helgeson, 1998; Helgeson & Fritz, 1996; Katz et al., 2002; see Helgeson, 1994 for a review of earlier work), although there have been exceptions (e.g., Bruch, 2002; Fritz, 2000).

Thus, women may be more likely than men to overvalue relationships as sources of self-worth, which interpersonal theories of depression have identified as a risk factor for depression (e.g., Barnett & Gotlib, 1988; Joiner & Coyne, 1999). Women who do overvalue relationships may seek reassurance from others to an extent that is excessive and annoys them (Joiner, Metalsky, Katz, & Beach, 1999). This can lead to rejection by others, or at least conflict in relationships, which then only feeds a woman's worries about the status of her relationships and presumably makes her more vulnerable to depression. There has not, however, been any direct test of whether unmitigated communion mediates the gender difference in depression.

COGNITIVE STYLES

Everyone has both bad events and good events that occur in their lives. In recent decades, several theories have proposed that there are individual differences in the characteristic ways that people think about the bad things that happen in their lives and that some ways of thinking about these negative life events make it more likely that people will become depressed (Abramson, Metalsky, & Alloy, 1989; Abramson, Seligman, & Teasdale, 1978; Beck, 1967). In a related vein, other investigators have explored the expectations that people have about whether good things or bad things will happen to them, hypothesizing that people who expect bad things to happen are more likely to become depressed (Alloy, Abramson, Metalsky, & Hartlage, 1988; Scheier & Carver, 1985). There are four major theories that have proposed that characteristic thought patterns, either about the bad things that happen in one's life or about whether good or bad things will happen in one's life, can lead to depression.

Optimism and Pessimism

Scheier and Carver (1985) have proposed that individuals characteristically have optimistic or pessimistic expectations of whether things will go well for them in the future. In this theory, optimism and pessimism are viewed as personality traits. That is, they probably have a genetic component (Plomin et al., 1992), they are established early in life, and they are relatively stable over the life course. (The latter two assumptions have generally not been tested, however.) The theory proposes that people who are pessimists and expect bad outcomes in the future are more likely to become depressed than people who are optimists and expect good outcomes in the future.

There is extensive and consistent data to support this view (Andersson, 1996; Carver, Lehman, & Antoni, 2003; Chang, 2002; Chang, Maydeu-Olivares, & D'Zurilla, 1997; Chang & Sanna, 2001; Devine, Forehand, Morse, Simon, Clark, & Kernis, 2000; Epping-Jordan et al., 1999; Isaacowitz & Seligman, 2002; Labbe, Lopez, Murphy, & O'Brien, 2002; Lightsey, 1997; Plomin et al., 1992; Scheier et al., 1994; Updegrafff, Taylor, Kemeny, & Wyatt, 2002; van Servellen, Aguirre, Sarna, & Brecht, 2002). These studies show strong concurrent correlations between optimism–pessimism and depressive symptoms for samples with a range of ethnic and class backgrounds, for people who are healthy and people who are ill, and for all ages from young adults through the elderly. Furthermore, as one might expect with a dispositional variable, longitudinal studies have consistently found that optimism–pessimism predicts depressive symptoms over and above prior levels of depressive symptoms (Brissette, Scheier, & Carver, 2002; Carver & Gaines, 1987; Devine et al., 2000; Isaacowitz & Seligman, 2002; Lightsey & Christopher, 1997; Vickers & Vogeltanz, 2000). None of these studies tested whether optimism–pessimism predicts increases in depression equally for men and women; although there is no reason to think that this might be the case, research directed to this point would be useful.

Overall there has been little research on the question of whether stressful or negative life events interact with optimism–pessimism to yield depression. It is certainly possible that, at low levels of stress, it does not matter very much whether one is optimistic or pessimistic but, at high levels of stress, optimistic people are less likely to become depressed than are pessimistic people. The small number of research reports that speak to this point provide mixed data (Carver & Gaines, 1987; Chang, 1998, 2002; Lightsey, 1997; Lightsey & Christopher, 1997).

The question of gender differences in optimism–pessimism has not been much explored: what data there are suggest that men and women are equally optimistic–pessimistic (Chang et al., 1997; Chang & Sanna, 2001; Isaacowitz & Seligman, 2002; Lightsey & Christopher, 1997; van Servellen et al., 2002), although a large community study in Sweden found women to be more pessimistic than men (Scott & Melin, 1998). Thus optimism–pessimism does not seem to contribute to the gender difference in depression.

Depressogenic Cognitive Styles

There are three major theories that have proposed that particular thought patterns make individuals more likely to become depressed when faced with stressful life events and circumstances. Beck (1967; Beck, Rush, Shaw, & Emery, 1979) was the first to propose that cognitive processes could be one of the causes of depression, arguing that some individuals are prone to certain kinds of errors when thinking about the bad things that happen in their lives. These errors include overgeneralization (sweeping conclusions on the basis of a single incident), selective abstraction (focusing on negative details and ignoring positive details), personalization (relating external events to oneself without justification), and catastrophizing (concluding that things are much worse than they really are). The result is that some individuals have a style of thinking that leads them to be unrealistically negative about themselves, their experiences, and the future (*the negative cognitive triad* in Beck's terminology). Such individuals believe that they are unworthy or inadequate, that negative life events constitute defeats and deprivations, and that any current difficulties will persist into the future. When stressful life events occur and these individuals view them through the lens of the negative cognitive triad, the result is depression.

The other two diathesis-stress theories are very similar to one another. Seligman, Abramson, and Alloy (Abramson et al., 1978; Abramson et al., 1989; Peterson & Seligman, 1984; Seligman, Abramson, Semmel, & von Baeyer, 1979) have proposed models in which some individuals are more likely than others to attribute negative life events to internal, global, and stable causes and positive life events to external, specific, and unstable causes. That is, some individuals, when faced with a stressful event, will think of it as a circumstance that is their fault and that is likely to affect many aspects of their lives for the foreseeable future. These individuals are said to have a pessimistic explanatory style for the events in their lives. The original reformulated learned helplessness theory (Abramson et al., 1978) proposed that this pessimistic way of thinking about negative life events leads a person to feel depressed and helpless about his or her ability to do anything about the stressful event. In an update to this theory, Abramson et al. (1989) argue that a pessimistic explanatory style leads to a particular type of depression, a hopelessness depression, which is characterized by the expectation that highly aversive outcomes are likely to occur but highly desired outcomes are unlikely to occur.

Longitudinal research is particularly important when considering cognitive style factors in depression, since it is well known and well documented that people who are depressed think negatively about themselves and their world (Coyne & Gotlib, 1986). Fortunately, there is a good deal of longitudinal research that has tested whether cognitive style variables predict increases in depressive symptoms or depressive episodes over time, with initial levels of symptoms or episodes taken into account (Abramson et al., 2000; Alloy et al., 1999; Alloy & Clements, 1998; Grazioli & Terry, 2000; Vickers & Vogeltanz, 1999). These studies have used time spans that vary from 1 month to 2.5 years and have generally found that a negative, or pessimistic, or vulnerable cognitive style predicts increases in depressive symptoms or episodes over time, although there are occasional exceptions (e.g., Vickers & Vogeltanz, 2000).

There are two different kinds of exceptions to this general result. First, although Sharpley and Yardley (1999) report the usual correlation between a pessimistic explanatory style and higher levels of depression for older adults in Australia, Isaacowitz and Seligman (2001, 2002) report that an optimistic rather than a pessimistic explanatory style for negative life events leads to increases in depressive symptoms in older adults. Isaacowitz and Seligman speculate that older adults experience more uncontrollable negative life events than do individuals in any other age group and that an optimistic explanatory style for negative life events encourages a problem-solving orientation, which, because it is an ineffective coping approach for uncontrollable events, will lead to more depression.

The second kind of exception arises in studies that have specifically tested the diathesis-stress component of these theories. In these studies, when the interaction of cognitive vulnerability and levels of negative life events predicts increases in depressive symptoms over time, cognitive vulnerability per se generally does not predict increases in depressive symptoms over time. The form of the interaction is always the same: those individuals who are cognitively vulnerable and experience high levels of stressful events show increases in depressive symptoms but individuals who experience low levels of negative life events (whether or not they are cognitively vulnerable), or who experience high levels of negative life events but are not cognitively vulnerable, do not experience increases in depressive symptoms (Kernis et al., 1998; Metalsky, Halberstadt, & Abramson, 1987; Metalsky & Joiner, 1992; Metalsky, Joiner, Hardin, & Abramson, 1993; Peterson & Seligman, 1984; Reilly-Harrington, Alloy, Fresco, & Whitehouse, 1999).

There has been almost no published explicit test of gender differ-ences in negative, or pessimistic, or vulnerable cognitive styles. However, researchers who have checked to see whether there are differences in cogni-tive style between men and women before combining the data in their sam-ples have generally found no gender differences (Isaacowitz & Seligman, 2002; Kernis et al., 1998; Metalsky et al., 1993; Reilly-Harrington et al., 1999; Sharpley & Yardley, 1999), so it seems unlikely that cognitive style can con-tribute to our understanding of the gender difference in depression.

COPING STRATEGIES

Life is full of stressful events. When stressful events occur, we must cope with them as best we can. But people cope with stressful events in different ways, and the various ways people deal with the circumstances of their lives may be more or less adaptive. It is also possible that different kinds of stressful events are best dealt with in different ways. That is, a response that is adaptive in one situation may actually be maladaptive in another. This may be particularly the case for controllable and uncontrollable events. Thus, what may matter most in terms of adaptive coping is whether an individual has multiple coping strategies and uses them flexibly (Folkman, Lazarus, Dunkel-Schetter, DeLongis, & Gruen, 2000).

Over the years, there have been many schemes for classifying the strate-gies that people use to cope with stressful circumstances and events. One of the earliest – and still one of the most widely used – divides coping strate-gies into approach versus avoidance (Folkman & Lazarus, 1988), proposing that a person either could work with the stressful situation to try to make it less stressful or could turn away and avoid the situation altogether. Oth-ers have divided approach strategies into problem-focused and emotion-focused strategies, with both contrasted to avoidance strategies (Holahan, Moos, & Sandler, 1996; Summerfeldt & Endler, 1996). These investigators argue that individuals faced with a stressful life event could try to make the situation less stressful either by solving the problem (either by changing it directly or by a cognitive reappraisal that changes its meaning) or by reg-ulating the negative emotion generated by the situation. In recent years, there has been a good deal of interest in emotion regulation (Barrett & Gross, 2001; Gross, 1998, 1999), which is currently one of the most promis-ing theoretical and research approaches to coping.

Interpreting the research on the relationship between coping strate-gies and depressive symptoms is made substantially more difficult by the range of definitions that are grouped together under *approach, problem-focused, emotion-focused,* or *avoidance* strategies. To interpret the results of most investigations, one must actually look at the kinds of questions that researchers have asked, and research reports frequently do not provide this level of detail. This is particularly a problem when emotion-focused strategies are being assessed, because the term *emotion focused* can refer to cognitive reappraisal strategies that are designed to redefine a situation and the emotion it generates (Gross, 1998, 1999) or to strategies such as rumination or suppression that are designed to deal with emotion that has been generated by a particular situation (Gross, 1998, 1999; Nolen-Hoeksema, 2003). Indeed, sometimes questions that reflect several possibilities are lumped together in a single measure, which makes the results essentially impossible to interpret.

Folkman and Lazarus (1988), along with many others, have proposed that, when people use approach strategies, they experience fewer depressive symptoms and, when people use avoidance strategies, they experience more depressive symptoms. The evidence for this is stronger and more consistent for avoidance strategies than for approach strategies.

The use of avoidance strategies is quite consistently correlated with higher levels of depressive symptoms (Aranda, Casteneda, Lee, & Sobel, 2001; Berghuis & Stanton, 2002; Clements & Sawhney, 2000; Cohen, 2002; Cronkite, Moos, Twohey, Cohen, & Swindle, 1998; Culver, Arena, Antoni, & Carver, 2002; Dekker & Oomen, 1999; Dunkley, Blankstein, Halsall, Williams, & Winkworth, 2000; Epping-Jordan et al., 1999; Felsten, 1998; Fillion, Kovacs, Gagnon, & Endler, 2002; Kukyen & Brewin, 1999; Macrodimitris & Endler, 2001; Nezu et al., 1999; Rudnicki, Graham, Habboushe, & Ross, 2001; Simoni & Ng, 2000; Stein & Nyamathi, 1999; Whatley, Foreman, & Richards, 1998). This relationship holds separately for men and women, for adults from their late teens through old age, and for people from a variety of socioeconomic classes, ethnicities, and nationalities. Some of these studies have asked participants to report how they cope when stressful things in general happen to them, whereas others have asked them how they have coped with specific stressful circumstances, usually a severe illness or injury. Although the vast majority of investigations have found that the use of avoidance coping strategies is related to higher levels of depressive symptoms, there are a few studies that have not found any relationship between avoidant coping strategies and depression (e.g., Da Costa, Larouche, Drista, & Brender, 2000; Flett, Blankstein, & Obertynski, 1996). More important, there is not very much longitudinal data to let us look at the extent to which avoidance coping strategies predict future depression, controlling for concurrent depression. What little there is sometimes finds evidence for increases in depressive symptoms over time (Blalock & Joiner, 2000) and sometimes does not (Culver et al., 2002; Osowiecki & Compas, 1999).

The data concerning the effectiveness of approach strategies are quite inconsistent, probably because approach measures can combine such disparate things as problem solving, positive reappraisal, support seeking, and acceptance or resignation, but perhaps also because studies usually do not differentiate between different kinds of stressful situations, and different kinds of approach strategies might be more or less effective depending on the situation. A review of the recent literature shows that sometimes use of approach strategies is related to fewer depressive symptoms (Berghuis & Stanton, 2002; Brissette et al., 2002; Clements & Sawhney, 2000; Macrodimitris & Endler, 2001; Nezu et al., 1999; Surmann, 1999; Whatley et al., 1998) and sometimes use of approach strategies is not correlated with depressive symptoms (Da Costa et al., Brender, 2000; Dunkley et al., 2000; Epping-Jordan et al., 1999; Felsten, 1998; Fillion et al., 2002; Flett et al., 1996; Kuykin & Brewin, 1999; Osowiecki & Compas, 1999; Stein & Nyamathi, 1998).

There has been some attention paid to the specific approach strategy of reappraisal of stressful situations, which usually involves thinking about the situation in a more positive way. In his extensive work on emotion regulation, Gross (1998) argues that cognitive reappraisal of a stressful situation changes the actual emotion that is then experienced. Insofar as this moves the experienced emotion from more to less negative, or even to positive, one would expect that individuals who engage in cognitive reappraisal would also experience fewer depressive symptoms. Although there is not a lot of evidence specifically directed at this speculation, what evidence there is, is mixed (Berghuis & Stanton, 2002; Brissette et al., 2002; Culver et al., 2002; Gross & John, 2003; Kuyken & Brewin, 1999). It should be noted, however, that a range of situations and measures has been used, which might make the results more variable.

Relatively few investigators have examined whether men and women differ in their use of approach or avoidance coping strategies. In general, however, there do not seem to be gender differences in the use of these kinds of strategies (Aranda et al., 2001; Berghuis & Stanton, 2002; Blalock & Joiner, 2000; Felsten, 1998), although Berghuis and Stanton (2002) did find that men used problem-solving coping more than women did. This suggests that gender differences in the use of approach and avoidance coping strategies probably do not mediate the gender difference in depression. As we discuss below, however, gender differences in emotion regulation strategies may be strong contributors to the gender difference in depression.

Most of the empirical work on emotion regulation has focused on the ways that individuals deal with their negative emotions. Nolen-Hoeksema and her colleagues have focused on individuals' tendencies to ruminate about current negative situations and feelings, which has been shown to be remarkably maladaptive (Nolen-Hoeksema, 2003).

When people ruminate, they have emotion-focused thoughts such as, "Why am I so unmotivated? I just can't get going. I'm never going to get my work done feeling this way." Although some rumination may be a natural response to distress and depression, there are stable individual differences in the tendency to ruminate (Nolen-Hoeksema & Davis, 1999).

People who ruminate a great deal in response to their sad or depressed moods have more depressive symptoms (Garnefski, Legerstee, Kraaij, van den Kommer, & Teerds, 2002; Just & Alloy, 1997; Kraaij, Pruyboom, & Garnefski, 2002; Mor & Winquist, 2002; Nolen-Hoeksema & Harrell, 2002; Rude & McCarthy, 2003), experience longer periods of depressive symptoms (Nolen-Hoeksema, 2000; Nolen-Hoeksema, McBride, & Larson, 1997; Nolen-Hoeksema, Morrow, & Frederickson, 1993, but see Just & Alloy, 1997), and are more likely to be diagnosed with major depressive disorder (Just & Alloy, 1997; Nolen-Hoeksema, 2000). The effects of rumination on depression generally remain significant even after controlling for baseline levels of depression (Nolen-Hoeksema & Morrow, 1991; Nolen-Hoeksema, Parker, & Larson, 1994), although there have been a few exceptions (see, for example, Lara, Klein, & Kasch, 2000).

Women are more likely than men to ruminate in response to sad, depressed, or anxious moods (Nolen-Hoeksema & Jackson, 2001; Nolen-Hoeksema et al., 1999). The gender difference in rumination is found both in self-report survey and interview studies and in laboratory studies in which women's and men's responses to sad moods are observed (Butler & Nolen-Hoeksema, 1994). In turn, when gender differences in rumination are statistically controlled, the gender difference in depression becomes nonsignificant, suggesting that rumination helps to account for the gender difference in depression (Nolen-Hoeksema et al., 1999; Treynor, Gonzalez, & Nolen-Hoeksema, 2003).

Recent work has suggested that there may be more than one kind of rumination, and different aspects of the overall tendency to ruminate may relate differently to depression (Robinson & Alloy, 2003; Treynor et al., 2003). Specifically, it would appear that reflection (e.g., "Analyze recent events to try to understand why you are depressed.") and brooding (e.g., "Think: What am I doing to deserve this?") are both aspects of rumination, but brooding, rather than reflection, is the aspect that predicts increases in depressive symptoms over time and that mediates the gender difference in depression (Treynor et al., 2003).

How does rumination contribute to depression? Laboratory studies show that when people ruminate in response to a depressed mood, their memories of their past, their interpretations of the present, and their expectations for the future become more negative and distorted (Lyubomirsky, Caldwell, & Nolen-Hoeksema, 1998; Lyubomirsky & Nolen-Hoeksema, 1995). Thus, ruminators become increasingly negative and hopeless in their thinking and show many of the cognitive errors described by Beck (1987) as contributing to depression. Moreover, the quality of solutions ruminators generate to solve their problems is lower, and they are less confident about implementing these solutions (Lyubomirsky & Nolen-Hoeksema, 1995; Ward, Lyubomirsky, Sousa, & Nolen-Hoeksema, 2003). Thus, ruminators are less likely to take positive action on their environment to overcome other factors contributing to their depression.

Instead of focusing on their negative emotions (and thereby amplifying them), people can take the opposite tack and try to suppress them. There is very little data on this as yet, but what there is suggests that the more people try to suppress negative emotions, the more they are likely to be depressed (Gross, 1998; Gross & John, 2003; Ravindran, Griffiths, Waddell, & Anisman, 1995; Rude & McCarthy, 2003). There is also some suggestion that men tend to suppress their emotions more than women do (Gross & John, 2003). This could be a promising line of research for the future.

SUMMARY AND CONCLUSIONS

There are four broad categories of cognitive variables that may be predisposing variables for depression: the characteristic ways that people think about themselves, the characteristic ways that people think about their relationships with others, the characteristic ways that people think about the things that happen to them and about what the future is likely to bring, and the characteristic ways that people deal with the stressful things that happen to them or with their depressed feelings. Within each of these categories there are specific variables that are reliably correlated with depressive symptoms. These include body image, self-esteem, self-efficacy and mastery, sociotropy, unmitigated communion, optimism–pessimism, cognitive and attributional styles, avoidance coping strategies, and the emotion regulation strategies of rumination and suppression. Although the correlational data in each instance are both extensive and consistent, there is much less data about whether these variables precede, and thus can be seen as risk factors for, depression. Such data must have the form of longitudinal studies that control for initial levels of depressive symptoms and ask whether a variable leads to increases in depressive symptoms over time. Only for the variables of self-esteem, self-efficacy and mastery, a pessimistic attributional style, and rumination are there sufficient longitudinal data to argue that they have been established as risk factors for depression.

Coping strategies are of course designed to buffer stressful life circumstances (and some seem to do so better than others), but some of the other variables also interact with stressful life events to yield higher or lower levels of depressive symptoms (e.g., self esteem, self-efficacy and mastery, sociotropy, cognitive and attributional styles). These interactions have the following form: under low levels of stress, there is little difference in depressive symptoms between people who are high on the variable of interest and people who are low; when stress is high, however, people who are high on the variable of interest either have more depression or less depression than people who are low (depending on whether the variable buffers the effect of stress or exacerbates it). Thus, it is good to have high levels of self-esteem, self-efficacy, or mastery because, when faced with stressful circumstances, people who have high levels of these variables have fewer depressive symptoms than people who have low levels of these variables. However, it is not helpful to have high levels of sociotropy, cognitive errors, or pessimistic attributions because, when faced with stressful circumstances, people whose thought processes can be characterized in this way have higher levels of depressive symptoms than people whose thought processes operate in a different fashion.

Because this is a handbook on women and depression, we conclude this chapter with some thoughts about how cognitive variables might relate to the well-established gender difference in depression. Only some of the variables that we have considered in this chapter and that are clearly related to, and in some instances are clearly risk factors for, depression exhibit gender differences. Only those variables that exhibit gender differences are candidates as mediators for the gender difference in depression. These variables include body image, sociotropy and unmitigated communion, and rumination (and possibly suppression).

Women are less satisfied with their bodies than men are. Women are more sociotropic and more likely to exhibit unmitigated communion than men are. In other words, women are more likely than men to exhibit a heightened concern about what others think of them and a greater dependence on the approval of others, which can lead them to focus on the needs of others to the exclusion of their own needs. Rumination is a coping strategy that is utilized by women much more than it is utilized by men (and the opposite may be true for suppression). We think it is worth noting here that most of the cognitive variables that are clearly related to depression do not exhibit gender differences, including self-esteem, self-efficacy and mastery, and the various cognitive and attributional style variables.

The two cognitive variables that seem most promising as contributors to the gender difference in depression are interpersonal orientation (sociotropy and unmitigated communion) and rumination. Women are more interpersonally oriented than men, and this may lead them to sacrifice their own wants and needs excessively to maintain positive relationships with others and to rely more on others' approval (which may or may not be forthcoming) for their own sense of self worth. This tendency toward sociotropy and unmitigated communion in turn may contribute to depression. Women are more prone than men to engage in rumination when sad, blue, or anxious, and this appears to contribute to their higher rates of depression. Rumination enhances negative thinking and interferes with good problem solving and motivation, making it more difficult for individuals to overcome problems that may be associated with their depression.

References

Abramson, L. Y., Alloy, L. B., Hankin, B. L., Clements, C. M., Zhu, L., Hogan, M. E., & Whitehouse, W. G. (2000). Optimistic cognitive styles and invulnerability to depression. In J. E. Gillham (Ed.), *The science of optimism and hope: Research essays in honor of Martin E. P. Seligman* (pp. 75–98). Philadelphia: Templeton Foundation Press.

Abramson, L. Y., Alloy, L. B., Hankin, B. L., Haeffel, G. J., MacCoon, D. G., & Gibb, B. E. (2002). Cognitive vulnerability-stress models of depression in a self-regulatory and psychobiological context. In I. H. Gotlib & C. L. Hammen (Eds.), *Handbook of depression.* (pp. 268–294). New York: Guilford Press.

Abramson, L. Y., Metalsky, G. I., & Alloy, L. B. (1989). Hopelessness depression: A theory-based subtype of depression. *Psychological Review, 96,* 358–372.

Abramson, L. Y., Seligman, M. E. P., & Teasdale, J. (1978). Learned helplessness in humans: Critique and reformulation. *Journal of Abnormal Psychology, 87,* 49–74.

Alford, B. A., & Gerrity, D. M. (1995). The specificity of socio-authonomy personality dimensions to depression vs. anxiety. *Journal of Clinical Psychology, 51,* 190–195.

Allen, N. B., Ames, D., Layton, T., Bennetts, K., & Kingston, K. (1997). The relationship between sociotropy/autonomy and patterns of symptomatology in the depressed elderly. *British Journal of Clinical Psychology, 36,* 121–132.

Alloy, L. B., Abramson, L. Y., Metalsky, G. I., & Hartlage, S. (1988). The hopelessness theory of depression: Attributional aspects. *British Journal of Clinical Psychology, 27,* 5–21.

Alloy, L. B., Abramson, L. Y., Whitehouse, W. G., Hogan, M. E., Tashman, N. A., Steinberg, D. L., Rose, D. T., & Donovan, P. (1999). Depressogenic cognitive styles: Predictive validity, information processing and personality characteristics, and developmental origins. *Behavior Research and Therapy, 37,* 503–531.

Alloy, L. B., & Clements, C. M. (1998). Hopelessness theory of depression: Tests of the symptom component. *Cognitive Therapy and Research, 22*(4), 303–335.

Andersson, G. (1996). The benefits of optimism: A meta-analytic review of the Life Orientation Test. *Personality and Individual Differences, 21,* 719–725.

Aranda, M. P., Castaneda, I., Lee, P-J., & Sobel, E. (2001). Stress, social support, and coping as predictors of depressive symptoms: Gender differences among Mexican Americans. *Social Work Research, 25,* 37–48.

Badger, T. A. (2001). Depression, psychological resources, and health-related quality of life in older adults 75 and above. *Journal of Clinical Geropsychology, 7,* 189–200.

Bandura, A., Pastorelli, C., Barbaranelli, C., & Caprara, G. V. (1999). Self-efficacy pathways to childhood depression. *Journal of Social and Personality Psychology, 76,* 258–269.

Barrett, L. F., & Gross, J. J. (2001). Emotional intelligence: A process model of emotion representation and regulation. In T. J. Mayne & G. A. Bonanno (Eds.), *Emotions: Current issues and future directions. Emotions and social behavior* (pp. 286–310). New York: Guilford Press.

Barnett, P. A., & Gotlib, I. H. (1988). Psychosocial functioning and depression: Distinguishing among the antecedents, concomitants, and consequences. *Psychological Bulletin, 104,* 97–126.

Barry, D. T., Grilo, C. M., & Masheb, R. M. (2002). Gender differences in patients with binge eating disorder. *International Journal of Eating Disorders, 31,* 63–70.

Bartelstone, J. H., & Trull, T. J. (1995). Personality, life events, and depression. *Journal of Personality Assessment, 64,* 279–294.

Beck, A. T. (1967). *Depression: Clinical, experimental, and theoretical aspects.* New York: Hoeber. (Republished as *Depression: Causes and treatment*). Philadelphia: University of Pennsylvania Press, 1972).

Beck, A. T. (1983). Cognitive therapy of depression: New perspectives. In P. J. Clayton and J. E. Barrett (Eds.), *Treatment of depression: Old controversies and new approaches.* New York: Raven Press.

Beck, A. T. (1987). Cognitive models of depression. *Journal of Cognitive Psychotherapy: An International Quarterly, 1,* 5–37.

Beck, A. T., Rush, A. J., Shaw, B. F., & Emery, G. (1979). *Cognitive therapy of depression.* New York: Guilford.

Beck, R., Robbins, M., Taylor, C., & Baker, L. (2001). An examination of sociotropy and excessive reassurance seeking in the prediction of depression. *Journal of Psychopathology and Behavioral Assessment, 23*(2), 101–105.

Beck, R., Taylor, C., & Robbins, M. (2003). Missing home: Sociotropy, and autonomy and their relationship to psychological distress and homesickness in college freshmen. *Anxiety, Stress, and Coping, 16*(2), 155–166.

Berghuis, J. P., & Stanton, A. (2002). Adjustment to a dyadic stressor: A longitudinal study of coping and depressive symptoms in infertile couples over an insemination attempt. *Journal of Counseling and Clinical Psychology, 70,* 433–438.

Berthiaume, M., David, H., Saucier, J-F., & Borgeat, F. (1998). Correltes of prepartum depressive symptomatology: A multivariate analysis. *Journal of Reproductive and Infant Psychology, 16,* 45–56.

Bieling, P. J., Beck, A. T., & Brown, G. K. (2000). The sociotropy-autonomy scales: Structure and implications. *Cognitive Therapy and Research, 24,* 763–780.

Bisconti, T. L., & Bergeman, C. S. (1999). Perceived social control as a mediator of the relationships amongsocial support, psychological well-being, and perceived health. *The Gerontologist, 39,* 94–103.

Blalock, J. A., & Joiner, T. E. (2000). Interaction of cognitive avoidance coping and stress in predicting depression/anxiety. *Cognitive Therapy and Research, 24,* 47–65.

Brissette, I., Scheier, M. F., & Carver, C. S. (2002). The role of optimism in social network development, coping, and psychological adjustment during a life transition. *Journal of Personality and Social Psychology, 82,* 102–111.

Bruch, M. A. (2002). The relevance of mitigated and unmitigated agency and communion for depression vulnerabilities and dysphoria. *Journal of Counseling Psychology, 49,* 449–459.

Bullers, S. (2000). The mediating role of perceived control in the relationship between social ties and depressive symptoms. *Women and Health, 31*(2/3), 97–116.

Butler, L. D., & Nolen-Hoeksema, S. (1994). Gender differences in responses to a depressed mood in a college sample. *Sex Roles, 30,* 331–346.

Campbell, D. C., & Kwon, P. (2001). Domain-specific hope and personal style: Toward an integrative understanding of dysphoria. *Journal of Social and Clinical Psychology, 20,* 498–520.

Carver, C. S., & Gaines, J. G. (1987). Optimism, pessimism, and postpartum depression. *Cognitive Therapy and Research, 11*, 449–462.

Carver, C. S., Lehman, J. M., & Antoni, M. H. (2003). Dispositional pessimism predicts illness-related disruption of social and recreational activities among breast cancer patients. *Journal of Personality and Social Psychology, 84*, 813–821.

Chang, E. C. (1998). Does dispositional optimism moderate the relation between perceived stress and psychological well-being?: A preliminary investigation. *Personality and Individual Differences, 25*, 233–240.

Chang, E. C. (2002). Optimism-pessimism and stress appraisal: Testing a cognitive interactive model of psychological adjustment in adults. *Cognitive Therapy and Research, 26*, 675–690.

Chang, E. C., Maydeu-Olivares, A., & D'Zurilla, T. J. (1997). Optimism and pessimism as partially independent constructs: Relationships to positive and negative affectivity and psychological well-being. *Personality and Individual Differences, 23*, 433–440.

Chang, E. C., & Sanna, L. J. (2001). Optimism, pessimism, positive and negative affectivity in middle-aged adults: A test of a cognitive-affective model of psychological adjustment. *Psychology and Aging, 16*, 524–531.

Chou, K.-L., & Chi, I. (2001). Social comparison in Chinese older adults. *Aging and mental Health, 5*(3), 242–252.

Clark, D. A., Beck, A. T., & Brown, G. K. (1992). Sociotropy, autonomy, and life events perceptions in dysphoric and nondysphoric individuals. *Cognitive Therapy and Research, 16*, 635–652.

Clements, C. M., & Sawhney, D. K. (2000). Coping with domestic violence: Control attributions, dysphoria, and hopelessness. *Journal of Traumatic Stress, 13*, 219–240.

Cohen, M. (2002). Coping and emotional distress in primary and recurrent breast cancer patients. *Journal of Clinical Psychology in Medical Settings, 9*(3), 245–251.

Cohen-Tovee, E. M. (1993). Depressed mood and concern with weight and shape in normal young women. *International Journal of Eating Disorders, 14*, 223–227.

Connor-Smith, J. K., & Compas, B. E. (2002). Vulnerability to social stress: Coping as a mediator or moderator of sociotropy and symptoms of anxiety and depression. *Cognitive Therapy and Research, 26*, 39–55.

Cotten, S. R. (1999). Marital status and mental health revisited: Examining the importance of risk factors and resources. *Family Relations, 48*, 225–233.

Coyne, J. C., & Gotlib, I. (1986). Studying the role of cognition in depression: Well-trodden paths and cul-de-sacs. *Cognitive Therapy and Research, 10*, 695–705.

Cronkite, R. C., Moos, R. H., Twohey, J., Cohen, C., & Swindle, R. Jr. (1998). Life circumstances and personal resources as predictors of the ten-year course of depression. *American Journal of Community Psychology, 26*, 255–280.

Culp, L. N., & Beach, S. R. H. (1998). Marriage and depressive symptoms: The role and bases of self-esteem differ by gender. *Psychology of Women Quarterly, 22*, 647–663.

Culver, J. L., Arena, P. L., Antoni, M. H., & Carver, C. S. (2002). Coping and distress among women under treatment for early stage breast cancer: Comparing African Americans, Hispanics and Non-Hispanic Whites. *Psycho-Oncology, 11*, 495–504.

Da Costa, D., Larouche, J., Dritsa, M., & Brender, W. (2000). Psychosocial correlates of prepartum and postpartum depressed mood. *Journal of Affective Disorders, 59*, 31–40.

Danziger, S. K., Carlson, M. J., & Henly, J. R. (2001). Post-welfare employment and psychological well-being. *Women and Health, 32*(1/2), 47–78.

Davila, J. (2001). Refining the association between excessive reassurance seeking and depressive symptoms: The role of related interpersonal constructs. *Journal of Social and Clinical Psychology, 20,* 538–559.

Davis, C., & Katzman, M. (1997). Charting new territory: Body esteem, weight satisfaction, depression, and self-esteem among Chinese males and females in Hong Kong. *Sex Roles, 36*(7/8), 449–459.

Davis, C., & Katzman, M. A. (1998). Chinese men and women in the United States and Hong Kong: Body and self-esteem ratings as a prelude to dieting and exercise. *International Journal of Eating Disorders, 23,* 99–102.

Dekker, J., & Oomen, J. (1999). Depression and coping. *European Journal of Psychiatry, 13*(3), 183–189.

Denniston, C., Roth, D., & Gilroy, F. (1992). Dysphoria and body image among college women. *International Journal of Eating Disorders, 12,* 449–452.

Devine, D., Forehand, R., Morse, E., Simon, P., Clark, L., & Kernis, M. (2000). HIV infection in inner-city African American women: The role of optimism in predicting depressive symptomatology. *International Journal of Rehabilitation and Health, 5*(3), 141–156.

Dunkley, D. M., Blankstein, K. R., Halsall, J., Williams, M., & Winkworth, G. (2000). The relation between perfectionism and distress: Hassles, coping, and perceived social support as mediators and moderators. *Journal of Counseling Psychology, 47,* 437–453.

Ennis, N. E., Hobfoll, S. E., & Schroder, K. E. E. (2000). Money doesn't talk, it swears: How economic stress and resistance resources impact inner-city women's depressive mood. *American Journal of Community Psychology, 28*(2), 149–173.

Epping-Jordan, J. E., Compas, B. E., Osowiecki, D. M., Oppedisano, G., Gerhardt, C., Primo, K., et al. (1999). Psychological adjustment in breast cancer: Processes of emotional distress. *Health Psychology, 18*(4), 315–326.

Fairbrother, N., & Moretti, M. (1998). Sociotropy, autonomy, and self-discrepancy: Status in depressed, remitted depressed, and control participants. *Cognitive Therapy and Research, 22*(3), 279–296.

Feingold, A. (1994). Gender differences in personality: A meta-analysis. *Psychological Bulletin, 116,* 429–456.

Feingold, A., & Mazzella, R. (1998). Gender differences in body image are increasing. *Psychological Science, 9,* 190–195.

Felsten, G. (1998). Gender and coping: Use of distinct strategies and associations with stress and depression. *Anxiety, Stress, and Coping, 11,* 289–309.

Fernandez, M. E., Mutran, E. J., & Reitzes, D. C. (1998). Moderating the effects of stress on depressive symptoms. *Research on Aging, 20*(2), 163–182.

Fillion, L., Kovacs, A. H., Gagnon, P., & Endler, N. S. (2002). Validation of the shortened COPE for use with breast cancer patients undergoing radiation therapy. *Current Psychology: Developmental, Learning, Personality, Social, 21,* 17–34.

Flett, G. L., Blankstein, K. R., & Obertynski, M. (1996). Affect intensity, coping styles, mood regulation expectancies, and depressive symptoms. *Personality and Individual Differences, 20,* 221–228.

Flett, G. L., Hewitt, P. L., Garshowitz, M., & Martin, T. R. (1997). Personality, negative social interactions, and depressive symptoms. *Canadian Journal of Behavioural Science*, *29*, 28–37.

Folkman, S., & Lazarus, R. S. (1988). Coping as a mediator of emotion. *Journal of Personality and Social Psychology*, *54*, 466–475.

Folkman, S., Lazarus, R. S., Dunkel-Schetter, C., DeLongis, A., & Gruen, R. J. (2000). The dynamics of a stressful encounter. In E. T. Higgins & A. W. Kruglanski (Eds.). *Motivational science: Social and personality perspectives. Key reading in social psychology*. (pp. 111–127). Philadelphia: Psychology Press.

Fresco, D. M., Sampson, W. S., Craighead, L. W., & Koons, A. N. (2001). The relationship of sociotropy and autonomy to symptoms of depression and anxiety. *Journal of Cognitive Psychotherapy*, *16*, 17–31.

Fritz, H. L. (2000). Gender-linked personality traits predict mental health and functional status following a first coronary event. *Health Psychology*, *19*, 420–428.

Fritz, H. L., & Helgeson, V. S. (1998). Distinctions of unmitigated communion from communion: Self-neglect and overinvolvement with others. *Journal of Personality and Social Psychology*, *75*, 121–140.

Furnham, A., & Greaves, N. (1994). Gender and locus of control correlates of body image dissatisfaction. *European Journal of Personality*, *8*, 183–200.

Garnefski, N., Legerstee, J., Kraaij, V., van den Kommer, T., & Teerds, J. (2002). Cognitive coping strategies and symptoms of depression and anxiety: A comparison between adolescents and adults. *Journal of Adolescence*, *25*, 603–611.

Ginexi, E. M., Howe, G. W., & Caplan, R. D. (2000). Depression and control beliefs in relation to reemployment: What are the directions of effect? *Journal of Occupational Health Psychology*, *5*, 323–336.

Grant, L. D., Long, B. C., & Willms, J. D. (2002). Women's adaptation to chronic back pain: Daily appraisals and coping strategies, personal characteristics and perceived spousal responses. *Journal of Health Psychology*, *7*, 545–563.

Grazioli, R., & Terry, D. J. (2000). The role of cognitive vulnerability and stress in the prediction of postpartum depressive symptomatology. *British Journal of Clinical Psychology*, *39*, 329–347.

Gross, J. J. (1998). The emerging field of emotion regulation: An integrative review. *Review of General Psychology*, *2*, 271–299.

Gross, J. J. (1999). Emotion regulation: Past, present, future. *Cognition and Emotion*, *13*, 551–573.

Gross, J. J., & John, O. P. (2003). Individual differences in two emotion regulation processes: Implications for affect, relationships, and well-being. *Journal of Personality and Social Psychology*, *85*, 348–362.

Hayaki, J., Friedman, M. A., & Brownell, K. D. (2002). Emotional expression and body dissatisfaction. *International Journal of Eating Disorders*, *31*, 57–62.

Heilemann, M. V., Lee. K. A., & Kury, F. S. (2002). Strengths and vulnerabilities of women of Mexican descent in relation to depressive symptoms. *Nursing Research*, *51*(3), 175–182.

Helgeson, V. (1994). Relation of agency and communion to well-being: Evidence and potential explanations. *Psychological Bulletin*, *116*, 412–428.

Helgeson, V., & Fritz, H. (1996). Implications of communion and unmitigated communion for adolescent adjustment to Type I diabetes. *Women's Health: Research on Gender, Behavior, and Policy*, *2*, 169–194.

Helgeson, V. S., & Fritz, H. L. (1998) A theory of unmitigated communion. *Personality and Social Psychology Review, 2*(3), 173–183.

Hobfoll, S. E., Bansal, A., & Schurg, R. (2002). The impact of perceived child physical and sexual abuse history on Native American women's psychological well-being and AIDS risk. *Journal of Consulting and Clinical Psychology, 70*, 252–257.

Hobfoll, S. E., Johnson, R. J., Ennis, N., & Jackson, A. P. (2003). Resource loss, resource gain, and emotional outcomes among inner city women. *Journal of Personality and Social Psychology, 84*, 632–643.

Holahan, C. J., Moos, R. H., & Schaefer, J. A. (1996). Coping, stress resistance, and growth: Conceptualizing adaptaive functioning. In M. Zeidner & N. S. Endler (Eds.), *Handbook of coping: Theory, research applications* (pp. 24–43). Oxford, England: Wiley.

Isaacowitz, D. M., & Seligman, M. E. P. (2001). Is pessimism a risk factor for depressive mood among community-dwelling older adults? *Behavior Research and Therapy, 39*, 255–272.

Isaacowitz, D. M., & Seligman, M. E. P. (2002). Cognitive style predictors of affect change in older adults. *International Journal of Aging and Human Development, 54*, 233–253.

Jack, D. C. (1991). *Silencing the Self: Women and Depression.* New York: Harper Perennial.

James, W., (1890). *Principles of psychology.* New York: Dover.

Jang, Y., Haley, W. E., Small, B. J., & Mortimer, J. A. (2002). The role of mastery and so ial resources in the associations between disability and depression in later life. *The Gerontologist, 42*, 807–813.

Joiner, T., & Coyne, J. C. (1999). *The interactional nature of depression: Advances in interpersonal approaches.* Washington, DC: American Psychological Association.

Joiner, T., Metalsky, G., Katz, J., & Beach, S. R. (1999). Depression and excessive reassurance-seeking. *Psychological Inquiry, 10*, 269–278.

Joiner, T. E., Wonderlich, S. A., Metalsky, G. I., & Schmidt, N. B. (1995). Body dissatisfaction: A feature of bulimia, depression, or both? *Journal of Social and Clinical Psychology, 14*, 339–355.

Just, N., & Alloy, L. B. (1997). The response styles theory of depression: Tests and an extension of the theory. *Journal of Abnormal Psychology, 106*, 221–229.

Katz, J., Joiner, T. E., Jr., & Kwon, P. (2002). Membership in a devalued social group and emotional well-being: Developing a model of personal self-esteem, collective self-esteem, and group socialization. *Sex Roles, 47*(9/10), 419–431.

Kernis, M. H., Whisenhunt, C. R., Waschull, S. B., Greenier, K. D., Berry, A. J., Herlocker, C. E., et al. (1998). Multiple facets of self-esteem and their relations to depressive symptoms. *Personality and Social Psychology Bulletin, 24*(6), 657–668.

Koenig, L. J., & Wasserman, E. L. (1995). Body image and dieting failure in college men and women: Examining links between depression and eating problems. *Sex Roles, 34*(3/4), 225–249.

Kraaij, V., Pruymboom, E., & Garnefski, N. (2002). Cognitive coping and depressive symptoms in the elderly: A longitudinal study. *Aging and Mental Health, 6*(3), 275–281.

Kuyken, W., & Brewin, C. R. (1999). The relation of early abuse to cognition and coping in depression. *Cognitive Therapy and Research, 23*, 665–677.

Kwon, P., & Whisman, M. A. (1998). Sociotropy and autonomy as vulnerabilities to specific life events: Issues in life event categorization. *Cognitive Therapy and Research, 22*, 353–362.

Labbe, E. E., Lopez, I., Murphy, L., & O'Brien, C. (2002). Optimism and psychosocial functioning in caring for children with Battens and other neurological diseases. *Psychological Reports, 90*, 1129–1135.

Lachman, M. E., & Weaver, S. L. (1998). The sense of control as a moderator of social class differences in health and well-being. *Journal of Personality and Social Psychology, 74*, 763–773.

Lara, M. E., Klein, D. N., & Kasch, K. L. (2000). Psychosocial predictors of the short-term course and outcome of major depression: A longitudinal study of a nonclinical sample with recent-onset episodes. *Journal of Abnormal Psychology, 109*, 644–650.

Lightsey, Jr., O. R. (1997). Stress buffers and dysphoria: A prospective study. *Journal of Cognitive Psychotherapy: An International Quarterly, 11*(4), 263–277.

Lightsey, Jr., O. R., & Christopher, J. C. (1997). Stress buffers and dysphoria in a non-Western population. *Journal of Counseling and Development, 75*, 451–459.

Lyubomirsky, S., Caldwell, N. D., & Nolen-Hoeksema, S. (1998). Effects of ruminative and distracting responses to depressed mood on retrieval of autobiographical memories. *Journal of Personality and Social Psychology, 75*, 166–177.

Lyubomirsky, S., & Nolen-Hoeksema, S. (1995). Effects of self-focused rumination on negative thinking and interpersonal problem-solving. *Journal of Personality and Social Psychology, 69*, 176–190.

Mable, H. M., Balance, W. D. G., & Galgan, R. J. (1986). Body-image distortion and dissatisfaction in university students. *Perceptual and Motor Skills, 63*, 907–911.

Maciejewski, P. K., Prigerson, H. G., & Mazure, C. M. (2000). Self-efficacy as a mediator between stressful life events and depressive symptoms. *British Journal of Psychiatry, 176*, 373–378.

Macrodimitris, S. D., & Endler, N. S. (2001). Coping, control, and adjustment in Type 2 diabetes. *Health Psychology, 20*(3), 208–216.

Maier, E. H., & Lachman, M. E. (2000). Consequences of early parental loss and separation for helth and well-being in midlife. *International Journal of Behavioral Development, 24*(2), 183–189.

Makaremi, A. (2000). Self-efficacy and depression among Iranian college students. *Psychological Reports, 86*, 386–388.

Marsella, A. J., Shizuru, L., Brennan, J., & Kameoka, V. (1981). Depression and body image satisfaction. *Journal of Cross-Cultural Psychology, 12*, 360–371.

Mazure, C. M., Bruce, M. L., Maciejewski, P. K., & Jacobs, S. C. (2000). Adverse life events and cognitive-personality characteristics in the prediction of major depression nd antidepressant response. *The American Journal of Psychiatry, 157*, 896–903.

Mazure, C. M., & Maciejewski, P. K. (2003). A model of risk for major depression: Effects of life stress and cognitive style vary by age. *Depression and Anxiety, 17*, 26–33.

Mazure, C. M., Raghavan, C., Maciejewski, P. K., Jacobs, S. C., & Bruce, M. L. (2001). Cognitive-personality characteristics as direct predictors of unipolar major depression. *Cognitive Therapy and Research, 25*, 215–225.

McCauley, M., Mintz, L., & Glenn, A. (1988). Body image, self-esteem, and depression-proneness: Closing the gender gap. *Sex Roles, 18*(7/8), 381–391.

Metalsky, G. I., Halberstadt, L. J., & Abramson, L. Y. (1987). Vulnerability to depressive mood reactions: Toward a more powerful test of the diathesis-stress and causal mediation components of the reformulated theory of depression. *Journal of Personality and Social Psychology, 52*, 386–393.

Metalsky, G. I., & Joiner, Jr., T. E. (1992). Vulnerability to depressive symptomatology: A prospective test of the diathesis-stress and causal mediation components of the hopelessness theory of depression. *Journal of Personality and Social Psychology, 63*, 667–675.

Metalsky, G. I., Joiner, Jr., T. E., Hardin, T. S., & Abramson, L. Y. (1993). Depressive reactions to failure in a naturalistic setting: A test of the hopelessness and self-esteem theories of depression. *Journal of Abnormal Psychology, 102*, 101–109.

Moore, R. G., & Blackburn, I-M. (1994). The relationship of sociotropy and autonomy to symptoms, cognition and personality in depressed patients. *Journal of Affective Disorders, 32*, 239–245.

Moore, R. G., & Blackburn, I-M. (1996). The stability of sociotropy and autonomy in depressed patients undergoing treatment. *Cognitive Therapy and Research, 20*, 69–80.

Mor, N., & Winquist, J. (2002). Self-focused attention and negative affect: A meta-analysis. *Psychological Bulletin, 128*, 638–662.

Mori, D. L., & Morey, L. (1991). The vulnerable body image of females with feelings of depression. *Journal of Research in Personality, 25*, 343–354.

Nezu, C. M., Nezu, A. M., Friedman, S. H., Houts, P. S., Dellicarpini, L. A., Bildner, C., & Faddis, S. (1999). Cancer and psychological distress: Two investigations regarding the role of social problem-solving. *Journal of Psychosocial Oncology, 16*(3/4), 27–40.

Nolen-Hoeksema, S. (1991). Responses to depression and their effects on the duration of depressive episodes. *Journal of Abnormal Psychology, 100*, 569–582.

Nolen-Hoeksema, S. (2000). The role of rumination in depressive disorders and mixed anxiety/depressive symptoms. *Journal of Abnormal Psychology, 109*, 504–511.

Nolen-Hoeksema, S. (2002). Gender differences in depression. In I. H. Gotlib & C. L. Hammen (Eds.), *Handbook of depression.* (pp. 492–509). New York: Guilford.

Nolen-Hoeksema, S. (2003). The response styles theory. In. C. Papageorgiou & A. Wells (Eds.), *Depressive rumination: Nature, theory, and treatment of negative thinking in depression* (pp. 107–123). New York: Wiley.

Nolen-Hoeksema, S., & Davis, C. G. (1999). "Thanks for sharing that": Ruminators and their social support networks. *Journal of Personality and Social Psychology, 77*, 801–814.

Nolen-Hoeksema, S., & Harrell, Z. A. (2002). Rumination, depression, and alcohol use: Tests of gender differences. *Journal of Cognitive Psychotherapy: An International Quarterly, 16*, 391–403.

Nolen-Hoeksema, S., & Jackson, B. (2001). Mediators of the gender differences in rumination. *Psychology of Women Quarterly, 25*, 37–47.

Nolen-Hoeksema, S., Larson, J., & Grayson, C. (1999). Explaining the gender difference in depression. *Journal of Personality and Social Psychology, 77*, 1061–1072.

Nolen-Hoeksems, S., McBride, A., & Larson, J. (1997). Rumination and psychological distress among bereaved partners. *Journal of Personality and Social Psychology, 72*, 855–862.

Nolen-Hoeksema, S., & Morrow, J. (1991). A prospective study of depression and posttraumatic stress symptoms after a natural disaster: The 1989 Loma Prieta earthquake. *Journal of Social and Personality Psychology, 61*, 115–121.

Nolen-Hoeksema, S., Morrow, J., & Frederickson, B. L. (1993). Response styles and the duration of episodes of depressed mood. *Journal of Abnormal Psychology, 102*, 20–28.

Nolen-Hoeksema, S., Parker, L. E., & Larson, J. (1994). Ruminative coping with depressed mood following loss. *Journal of Personality and Social Psychology, 67*, 92–104.

Nunn, J. D., Mathews, A., & Trower, P. (1997). Selective processing of concern-related information in depression. *British Journal of Clinical Psychology, 36*, 489–503.

Oates-Johnson, T., & DeCourville, N. (1999). Weight preoccupation, personality, and depression in university students: An interactionist perspective. *Journal of Clinical Psychology, 55*, 1157–1166.

Osowiecki, D. M., & Compas, B. E. (1999). A prospective study of coping, perceived control, and psychological adaptation to breast cancer. *Cognitive Therapy and Research, 23*, 169–180.

Pakriev, S., Poutanen, O., & Salokangas, R. K. R. (2002). Causal and pathoplastic risk factors of depression: Findings of the Tampere Depression Project. *Nordic Journal of Psychiatry, 56*, 29–32.

Pearlin, L. I., & Schooler, C. (1978). The structure of coping. *Journal of Health and Social Behavior, 19*, 2–21.

Peterson, C., & Seligman, M. E. P. (1984). Causal explanations as a risk factor for depression: Theory and evidence. *Psychological Review, 91*, 347–374.

Plomin, R., Scheier, M. F., Bergeman, C. S., Pedersen, N. L., Nesselroade, J. R., & McClearn, G. E. (1992). Optimism, pessimism and mental health: A twin/adoption analysis. *Personality and Individual Differences, 13*, 921–930.

Quinn, D. M., & Crocker, J. (1998). Vulnerability to the affective consequences of the stigma of overweight. In J. K. Swim & C. Stangor (Eds.), *Prejudice: The target's perspective* (pp. 125–143). San Diego, CA: Academic Press.

Raghavan, C., Le, H-N., & Berenbaum, H. (2002). Predicting dysphoria and hostility using the diathesis-stress model of sociotropy and autonomy in a contextualized stress setting. *Cognitive Therapy and Research, 26*, 231–244.

Ravindran, A. R., Griffiths, J., Waddell, J., & Anisman, H. (1995). Stressful life events and coping styles in relation to dysthymia and major depressive disorder: Variations associated with alleviation of symptoms following pharmacotherapy. *Progress in Neuro-Psychopharmacology & Biological Psychiatry, 19*, 637–653.

Regehr, C., Cadell, S., & Jansen, K. (1999). Perceptions of control and long-term recovery from rape. *American Journal of Orthopsychiatry, 69*, 110–115.

Reilly-Harrington, N. A., Alloy, L. B., Fresco, D. M., & Whitehouse, W. G. (1999). Cognitive styles and life events interact to predict bipolar and unipolar symptomatology. *Journal of Abnormal Psychology, 108*, 567–578.

Reinherz, H. Z., Giaconia, R. M., Hauf, A. M. C., Wasserman, M. S., & Silverman, A. B. (1999). Major depression in the transition to adulthood: Risks and impairments. *Journal of Abnormal Psychology, 108*, 500–510.

Ritter, C., Hobfoll, S. E., Lavin, J., Cameron, R. P., & Hulsizer, M. R. (2000). Stress, psychosocial resources, and depressive symptomatology during pregnancy in low-income, inner-city women. *Health Psychology, 19*, 576–585.

Roberts, S. B., & Kendler, K. S. (1999). Neuroticism and self-esteem as indices of the vulnerability to major depression in women. *Psychological Medicine, 29*, 1101–1109.

Roberts, J. E., Shapiro, A. M., & Gamble, S. A. (1999). Level and perceived stability of self-esteem prospectively predict depressive symptoms during psychoeducational group treatment. *British Journal of Clinical Psychology, 38*, 425–429.

Robins, C. J., & Block, P. (1988). Personal vulnerability, life events, and depressive symptoms: A test of a specific interactional model. *Journal of Personality and Social Psychology, 54*, 847–852.

Robins, C. J., Hayes, A. M., Block, P., Kramer, R. J., & Villena, M. (1995). Interpersonal and achievement concerns and the depressive vulnerability and symptoms specificity hypotheses: A prospective study. *Cognitive Therapy and Research, 19*, 1–20.

Robinson, M. S., & Alloy, L. B. (2003). Negative cognitive styles and stress-reactive rumination interact to predict depression: A prospective study. *Cognitive Therapy and Research, 27*, 275–292.

Rosenberg, M. (1965). *Society and the adolescent self-image*. Princeton, NJ: Princeton University Press.

Rosenberg, M. (1979). *Conceiving the self*. New York: Basic Books.

Rosenberg, M., Schooler, C., Schoenbach, C., & Rosenberg, F. (1995). Global self-esteem and specific self-esteem: Different concepts, different outcomes. *American Sociological Review, 60*, 141–156.

Rude, S. S., & Burnham, B. L. (1993). Do interpersonal and achievement vulnerabilities interact with congruent events to predict depression? Comparison of DEQ, SAS, DAS, and combined scales. *Cognitive Therapy and Research, 17*, 531–548.

Rude, S. S., & McCarthy, C. T. (2003). Emotional functioning in depressed and depression-vulnerable college students. *Cognition and Emotion, 17*, 799–806.

Rudnicki, S. R., Graham, J. L., Habboushe, D. F., & Ross, R. D. (2001). Social support and avoidant coping: Correlates of depressed mood during pregnancy in minority women. *Women and Health, 34*(3), 19–35.

Sarwer, D. B., Wadden, T. A., & Foster, G. D. (1998). Assessment of body image dissatisfaction in obese women: Specificity, severity, and clinical significance. *Journal of Consulting and Clinical Psychology, 66*, 651–654.

Sato, T., & McCann, D. (1997). Vulnerability factors in depression: The facets of sociotropy and autonomy. *Journal of Psychopathology and Behavioral Assessment, 19*, 41–62.

Sato, T., & McCann, D. (1998). Individual differences in relatedness and individuality: A exploration of two constructs. *Personality and Individual Differences, 24*, 847–859.

Scheier, M. F., & Carver, C. S. (1985). Optimism, coping, and health: Assessment and implications of generalized outcome expectancies. *Health Psychology, 4*, 219–247.

Scheier, M. F., Carver, C. S., & Bridges, M. W. (1994). Distinguishing optimism from neuroticism (and trait anxiety, self-mastery, and self-esteem): A reevaluation of the Life Orientation Test. *Journal of Personality and Social Psychology, 67*, 1063–1078.

Schieman, S., & Turner, H. A. (2001). "When feelilng other people's pain hurts": The influence of psychosocial resources on the association between self-reported empathy and depressive symptoms. *Social Psychology Quarterly, 64,* 376–389.

Schieman, S., Van Gundy, K., & Taylor, J. (2002). The relationship between age and depressive symptoms: A test of competing explanatory and suppression influences. *Journal of Aging and Health, 14,* 260–285.

Scott, B., & Melin, L. (1998). Psychometric properties and standardized data for questionnaires measuring negative affect, dispositional style and daily hassles. A nationwide sample. *Scandinavian Journal of Psychology, 39,* 301–307.

Seligman, M. E., Abramson, L. Y., Semmel, A., & von Baeyer, C. (1979). Depressive attributional style. *Journal of Abnormal Psychology, 88,* 242–247.

Sharpley, C. F., & Yardley, P. (1999). The relationship between cognitive hardiness, explanatory style, and depression-happiness in post-retirement men and women. *Australian Psychologist, 34*(3), 198–203.

Simoni, J. M., & Ng, M. T. (2000). Trauma, coping, and depression among women with HIV/AIDS in New York City. *AIDS Care, 12,* 567–580.

Smith, H. M., & Betz, N. E. (2002). An examination of efficacy and esteem pathways to depression in young adulthood. *Journal of Counseling Psychology, 49,* 438–448.

Steer, R. A., & Clark, D. A. (1997). Psychometric characteristics of the Beck Depression Inventory-II with college students. *Measurement and Evaluation in Counseling and Development, 30*(3), 128–136.

Stein, J. A., & Nyamathi, A. (1999). Gender differences in relationships among stress, coping, and health risk behaviors in impoverished, minority populations. *Personality and Individual Differences, 26,* 141–157.

Strong, S. M., Williamson, D. A., Netemeyer, R. G., & Geer, J. H. (2000). Eating disorder symptoms and concerns about body differ as a function of gender and sexual orientation. *Journal of Social and Clinical Psychology, 19,* 240–255.

Summerfeldt, L. J., & Endler, N. S. (1996). Coping with emotion and psychopathology. In M. Zeidner & N. S. Endler (Eds.), *Handbook of coping: Theory, research applications* (pp. 602–639). Oxford, England: Wiley.

Surmann, A. T. (1999). Negative mood regulation expectancies, coping, and depressive symptoms among American nurses. *Journal of Social Psychology, 139,* 540–543.

Thomas, S., & Vindhya, U. (2000). Women and stress: A study of stressful life events, depression and the moderating influence of self-esteem. *Journal of Indian Psychology, 18,* 38–51.

Thompson, J. K., & Psaltis, K. (1988). Multiple aspects and correlates of body figure ratings: A replication and extension of Fallon and Rozin (1985). *International Journal of Eating Disorders, 7,* 813–817.

Treynor, W., Gonzalez, R., & Nolen-Hoeksema, S. (2003). Rumination reconsidered: A psychometric analysis. *Cognitive Therapy and Research, 27,* 247–259.

Turner, H. A., & Butler, M. J. (2003). Direct and Indirect effects of childhood adversity on depressive symptoms in young adults. *Journal of Youth and Adolescence, 32*(2), 89–103.

Turner, R. J., Lloyd, D. A., & Roszell, P. (1999). Personal resources and the social distribution of depression. *American Journal of Community Psychology, 27,* 643–672.

Updegraff, J. A., Taylor, S. E., Kemeny, M. E., & Wyatt, G. E. (2002). Positive and negative effects of HIV infection in women with low socioeconomic resources. *Personality and Social Psychology Bulletin, 28,* 382–394.

Van Servellen, G., Aguirre, M., Sarna, L., & Brecht, M-L. (2002). Differential predictors of emotional distress in HIV-infected men and women. *Western Journal of Nursing Research, 24,* 49–72.

Vickers, K. S., & Vogeltanz, N. D. (2000). Dispositional optimism as a predictor of depressive symptoms over time. *Personality and Individual Differences, 28,* 259–272.

Walker, L., Timmerman, G. M., Kim, M., & Sterling, B. (2002). Relationships between body image and depressive symptoms during postpartum in ethnically diverse, low income women. *Women and Health, 36,* 101–121.

Ward, A., Lyubomirsky, S., Sousa, L., & Nolen-Hoeksema, S. (2003). Can't quite commit: Ruminators and uncertainty. *Personality and Social Psychology Bulletin, 29,* 96–107.

Whatley, S. L., Foreman, A. C., & Richards, S. (1998). The relationship of coping style to dysphoria, anxiety, and anger. *Psychological Reports, 83,* 783–791.

Wilhelm, K., Parker, G., Dewhurst-Savellis, J., & Asghari, A. (1999). Psychological predictors of single and recurrent major depressive episodes. *Journal of Affective Disorders, 54,* 139–147.

Zhang, J., & Norvilitis, J. M. (2002). Measuring Chinese psychological well-being with Western developed instruments. *Journal of Personality Assessment, 79,* 492–511.

8

Personality and Depression in Women

Thomas A. Widiger, Stephanie Mullins-Sweatt, and Kristen G. Anderson

Everybody has a personality, or a characteristic manner of thinking, feeling, behaving, and relating to others. Some persons are typically introverted and withdrawn; others are more extraverted and outgoing. Some persons are characteristically agreeable, cooperative, and even altruistic, whereas others might be consistently argumentative, self-centered, and exploitative. The personality traits that distinguish one person from another have been shown to have substantial temporal stability, heritability, cross-situational consistency, and functional relevance to a wide variety of important life outcomes, including (but not limited to) career success, well-being, marital stability, and physical health (Pervin & John, 1999).

Personality traits are important to consider when attempting to explain gender differences in depression for two fundamental reasons. First, there is a considerable amount of research indicating that some personality traits do provide a vulnerability to the occurrence of mood disorders, and, second, these personality traits have been shown to obtain consistent gender differences.

DEPENDENCY, DEPRESSION, AND WOMEN

One of the more heavily researched personality dispositions for the development of depressive episodes is the personality trait of dependency (Clark, Watson, & Mineka, 1994). Dependency has been studied extensively within general personality research, assessed typically by the Depressive Experience Questionnaire (Blatt & Zuroff, 1992; Santor, Zuroff, & Fielding, 1997), the Sociotropy-Autonomy Scale (SAS; Robins, Hayes, Block, Kramer, & Villena, 1995), the Personality Style Inventory (Robins et al., 1995), or the Interpersonal Dependency Inventory (IDI; Hirschfield et al., 1977). Extreme forms of dependency are even diagnosed as a mental disorder and included within the American Psychiatric Association's (APA's) *Diagnostic and Statistical Manual of Mental Disorders* (*DSM-IV*; APA, 2000) as a dependent personality disorder. Dependent personality disorder is defined in *DSM-IV* as "a pervasive and excessive need to be taken care of that leads to submissive and clinging behavior and fears of separation" (APA, 2000, p. 721).

Dependent persons are said "to have a desperate need to keep in close physical contact with need-gratifying others, and they experience deep longings to be loved, cared for, nurtured, and protected" (Blatt & Zuroff, 1992, p. 528). Beck (1983) discusses from a more cognitive-behavioral perspective a constellation of sociotropic personality traits that "includes passive-receptive wishes (acceptance, intimacy, understanding, support guidance)" (p. 273). Highly sociotropic persons are "particularly concerned about the possibility of being disapproved of by others, and often acting in ways designed to please those others and thereby securing their attachments" (Robins & Block, 1988, p. 848). "These dependent individuals rely intensely on others to provide and maintain a sense of well-being, and therefore they have great difficulty expressing anger for fear of losing the need gratification others can provide" (Blatt & Zuroff, 1992, p. 528). Separation, rejection, and interpersonal loss are thought to be particularly stressful for such persons and "depression is most likely to occur in such individuals in response to perceived loss or rejection in social situations" (Blatt & Zuroff, 1992, p. 528).

Many studies have documented that dependent personality traits provide a vulnerability to the development of mood disorders in women (Blatt & Zuroff, 1992; Bornstein, 1992, 1993; Nietzel & Harris, 1990; Overholser, 1996). Nietzel and Harris (1990) conducted a meta-analysis of 21 studies that yielded 24 independent effect sizes for the relationship between dependency and depression. They concluded that "the interaction of elevated dependency needs with negative social events is a uniquely pernicious combination" (p. 291). They speculated that the eventual depressive episode is the result of a "sequence of unintentional self-defeating behaviors motivated by a deep desire to restore lost love and intimacy and bolster self-esteem" (Nietzel & Harris, 1990, p. 293).

Hammen et al. (1995) obtained 6-month and 12-month follow-up assessments of 129 high school women. They conducted multiple regression analyses to predict depression on the basis of dependency cognitions, prior interpersonal stress, and the interaction between them, controlling for initial levels of depression. All of these young women experienced stressful life events during this period of their lives, including moving away from home, separation from an important relationship, and loss of a romantic partner, but most of them did not become depressed. "It was the women with cognitions about relationships representing concerns about rejection or untrustworthiness of others who were especially challenged by normative changes" (Hammen et al., 1995, p. 441). Hammen et al. concluded that "overall, the results suggest that dysfunctional attachment cognitions contribute to both onset and severity of symptomatology" (p. 441).

Ayduk, Downey, and Kim (2001) conducted a 6-month longitudinal study of college women. They reported that the women high in rejection sensitivity (i.e., disposition to anxiously expect, readily perceive, and overreact to rejection) became more depressed when they experienced a partner-initiated breakup during the follow-up period than women low in rejection sensitivity. No differences were obtained when they experienced a self-initiated or mutually initiated breakup nor when the stressor was not interpersonal in nature.

Mazure, Bruce, Maciejewski, and Jacobs (2000) used a multivariate approach to test how adverse life events and cognitive-personality style (including need for approval from the SAS) were related to an onset of depression. They reported that the "results of our study indicated that depression was nine times more likely after a major adverse event and was almost three times more likely in the presence of cognitive-perceptual characteristics that emphasized either concern about disapproval or need for control" (Mazure et al., 2000, pp. 900–901). Mazure, Raghavan, Maciejewski, Jacobs, and Bruce (2001) subsequently reported that the needs for approval (as well as autonomy) "were strong predictors of depressive status independent of the occurrence of stressful life events" (p. 215).

Sanathara, Gardner, Prescott, and Kendler (2003) administered the subscale of the IDI concerned with an emotional reliance on another person within a multiwave population-based twin study, involving 7,174 participants. They reported that dependency scores were strongly associated with a lifetime risk for major depressive episodes. Premorbid dependency scores were also predictive of future onsets of depression, females obtained substantially higher dependency scores than males, and sex differences in risk for depression were explained largely by individual differences in dependency. They concluded that "these results suggest that a nontrivial proportion of the gender differences in risk for major depression might result from gender differences in interpersonal dependency" (Sanathara et al., 2003, p. 930).

GENDER DIFFERENCES IN DEPENDENCY

Many of the dependency studies have been confined entirely to women (e.g., Ayduk et al., 2001; Hammen et al., 1995). Although some of the above studies did include male participants, females consistently obtain higher scores than men on objective measures of sociotropy and dependency (Bornstein, 1996; Corbitt & Widiger, 1995; Turner & Turner, 1999). However, the gender differences in the assessment of dependency have been controversial (Widiger, 1998). Gender bias concerns have been raised with respect to the conceptualization of dependency (e.g., a false assumption that attachment needs are excessive), the wording of diagnostic criteria, the application of diagnostic criteria by clinicians (e.g., clinicians being overly sensitive to the appearance of dependent behaviors in women), thresholds for diagnosis, clinical presentation, research sampling, the self-awareness and openness of respondents, and the items included within self-report inventories (Bornstein, 1996; Morey, Warner, & Boggs, 2002; Sprock, Crosby, & Nielsen, 2001; Widiger, 1998).

For example, studies have suggested that some self-report personality disorder inventories are providing gender-biased assessments (Widiger, 1998). Some self-report inventories include gender-related items that are keyed in the direction of adaptive rather than maladaptive functioning. An item need not assess for dysfunction to provide a valid assessment of personality disorder. Items assessing gregariousness can identify histrionic persons, items assessing self-confidence can identify narcissistic persons, and items assessing conscientiousness can identify obsessive-compulsive persons (Millon, Millon, & Davis, 1997). Items keyed in the direction of adaptive rather than maladaptive functioning can also be helpful in countering the tendency of some respondents to deny or minimize personality disorder symptomatology. However, these items will not be useful in differentiating abnormal from normal personality functioning and are likely to contribute to the overdiagnosis of personality disorders in normal or minimally dysfunctional populations, such as student counseling centers, child custody disputes, or personnel selection (Boyle & Le Dean, 2000). When these items are related to the sex or gender of respondents, as many are in the case of the dependent and histrionic personality disorder scales of the Millon Clinical Multiaxial Inventory-III (MCMI-III; Millon et al., 1997) and the Minnesota Multiphasic Personality Inventory-2 (Colligan et al., 1994), they may contribute to gender-biased assessments (Lampel, 1999; Lindsay, Sankis, & Widiger, 2000). Lindsay and Widiger (1995) reported that some self-report inventories for the assessment of dependency tend to identify adaptive, normative behavior that occurs more often in women. For example, one of the items on the Millon Clinical Multiaxial Inventory-II (Millon, 1987) dependent personality disorder scale was, "In the past, I've gotten involved sexually with many people who didn't matter much to me" (keyed false for the dependent personality disorder). This item (keyed false) correlated positively with subject sex and with femininity, but negatively with measures of maladaptive personality functioning. There is little that is maladaptive in not becoming sexually involved with many persons that do not matter much to oneself, yet this apparent sexual responsibility was being used as an indicator for the presence of a dependent personality disorder.

Kaplan (1983) argued in her influential critique of the APA *Diagnostic and Statistical Manual* that "whereas women's expression of dependency merits clinicians' labeling and concern, men's expression of dependency does not" (p. 790). She suggested that the criteria for dependent personality disorder should provide an equal representation of masculine forms of dependency and a number of researchers have agreed with her proposal. Walker (1994) suggested that "men who rely on others to maintain their homes and take care of their children are . . . expressing personality-disordered dependency behaviors" (p. 25). Influential members of the *DSM-IV* task force have also agreed with Kaplan (1983): Frances, First, and Pincus (1995) stated that "clinicians should be aware of a possible sex bias . . . so as not to miss stereotypically masculine forms of dependency expressed through domineering behavior, ordering others to help rather than demanding or pleading" (p. 377).

Bornstein (1995, 1996) suggests that men are in fact as dependent as women, and that differences obtained with self-report inventories and semistructured interviews is due largely to a greater willingness of women to acknowledge their feelings of dependency. Bornstein (1995) conducted a meta-analysis of all studies published since 1950 that provided data on sex differences in dependency. A total of 97 studies were identified, and the findings "revealed that (a) women of all ages consistently obtain higher dependency scores than do men on objective dependency tests; [but] (b) adult men obtain slightly higher scores than do adult women on projective dependency tests" (Bornstein, 1995, p. 319). The validity of Bornstein's (1995) hypothesis that men are as dependent as women, however, does depend in part on the relative validity of the projective and objective measures of dependency. Bornstein (1993, 1995) provides a considerable amount of research to support the validity of projective assessments of dependency, but the validity of projective tests remains highly controversial (Wood, Garb, Lilienfeld, & Nezworski, 2002) and it is possible that the absence of sex differences on projective measures of dependency reflects an inadequate sensitivity or validity in their assessment of dependency.

MECHANISMS OF EFFECT

An important focus of future research on the contribution of personality traits to depression in women will be the process or mechanisms through which this association occurs (Klein, Wonderlich, & Shea, 1993). Dependent personalty traits could provide a vulnerability for mood disorders in women through a variety of different means (Bornstein, 1992; Widiger & Bornstein, 2001).

One of the new *DSM-IV* diagnostic criteria for dependent personality disorder was "urgently seeks another relationship as a source of care and support when a close relationship ends" (APA, 1994, p. 725; Hirschfeld, Shea, & Weise, 1995). Dependent persons may at times be indiscriminate in partner selection, becoming quickly and intensely involved with persons who are unreliable, undependable, and perhaps even abusive. Dependent persons can be their own worst enemies, driving their partners away by repeated demands for reassurance and support. A number of studies have indicated that dependent persons' weak, ineffectual self-image, and excessive need to please others contribute to a variety of interpersonal problems and maladaptive consequences (Overholser, 1996). Santor and Zuroff (1997), for example, indicated how dependent persons were excessively concerned about maintaining interpersonal relatedness, adopted the responses of friends who outperformed them, praised the persons who criticized them, and minimized their disagreements. Joiner and Metalsky (1995) indicated how persons characterized by excessive reassurance seeking can in fact contribute to their worst fears of interpersonal rejection. "A dysphoric individual who seeks excessive reassurance in response to perceived threat in one domain (e.g., fear of being fired) may, by excessive reassurance seeking, generate stress in another domain (e.g., his or her spouse may withdraw after failing to assuage the individual's worries)" (Joiner & Metalsky, 2001, p. 378).

However, the mechanism by which dependency leads to depression may also be due more to an affective dysregulation than an interpersonal interaction. This perspective is more readily evident when dependency is understood from the perspective of general personality functioning. Pincus (2002) indicates how dependency can be understood from a broader perspective of general personality functioning; specifically, the five-factor model (FFM). The FFM is a well-validated model of general personality functioning that consists of five broad dimensions: neuroticism, extraversion, openness, agreeableness, and conscientiousness (John & Srivastava, 1999). Each of these broad domains of personality functioning can be differentiated into more specific facets (McCrae & Costa, 1999). For example, the domain of neuroticism (or negative affectivity) consists of anxiousness, depressiveness, angry hostility, vulnerability, self-consciousness, and impulsivity. The domain of agreeableness vs. antagonism consists of such facets as trusting gullibility vs. cynical suspiciousness, tender-minded sympathy vs. tough-minded callousness, meek modesty vs. confident arrogance, altruistic self-sacrifice vs. selfish exploitativeness, docile compliant submissiveness vs. oppositional, combative aggressiveness, and straightforward honesty vs. manipulative deceptiveness.

Each of the DSM-IV personality disorders can be readily understood as maladaptive variants of the domains and facets of the FFM (Widiger & Costa, 2002). For example, dependent personality disorder might be essentially facets of neuroticism (more specifically anxiousness, depressiveness, and feelings of vulnerability), agreeableness (excessive compliance, trust, gullibility, and meekness) and extraversion (excessive needs for warmth and attachment) (Widiger, Trull, Clarkin, Sanderson, & Costa,

2002). Empirical support for this conceptualization of dependency is provided in studies by Bagby et al. (2001), Haigler and Widiger (2001), Mongrain (1993), Pincus and Gurtman (1995) and Zuroff (1994).

There are a number of advantages of conceptualizing dependency in terms of the FFM. One is that the considerable amount of research on the development, course, heritability, and neurobiological mechanisms of general personality functioning can be applied to the study of dependency. In addition, the FFM conceptualization of dependency dismantles the construct into more specific components, differentiating (for example) the affective dysregulation (facets of neuroticism) from the manner of interpersonal relatedness (facets of agreeableness and extraversion). One implication of this reformulation is that its vulnerability to depressive mood disorders may not be specific to attachment needs but may reflect instead a more general emotional instability or insecurity (e.g., neuroticism) that is shared with other disorders of personality (Bagby et al. 2001; Bornstein & Cecero, 2000; Mongrain, 1993; Zuroff, 1994).

Persons who are characteristically high in neuroticism are "prone to feelings of guilt, sadness, hopelessness, and loneliness. They are easily discouraged and often dejected" (Costa & McCrae, 1992, p. 16). Neuroticism is a fundamental domain of personality functioning for which consistent gender differences have been obtained (Clark, Watson, & Mineka, 1994; John & Srivastava, 1999; McCrae & Costa, 1999). Feingold (1994) conducted a meta-analysis of personality trait gender differences by examining 36 independent normative samples provided by 13 different personality inventories, the findings for which were organized with respect to the FFM (McCrae & Costa, 1999). Feingold reported that women score higher than men on personality measures of depressiveness, anxiousness, vulnerability, and other components of neuroticism, consistent with the findings obtained by the predominant measure of the five factor model, the NEO Personality Inventory–Revised (NEO PI-R; Costa & McCrae, 1992). Costa, Terracciano, and McCrae (2001) reported consistent gender differences in the five factors when examined with the NEO PI-R across 26 cultures, ranging from very traditional (Pakistan) to modern (The Netherlands). Consistent with previous results, they found that women scored higher than men in the domains of neuroticism (including the facets of anxiousness, depressiveness, and vulnerability) and agreeableness (including the facets of trust, altruism, and compliance).

Neuroticism has also obtained substantial empirical support for providing a general vulnerability to the development of clinically significant depressive mood disorders (e.g., Krueger, Caspi, Moffitt, Silva, & McGee, 1996; Surtees & Wainwright, 1996; Trull & Sher, 1994). The role of neuroticism in the development of mood disorders within women has been a specific focus of investigation in a longitudinal twin study by Kendler and his colleagues (Kendler, Gardner, & Prescott, 2002). In one of the earlier investigations, Kendler, Neale, Kessler, Heath, and Eaves (1993) studied 2,035 female twins who completed personality tests at two times, separated by 12 months. They found that approximately 55% of the genetic liability of major depression in women was shared with neuroticism, whereas 45% of the liability was unique to major depression. Kendler et al. (1993) concluded that "in women, the relationship between neuroticism and the liability to major depression is substantial and largely the result of genetic factors that predispose to both neuroticism and major depression" (p. 853). Subsequent studies have not been able to verify that the higher levels of neuroticism can explain the gender differences in rate of depression (Fanous, Gardner, Prescott, Cancro, & Kendler, 2002), but it is apparent that a fundamental neurobiological temperament of neuroticism does appear to provide a foundation for an eventual development of depression in women (Kendler et al., 2002).

The one personality trait to emerge from the National Institutes of Mental Health (NIMH) Collaborative Program on the Psychobiology of Depression for providing a contribution to the etiology of depression was neuroticism. Hirschfeld et al. (1989) reported the results of a follow-up study of 399 never previously ill subjects, 29 of whom eventually experienced an episode of major depression within 6 years of previously assessed personality traits. "The personality features most predictive of first onset were those indicating decreased emotional strength, such as emotional stability and less ability to react to stressful situations" (Hirschfeld et al., 1989, p. 348). Other studies have also reported that high levels of neuroticism prospectively predict later onsets of depression in persons with no history of depression (Boyce, Parker, Barnett, Cooney, & Smith, 1991; Nystrom & Lindgard, 1975).

TRAITS AND STATES

A methodological issue of considerable importance to research on the relationship of personality to depression in women is the pathoplastic effect of depressive mood disorders on the appearance, expression, or assessment of personality traits (Coyne & Whiffen, 1995; Widiger, Verheul, & van den Brink, 1999). Clinicians will typically assess a patient's personality during an initial intake procedure, yet this can be the worst time to do so. Women who are depressed at the time of the assessment will fail to provide an accurate description of their usual way of thinking, feeling, behaving, and relating to others. Low self-esteem and negativism are central features of a depressive mood disorder (APA, 2000), and women who are depressed will describe themselves as being more dependent, introverted, self-conscious, vulnerable, or pessimistic than is in fact the case (Widiger, 1993).

An alternative perspective on this issue has been to hypothesize that the depressogenic cognitions are latent and require a stressful life event to be activated (Miranda, Persons, & Byers, 1990). From this perspective, persons do not vary in the presence of negative cognitive schemas but vary instead with respect to the availability or accessibility of latent cognitive schemas. "These structures are stored in the brain and continue to exist regardless of the clinical state" (Zuroff, Blatt, Sanislow, Bondi, & Pilkonis, 1999, p. 77). However, this reasoning is perhaps comparable to inferring the presence of an unconscious conflict that becomes accessible only when the person develops the mental disorder for which it is hypothesized to be its cause. If the cognitions cannot be accessed until the person is already depressed, then the dispositions cannot be used to predict or prevent depressive episodes and may perhaps be more validly understood as a symptom, complication, or effect of the depressive mood disorder.

One of the more well established and consistently replicated findings in research on personality and psychopathology is the pathoplastic effect of psychopathology on the presentation, manifestation, and perception of personality. Persons who are significantly anxious, depressed, angry, or distraught will fail to provide accurate descriptions of their usual way of thinking, feeling, behaving, and relating to others (Widiger & Trull, 1992). The pathoplastic effects of mood may not be as problematic for particular instruments or within settings that involve relatively less psychopathology (Bagby et al., 1998; Loranger et al., 1991; Trull & Goodwin, 1993) but one should not become too sanguine about the resilience of any particular measure or population to the potential distortions of depressive mood states. Self-report inventories are especially vulnerable to the artifactual distortions in self-descriptions secondary to mood states (e.g., Bagby, Joffe, Parker, Kalemba, & Harkness, 1995; Griens, Jonker, Spinhoven, & Blom, 2002; Harkness, Bagby, Joffe, & Levitt, 2002). One of the conclusions of the Basic Behavioral Science Task Force of the National Advisory Mental Health Council (1996) was that "future research in personality evaluation should move beyond the use of self-report questionnaires to the more frequent inclusion of judgments by other informants and observations of behavior in natural settings and in laboratory settings that can provide concurrent psychophysiological recordings" (p. 28). For example, a popular self-report inventory for the assessment of personality disorders is the MCMI-III (Millon et al., 1997). However, the instructions to MCMI-III test respondents are to describe "feelings and attitudes . . . to help your doctor in learning about your problems and in planning to help you." There is no instruction for respondents to describe their characteristic manner of thinking, feeling, behaving, or relating to others prior to the onset of the current mental disorder. In fact, the instructions focus the respondents' attention on their current problems. In addition, many of the items within MCMI-III personality disorder scales concern the symptomatology of other mental mental disorders as much as, if not more than, maladaptive personality traits. For example, "I have tried to commit suicide" (Millon et al., 1997, p. 99) is one of the items within the MCMI-III scale for borderline personality disorder. It is unlikely that test respondents would or could differentiate a suicidality secondary to a recent depressive mood disorder from the self-destructive behavior evident within a borderline personality disorder when they respond to this item, particularly in the absence of any instruction to do so or any guidelines for how to do so. Indeed, the test authors themselves use affirmative responses to this item to assess for the presence of a depressive mood disorder as well as for the presence of a borderline personality disorder.

Semistructured interviews also fail to be immune to these pathoplastic distortions (e.g., Johnson et al., 1997; Peselow, Sanfilipo, & Fieve, 1994). An interviewer can easily fail to appreciate the extent to which persons' self-descriptions are being distorted by their mood. For example, Loranger et al. (1991) compared assessments obtained by a semistructured interview at the beginning of an inpatient admission to those obtained 1 week to 6 months later and reported that "there was a significant reduction in the mean number of criteria met on all of the personality disorders except schizoid and antisocial" (p. 726). Loranger et al. argued that the reduction was not due to an inflation of scores secondary to depressed or anxious mood because changes in personality disorder scores were not correlated with changes in anxiety or depression. However, an alternative perspective is that the study lacked sufficiently sensitive or accurate measures to explain why there was a substantial decrease on 10 of the 12 personality disorder scales. It is unlikely that 1 week to 6 months of treatment resulted in the extent of changes to personality that were indicated by the semistructured interview (the change scores also failed to correlate with length of treatment). In fact, four of the patients were diagnosed with a histrionic personality disorder by the instrument at admission whereas eight patients were diagnosed with this disorder at discharge. If the change in scores represented valid changes in personality functioning then the brief inpatient treatments apparently created histrionic personality disorders in some of the patients.

In sum, researchers should expect that self-report assessments of personality will be affected by depressed mood, anger, or anxiousness or at least not assume that they will not be, and they should include within any particular study concerned with the relationship of personality to psychopathology an effective means by which to address or at least identify the extent of this potential distortion of the personality assessments. There are a variety of methods by which researchers have attempted to offset, address, or assess pathoplastic mood state distortions, including (1) the administration of the personality instrument to persons other than or in addition to the depressed patient (e.g., Bagby et al., 1998; Molinari, Kunik, Mulsant, & Rifai, 1998; Peselow et al., 1994), (2) the inclusion of validity or moderating scales that assess for potential sources of response distortions (Butcher & Rouse, 1996), and (3) delaying the administration of the instrument to a time when the person no longer appears to be susceptible to the effects of the disorder (e.g., Trull & Goodwin, 1993). These approaches have their respective limitations and none are necessarily sufficient, but they should be given serious consideration in any clinical study concerned with the relationship of personality to depression.

Just as there is no infallible instrument for the assessment of personality traits, there is also no conclusive research design. The all too often approach is to conduct a study of convenience, in which a readily available sample of patients within a particular clinical setting is administered a familiar instrument for the assessment of a topical personality construct. Cross-sectional studies do provide informative results, particularly when generating hypotheses regarding constructs that have not yet been considered (e.g., Ball et al., 1997; Trull & Sher, 1994) or raising concerns regarding the contribution of pathoplastic relationships (e.g., Bagby et al., 1995), but longitudinal studies should perhaps be the predominant and standard means with which hypotheses concerning causal relationship of personality to depression are studied (Costa & McCrae, 1998; Wetherell, Gatz, & Pederson, 2001).

There is perhaps no easy or ready solution to the problem of disentangling the effects of mood, cognition, and personality, outside of longitudinal studies that follow persons from early childhood into old age. Any study that assesses personality traits in women who are currently or have even historically suffered from a major depressive episode is susceptible to methodological concerns regarding the impact of the mood disorder on the personality assessment; however, any study confined to women with no prior history of a major depressive episode would likely be excluding the very persons who would most likely have a personality trait vulnerability for depression (Widiger et al., 1999).

TRAITS AND SITUATIONS

An additional methodological issue of particular importance to the contribution of personality to depression in women is disentangling the effects of personality from the effects of unstable relationships (Coyne & Whiffen, 1995). In theory, dependent and sociotropic personality traits contribute to the instability of intimate and supportive relationships through the expression of excessive needs for reassurance and contribute to the onset of depressive mood disorders through a premorbid emotional instability and pathogenic attitudes. However, it is also possible that the emotional instability and pathologic attitudes are themselves the result of unstable interpersonal relationships. Coyne and Whiffen (1995) noted that some self-report measures of personality traits include items that concern the stability of recent relationships. These measures could then be assessing the actual instability of the relationships rather than dependent perceptions of the relationships. "Intimate relationships that are insecure or have an uncertain future may engender dependency and reassurance seeking" (Coyne & Whiffen, 1995, p. 367). A broader and more provocative reformulation of dependency in women is that the apparent feelings of insecurity in women may say less about the women than about the persons with whom the women are involved. "Men and women may differ in what they seek from relationships, but they may also differ in what they provide to each other" (Coyne & Whiffen, 1995, p. 368). In other words, "women might appear (and be) less dependent if they weren't involved with such undependable men" (Widiger & Anderson, 2003, p. 63).

Personality disorder diagnoses can be used to inappropriately or inaccurately blame women for the troubles in their lives. For example, repeated proposals to include a masochistic or self-defeating personality disorder in the APA diagnostic manual have been opposed in part because of the concern that the presence of such a diagnosis might contribute to an overdiagnosis of maladaptive personality traits in women within abusive relationships (Widiger, 1995). The intention of the masochistic or self-defeating diagnosis was to identify persons who are characteristically pessimistic, self-blaming, and self-defeating, an interest that has since been supplanted by the depressive personality disorder diagnosis (e.g., Kwon et al., 2000; Phillips, Gunderson, Hirschfeld, & Smith, 1990; Phillips et al., 1998). It was apparent that a proportion of women diagnosed with a self-defeating or masochistic personality disorder would be within physically abusive relationships and their self-defeating behaviors might say more about the constant threat of being physically harmed than any particular personality trait pathology that might be present (Widiger, 1995). A diagnosis of masochistic or self-defeating personality disorder could in fact be used to blame the victim rather than the perpetrator. A historical example is provided by Snell, Ronsenwald, and Robey (1964), who explained the occurrence of 12 husbands charged with assault as "filling masochistic needs of the wife" (p. 110). Women who have been repeatedly physically, sexually, or psychologically abused during childhood do have an increased risk of developing cognitive, neurobiological, and psychodynamic pathologies that may increase the likelihood of being victimized again as adults. Coolidge and Anderson (2002) reported a significantly higher rate of dependent personality disorder in women with a history of multiple abusive relationships. However, many current victims of abuse who seem unwilling or unable to extricate themselves from a relationship could be acting realistically in response to threats of physical harm and to the absence of a safe or meaningful alternative (Walker, 1994). It can be very difficult to leave a relationship in which one has a significant emotional involvement and at times it might seem preferable to suffer occasional assaults than to be perpetually harassed, stalked, and perhaps eventually killed.

Future studies concerned with the contribution of dependent personality traits to depression in women should include an objective assessment of the contribution of the women's partners to the depression and to the dependent personality traits (Besser & Priel, 2003). Longitudinal studies that explore the relationship of personality to marital instability and dissatisfaction would be particularly informative. Kelly and Conley (1987) investigated a panel of 300 couples followed over a period of 50 years (1930–1980). Twenty-two of the 300 original couples subsequently broke their engagements, and 50 eventually became divorced sometime between 1935 and 1980. Personality traits of both spouses were assessed during the initial 1930s data collection through aggregated ratings of each spouse provided by five acquaintances. Additional predictor variables were social environment, attitudes concerning marriage, sexual history, and stressful life events that occurred subsequently during the marriage. Three personality variables accounted for more than half of the predictable variance in marital instability and dissatisfaction: the neuroticism of the husband, the neuroticism of the wife, and the impulsivity in the husband. Neuroticism of the spouses in the 1930s was linked with marital dissatisfaction in 1955 and again in 1980. Impulsivity of the husband in the 1930s was linked with the subsequent occurrence of extramarital affairs during the marriage, which were then subsequently emphasized by the wives in their descriptions of their reasons for divorce.

More recent studies have been conducted by Besser and Priel (2003), Botwin, Buss, and Shackelford (1997) and Bouchard, Lussier, and Sabourin (1999). Botwin et al., for example, studied the relationship of the partners' personality traits, their preferences for an ideal partner, and marital and sexual dissatisfaction. Personality assessments consisted of scores aggregated across self, partner, and acquaintance ratings. Women expressed a greater preference than men for a wide array of socially desirable personality traits, and both men and women reported wanting a partner who was relatively close to them in personality traits. However, relationship satisfaction was primarily associated with the personalty traits of the partner, irrespective of the degree of matching with self or with the ideal partner. Women were especially happy with their relationships if their male partners were high in agreeableness, low in neuroticism, and high in openness. Happiness in women is perhaps as much the result of the personality traits of their (often male) partners as it is a reflection of their own personality functioning.

IMPLICATIONS FOR TREATMENT OF DEPRESSION IN WOMEN

The role of personality in the treatment of depression in women might be considered to be an oxymoron, as personality traits are notoriously difficult to treat. Personality traits are, by definition, chronic and stable behavior patterns that have been evident since young adulthood (APA, 2000). Persons consider many of their personality traits to be integral to their sense of self, and persons may value particular aspects of their personality that a clinician would consider to be an important target of treatment (Stone, 1993). Maladaptive personality traits (e.g., those traits that would provide a vulnerability for depressive episodes) are even said to be relatively more inflexible than normal personality traits (Millon et al., 1996).

Personality disorders were placed on a separate axis in *DSM-III* (APA, 1980) because of their potential importance to the clinical treatment of other mental disorders (Frances, 1980). Maladaptive personality functioning can go far in explaining why some women have failed to respond as expected to a treatment of their depression (Shea, Widiger, & Klein, 1992). Researchers who are concerned with the treatment of depression in women are then advised to include a measure of personality disorder symptomatology if they intend to account fully for variation in treatment responsivity. In a complementary fashion, other personality traits (e.g., high self-discipline, conscientiousness, and openness to experience) may facilitate treatment responsivity (Sanderson & Clarkin, 2002). "The last 40 years of individual differences research require the inclusion of personality trait assessment for the construction and implementation of any treatment plan that would lay claim to scientific status" (Harkness & Lilienfeld, 1997, p. 349).

However, there is an unfortunate perception that there is currently no effective treatment for personality disorders and that personality traits are essentially immutable (Costa & McCrae, 1994; Frances et al., 1995). Personality disorders are among the more difficult of disorders to treat, in part because they involve pervasive and entrenched behavior patterns that have been present throughout much of a person's life, but this does not imply that clinically and socially meaningful treatments of personality disorder do not occur. There is compelling empirical support to indicate that meaningful responsivity to treatment does occur (Perry, Banon, & Ianni, 1999; Sanislow & McGlashan, 1998). Perry et al. (1999) conducted a meta-analysis of 15 published personality disorder psychotherapy studies. Among their results was the finding that approximately 50% of patients with a personality disorder tend to recover (i.e., no longer meet diagnostic criteria for the respective disorder) after 1.3 years of focused treatment (approximately 93 sessions), whereas a 50% recovery rate would not occur until approximately 10.5 years over the natural course of the disorder (which may have included brief periods of unknown treatments). Perry et al. (1999) concluded that "psychotherapy is an effective treatment for personality disorders and may be associated with up to a sevenfold faster rate of recovery in comparison with the natural history of the [personality] disorder" (p. 1312). Similar conclusions were reached by Sanislow and McGlashan (1998) in their comprehensive review of pharmacologic and psychosocial personality disorder treatment studies. They indicated that patients do not reach a level of "normalcy" but there is compelling "evidence that effective treatments exist to alleviate symptoms and reduce symptomatic behavior" (Sanislow & McGlashan, 1998, p. 237).

Knutson et al. (1998) even demonstrated the potential effects of pharmacotherapy on ostensibly normal personality functioning. Knutson et al. administered for four weeks in a double-blind manner paroxetine, a selective serotonin reuptake inhibitor (SSRI), to 23 of 48 normal volunteers. They reported that this SSRI administration (relative to placebo) reduced negative affectivity and increased social facilitation. The magnitude of changes was even correlated with the plasma levels of SSRI within the SSRI group. "This is the first empirical demonstration that chronic administration of a selective serotonin reuptake blockade can have significant personality and behavioral effects in normal humans in the absence of baseline depression or other psychopathology" (Knutson et al., 1998, p. 378).

Linehan (1993) has developed an extensive and comprehensive treatment program, titled Dialectical Behavior Therapy (DBT), for the many problematic traits and behaviors evident within borderline personality disorder, a constellation of maladaptive personality traits that is diagnosed substantially more frequently in women. DBT was constructed originally for the treatment of parasuicidal women, and Linehan and her colleagues have amassed compelling empirical support for the effectiveness of DBT in treating facets of borderline personality disorder (Koerner & Dimeff, 2000; Linehan, 2000). DBT is unlikely to result in the development of a fully healthy or ideal personality structure (whatever that may entail), but clinically and socially meaningful change to personality structure and functioning does occur. Scheel (2000) provided a thorough and detailed critique of most every empirical study of the effectiveness of DBT for borderline personality disorder. She was concerned that the effectiveness of DBT has been exaggerated and that its implementation has surpassed its empirical support. Nevertheless, despite her critical perspective she also reached the conclusion that "summarizing published empirical results across studies, standard outpatient dialectical behavior therapy has been associated with lesser parasuicidal behavior, psychiatric hospitalization, anger, and psychotropic medication usage, and with increased client retention, overall level of functioning, overall social adjustment, and employment performance" (Scheel, 2000, p. 76).

A lessen to be learned from DBT research for the treatment of depression in women with maladaptive personality traits is that even though these traits will hinder and complicate (often substantially) the treatment of depression, this does not necessarily mean that these personality traits are themselves untreatable or are unresponsive to systematic and comprehensive clinical interventions. A useful focus of investigation for future research on the treatment of depression in women would be the long-term costs and benefits of sustained or comprehensive treatment programs relative to the more brief or limited interventions that are currently preferred by managed care programs. Insurance provider programs are often reluctant to fund the treatment of maladaptive personality traits and may even consider the presence of a personality disorder to indicate that the patient is largely untreatable. However, one of the outcome variables emphasized by Linehan and her colleagues has been health care utilization and cost effectiveness (Linehan & Heard, 1999). The American Psychiatric Association provided a 1998 Gold Achievement Award to the DBT program at the Mental Health Center of Greater Manchester, New Hampshire.

Data on the first 14 clients to complete the first year of contracted DBT showed a 77% decrease in hospital days (from 479 days to 85 days), a 76% decrease in partial hospital days (from 173 days to 42 days), a 56% decrease in crisis bed days (from 170 days to 73 days), an 80% decrease in face-to-face contacts with emergency services (from 61 to 12), a 43% increase in number employed (from two to eight) and a 58% decrease in total treatment costs (from $645,000 to $273,000) (Linehan, 2000, p. 114).

CONCLUSIONS

Personality functioning is of significant importance in the understanding of depression in women. Personality functioning should be assessed in most studies concerned with the treatment of depression in women, as some maladaptive personality traits will hinder treatment efficacy whereas others may facilitate treatment effectiveness. The optimal research design for an understanding of the etiology of depression in women would also include a reasonably comprehensive sample of personality functioning because the traits that will be discovered to be significantly beneficial or problematic could also fail to have been the traits of initial importance or interest to the researcher or clinician.

The most informative research design for identifying the contribution of personality traits to the development and treatment of depression will be longitudinal studies of women selected on the basis of their possession of relevant personality trait dispositions, followed into and through their episodes and treatment of depression. Relationship and marital instability will be stressors of particular importance in a study of depression in women, and the personality traits of the women's partners might be as important in predicting the relationship instability and the women's eventual depression as the personality traits of the women themselves. Hammen et al. (1995), Johnson, Cohen, Brown, Smailes, and Bernstein (1999), Lenzenweger, Loranger, Korfine, and Neff (1997), Trull, Useda, Conforti, and Doan (1997), and others are conducting prospective longitudinal studies of personality disorder symptomatology in adolescents and young adults, many of whom have not yet developed a clinically significant depression but are presumably at risk for doing so. "These initial studies...mark an important step in identifying young adults with significant levels of borderline features who may go on to experience significant dysfunction in later years" (Trull et al., 1997, p. 308).

References

American Psychiatric Association. (1980). *Diagnostic and statistical manual of mental disorders* (3rd ed.). Washington, DC: American Psychiatric Association.

American Psychiatric Association. (1994). *Diagnostic and statistical manual of mental disorders* (4th ed.). Washington, DC: American Psychiatric Association.

American Psychiatric Association. (2000). *Diagnostic and statistical manual of mental disorders* (4th ed., rev. ed.). Washington, DC: American Psychiatric Association.

Ayduk, O., Downey, G., & Kim, M. (2001). Rejection sensitivity and depressive symptoms in women. *Personality and Social Psychology Bulletin, 27*, 868–877.

Bagby, R. M., Gilchrist, E. J., Rector, N. A., Dickens, S. E., Joffe, R., Levitt, A., et al. (2001). The stability and validity of the sociotropy and autonomy personality dimensions as measured by the Revised Personal Style Inventory. *Cognitive Therapy and Research, 25*, 765–779.

Bagby, R. M., Joffe, R. T., Parker, J. D. A., Kalemba, V., & Harkness, K. L. (1995). Major depression and the five-factor model of personality. *Journal of Personality Disorders, 9*, 224–234.

Bagby, R. M., Rector, N. A., Kirstin, B., Dickens, S. E., Levitan, R. D., & Kennedy, S. H. (1998). Self-report ratings and informants' ratings of personalities of depressed outpatients. *American Journal of Psychiatry, 155*, 437–438.

Ball, S. A., Tennen, H., Poling, J. L., Kranzler, H. W., & Rounsaville, B. J. (1997). Personality, temperament, and character dimensions. *Journal of Abnormal Psychology, 106*, 545–553.

Basic Behavioral Science Task Force of the National Advisory Mental Health Council. (1996). Basic behavioral science research for mental health. Vulnerability and resilience. *American Psychologist, 51*, 22–28.

Beck, A. T. (1983). Cognitive therapy of depression: New perspectives. In P. J. Clayton & J. E. Barrett (Eds.). *Treatment of depression: Old controversies and new approaches* (pp. 265–290). New York: Raven Press.

Besser, A., & Priel, B. (2003). A multisource approach to self-critical vulnerability to depression: The moderating role of attachment. *Journal of Personality, 71*, 515–555.

Blatt, S. J., & Zuroff, D. (1992). Interpersonal relatedness and self-definition: Two prototypes for depression. *Clinical Psychology Review, 12*, 527–562.

Bornstein, R. F. (1992). The dependent personality: Developmental, social, and clinical perspectives. *Psychological Bulletin, 112*, 3–23.

Bornstein, R. F. (1993). *The dependent personality*. New York: Guilford.

Bornstein, R. F. (1995). Sex differences in objective and projective dependency tests: a meta-analytic review. *Assessment, 2*, 319–331.

Bornstein, R. F. (1996). Sex differences in dependent personality disorder prevalence rates. *Clinical Psychology: Science and Practice, 3*, 1–12.

Bornstein, R. F., & Cecero, J. J. (2000). Deconstructing dependency in a five-factor world: A meta-analytic review. *Journal of Personality Assessment, 74*, 324–343.

Botwin, M. D., Buss, D. M., & Shackelford, T. K. (1997). Personality and mate preferences: Five factors in mate selection and marital satisfaction. *Journal of Personality, 65*, 107–136.

Bouchard, G., Lussier, Y., & Sabourin, S. (1999). Personality and marital adjustment: Utility of the five-factor model of personality. *Journal of Marriage and Family, 61*, 651–660.

Boyce, P., Parker, G., Barnett, B., Cooney, M., & Smith, F. (1991). Personality as a vulnerability factor to depression. *British Journal of Psychiatry, 159*, 106–114.

Boyle, G. J., & Le Dean, L. (2000). Discriminant validity of the Illness Behavior Questionnaire and Millon Clinical Multiaxial Inventory-III in a heterogeneous sample of psychiatric outpatients. *Journal of Clinical Psychology, 56*, 779–791.

Butcher, J. N., & Rouse, S. V. (1996). Personality: Individual differences and clinical assessment. *Annual Review of Psychology, 47*, 87–111.

Clark, L. A., Watson, D., & Mineka, S. (1994). Temperament, personality, and the mood and anxiety disorders. *Journal of Abnormal Psychology, 103*, 103–116.

Colligan, R. C., Morey, L. C., & Offord, K. P. (1994). MMPI/MMPI-2 personality disorder scales. Contemporary norms for adults and adolescents. *Journal of Clinical Psychology, 50*, 168–200.

Coolidge, F. L., & Anderson, L. W. (2002). Personality profiles of women in multiple abusive relationships. *Journal of Family Violence, 17*, 117–131.

Corbitt, E. M., & Widiger, T. A. (1995). Sex differences among the personality disorders: An exploration of the data. *Clinical Psychology: Science & Practice, 2*, 225–238.

Costa, P. T., & McCrae, R. R. (1992). *Revised NEO Personality Inventory (NEO-PI-R) and NEO Five-Factor Inventory (NEO-FFI): Professional manual*. Odessa, FL: Psychological Assessment Resources.

Costa, P. T., & McCrae, R. R. (1994). Set like plaster? Evidence for the stability of adult personality. In T. Heatherton & J. L. Weinberger (Eds.), *Can personality change?* (pp. 21–40). Washington, DC: American Psychological Association.

Costa, P. T., & McCrae, R. R. (1998). Trait theories of personality. In D. F. Barone, M. Hersen, & V. B. van Hasselt (Eds.), *Advanced personality* (pp. 103–121). New York: Plenum.

Costa, P. T., Terracciano, A., & McCrae, R. R. (2001). Gender differences in personality traits across cultures: Robust and surprising findings. *Journal of Personality and Social Psychology, 81*, 322–331.

Coyne, J. C., & Whiffen, V. E. (1995). Issues in personality as diathesis for depression: The case of sociotropy-dependency and autonomy-self-criticism. *Psychological Bulletin, 118*, 358–378.

Fanous, A., Gardner, C. O., Prescott, C. A., Cancro, R., & Kendler, K. S. (2002). Neuroticism, major depression, and gender: A population-based twin study. *Psychological Medicine, 32*, 719–728.

Feingold, A. (1994). Gender differences in personality: A meta-analysis. *Psychological Bulletin, 116*, 429–456.

Frances, A. J. (1980). The DSM-III personality disorders sections: A commentary. *American Journal of Psychiatry, 137*, 1050–1054.

Frances, A. J., First, M. B., & Pincus, H. A. (1995). *DSM-IV guidebook*. Washington, DC: American Psychiatric Press.

Griens, A. M. G. F., Jonker, K., Spinhoven, P., & Blom, M. B. J. (2002). The influence of depressive state features on trait measurement. *Journal of Affective Disorders*, *70*, 95–99.

Haigler, E. D., & Widiger, T. A. (2001). Experimental manipulation of NEO PI-R items. *Journal of Personality Assessment*, *77*, 339–358.

Hammen, C. L., Burge, D., Daley, S. E., Davila, J., Paley, B., & Rudolph, K. D. (1995). Interpersonal attachment cognitions and predictions of symptomatic responses to interpersonal stress. *Journal of Abnormal Psychology*, *104*, 436–443.

Harkness, A. R., & Lilienfeld, S. O. (1997). Individual differences science for treatment planning: Personality traits. *Psychological Assessment*, *9*, 349–360.

Harkness, K. L., Bagby, M. R., Joffe, R. T., & Levitt, A. (2002). Major depression, chronic minor depression, and the five-factor model of personality. *European Journal of Personality*, *16*, 271–281.

Hirschfeld, R. M. A., Klerman, G. L., Gough, H. G., Barrett, J., Korchin, S. J., & Chodoff, P. (1977). A measure of interpersonal dependency. *Journal of Personality Assessment*, *41*, 610–618.

Hirschfeld, R., Klerman, G., Lavori, P., Keller, M., Griffith, P., & Coryell, W. (1989). Premorbid personality assessments of first onset of major depression. *Archives of General Psychiatry*, *46*, 345–350.

Hirschfeld, R., Shea, M. T., & Weise, R. (1995). Dependent personality disorder. In W. J. Livesley (Ed.), *The DSM-IV personality disorders* (pp. 239–256). New York: Guilford.

John, O. P., & Srivastava, S. (1999). The big five trait taxonomy: History, measurement, and theoretical perspectives. In L. A. Pervin & O. P. John (Eds.), *Handbook of personality: Theory and research* (pp. 102–138). New York: Guilford.

Johnson, J. G., Cohen, P., Brown, J., Smailes, E. M., & Bernstein, D. P. (1999). Childhood maltreatment increases risk for personality disorders during early adulthood. *Archives of General Psychiatry*, *56*, 600–606.

Johnson, J. G., Williams, J. B. W., Goetz, R. R., Rabkin, J. G., Lipsitz, J. D., & Remien, R. H. (1997). Stability and change in personality disorder symptomatology: Findings from a longitudinal study of HIV+ and HIV− men. *Journal of Abnormal Psychology*, *106*, 154–158.

Joiner, T. J., & Metalsky, G. I. (1995). A prospective test of an integrative interpersonal theory of depression: A naturalistic study of college roommates. *Journal of Personality and Social Psychology*, *69*, 778–788.

Joiner, T. J., & Metalsky, G. I. (2001). Excessive reassurance seeking. Delineating a risk factor involved in the development of depressive symptoms. *Psychological Science*, *12*, 371–378.

Kaplan, M. (1983). A woman's view of DSM-III. *American Psychologist*, *38*, 786–792.

Kelly, E. L., & Conley, J. J. (1987). Personality and compatibility: A prospective analysis of marital stability and marital satisfaction. *Journal of Personality and Social Psychology*, *52*, 27–40.

Kendler, K. S., Gardner, C. L., & Prescott, C. A. (2002). Toward a comprehensive developmental model for major depression in women. *American Journal of Psychiatry*, *159*, 1133–1145.

Kendler, K. S., Neale, M. C., Kessler, R. C., Heath, A. C., & Eaves, L. J. (1993). A longitudinal twin study of personality and major depression in women. *Archives of General Psychiatry*, *50*, 853–862.

Klein, M. H., Wonderlich, S., & Shea, M. T. (1993). Models of relationship between personality and depression: Toward a framework for theory and research. In M. H. Klein, D. J. Kupfer, & M. T. Shea (Eds.), *Personality and depression* (pp. 1–54). New York: Guilford.

Knutson, B., Wolkowitz, O. M., Cole, S. W., Chan, T., Moore, E. A., Johnson, R. C., et al. (1998). Selective alteration of personality and social behavior by serotonergic intervention. *American Journal of Psychiatry, 155*, 373–379.

Koerner, K., & Dimeff, L. A. (2000). Further data on Dialectical Behavior Therapy. *Clinical Psychology: Science and Practice, 7*, 104–112.

Krueger, R. F., Caspi, A., Moffitt, T. E., Silva, P. A., & McGee, R. (1996). Personality traits are differentially linked to mental disorders: A multitrait-multidiagnosis study of an adolescent birth cohort. *Journal of Abnormal Psychology, 105*, 299–312.

Kwon, J. S., Kim, Y. M., Chang, C. G., Park, B. J., Kim, L., Yoon, D. J., et al. (2000). Three-year follow-up of women with the sole diagnosis of depressive personality disorder: Subsequent development of dysthymia and major depression. *American Journal of Psychiatry, 157*, 1966–1972.

Lampel, A. K. (1999). Use of the Millon Clinical Multiaxial Inventory-III in evaluating child custody litigants. *American Journal of Forensic Psychology, 17*, 19–31.

Lenzenweger, M. F., Loranger, A. W., Korfine, L., & Neff, C. (1997). Detecting personality disorders in a nonclinical population. *Archives of General Psychiatry, 54*, 345–351.

Lindsay, K. A., Sankis, L. M., & Widiger, T. A. (2000). Gender bias in self-report personality disorder inventories. *Journal of Personality Disorders, 14*, 218–232.

Lindsay, K., & Widiger, T. A. (1995). Sex and gender bias in self-report personality disorder inventories: Items analyses of the MCMI-II, MMPI, and PDQ-R. *Journal of Personality Assessment, 65*, 1–20.

Linehan, M. M. (1993). *Cognitive-behavioral treatment of borderline personality disorder*. New York: Guilford.

Linehan, M. M. (2000). The empirical basis of Dialectical Behavior Therapy: Development of new treatments versus evaluation of existing treatments. *Clinical Psychology: Science and Practice, 7*, 113–119.

Linehan, M. M., & Heard, H. (1999). Borderline personality disorder: Costs, course, and treatment outcome. In N. Miller & K. Magruder (Eds.), *The cost-effectiveness of psychotherapy: A guide for practitioners, researchers, and policy-makers* (pp. 291–305). New York: Oxford University Press.

Loranger, A. W., Lenzenweger, M. F., Gartner, A. F., Susman, V. L., Herzig, J., Zammit, G. K., et al. (1991). Trait-state artifacts and the diagnosis of personality disorders. *Archives of General Psychiatry, 48*, 720–729.

Mazure, C. M., Bruce, M. L., Maciejewski, P. K., & Jacobs, S. C. (2000). Adverse life events and cognitive personality characteristics in the predictor of major depression and antidepressant response. *American Journal of Psychiatry, 157*, 896–903.

Mazure, C. M., Raghavan, C., Maciejewski, P. K., Jacobs, S. C., & Bruce, M. L. (2001). Cognitive-personality characteristics as direct predictors of unipolar major depression. *Cognitive Therapy and Research, 25*, 215–225.

McCrae, R. R., & Costa, P. T. (1999). A five-factor theory of personality. In L. A. Pervin & O. P. John (Eds.), *Handbook of personality: Theory and research* (pp. 139–153). New York: Guilford.

Millon, T. (1987). *Manual for the MCMI-II* (2nd ed.). Minneapolis, MN: National Computer Systems.

Millon, T., Davis, R. D., Millon, C. M., Wenger, A., Van Zullen, M. H., Fuchs, M., et al. (1996). *Disorders of personality. DSM-IV and beyond* (2nd ed.). New York: Wiley.

Millon, T., Millon, C., & Davis, R. (1997). *MCMI-III manual* (2nd ed.). Minneapolis, MN: National Computer Systems.

Miranda, J., Persons, J., & Byers, C. (1990). Endorsement of dysfunctional beliefs depends on current mood state. *Journal of Abnormal Psychology, 99,* 237–241.

Molinari, V., Kunik, M. E., Mulsant, B., & Rifai, A. H. (1998). The relationship between patient, informant, social worker, and consensus diagnoses of personality disorder in elderly depressed patients. *American Journal of Geriatric Psychiatry, 6,* 136–144.

Mongrain, M. (1993). Dependency and self-criticism located within the five-factor model of personality. *Personality and Individual Differences, 15,* 455–462.

Morey, L. L., Warner, M. B., & Boggs, C. D. (2002). Gender bias in the personality disorder criteria: An investigation of five bias indicators. *Journal of Psychopathology and Behavioral Assessment, 24,* 55–65.

Nietzel, M. T., & Harris, M. J. (1990). Relationship of dependency and achievement/autonomy to depression. *Clinical Psychology Review, 10,* 279–297.

Nystrom, S., & Lindegard, B. (1975). Predisposition for mental syndromes: A study comparing predisposition for depression, neurasthenia, and anxiety state. *Acta Psychiatrica Scandinavica, 51,* 69–76.

Overholser, J. C. (1996). The dependent personality and interpersonal problems. *Journal of Nervous and Mental Disease, 184,* 8–16.

Perry, J. C., Banon, E., & Ianni, F. (1999). Effectiveness of psychotherapy for personality disorders. *American Journal of Psychiatry, 156,* 1312–1321.

Pervin, L. A., & John, O. P. (Eds.) (1999). *Handbook of personality: Theory and research* (2nd ed.). New York: Guilford.

Peselow, E. D., Sanfilipo, M. P., & Fieve, R. R. (1994). Patients' and informants' reports of personality traits during and after major depression. *Journal of Abnormal Psychology, 103,* 819–824.

Phillips, K. A., Gunderson, J. G., Hirschfeld, R. M. A., & Smith, L. E. (1990). A review of the depressive personality. *American Journal of Psychiatry, 147,* 830–837.

Phillips, K. A., Gunderson, J. G., Triebwasser, J., Kimble, C. R., Faedda, G., Lyoo, I. K., et al. (1998). Reliability and validity of depressive personality disorder. *American Journal of Psychiatry, 155,* 1044–1048.

Pincus, A. L. (2002). Constellations of dependency within the five factor model of personality. In P. T. Costa and T. A. Widiger (Eds.), *Personality disorders and the five factor model of personality* (pp. 203–214). Washington, DC: American Psychological Association.

Pincus, A. L., & Gurtman, M. B. (1995). The three faces of interpersonal dependency: Structural analysis of self-report dependency measures. *Journal of Personality and Social Psychology, 69,* 744–758.

Robins, C. J., & Block, P. (1988). Personal vulnerability, life events, and depressive symptoms: A test of a specific interactional model. *Journal of Personality and Social Psychology, 54,* 847–852.

Robins, C. J., Hayes, A. H., Block, P., Kramer, R. J., & Villena, M. (1995). Interpersonal and achievement concerns and the depressive vulnerability and symptom specificity hypothesis: A prospective study. *Cognitive Therapy and Research, 19,* 1–20.

Sanathara, V. A., Gardner, C. O., Prescott, C. A., & Kendler, K. S. (2003). Interpersonal dependence and major depression: Aetiological inter-relationship and gender differences. *Psychological Medicine, 33,* 927–931.

Sanderson, C. J., & Clarkin, J. F. (2002). Further use of the NEO PI-R personality dimensions in differential treatment planning. In P. T. Costa & T. A. Widiger (Eds.), *Personality disorders and the five-factor model of personality* (pp. 351–376). Washington, DC: American Psychological Association.

Sanislow, C. A., & McGlashan, T. H. (1998). Treatment outcome of personality disorders. *Canadian Journal of Psychiatry, 43,* 237–250.

Santor, D. A., & Zuroff, D.C. (1997). Interpersonal responses to threats of status and interpersonal relatedness: effects of dependency and self-criticism. *British Journal of Clinical Psychology, 36,* 521–541.

Santor, D. A., Zuroff, D.C., & Fielding, A. (1997). Analysis and revision of the Depressive Experiences Questionnaire: Examining scale performance as a function of scale length. *Journal of Personality Assessment, 69,* 145–163.

Scheel, K. R. (2000). The empirical basis of dialectical behavior therapy: Summary, critique, and implications. *Clinical Psychology: Science and Practice, 7,* 68–86.

Shea, M. T., Widiger, T. A., & Klein, M. H. (1992). Comorbidity of personality disorders and depression: Implications for treatment. *Journal of Consulting and Clinical Psychology, 60,* 857–868.

Snell, J., Rosenwald, R., & Robey, A. (1964). The wifebeater's wife. A study of family interaction. *Archives of General Psychiatry, 11,* 107–112.

Sprock, J., Crosby, J. P., & Nielsen, B. A. (2001). Effects of sex and sex roles on the perceived maladaptiveness of DSM-IV personality disorder symptoms. *Journal of Personality Disorders, 15,* 41–59.

Stone, M. H. (1993). *Abnormalities of personality. Within and beyond the realm of treatment.* New York: W. W. Norton.

Surtees, P. G., & Wainwright, N. W. J. (1996). Fragile states of mind: Neuroticism, vulnerability and long-term outcome in depression. *British Journal of Psychiatry, 169,* 338–347.

Trull, T. J., & Goodwin, A. H. (1993). Relationship between mood changes and the report of personality disorder symptoms. *Journal of Personality Assessment, 61,* 99–111.

Trull, T. J., & Sher, K. J. (1994). Relationship between the five-factor model of personality and Axis I disorders in a nonclinical sample. *Journal of Abnormal Psychology, 103,* 350–360.

Trull, T. J., Useda, D., Conforti, K., & Doan, B-T. (1997). Borderline personality disorder features in nonclinical young adults: 2. Two-year outcome. *Journal of Abnormal Psychology, 106,* 307–314.

Turner, H. A., & Turner, R. J. (1999). Gender, social status, and emotional reliance. *Journal of Health and Social Behavior, 40,* 360–373.

Walker, L. E. A. (1994). Are personality disorder gender biased? In S. A. Kirk & S. D. Einbinder (Eds.), *Controversial issues in mental health* (pp. 22–29). New York: Allyn & Bacon.

Watson, D., & Clark, L. A. (1994). Introduction to the special series on personality and psychopathology. *Journal of Abnormal Psychology, 103,* 3–5.

Wetherell, J. L., Gatz, M., & Pederson, N. L. (2001). A longitudinal analysis of anxiety and depressive symptoms. *Psychology and Aging, 16,* 187–195.

Widiger, T. A. (1993). Personality and depression: Assessment issues. In M. H. Klein, D. J. Kupfer, & M. T. Shea (Eds.), *Personality and depression: A current view* (pp. 77–118). New York: Guilford.

Widiger, T. A. (1995). Deletion of the self-defeating and sadistic personality disorder diagnoses. In W. J. Livesley (Ed.), *The DSM-IV personality disorders* (pp. 359–373). New York: Guilford.

Widiger, T. A. (1998). Sex biases in the diagnosis of personality disorders. *Journal of Personality Disorders, 12,* 95–118.

Widiger, T. A., & Anderson, K. G. (2003). Personality and depression in women. *Journal of Affective Disorders, 74,* 59–66.

Widiger, T. A., & Bornstein, R. F. (2001). Histrionic, narcissistic, and dependent personality disorders. In H. E. Adams & P. Sutker (Eds.), *Comprehensive handbook of psychopathology* (3rd ed., pp. 507–529). New York: Plenum.

Widiger, T. A., & Costa, P. T. (2002). Five factor model personality disorder research. In P. T. Costa & T. A. Widiger (Eds.), *Personality disorders and the five factor model of personality* (2nd ed., pp. 59–87). Washington, DC: American Psychological Association.

Widiger, T. A., & Trull, T. S. (1992). Personality and psychopathology: An application of the five-factor model. *Journal of Personality, 60,* 363–393.

Widiger, T. A., Trull, T. J., Clarkin, J. F., Sanderson, C., & Costa, P. T. (2002). A description of the DSM-IV personality disorders with the five-factor model of personality. In P. T. Costa & T. A. Widiger (Eds.), *Personality disorders and the five factor model of personality* (2nd ed., pp. 89–99). Washington, DC: American Psychological Association.

Widiger, T. A., Verheul, R., & van den Brink, W. (1999). Personality and psychopathology. In L. Pervin & O. John (Eds.), *Handbook of personality* (2nd ed., pp. 347–366). New York: Guilford.

Wood, J. M., Garb, H. N., Lilienfeld, S. O., & Nezworski, M. T. (2002). Clinical assessment. *Annual Review of Psychology, 53,* 519–543.

Zuroff, D. C. (1994). Depressive personality styles and the five-factor model of personality. *Journal of Personality Assessment, 63,* 453–472.

Zuroff, D. C., Blatt, S. J., Sanislow, C. A., Bondi, C. M., & Pilkonis, P. A. (1999). Vulnerability to depression: Reexamining state dependence and relative stability. *Journal of Abnormal Psychology, 108,* 76–89.

9

The Social Costs of Stress

How Sex Differences in Stress Responses Can Lead to Social Stress Vulnerability and Depression in Women

Laura Cousino Klein, Elizabeth J. Corwin, and Rachel M. Ceballos

Men and women differ in their mental and physical disease vulnerability, and sex differences[1] appear to matter in the biological contributions to human health (Institute of Medicine, 2001). With respect to depression, approximately 12 million women in the United States experience depression annually, a rate of diagnosis that is at least twice the frequency of that found among men (National Mental Health Association, 2004). Although genetic factors, including sex, certainly play a role in the development of psychiatric illness, we now know that genotype alone does not singly determine whether an individual will develop a given psychiatric disorder (Plomin, DeFries, Craig, & McGuffin, 2003). For example, although there are well-known familial risks and sex differences in the prevalence rates of schizophrenia, depression, anxiety, and attention deficit/hyperactivity disorder (ADHD) (American Psychiatric Association, 1994), recent data suggest that environmental and biological factors, including stress, age, social support, and socioeconomic status, also contribute to the expression of these disorders (e.g., depression, anxiety, and posttraumatic stress disorder (PTSD); Kubzansky, Berkman, & Seeman, 2000; McEwen, 1998; Sapolsky, 1994; Vinokur, Price, & Caplan, 1996). A recent report proposes that there are sex differences in biological and behavioral responses to stress – the tend-and-befriend response (Taylor, Klein, Lewis, Gruenewald, Gurung, & Updegraff, 2000) – and suggests that social stressors may influence the

[1] The term *sex* refers to the classification of human and nonhuman animals as male or female according to the function of their reproductive organs. In this chapter we chose to use this term in comparison to the term *gender*, which refers to a person's self-representation as a male or female (Institute of Medicine, 2001).

The ideas developed for portions of this manuscript were supported by a Penn State University seed grant awarded to LCK and EJC. We thank Meredith S. Shiels and Lyn M. Weinberg for their assistance with preparing this chapter.

manifestation of some psychiatric disorders, such as depression, differently in men and women (Klein & Corwin, 2002).

This chapter examines how sex differences in the biobehavioral responses to stress may make women particularly vulnerable to social stressors and, consequently, depression. We begin with a discussion of stress physiology including an overview of a recent theory on sex differences in stress responses – the tend-and befriend response (Taylor et al., 2000) – which we think may provide a basis for understanding why women may be at risk for the depression-inducing effects of social stressors compared to men. We then move to a discussion of implications of the tend-and-befriend theory for vulnerability to depression among women, with a particular emphasis on oxytocin, and how this theory suggests differential biological and behavioral sensitivity to social stressors among women and men. We then discuss implications for research in the area of depression among women, including postpartum depression, and in novel pharmacologic treatment approaches for depression. We conclude this chapter with a brief discussion of the implications of this recent tend-and befriend theory on future research goals.

BIOLOGY OF STRESS

Stress is a process in which stressors threaten an organism's safety and well-being (Baum, Grunberg, & Singer, 1982). Stressors include a wide scope of events ranging from psychological (e.g., speech anxiety) and environmental (e.g., natural disasters, financial strain) to physical (e.g., exercise) and immunological (e.g., illness). Stressors may be pleasant ("eustress"), as in the case of a wedding or graduation, or unpleasant ("distress"), such as losing a job. Stressors also vary in duration (i.e., acute versus chronic) and frequency (e.g., daily vs. monthly). It is not always necessary for the stressor to be physically present to experience its biobehavioral effects, as in the case of PTSD. Although they occurred in the past, remembered stressors, such as the events of September 11 or the death of a spouse, are traumatic enough to provoke a stress response just from being recalled.

Stressors trigger a coordinated cascade of biological and behavioral responses that are designed to ensure the safety and well-being of the organism. This biobehavioral cascade begins with activation of the sympathetic (SNS) branch of the autonomic nervous system (ANS) (see Figure 9.1).

SNS stimulation results in release of the catecholamines epinephrine (EPI) and norepinephrine (NE) from the adrenal medulla, which in turn triggers a fight-or-flight response (Cannon, 1932). Designed to stimulate and sustain a rapid reaction to threat, these catecholamines and SNS activation result in elevated blood pressure, heart rate, and respiration and

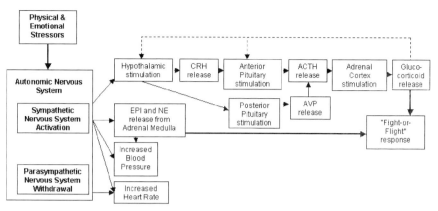

FIGURE 9.1. A simplified representation of the components of the autonomic nervous system (ANS) and hypothalamic-pituitary-adrenal (HPA) axis stress system. CRH = corticotrophin releasing hormone, ACTH = adrenocorticotropin hormone, EPI = epinephrine, NE = norepinephrine, AVP = arginine vasopressin. Solid lines represent direct or indirect stimulatory pathways. Dashed lines represent direct or indirect inhibitory pathways. (Adapted from Klein and Corwin, 2002.)

trigger arousal within seconds of stress exposure. Physiological reactions also include pupil dilation, blood vessel constriction, sweating, and dry mouth. Proposed by the physiologist Walter Cannon in the early 1900s, this fight-or-flight response is a prototypical mammalian stress response related to species survival in which an organism (e.g., a human) either fights or flees when faced with a threat (e.g., a tiger) (Cannon, 1932).

SNS activation by stress also stimulates the hypothalamic-pituitary-adrenal (cortex) axis (HPA axis) (see Stratakis & Chrousos, 1995, for review). HPA-axis activation (see Figure 9.1) results in release of corticotropin-releasing hormone (CRH) from the paraventricular nucleus of the hypothalamus, which stimulates release of adrenocorticotropin hormone (ACTH) from the anterior pituitary, as well as arginine vasopressin (AVP) from the posterior pituitary gland (de Wied, Diamant, & Fodor, 1993; Gibbs, 1986; Russell, 2002). AVP acts centrally to support the fight-or-flight response, whereas ACTH circulates to the adrenal cortex, located on top of the kidneys, to stimulate glucocorticoid release, including corticosteroids (e.g., cortisol, corticosterone). Corticosteroids themselves regulate continued HPA-axis function through a negative feedback loop by dampening further CRH release from the hypothalamus and ACTH release from the anterior pituitary gland. Ultimately, these stress hormones mobilize energy stores so that an organism can adapt to the stressor.

Unfortunately, repeated and chronic activation of the stress response can result in negative health consequences through dysregulation of the

HPA-axis negative feedback loop and alterations in immune function (see Segerstrom & Miller, 2004, for review) (see Figure 9.2). For example, prolonged stress exposure appears to uncouple corticosteroids from their ability to inhibit the further secretion of CRH and ACTH. This HPA-axis dysregulation leads to pathological overproduction of corticosteroids, which has been associated with destruction of hippocampal neurons, thereby leading to problems in learning, memory, and attention, as well as the development of psychiatric disorders such as episodes of repeated and severe depression (Heffelfinger & Newcomer, 2001; McEwen, 1998; Sapolsky, 1996, 2000a, 2000b). Similarly, dysregulation of the SNS- and HPA-axis stress response systems can lead to dysregulated centralized NE levels, a condition that has been associated with anxiety disorders such as generalized anxiety disorder and PTSD (Geracioti et al., 2001; Nutt, 2001; Ressler & Nemeroff, 2000), as well as depression (O'Connor, O'Halloran, & Shanahan, 2000; Ressler & Nemeroff, 2000). Pharmacotherapeutic agents that alter this neurotransmitter system are effective in reducing symptoms of generalized anxiety disorder (Ballenger, 2001) and depression (Goldstein, Lu, Detke, Wiltse, Mallinckrodt, & Demitrack, 2004; Mallinckrodt, Goldstein, Detke, Lu, Watkin, & Tran, 2003; Nemeroff & Owens, 2004).

With regard to stress and immune function, chronic stress also is associated with an increased susceptibility to infectious disease (e.g., influenza, common cold) and a decreased immune responsiveness to a variety of antigenic stimuli (e.g., flu vaccine, pneumococcal pneumonia vaccine; Chrousos, 1995; Cohen, Tyrell, & Smith, 1991; Glaser, Kiecolt-Glaser, Malarkey, & Sheridan, 1998; Glaser, Sheridan, Malarkey, MacCallum, & Kiecolt-Glaser, 2000; Kiecolt-Glaser, Glaser, Gravenstein, Malarkey, & Sheridan, 1996). Chronic stress functions to decrease immune responsiveness at least partially as a result of an increase in corticosteroid-mediated (via HPA-axis activation) inhibition of virtually all facets of immune responsiveness including B-cell and T-cell function, macrophage phagocytosis, and mast cell degranulation (see Chrousos, 1995, for review).

Stress also has been reported to alter plasma levels of various cytokines (Watkins, Nguyen, Lee, & Maier, 1999), especially interleukin (IL)-1 and IL-6 (Liao, Keiser, Scales, Kunkel, Kluger, 1995), two cytokines known to have predominately pro-inflammatory effects. These changes in cytokine levels are important because cytokines may play a role in depressive illness (see Anisman & Merali, 2002, 2003, for review). For example, IL-1β stimulates CRH (Gabay & Kushner, 1999), thereby contributing to depression (e.g., Dantzer et al., 1999), and IL-6 appears to be associated with symptoms of depression in humans (Glaser, Robles, Sheridan, Malarkey, & Kiecolt-Glaser, 2003, Maes et al., 1995, 1999). Taken together, experimental evidence suggests that neuroendocrine and immune responses to stress may contribute to the development of some psychiatric disorders, including depression (e.g., O'Connor et al., 2000).

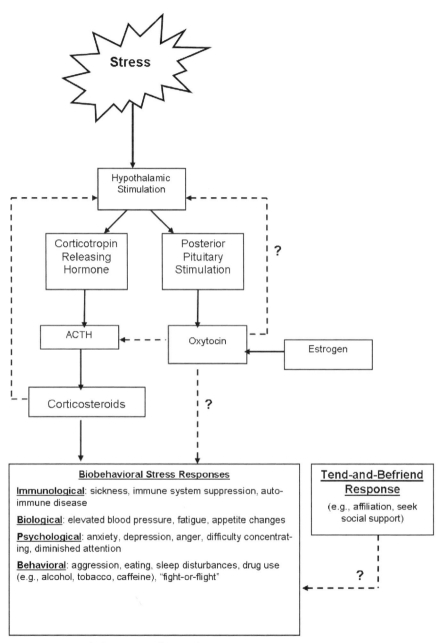

FIGURE 9.2. Biobehavioral effects of HPA-axis hormones released during stress, including effects of oxytocin released from the posterior pituitary gland. Solid lines represent direct or indirect stimulatory pathways. Dashed lines represent direct or indirect inhibitory pathways. ACTH = adrenocorticotropin hormone; "?" = hypothesized effects.

SEX DIFFERENCES IN BIOBEHAVIORAL RESPONSES TO STRESS

Studies with animals and humans indicate that males and females differ in their biological (e.g., neuroendocrine, immunological) and behavioral (e.g., eating, illicit drug consumption) responses to stressors (e.g., Cannon & St. Pierre, 1997; Gray, 1971, Greeno & Wing, 1994; Jezova, Jurankova, Mosnarova, Kriska, & Skultetyova, 1996; Kant et al., 1983; Klein, Faraday, Quigley, & Grunberg, 2004; Klein, Popke, & Grunberg, 1997). Taylor, Klein, and colleagues recently suggested that although both males and females display the traditional fight-or-flight response to some stressors, a behavioral pattern of tend-and-befriend might better explain females' stress responses in general (Taylor et al., 2000) (see Figure 9.2). The tend-and-befriend response to stress suggests that for females, a more adaptive response to some stressors may not be either fight or flight, but rather the use of social interactions to provide physical and perhaps psychological protection against a stressor. Designed to increase the likelihood of survival when faced with a threat, tending and befriending promotes safety and diminishes distress through creating and maintaining social networks (i.e., befriending), along with nurturing activities that protect the female and her offspring (i.e., tending). In this regard, the tend-and-befriend response differs distinctly from the traditional fight-or-flight response, which in reality would often leave the female and any dependent offspring unable to defeat or escape from the threat (Taylor et al., 2000).

The underlying neurobiological system of the tend-and-befriend response is proposed to differ from the fight-or-flight response in that it appears to be modulated by the posterior pituitary hormone oxytocin (see Figure 9.2) (Taylor et al., 2000). Oxytocin, although present in both males and females, is elevated in females throughout their reproductive years, and, at times, dramatically so (e.g., after giving birth). It has recently been demonstrated that oxytocin is associated with increased affiliative, maternal, bonding, and monogamy behavior (e.g., Bartels & Zeki, 2004; Carter & Altemus, 1997; Cho, DeVries, Williams, & Carter, 1999; McCarthy, 1995; Insel, Winslow, Wang, & Young, 1998; McCarthy & Altemus, 1997; Uvnas-Moberg, 1997) (for review, see Carter, 2003; Kendrick, 2000; Young, Lim, Gingrich, & Insel, 2001).

In regard to the role of oxytocin in the biobehavioral stress response, stress stimulates the hypothalamo-neurohypophysial system, which results in the concurrent release of oxytocin and AVP from the posterior pituitary gland (Engelmann, Ebner, Landgraf, Holsboer, & Wotjak, 1999; Engelmann, Wotjak, Ebner, & Landgraf, 2000; Gash & Boer, 1987; Hatton, 1990; Swaab et al., 1993). Oxytocin and AVP are synthesized in separate magnocellular neurons of the supraoptic nuclei (SON), paraventricular nuclei (PVN), and accessory nuclei of the hypothalamus (de Wied et al., 1993; Gash & Boer, 1987; Hatton, 1990; Swaab et al., 1993). Structurally similar, these nonapeptides differ from one another by only two amino acids (AVP contains phenylalanine and arginine, oxytocin does not), yet their biobehavioral stress effects appear to differ profoundly. Whereas AVP plays a stimulatory role in the HPA-axis response to stress (see Figure 9.1), oxytocin appears to dampen the HPA-axis stress response by inhibiting ACTH and, perhaps, CRH release (see Figure 9.2) (Gibbs, 1986; Legros, 2001), which could result in lowered corticosteroid levels (e.g., Heinrichs, Baumgartner, Kirschbaum, & Ehlert, 2003). Because oxytocin levels are higher in females than in males and the biobehavioral effects of oxytocin are enhanced in the presence of estrogen (McCarthy, 1995; McCarthy & Altemus, 1997), it is hypothesized that oxytocin's stress-reducing effects are more prominent in females (see Taylor et al., 2000, for review). In contrast, basal plasma and urine AVP concentrations are higher in males than in females, regardless of estrogen and progesterone changes across the estrous cycle phase (Crofton, Ratliff, Brooks, & Share, 1986; Crofton, Share, & Brooks, 1988). These AVP differences appear to be a function of greater AVP release from PVN and SON neurons Rhodes and Rubin 1999, although sex differences in the correlation between peripheral and central AVP levels in humans have not been reported in the literature (Rhodes & Rubin, 1999). An intriguing body of literature examining the role of AVP and oxytocin in the modulation of sympathetic and HPA-axis activation in response to stress is emerging and suggests important sex differences in biobehavioral responses to stress (e.g., Neumann, Torner, & Wigger, 2000; Neumann, Wigger, Torner, Holsboer, & Landgraf, 2000; Windle, Shanks, Lightman, & Ingram, 1997).

Increasing evidence with animals and humans suggests that oxytocin release actually may mediate many of the biobehavioral health effects of stress (see Figure 9.2) (Carter & Altemus, 1997; McCarthy, 1995; McCarthy & Altemus, 1997; Uvnas-Moberg, 1997), including reduced blood pressure (Light et al., 2000; although see Altemus et al., 2001, for further discussion), perceived stress levels (Mezzacappa & Katlin, 2002), anxiety (Heinrichs et al., 2003; McCarthy, 1995; McCarthy & Altemus, 1997; Turner, Altemus, Enos, Cooper, & McGuinness, 1999), aggression (Harmon, Huhman, Moore, & Albers, 2002; Lubin, Elliott, Black, & Johns, 2003), depression (Anderberg & Uvnas-Moberg, 2000), and, perhaps, improved attention and social memory (Brett & Baxendale, 2001; see Ferguson, Young, & Insel, 2002, for review).

IMPLICATIONS OF THE TEND-AND-BEFRIEND RESPONSE FOR WOMEN AND DEPRESSION

The potential implications of these biobehavioral sex differences in response to stress are exciting and open up several new directions in the behavioral medicine field. The tend-and-befriend theory suggests differential biological and behavioral sensitivity to social stressors and disease manifestation in response to stress. This differential stress sensitivity has implications for research in the area of depression among women, including postpartum depression (PPD), and in novel pharmacologic treatment approaches for depression.

Sensitivity to Social Stressors

One implication of these sex differences in biobehavioral stress responses is that men and women may differ in their vulnerability to social stressors such as family, work, and social relationships (Klein & Corwin, 2002; Taylor et al., 2000; Taylor, Dickerson, & Klein, 2002). More specifically, women may be particularly vulnerable to the health consequences (e.g., depression) of social stressors ranging from social isolation and interpersonal conflict to romantic and marital relationships (Hammen, 2003a, 2003b; Klein & Corwin, 2002; see Robles & Kiecolt-Glaser, 2003, for review). Indeed, women are more vulnerable than are men to developing classic signs of depression in response to stressful life events (Maciejewski, Prigerson, & Mazure, 2001; Sherrill et al., 1997; Weich, Sloggett, & Lewis, 2001; although see Sherrill et al., 1997, for additional review). These life stressors can include the loss of companionable social relationships (i.e., friend or other relative) or exposure to social stressors such as a home relocation or physical assault (Maciejewski et al., 2001). Social support also can moderate the stress–depression relationship. For other stressors, however, such as the death of a spouse or child, men and women do not differ in their risk for depression. These reports appear to be consistent with the tend-and-befriend response (Taylor et al., 2000, 2002), which suggests that sex-specific vulnerability may be linked to the social context of the stressor.

Research in the area of stress (e.g., life events), social support, and the course of depression in women and men continues to grow with mixed results; some studies report profound gender differences in the relationship among social support, stress, and depression, whereas others do not. From the perspective of the tend-and-befriend theory, we propose that one potential factor that could moderate these stress effects is the quality of the stressor, rather than the level or quality of social support available to the individual. More specifically, whether the stressor is related to the social domain may influence how any individual may respond to that challenge. For example, Stroud and colleagues reported that men display heightened HPA-axis reactivity (i.e., cortisol) to achievement stress in the laboratory compared to women, but that women display elevated HPA-axis reactivity to social (i.e., rejection) stress compared to men (Stroud, Salovey, & Epel, 2002). Likewise, relational aggression, which often results in social isolation, appears to be a particular stressor for women that can result in poor health outcomes such as bulimia (Werner & Crick, 1999). For example, adolescent girls who display high levels of relational aggression appear to be at particular risk for depression and loneliness (Crick & Grotpeter, 1995). Although there are behavioral and psychological explanations available to understand this finding, one could speculate that underlying differences in oxytocin and other stress-related hormones may influence relational aggressive behavior and associated levels of depression.

We could not find published data on the role that oxytocin or the tend-and-befriend response may play in moderating social stress and depression in women, though several scientific reviews and a few laboratory studies are beginning to implicate this system (e.g., Altemus, Roca, Galliven, Romanos, & Deuster, 2001; Swaab et al., 2000; Teicher, Andersen, Polcari, Anderson, & Navalta, 2002). Pilot research in our laboratory suggests that plasma levels of oxytocin are negatively correlated with self-reported levels of hostility, anxiety, and social isolation. In contrast, oxytocin levels are positively correlated with perceived social support among healthy women.

Interestingly, the effects of oxytocin and social support may not be limited to women. Taylor and colleagues suggest that although men have lower levels of oxytocin than do women, they too may benefit from the presence of oxytocin and display a tend-and-befriend response to certain stressors (Taylor et al., 2000, 2002). A clever paper by Heinrichs and colleagues found that social support and nasally administered oxytocin appear to suppress laboratory stress-induced cortisol and perceived stress levels among men (Heinrichs et al., 2003). It would be interesting to examine women's cortisol and perceived stress responses in a similar laboratory paradigm to explore potential sex differences in these effects. We encourage further research on gender differences in biobehavioral responses to social stress and effects on depression to better elucidate the potential mechanisms that underlie vulnerability to depression. One result of understanding these mechanisms is the potential to develop targeted interventions, such as novel pharmacotherapies.

Novel Pharmacotherapeutic Interventions

The two posterior pituitary hormones implicated in sex differences in stress reactivity and in the development of stress-related psychiatric illnesses, AVP and oxytocin, have received recent attention as an avenue of pharmacotherapeutic treatments for depression and anxiety (Holmes et al., 2003). Griebel and colleagues (2002) suggest that a novel vasopressin receptor antagonist, which blocks the AVP V1b receptor located throughout the limbic region, has strong anxiolytic effects in rats exposed to traumatic stress. Further, oxytocin appears to mediate the antidepressant effects of the serotonin-specific reuptake inhibitor, citalopram, in rats (Uvnas-Moberg, Bjokstrand, Hillegaart, & Ahlenius, 1999). Oxytocin may moderate depressive symptoms through interactions with the HPA axis, thus influencing levels of cortisol and immune cell function, both factors that are known to be associated with depression. This hypothesis partially is supported by a recent study that found significantly lower basal levels of oxytocin among women diagnosed with depression compared to those without a depression diagnosis (Anderberg & Uvnas-Moberg, 2000). Interestingly, lower levels of oxytocin also were found among women with high levels of self-reported stress and anxiety. Although these women also suffered from fibromyalgia, a syndrome believed to have immunological, neuroendocrinological, and psychological components, the data provide some insight into potential targets for therapeutic interventions. These data, coupled with Heinrichs and colleagues' recent report that oxytocin administration and social support may reduce biobehavioral stress reactivity among men (Heinrichs et al., 2003), suggest important individual differences in oxytocin functioning and an exciting new line of research investigation that appears to be uncharted to date.

Postpartum Depression and Maternal Health

In addition to supporting the tend-and-befriend response to stress, oxytocin plays a key role in maternal bonding and in breast-feeding (Bartels & Zeki, 2004; Matthiesen, Ransjo-Arvidson, Nissen, & Uvnas-Moberg, 2001; for review, see Carter, 2003; Kendrick, 2000). Pregnancy, labor, delivery, and the postpartum are periods of dynamic hormonal and immunological fluctuations that involve changes in the HPA axis and posterior pituitary regulation of oxytocin (e.g., Hendrick, Altshuler, & Suri, 1998; McCoy, Beal, & Watson, 2003; Schmeelk, Granger, Susman, & Chrousos, 1999; Tsigos & Chrousos, 2002). It has been proposed that dysregulation of these systems during these dramatic reproductive periods may contribute to the development of PPD (e.g., Hendrick et al., 1998; McCoy et al., 2003; Schmeelk et al., 1999; Tsigos & Chrousos, 2002; though see O'Hara, Schlechte, Lewis & Varner, 1991, for further review).

   PPD is a relatively common and potentially devastating disorder that develops in women within the first year after having given birth (Steiner, 1998; Wood, Thomas, Droppleman, & Meighan, 1997). Prevalence rates are high, with estimates of approximately 12% for major and 19% for minor PPD (Beck & Gable, 2001; Hopkins, Marcus, & Campbell, 1984; Whiffen, 1992). The implications of PPD extend beyond simply dampening the pleasure a new mother may feel after giving birth and in fact may result in serious risk to mother and infant alike. Even without overt risks to health, subtle interruptions in maternal–infant bonding may occur, with negative effects on infant behavioral and cognitive development (Beck, 1995;

Cooper & Murray-Kolb, 1998; Holden, 1991). These effects may last years beyond infancy (Cox, Puckering, Pound, & Mills, 1987; Field et al., 1988; Holden, 1991; Stein et al., 1991; Whiffen & Gotlib, 1989) and other family members, including partners and older children, may also suffer (Holden, 1991; Boath, Pryce, & Cox, 1998).

A number of psychosocial risk factors specific or common to women, including prenatal depression, child-care stress, low self-esteem, life events, and poor social support have been identified as contributing to the development of PPD (Beck & Gable, 2001; O'Hara et al., 1991). Some of these factors can be reduced with interventions; others, for example, poor social support and life events, are more difficult to address. Recently, we identified self-report of severe postpartum fatigue (PPF) as a physiological variable whose presence early in the postpartum significantly increases a woman's risk of developing symptoms of PPD (Bozoky & Corwin, 2002; Corwin, Brownstead, Barton, Heckard, & Morin, in press). Our results indicate that although all new mothers experience fatigue to some extent after giving birth, mothers who report continuing and unremitting fatigue as early as 1–2 weeks postpartum are significantly more likely to develop symptoms of postpartum depression than are mothers whose fatigue begins to fall as they moved beyond the first week postpartum. Although fatigue is known to be a symptom of depression, the two are not identical and women who suffer from PPD can differentiate between the two (Milligan, Lenz, Parks, Pugh, & Kitzman, 1996). These studies and others examining fatigue and postpartum depression (Troy, 1999; Whiffen, 1992) provide incentive for the development of interventions aimed at helping women reduce their postpartum fatigue level and identifying those who are at risk of depression by an early query as to their fatigue level. Although thyroid dysfunction and anemia may contribute to PPF and PPD as well (Cooper & Murray-Kolb, 1998; Corwin, Murray-Kolb, & Beard, 2003), we and others propose that dysregulated stress neuroendocrine and immune physiology might also be interrelated during the postpartum period to stimulate the development of depression (Hendrick et al., 1998; McCoy et al., 2003; Schmeelk et al., 1999; Tsigos & Chrousos, 2002). The tend-and-befriend model would suggest that social support interventions and therapies designed to increase oxytocin and facilitate infant bonding could help alleviate the onset and progression of PPD and PPF. Interestingly, the few published studies of social support, infant massage, and guided infant bonding, which should also increase oxytocin levels and reduce HPA-axis activity, demonstrate improved maternal and infant outcomes (Field, 1998, 2002; Matthiesen et al., 2001), including reduced reports of maternal depression (Field et al., 2000; Glover, Onozawa, & Hodgkinson, 2002; Scott, Klaus, & Klaus, 1999; see Federenko & Wadhwa, 2004, for review). It may be that oxytocin mediates these effects. Additional research is needed to further explore this relationship.

Tend-and-Befriend, Social Support, Immune Function, and Health Outcomes

One implication of the tend-and-befriend theory is that oxytocin dampens stress reactivity, thereby improving health outcomes for women compared with men (Carter, 2003; Klein & Corwin, 2002; Taylor et al., 2000). It also is possible that social support and affiliation under stress influence oxytocin and the immune system and, consequently, health outcomes. The idea that social support and social isolation can modify immune, neuroendocrine, physiological, and psychological reactivity recently has been proposed (for review, see Hawkley & Cacioppo, 2003; Robles & Kiecolt-Glaser, 2003), although the exact mechanisms are not well understood. A literature search revealed very limited research on the role that oxytocin may play in mediating the health benefits of social support, or health consequences of social isolation (e.g., loneliness), although links between oxytocin and the immune system have been reported (e.g., Savino, Arzt, & Dardenne, 1999). In fact, a MedLine search using the key words "oxytocin AND immune AND social support" revealed no publications at all. Substituting "social isolation," "loneliness," or "depression" for "social support" yielded similar results. This limited search suggests minimal research investigations of the interactions among stress, social support, depression, and immunity with regard to gender differences in stress reactivity. Given the particular role that oxytocin may play in mediating social and affiliative behaviors, stress reactivity, and integrative functioning (e.g., Carter, 2003), it seems reasonable to propose further research investigations into these areas to better understand social and biological contributors to mental and physical health functioning among women.

CONCLUSIONS

Although speculative at this time, the observation that there are sex differences in the biobehavioral effects of stress may help to understand women's particular vulnerability to social stress and depression. Examining sex differences in biobehavioral responses to stress provides a unique opportunity for the development of new hypotheses with respect to understanding the etiology of depression, the role that stress may play in the development of depression among women, and novel pharmacologic treatments for depression. On a broader level, assuming that social stressors may affect women more than they do men, perhaps diagnostic criteria for other psychiatric disorders should include questions about disruptions in social relationships. In contrast, this new perspective also may provide insight regarding the manifestation of psychiatric disorders that are less frequently diagnosed in men, such as depression. For example, perhaps depression is manifested differently among men and is revealed through behaviors such as fighting with others, including coworkers, friends, or spouse (e.g., fighting response). These fight-or-flight behaviors are not captured in current diagnostic criteria and are worth considering and investigating further. Further inquiry into the importance of sex-based differences in the behavioral manifestation of some psychiatric illnesses, including depression, may provide the opportunity for earlier diagnosis and treatment for both women and men.

Stress is a relevant construct to consider in the study of sex-based differences in the manifestation of depression and the role that social relationships may play in the progression of this illness. Stress also is a useful tool to determine biological and behavioral underpinnings of depression and social relationships in women and men. A better understanding of these sex differences could lead to more effective diagnosing and treatment approaches by shifting our perspective on the types of behaviors and biological responses we should expect to see from women and men. Although the tend-and-befriend response may occur more frequently among women (Taylor et al., 2000), it is important to determine the circumstances in which men display this response and vice versa for women responding to a stressor with a fight-or-flight response. It also may be that some men might experience health benefits from engaging in a tend-and-befriend response over a fight-or-flight response, whereas women might benefit from coping with some stressors with a fight-or-flight response when appropriate or necessary. Both domains of biobehavioral responses to stress are important to consider, and new research already suggests differential biological mechanisms for psychiatric illnesses such as major depression (Swaab, Fliers, Hoogendijk, Veltman, & Zhou, 2000). We encourage further biobehavioral studies that include women and men, as well as stress as a comparison behavior, to better understand mechanisms that may underlie sex differences in the biological, social, and behavioral contributors to depression. The present chapter provides a framework and some examples to highlight and encourage discussion about sex-based prevalence rates in depression.

References

Altemus, M., Redwine, L. S., Leong, Y. M., Frye, C. A., Porges, S. W., & Carter, C. S. (2001). Responses to laboratory psychosocial stress in postpartum women. *Psychosomatic Medicine, 63*, 814–821.

Altemus, M., Roca, C., Galliven, E., Romanos, C., & Deuster, P. (2001). Increased vasopressin and adrenocorticotropin responses to stress in the midluteal phase of the menstrual cycle. *Journal of Clinical Endocrinology and Metabolism, 86*, 2525–2530.

American Psychiatric Association. (1994). *Diagnostic and statistical manual of mental disorders* (4th ed.). Washington, DC: American Psychiatric Association.

Anderberg, U. M., & Uvnas-Moberg, K. (2000). Plasma oxytocin levels in female fibromyalgia syndrome patients. *Zeitschrift fur Rheumatologie, 59*, 373–379.

Anisman, H., & Merali, Z. (2002). Cytokines, stress, and depressive illness. *Brain, Behavior, and Immunity, 16*, 513–524.

Anisman, H., & Merali, Z. (2003). Cytokines, stress and depressive illness: Brain-immune interactions. *Annals of Medicine, 35*, 2–11.

Ballenger, J. C. (2001). Overview of different pharmacotherapies for attaining remission in generalized anxiety disorder. *Journal of Clinical Psychiatry, 62* (Suppl. 19), 11–19.

Bartels, A., & Zeki, S. (2004). The neural correlates of maternal and romantic love. *NeuroImage, 21*, 1155–1166.

Baum, A., Grunberg, N. E., & Singer, J. E. (1982). The use of psychological and neuroendocrinological measurements in the study of stress. *Health Psychology, 1*, 217–236.

Beck, C. T. (1995). The effects of postpartum depression on maternal-infant interaction: A meta-analysis. *Nursing Research, 44*, 298–304.

Beck, C. T., & Gable, R. K. (2001). Further validation of the postpartum depression screening scale. *Nursing Research, 50*, 201–209.

Boath, E. H., Pryce, A. J., & Cox, J. L. (1998). Postnatal depression: The impact on the family. *Journal of Reproductive and Infant Psychology, 16*, 199–203.

Bozoky, I., & Corwin, E. J. (2002). Fatigue as a predictor of postpartum depression. *Journal of Obstetric Gynecologic, and Neonatal Nursing, 31*, 436–443.

Brett, M., & Baxendale, S. (2001). Motherhood and memory: A review. *Psychoneuroendocrinology, 26*, 339–362.

Cannon, W. B. (1932). *The wisdom of the body*. New York: Norton.

Cannon, J. G., & St. Pierre, B. (1997). Gender differences in host defense mechanisms. *Journal of Psychiatric Research, 31*, 99–113.

Carter, C. S. (2003). Developmental consequences of oxytocin. *Physiology & Behavior, 79*, 383–397.

Carter, C. S., & Altemus, M. (1997). Integrative functions of lactational hormones in social behavior and stress management. *Annals of the New York Academy of Sciences, 807*, 164–174.

Cho, M. M., DeVries, A. C., Williams, J. R., & Carter, C. S. (1999). The effects of oxytocin and vasopressin on partner preferences in male and female prairie voles (Microtus ochrogaster). *Behavioral Neuroscience, 113*, 1071–1079.

Chrousos, G. P. (1995). The hypothalamic-pituitary-adrenal axis and immune-mediated inflammation. *New England Journal of Medicine, 332*, 1351–1362.

Cohen, S., Tyrell, D. A., & Smith, A. P. (1991). Psychological stress and susceptibility to the common cold. *New England Journal of Medicine, 325*, 606–612.

Cooper, P., & Murray-Kolb, L. (1998) Postnatal depression. *British Medical Journal, 316*, 1884–1886.

Corwin, E. J., Brownstead, J., Barton, N., Heckard, S., & Morin, K. (in press). The impact of fatigue on the development of postpartum depression. *Journal of Obstetric, Gynecological, and Neonatal Nursing*.

Corwin, E. J., Murray-Kolb, L., & Beard, J. L. (2003). Low hemoglobin level is a risk factor for postpartum depression. *Journal of Nutrition, 133*, 4139–4142.

Cox, A. D., Puckering, C., Pound, A., & Mills, M. (1987). The impact of maternal depression in young children. *Journal of Child Psychology and Psychiatry, and Allied Disciplines, 28*, 917–928.

Crick N. R., & Grotpeter J. K. (1995). Relational aggression, gender, and social-psychological adjustment. *Child Development, 66*, 710–722.

Crofton, J. T., Ratliff, D. L., Brooks, D. P., & Share, L. (1986). The metabolic clearance rate of and pressor responses to vasopressin in male and female rats. *Endocrinology, 118*, 1777–1781.

Crofton, J. T., Share, L., & Brooks, D. P. (1988). Pressor responsiveness to and secretion of vasopressin during the estrous cycle. *American Journal of Physiology, 255*, R1041–R1048.

Dantzer, R., Aubert, A., Bluthe, R., Gheusi, G., Cremona, S., Laye, S., et al. (1999). Mechanisms of the behavioural effects of cytokines. *Advances in Experimental Medicine and Biology, 461*, 83–106.

de Wied, D., Diamant, M., & Fodor, M. (1993). Central nervous effects of neurohypophysial hormones and related peptides. *Neuroendocrinology, 14*, 251–302.

Engelmann, M., Ebner, K., Landgraf, R., Holsboer, F., & Wotjak, C. T. (1999). Emotional stress triggers intrahypothalamic but not peripheral release of oxytocin in male rats. *Journal of Neuroendocrinology, 11*, 867–872.

Engelmann, M., Wotjak, C. T., Ebner, K., & Landgraf, R. (2000). Behavioural impact of intraseptally released vasopressin and oxytocin in rats. *Experimental Physiology, 85S*, 125S–130S.

Federenko, I. S., & Wadhwa, P. D. (2004). Women's mental health during pregnancy influences fetal and infant developmental and health outcomes. *CNS Spectrums, 9*, 198–206.

Ferguson, J. N., Young, L. J., & Insel, T. R. (2002). The neuroendocrine basis of social recognition. *Frontiers in Neuroendocrinology, 23*, 200–224.

Field, T. (1998). Early interventions for infants of depressed mothers. *Pediatrics, 102*, 1305–1310.

Field, T. (2002). Massage therapy. *The Medical Clinics of North America, 86*, 163–171.

Field, T., Healy, B., Goldstein, S., Perry, S., Bendell, D., Schanberg, S., et al. (1988). Infants of depressed mothers show "depressed" behavior even with nondepressed adults. *Child Development, 59*, 1569–1579.

Field, T., Pickens, J., Prodromidis, M., Malphurs, J., Fox, N., Bendell, D., et al. (2000). Targeting adolescent mothers with depressive symptoms for early intervention. *Adolescence, 35*, 381–414.

Gabay, C., & Kushner, I. (1999). Acute-phase proteins and other systemic responses to inflammation. *The New England Journal of Medicine, 340*, 448–454.

Gash, D. M., & Boer, G. J. (1987). *Vasopressin, principles and properties*. New York: Plenum.

Geracioti, T. D., Baker, D. G., Ekhator, N. N., West, S. A., Hill, K. K., Bruce, A. B., et al. (2001). CSF norepinephrine concentrations in posttraumatic stress disorder. *American Journal of Psychiatry, 158*, 1227–1230.

Gibbs, D. M. (1986). Vasopressin and oxytocin: Hypothalamic modulators of the stress response: A review. *Psychoneuroendocrinology, 11*, 131–140.

Glaser, R., Kiecolt-Glaser, J. K., Malarkey, W. B., & Sheridan, J. F. (1998). The influence of psychological stress on the immune response to vaccines. *Annals of the New York Academy of Sciences, 840*, 649–655.

Glaser, R., Robles, T. F., Sheridan, J., Malarkey, W. B., & Kiecolt-Glaser, J. K. (2003). Mild depressive symptoms are associated with amplified and prolonged inflammatory responses after influenza virus vaccination in older adults. *Archives of General Psychiatry, 60,* 1009–1014.

Glaser, R., Sheridan, J., Malarkey, W. B., MacCallum, R. C., & Kiecolt-Glaser, J. K. (2000). Chronic stress modulates the immune response to a pneumococcal pneumonia vaccine. *Psychosomatic Medicine, 62,* 804–807.

Glover, V., Onozawa, K., & Hodgkinson, A. (2002). Benefits of infant massage for mothers with postnatal depression. *Seminars in Neonatology, 7,* 495–500.

Goldstein, D. J., Lu, Y., Detke, M. J., Wiltse, C., Mallinckrodt, C., & Demitrack, M. A. (2004). Duloxetine in the treatment of depression: A double-blind placebo-controlled comparison with Paroxetine. *Journal of Clinical Psychopharmacology, 24,* 389–399.

Gray, J. (1971). Sex differences in emotional behavior in mammals including man: Endocrine bases. *Acta Psycholgie, 35,* 29–46.

Greeno, C. G., & Wing, R. R. (1994). Stress-induced eating. *Psychological Bulletin, 115,* 444–464.

Griebel, G., Simiand, J., Serradeil-Le Gal, C., Wagnon, J., Pascal, M., Scatton, B., et al. (2002). Anxiolytic- and antidepressant-like effects of the non-peptide vasopressin V1b receptor antagonist, SSR149415, suggest an innovative approach for the treatment of stress-related disorders. *Proceedings of the National Academy of Sciences, 99,* 6370–6375.

Hammen, C. (2003a). Interpersonal stress and depression in women. *Journal of Affective Disorders, 74,* 49–57.

Hammen, C. (2003b). Social stress and women's risk for recurrent depression. *Archives of Womens Mental Health, 6,* 9–13.

Harmon, A. C., Huhman, K. L., Moore, T. O., & Albers, H. E. (2002). Oxytocin inhibits aggression in female Syrian hamsters. *Journal of Neuroendocrinology, 14,* 963–969.

Hatton, G. I. (1990). Emerging concepts of structure-function dynamics in adult brain: The hypothalamo-neurohypophysial system. *Progressive Neurobiology, 34,* 437–504.

Hawkley, L. C., & Cacioppo, J. T. (2003). Loneliness and pathways to disease. *Brain Behavior and Immunity, 17,* S98–105.

Heffelfinger, A. K., & Newcomer, J. W. (2001). Glucocorticoid effects on memory function over the human life span. *Development and Psychopathology, 13,* 491–513.

Heinrichs, M., Baumgartner, T., Kirschbaum, C., & Ehlert, U. (2003). Social support and oxytocin interact to suppress cortisol and subjective responses to psychosocial stress. *Biological Psychiatry, 54,* 1389–1398.

Hendrick, V., Altshuler, L. L., & Suri, R. (1998). Hormonal changes in the postpartum and implications for postpartum depression. *Psychosomatics, 39,* 93–101.

Holden, J. M. (1991). Postnatal depression: its nature, effects, and identification using the Edinburgh Postnatal Depression scale. *Birth, 18,* 211–221.

Holmes, A., Heilig, M., Rupniak, N. M., Steckler, T., & Griebel, G. (2003). Neuropeptide systems as novel therapeutic targets for depression and anxiety disorders. *Trends in Pharmacological Sciences, 24,* 580–588.

Hopkins, J., Marcus, M., & Campbell, S. B. (1984). Postpartum depression: A critical review. *Psychological Bulletin, 95,* 498–515.

Insel, T. R., Winslow, J. T., Wang, Z., & Young, L. J. (1998). Oxytocin, vasopressin, and the neuroendocrine basis of pair bond formation. In H. H. Zingg, C. W. Bourque, & D. G. Bichet (Eds.). *Vasopressin and oxytocin: Molecular, cellular, and clinical advances* (pp. 215–224). New York: Plenum Press.

Institute of Medicine. (2001). *Exploring the biological contributions to human health: Does sex matter?* Washington, DC: National Academy Press.

Jezova, D., Jurankova, E., Mosnarova, A., Kriska, M., & Skultetyova, I. (1996). Neuroendocrine response during stress with relation to gender differences. *Acta Neurobiologae Experimentalis, 56*, 779–785.

Kant, G. J., Lenox, R. H., Bunnell, B. N., Mougey, E. H., Pennington, L. L., & Meyerhoff, J. L. (1983). Comparison of the stress response in male and female rats: Pituitary cyclic AMP and plasma prolactin, growth hormone and corticosterone. *Psychoneuroendocrinology, 8*, 421–28.

Kendrick, K. M. (2000). Oxytocin, motherhood and bonding. *Experimental Physiology, 85*, 111S–124S.

Kiecolt-Glaser, J. K., Glaser, R., Gravenstein, S., Malarkey, W. B., & Sheridan, J. (1996). Chronic stress alters the immune response to influenza virus vaccine in older adults. *Proceedings of the National Academy of Sciences, 93*, 3043–3047.

Klein, L. C., & Corwin, E. J. (2002). Seeing the unexpected: How sex differences in stress responses may provide a new perspective on the manifestation of psychiatric disorders. *Current Psychiatry Reports, 4*, 441–448.

Klein, L. C., Faraday, M. M., Quigley, K. S., & Grunberg, N. E. (2004). Gender differences in biobehavioral aftereffects of stress on eating, frustration, and cardiovascular responses. *Journal of Applied Social Psychology, 34*, 538–562.

Klein, L. C., Popke, E. J., & Grunberg, N. E. (1997). Sex differences in effects of predictable and unpredictable footshock on fentanyl self-administration in rats. *Experiments in Clinical Psychopharmacology, 5*, 99–106.

Kubzansky, L. D., Berkman, L. F., & Seeman, T. E. (2000). Social conditions and distress in elderly persons: Findings from the MacArthur Studies of Successful Aging. *Journal of Gerontology: Psychological Sciences, 55*, 238–246.

Legros, J. J. (2001). Inhibitory effect of oxytocin on corticotrope function in humans: Are vasopressin and oxytocin ying-yang neurohormones? *Psychoneuroendocrinology, 26*, 649–655.

Liao, J., Keiser, J., Scales, W. E., Kunkel, S. L., & Kluger, M. J. (1995). Role of corticosterone in TNF and IL-6 production in isolated perfused rat liver. *American Journal of Physiology, 268*, R6999–R7006.

Light, K. C., Smith, T. E., Johns, J. M., Brownley, K. A., Hofheimer, J. A., & Amico, J. A. (2000). Oxytocin responsivity in mothers of infants: A preliminary study of relationships with blood pressure during laboratory stress and normal ambulatory activity. *Health Psychology, 19*, 560–567.

Lubin, D. A., Elliott, J. C., Black, M. C., & Johns, J. M. (2003). An oxytocin antagonist infused into the central nucleus of the amygdala increases maternal aggressive behavior. *Behavioral Neuroscience, 117*, 195–201.

Maciejewski, P. K., Prigerson, H. G., & Mazure, C. M. (2001). Sex differences in event-related risk for major depression. *Psychological Medicine, 31*(4), 593–604.

Maes, M., Meltzer, H. Y., Bosmans, E., Bergmans, R., Vandoolaeghe, E., Ranjan, R., et al. (1995). Increased plasma concentrations of interleukin-6, soluble interleukin-6, soluble interleukin-2, and transferrin receptor in major depression. *Journal of Affective Disorders, 34,* 301–309.

Maes, M., Van Bockstaele, D. R., Gastel, A., Song, C., Schotte, C., Neels, H., et al. (1999). The effects of psychological stress on leukocyte subset distribution in humans: evidence of immune activation. *Neuropsychobiology, 39,* 1–9.

Mallinckrodt, C. H., Goldstein, D. J., Detke, M. J., Lu, Y., Watkin, J. G., & Tran, P. V. (2003). Duloxetine: A new treatment for the emotional and physical symptoms of depression. *Primary Care Companion to the Journal of Clinical Psychiatry, 5,* 19–28.

Matthiesen, A. S., Ransjo-Arvidson, A. B., Nissen, E., & Uvnas-Moberg, K. (2001). Postpartum maternal oxytocin release by newborns: Effects of infant hand massage and sucking. *Birth, 28,* 13–19.

McCarthy, M. M. (1995). Estrogen modulation of oxytocin and its relation to behavior. In R. Ivell, & J. Russell (Eds.), *Oxytocin: Cellular and molecular approaches in medicine and research* (pp. 235–242). New York: Plenum Press.

McCarthy, M. M., & Altemus, M. (1997). Central nervous system actions of oxytocin and modulation of behavior in humans. *Molecular Medicine Today, 3,* 269–275.

McCoy, S. J., Beal, J. M., & Watson, G. H. (2003). Endocrine factors and postpartum depression. A selected review. *Journal of Reproductive Medicine, 48,* 402–408.

McEwen, B. S. (1998). Protective and damaging effects of stress mediators. *New England Journal of Medicine, 338,* 171–179.

Mezzacappa, E. S., & Katlin, E. S. (2002). Breast-feeding is associated with reduced perceived stress and negative mood in mothers. *Health Psychology, 21,* 187–193.

Milligan, R., Lenz, E. R., Parks, P. L., Pugh, L. C., & Kitzman, H. (1996). Postpartum fatigue: clarifying a concept. *Scholarly Inquiry for Nursing Practice, 10,* 279–291.

National Mental Health Association. (2004). Depression in women. Retrieved July 2004, from http://www.nmha.org/infoctr/factsheets/23.cfm.

Nemeroff, C. B., & Owens, M. J. (2004). Pharmacologic differences among the SSRIs: Focus on monoamine transporters and the HPA axis. *CNS Spectrums, 9,* 23–31.

Neumann, I. D., Torner, L., & Wigger, A. (2000). Brain oxytocin: Differential inhibition of neuroendocrine stress responses and anxiety-related behaviour in virgin, pregnant and lactating rats. *Neuroscience, 95,* 567–575.

Neumann, I. D., Wigger, A., Torner, L., Holsboer, F., & Landgraf, R. (2000). Brain oxytocin inhibits basal and stress-induced activity of the hypothalamo-pituitary-adrenal axis in male and female rats: Partial action within the paraventricular nucleus. *Journal of Neuroendocrinology, 12,* 235–243.

Nutt, D. J. (2001). Neurobiological mechanisms in generalized anxiety disorder. *Journal of Clinical Psychiatry, 62,* 22–27.

O'Connor, T. M., O'Halloran, D. J., & Shanahan, F. (2000). The stress response and the hypothalamic-pituitary-adrenal axis: from molecule to melancholia. *QJM: Monthly Journal of the Association of Physicians, 93,* 323–333.

O'Hara, M. W., Schlechte, J. A., Lewis, D. A., & Varner, M. W. (1991). Controlled prospective study of postpartum mood disorders: psychological, environmental, and hormonal variables. *Journal of Abnormal Psychology, 100,* 63–73.

Plomin, R., DeFries, J. C., Craig, I. W., & McGuffin, P. (2003). *Behavioral genetics in the postgenomic era.* Washington, DC: American Psychological Association.

Ressler, K. J., & Nemeroff, C. B. (2000). Role of serotonergic and noradrenergic systems in the pathophysiology of depression and anxiety disorders. *Depression and Anxiety, 12,* 2–19.

Rhodes, M. E., & Rubin, R. T. (1999). Functional sex differences (sexual diergism) of central nervous system cholinergic systems, vasopressin, and hypothalamic-pituitary-adrenal axis activity in mammals: A selective review. *Brain Research Reviews, 30,* 135–152.

Robles, T. F., & Kiecolt-Glaser, J. K. (2003). The physiology of marriage: pathways to health. *Physiology & Behavior, 79,* 409–416.

Russell, J. A. (2002). Editorial: Neuroendocrinology with feeling. *Journal of Neuroendocrinology, 14,* 1–3.

Sapolsky, R. M. (1994). *Why zebras don't get ulcers: A guide to stress, stress-related disease, and coping.* New York: Freeman.

Sapolsky R. M. (1996). Why stress is bad for your brain. *Science, 273,* 749–750.

Sapolsky, R. M. (2000a). Glucocorticoids and hippocampal atrophy in neuropsychiatric disorders. *Archives of General Psychiatry, 57,* 925–935.

Sapolsky, R. M. (2000b). The possibility of neurotoxicity in the hippocampus in major depression: a primer on neuron death. *Biological Psychiatry, 48,* 755–765.

Savino, W., Arzt, E., & Dardenne, M. (1999). Immunoneuroendocrine connectivity: The paradigm of the thymus-hypothalamus/pituitary axis. *Neuroimmunomodulation, 6,* 126–136.

Schmeelk, K. H., Granger, D. A., Susman, E. J., & Chrousos, G. P. (1999). Maternal depression and risk for postpartum complications: Role of prenatal corticotropin-releasing hormone and interleukin-1 receptor antagonist. *Behavioral Medicine, 25,* 88–94.

Scott, K. D., Klaus, P. H., & Klaus, M. H. (1999). The obstetrical and postpartum benefits of continuous support during childbirth. *Journal of Women's Health and Gender-Based Medicine, 8,* 1257–1264.

Segerstrom, S. C., & Miller, G. E. (2004). Psychological stress and the human immune system: A meta-analytic study of 30 years of inquiry. *Psychological Bulletin, 130,* 601–630.

Sherrill, J. T., Anderson, B., Frank, E., Reynolds, C. F. 3rd, Tu, X. M., Patterson, D., et al. (1997). Is life stress more likely to provoke depressive episodes in women than in men? *Depression and Anxiety, 6,* 95–105.

Stein, A., Gath, D. H., Bucher, J., Bond, A., Day, A., & Cooper, P. J. (1991). The relationship between post-natal depression and mother-child interaction. *British Journal of Psychiatry, 158,* 46–52.

Steiner, M. (1998). Perinatal mood disorders: Position paper. *Psychopharmacology Bulletin, 34,* 301–306.

Stratakis, C. A., & Chrousos, G. P. (1995). Neuroendocrinology and pathophysiology of the stress system. In G. P. Chrousos, R. McCarty, K. Pacák, G. Cizza, E. Sternberg, P. W. Gold, & R. Kvetňanský (Eds.), *Stress: Basic mechanisms and clinical implications* (Vol. 771, pp. 1–18). New York: New York Academy of Sciences.

Stroud, L. R., Salovey, P., & Epel E. S. (2002). Sex differences in stress responses: social rejection versus achievement stress. *Biological Psychiatry, 15,* 318–327.

Swaab, D. F., Fliers, E., Hoogendijk, W. J. G., Veltman, D. J., & Zhou, J. N. (2000). Interaction of prefrontal cortical and hypothalamic systems in the pathogenesis of depression. *Progress in Brain Research, 126,* 369–396.

Swaab, D. F., Hofman, M. A., Lucassen, P. J., Purba, J. S., Raadsheer, F. C., & Van de Nes, J. A. P. (1993). Functional neuroanatomy and neuropathology of the human hypothalamus. *Anatomy and Embryology, 187,* 317–330.

Taylor, S. E., Dickerson, S. S., & Klein, L. C. (2002). Toward a biology of social support. In C. R. Snyder, & S. J. Lopez (Eds). *Handbook of positive psychology* (pp. 556–569). New York: Oxford University Press.

Taylor, S. E., Klein, L. C., Lewis, B. P., Gruenewald, T. L., Gurung, R. A. R., & Updegraff, J. A. (2000). Female responses to stress: Tend-and-befriend, not fight-or-flight. *Psychological Review, 107,* 411–429.

Teicher, M. H., Andersen, S. L., Polcari, A., Anderson, C. M., & Navalta, C. P. (2002). Developmental neurobiology of childhood stress and trauma. *Psychiatric Clinics of North America, 25,* 397–426.

Troy, N. W. (1999). A comparison of fatigue and energy levels at 6 weeks and 14 to 19 months postpartum. *Clinical Nursing Research, 8,* 135–152.

Tsigos, C., & Chrousos, G. P. (2002). Hypothalamic-pituitary-adrenal axis, neuroendocrine factors and stress. *Journal of Psychosomatic Research, 53,* 865–871.

Turner, R. A., Altemus, M., Enos, T., Cooper, B., & McGuinness, T. (1999). Preliminary research on plasma oxytocin in healthy, normal cycling women investigating emotion and interpersonal distress. *Psychiatry, 62,* 97–113.

Uvnas-Moberg, K. (1997). Oxytocin linked antistress effects–the relaxation and growth response. *Acta Psychologica Scandinavica, 640,* 38–42.

Uvnas-Moberg, K., Bjokstrand, E., Hillegaart, V., & Ahlenius, S. (1999). Oxytocin as a possible mediator of SSRI-induced antidepressant effects. *Psychopharmacology, 142,* 95–101.

Vinokur, A. D., Price, R. H., & Caplan, R. D. (1996). Hard times and hurtful partners: How financial strain affects depression and relationship satisfaction of unemployed persons and their spouses. *Journal of Personality and Social Psychology, 71,* 166–179.

Watkins, L. R., Nguyen, K. T., Lee, J. E., & Maier, S. F. (1999). Dynamic regulation of proinflammatory cytokines. *Advances in Experimental Medicine and Biology, 461,* 153–178.

Weich, S., Sloggett, A., & Lewis, G. (2001). Social roles and the gender difference in rates of the common mental disorders in Britain: A 7-year, population-based cohort study. *Psychological Medicine, 31,* 1055–1064.

Werner, N. E., & Crick, N. R. (1999). Relational aggression and social-psychological adjustment in a college sample. *Journal of Abnormal Psychology, 108,* 615–623.

Whiffen, V. E. (1992). Is postpartum depression a distinct diagnosis? *Clinical Psychology Review, 12,* 495–508.

Whiffen, V. E., & Gotlib, I. H. (1989). Infants of postpartum depressed mothers: Temperament and cognitive status. *Journal of Abnormal Psychology, 98,* 274–279.

Windle, R. J., Shanks, N., Lightman, S. L., & Ingram, C. D. (1997). Central oxytocin administration reduces stress-induced corticosterone release and anxiety behavior in rats. *Endocrinology, 138,* 2829–2834.

Wood, A. F., Thomas, S. P., Droppleman, P. G., & Meighan, M. (1997). The downward spiral of postpartum depression. *MCN: The American Journal of Maternal Child Nursing, 22,* 308–316.

Young, L. J., & Lim, M. M., Gingrich, B., & Insel, T. R. (2001). Cellular mechanisms of social attachment. *Hormones and Behaviors, 40,* 133–138.

10

Marriage and Depression

Mark A. Whisman, Lauren M. Weinstock,
and Natalie Tolejko

As has been noted throughout this volume, women are nearly twice as likely as men to experience depression in their lifetime (Nolen-Hoeksema, 1987). Cross-national studies have demonstrated that this is true in both Western and non-Western countries (Weissman et al., 1993). As has been discussed elsewhere (e.g., McGrath, Keita, Strickland, & Russo, 1990; Nolen-Hoeksema, 1987) and throughout the chapters of this volume, there are numerous explanations that have been advanced to account for this gender difference.

This chapter is devoted to evaluating a social explanation for gender differences in depression; namely, that this gender difference could be due, in part, to gender differences in experiences in marriage. It has been shown that the potential exposure to marital distress and other negative relationship outcomes such as divorce is very widespread. For example, it has been estimated that within the United States, 9 out of every 10 Americans marry at least once in their lifetime (Kreider & Fields, 2001), that many people who do not marry will live with a partner in nonmarital familial relationships (Seltzer, 2000), that approximately 16–20% of individuals are dissatisfied with their marriages or cohabiting relationships at any given time (e.g., Hjemboe & Butcher, 1991), and that nearly 50% of recent marriages may end in divorce (Kreider & Fields, 2001). Insofar as achieving a satisfying intimate relationship has been identified as one of the most important goals in life (e.g., Roberts & Robins, 2000), then difficulty in achieving or maintaining this goal is likely to result in negative outcomes, including depression.

The chapter is divided into three sections. The first section discusses gender differences in depression that could be due to gender differences in the association between marital status and depression. The second section discusses gender differences in specific aspects of marital functioning that could account for gender differences in depression. The third and final section discusses the impact of a spouse's behavior on the course of depression and suggests that potential gender differences in a spouse's behavior could contribute to gender differences in depression. Because the majority of the studies reviewed in this chapter have been conducted on married individuals, we generally refer to marital status, marital functioning, and spouses throughout the chapter. It remains to be determined if similar associations would be obtained for more general relationship status and functioning, as, for example, in the case of cohabiting couples. Finally, given the immense literature on marriage and depression, the chapter focuses on the most recent, the most accepted, and the most empirically supported perspectives on marriage and depression.

MARITAL STATUS AND DEPRESSION

The Association between Marital Status and Depression

There is a large body of literature evaluating the association between marital status and depression. Overall, the prevailing perspective concerning this association is that marriage may protect individuals against depression, and that this effect may be greater for men than for women. Evidence on this perspective, however, is mixed, and as such, alternate explanations need to be considered. We begin our review with the empirical and theoretical findings regarding the hypothesized protective nature of marriage more generally and conclude this section with a review of potential gender differences for this effect as well as possible explanations for any such differences.

Many researchers have found evidence that married individuals exhibit less numerous, less intense, or less persistent depressive symptoms than the unmarried (e.g., A. E. Barrett, 2000; Ensel, 1982; Lehtinen & Joukamaa, 1994). In fact, those who are not married have been found to be twice as likely than those who are married to experience depression (Gutierrez-Lobos, Woelfl, & Scherer, 2000). Several theories have been proposed to account for differences in depression due to marital status. One theory that has been advanced is that people who are naturally more resilient or resistant to depression are more likely to marry, resulting in a disproportionate number of healthy individuals in marriages compared to other marital statuses. This theory may not be sufficient, insofar as there is evidence that people who are naturally less depressed are not selected into marriage regardless of their gender (Horwitz & White, 1991). Another theory for the association between marital status and depression is that marriage supplies individuals with more social resources and social support. This theory has some support, insofar as lack of confidants is at least partially responsible for the rise in depression following divorce (Menaghan & Lieberman, 1986). However, a lack of social resources such as contact with close friends and the availability of confidants is an unlikely explanation beyond the disruption of the intimate marital relationship, as broader social networks seem to have very little impact on how people are able to cope with life stress, and marriage provides protection above and beyond variations in intimacy among those who are and are not married (Kessler & Essex, 1982). A third theory regarding the association between marital status and depression is that married individuals experience less stress than nonmarried individuals (i.e., never married or previously married individuals), which would serve to decrease depression vulnerability among married individuals. Once again the supporting evidence is somewhat mixed. Some studies have found that unmarried people are subjected to higher levels of stressful life events such as economic hardship and social isolation (e.g., Menaghan & Lieberman, 1986; Turner, Wheaton, & Lloyd, 1995), whereas other studies find that differences in the amount of stress the two groups experience are only marginally related to depression (Kessler & Essex, 1982). A fourth theory regarding an association between marital status and depression is that compared to unmarried individuals, married individuals are less reactive to the stress they do encounter. This explanation has received empirical support insofar as it has been found that married people are less reactive to the economic, housework-related, and parental stress they do experience than those who are not married (Kessler & Essex, 1982).

Upon closer examination, however, it appears the negative consequences associated with nonmarried status may be largely attributed to the dissolution of marriage rather than never having been married. For example, people who are divorced, separated, or widowed have more depressive symptoms and for specific symptoms, endorse more elevated and persistent impairment than people who are married or single (e.g., A. E. Barrett, 2000; J. Barrett, Oxman, & Gerber, 1987; Craig & Van Natta, 1979). In addition, results from studies using representative community samples suggest that separated or divorced individuals are more likely than married individuals to meet criteria for major depressive disorder (e.g., Kessler et al., 2003; Weissman, Bruce, Leaf, Florio, & Holzer, 1991). Similarly, prospective studies have shown that marital dissolution is associated with increased risk for depression (e.g., Aseltine & Kessler, 1993; Bruce, 1998; Coryell, Endicott, & Keller, 1992; Maciejewski, Prigerson, & Mazure, 2001). However, studies have found that depression does not differ between never married and married people (e.g., Horwitz & White, 1991; Kessler et al., 2003; Weissman et al., 1991) and that the never married are actually less depressed than those who were currently married (Romanoski et al., 1992).

There are several major explanations for why divorced people may be more vulnerable to depression than those who have never married. First, the dissolution of marriage itself is often a stressful event that can cause emotional distress. People in this situation are typically faced with changes in self-image, social life, personal habits and routines, and social roles that may be very distressing. Following divorce, there may also be a loss of emotional support, economic decline, continuing conflict with one's ex-spouse, and sole parenting responsibility or loss of custody of children (Amato, 2000). Second, there are powerful cultural norms that view marriage as desirable and necessary to happiness, so the loss of this accomplishment may be seen as a personal failure, which could in turn increase the risk of depression. In discussing the impact of divorce on depression, it should also be noted that marital dissolution has less of an effect on mental health when a couple was experiencing high levels of marital distress prior to divorce (Aseltine & Kessler, 1993; Wheaton, 1990).

In summary, the dissolution of a marital relationship seems to put people at a higher risk for depression than those who remain single or married. Empirical findings regarding differences in depression between married people and those who have never been married are less conclusive in part due to the fact that single people have not been widely examined separately from divorced and separated participants. The limited evidence available is mixed, and more research is needed before firm conclusions can be drawn.

Gender Differences in the Association between Marital Status and Depression

In addition to evaluating the association between marital status and depression across individuals, investigators have also evaluated gender differences in this association. One commonly held perspective is that men benefit more from marriage than women and, as such, can be expected to suffer more intensely when marriage ends. A common source given for the belief that marriage is good for men and bad for women is a book by Jessie Bernard (1975) titled *The Future of Marriage*. In this book, Bernard argues that the restrictions imposed on women by traditional marriage contribute to women's unhappiness and ill health. Other researchers have advanced similar perspectives on the greater benefits of marriage for men (e.g., Hafner, 1993). Insofar as men benefit more from marriage than women, this perspective would suggest that men would be more likely than women to become depressed following marital dissolution. An opposing perspective on gender differences in response to divorce comes from research indicating that economic consequences of divorce are greater for women than men (for a review, see Amato, 2000). The effect of economic disadvantage for women is particularly pronounced for women who have custody of children. For example, Bianchi, Subaiya, and Kahn (1999) found that custodial mothers experienced a 36% decline in standard of living following separation, whereas noncustodial fathers experienced a 28% increase. The net effect of this difference was that mothers' standard of living following separation was only about one half that of fathers. Insofar as women may be more likely than men to experience economic decline following marital dissolution, this perspective would suggest that women would be more likely than men to become depressed following marital dissolution.

In contrast to the theories that suggest that one gender benefits more from marriage than the other gender (or that one gender is more adversely affected by marital dissolution), another possible explanation for gender differences in the association between marital status and depression is that men and women simply react to the experience of marriage and its dissolution in different ways (Mirowsky, 1996). Men and women may experience comparable but different benefits from marriage and may therefore diverge in how they manifest their distress at the loss of these benefits. In support of this notion, women's risk of depression rises after a marital dissolution, whereas men experience a rise in the risk of alcoholism (e.g., Horwitz, White, & Howell-White, 1996). In addition, not only are women and men likely to react to dissolution in different ways, but also they may actually be experiencing very different sources of relationship stress based on the roles they occupy during marriage. For example, men are more likely than women to have problems managing a household, following the death of a spouse, whereas women are more likely to experience greater financial strain (Umberson, Wortman, & Kessler, 1992).

Turning from the theoretical to the empirical literature, results have been inconsistent with respect to gender effects of divorce on depression. Some studies suggest that the effects of marital disruption on depression are stronger for men than for women (e.g., Bruce & Kim, 1992; Kendler, Thornton, & Prescott, 2001), whereas other studies have found that relative to men, women are more likely to be depressed following marital dissolution (e.g., Aseltine & Kessler, 1993; Coryell et al., 1992; Doherty, Su, & Needle, 1989; Simon & Marcussen, 1999) and have more persistent symptoms of depression (Bracke, 2000). Still other studies have failed to find gender differences in the association between marital status (or related variables such as relationship dissolution) and depression, when gender differences in employment, lifestyle, and other demographics are statistically controlled (e.g., Aneshensel, Frerichs, & Clark, 1981; Gore & Mangione, 1983; Maciejewski et al., 2001).

In conclusion, there are no definitive gender differences in the association between marital status and depression. There is some evidence for women being at greater risk than men for depression following marital dissolution, which might help to explain the gender differences in rates of depression. However, there is also evidence for men being at greater risk than women and still other evidence for no gender differences in this association. Given the inconsistencies in prior studies, future research should focus on teasing apart the association between marital status and depression for never married vs. previously married individuals and more research should focus on clarifying gender differences in response to marriage dissolution. Furthermore, future research should evaluate how gender interacts with other variables such as economic well-being in predicting depression following marital dissolution.

MARITAL FUNCTIONING AND DEPRESSION

In this section, we review the theoretical and empirical literature on marital functioning and depression, with a focus on gender differences. Much of the literature on marital functioning and depression has been based on the marital discord model of depression (S. R. H. Beach, Sandeen, & O'Leary, 1990), which focuses on the roles of marital functioning in the onset, course, and treatment of depression. Specifically, this theory suggests that marital functioning impacts depression through the presence of negative facets of the marital relationship that are capable of inducing stress and strain and by the absence of positive facets of the marital relationship that are capable of providing social support and enhanced coping. In support of this theory, the association between marital distress and depression has been well documented (for a review, see Whisman, 2001). For example, compared to people with no current depression, marital distress is greater among individuals with major depression, as seen in both epidemiological (e.g., Whisman, 1999) and treatment-seeking samples (e.g., Vega et al., 1993). In addition, results from longitudinal studies indicate that marital distress predicts subsequent level of depressive symptoms (e.g., S. R. Beach & O'Leary, 1993) and onset of major depression (Whisman & Bruce, 1999). Furthermore, the association between marital functioning and depressive disorders remains significant when controlling for potential confounds of other dimensions of general interpersonal distress (Whisman, Sheldon, & Goering, 2000) or other Axis I psychiatric disorders (Whisman, 1999).

As initially proposed, the marital discord model does not specifically address gender differences in depression, although the need to incorporate such explanations for gender differences has recently been acknowledged (Whisman, 2001). There are at least two potential ways in which marital functioning may contribute to gender differences in depression. First, gender may moderate the association between marital functioning and depression, such that negative aspects of marital functioning (e.g., marital distress, conflict) exhibit a stronger association with depression for women than for men. Indeed, there are several theoretical perspectives that suggest that women's relational traits, identities, and self-representations may make them more responsive than men to disruptive relationship events (e.g., Cross & Madson, 1997; Jordon, Kaplan, Miller, Stiver, & Surrey, 1991). As applied to the study of depression, these theories would support the perspective that gender differences in depression could be due, in part, to gender differences in response to aspects of relationship functioning. A second way in which marital functioning could contribute to gender differences in depression is that compared to men, women may be exposed to more frequent or more severe relationship stressors (or less frequent or less intense relationship supports). In sum, gender differences in depression could be the result of gender differences in reaction to relationship functioning, gender differences in exposure to relationship stressors or supports, or both. In this section, we review research addressing both pathways by which marital functioning may contribute to gender differences in depression.

Gender Differences in Response to Marital Distress

Gender differences in the cross-sectional association between marital distress and depressive symptoms have been evaluated in several studies. In a recent meta-analysis of these studies, Whisman (2001) reported a weighted mean effect size (r) of .42 for women (based on 26 studies and 3,745 participants) and .37 for men (based on 21 studies and 2,700 participants). Although the difference between these two correlations was statistically significant, $z = 2.19$, $p < .05$, the effect size was not large. For example, the difference between the two correlations yields an effect size (q) of .06, which falls in the small effect size range (i.e., $q = .10$), using the conventions proposed by Cohen (1988). Another way of considering the magnitude of this difference is to consider the amount of change in the proportion of variance accounted for by these two correlations, using the formula $r_1^2 - r_2^2$ (Cohen, 1988, p. 114). Evaluated from this perspective, marital distress accounts for only 4% ($= .42^2 - .37^2$) more variance in depressive symptoms for women than for men.

Gender differences in the cross-sectional association between marital distress and major depression have less frequently been evaluated. The studies that have evaluated this association, however, have been based on large randomly selected and representative samples, which suggest that the results are likely to be highly generalizable. Whisman et al. (2000) and Uebelacker and Whisman (in press) found that gender did not moderate the magnitude of the association between marital distress and major depression in large population-based samples of participants from Canada and the United States, respectively. Similarly, Weissman (1987) found similar associations between marital distress and major depression for women and men in a large community sample, although the effects of gender were not formally tested.

In addition to evaluating gender differences in the magnitude of the cross-sectional association between marital distress and depression, investigators have also evaluated gender differences in the *longitudinal* association between the two. These studies have sought to evaluate whether greater marital distress at baseline is associated with increases in depression from baseline to follow-up. Investigations of this longitudinal association have been limited to studies evaluating longitudinal effects on change in depressive symptoms, the results of which have been inconsistent. Fincham and Bradbury (1993) used regression analyses to evaluate change in symptoms over time and found that baseline marital distress predicted residual change in depressive symptoms for women but not for men. That is to say, compared to women who were less distressed at baseline, women who reported greater marital distress at baseline reported greater increases in depression between baseline and follow-up. It should be noted, however, that this study did not formally evaluate gender differences (i.e., interactions between gender and baseline level of marital distress) and therefore should not be interpreted in terms of significant gender differences. Gender differences were, however, directly tested by Beach and O'Leary (1993) and Dehle and Weiss (1998), who used regression analyses to evaluate interaction effects between gender and marital distress on changes in depressive symptoms. Beach and O'Leary (1993) failed to find evidence for gender differences, whereas Dehle and Weiss (1998) found that gender was a significant moderator of the longitudinal association between marital distress and change in depressive symptoms. Specifically, Dehle and Weiss (1998) found that higher baseline marital distress was associated with greater increases in depressive symptoms for women than for men. More recently, investigators have used structural equation modeling to evaluate the longitudinal association between marital distress and depressive symptoms. Again, results have been mixed, with some studies finding a stronger association between baseline marital distress and depressive symptoms for women than for men (Fincham, Beach, Harold, & Osborne, 1997), whereas other studies have failed to find evidence of gender differences (S. R. H. Beach, Katz, Kim, & Brody, 2003). Finally, most recently, hierarchical linear modeling (also known as growth curve analysis) has been used to evaluate gender differences in the longitudinal association between marital distress and depressive symptoms. Both Kurdek (1998) and Davila, Karney, Hall, and Bradbury (2003) found little evidence for gender differences in within-subject associations between these variables over time. In comparison, Beach, Davey, and Fincham (1999) found evidence of gender differences when they reanalyzed the data from the Fincham et al. (1997) study using these types of analyses.

In summary, there is some evidence that marital distress is more strongly associated with depressive symptoms for women than for men in cross-sectional studies, although the magnitude of this difference is relatively small. In comparison, available evidence indicates that the cross-sectional association between marital distress and major depression is similar for women and men. Finally, studies evaluating the longitudinal association between marital distress and depressive symptoms has yielded inconsistent findings, which may be due in part to different time lags, different analytic strategies, and different types of samples. Therefore, available evidence suggests that the gender differences in depression are unlikely to be largely influenced by gender differences in the magnitude of the association between marital distress and depression. Thus, the first pathway by which marital functioning may contribute to gender differences in depression – that women are more reactive than men to marital distress – has generally not been supported. The possibility remains, however, that there are gender differences in reactivity to measures of marital functioning other than global marital distress. For example, Bolger, DeLongis, Kessler, and Schilling (1989) found that daily ratings of arguments with spouse were more strongly associated with daily mood for women than for men.

Gender Differences in Exposure to Marital Distress and Other Relationship Stressors

In this section, we review research on gender differences in exposure to marital distress and other relationship stressors. If women are more distressed than men with their relationship, or if women are more likely than men to be exposed to relationship stressors, then the gender differences in depression may be due in part to gender differences in exposure to negative aspects of relationship functioning.

We begin this section with a review of gender differences in marital distress. Results from recent large-scale studies suggest there is a small but reliable sex difference in marital distress, with women reporting higher levels of distress than men (e.g., Fowers, 1991; Schumm, Webb, & Bollman, 1998). Thus, to the extent that marital distress is associated with the onset and course of depression, higher levels of marital distress in women relative to men could result in greater rates of depression for women.

We now shift our focus from global evaluations of the relationship (i.e., marital distress) to a review of research on specific marital stressors and supports that are believed to be associated with depression. First, there may be gender differences in stress exposure that occur within the context of marriage that may result in gender differences in depression. Despite increasing participation in the paid labor force, women have continued to bear a significantly larger share of household and childrearing responsibilities (Cowan & Cowan, 1988; Hochschild, 1989). Insofar as greater responsibilities are often associated with greater stress and, in turn, greater risk for depression, then gender differences in these responsibilities could contribute to gender differences in depression. Consistent with this perspective, Glass and Fujimoto (1994) found that hours of household work were associated with level of depression for both wives and husbands and that wives reported significantly more hours of household work than husbands. Similarly, Strazdins and Broom (2004) found that emotional work (i.e., behavior directed at understanding and having empathy for others) was related to depression for both wives and husbands and that wives reported significantly more emotional work than husbands. Recently,

Nolen-Hoeksema, Larson, and Grayson (1999) reported that chronic strains in household and interpersonal domains (along with low mastery and rumination) were more common in women than in men and mediated the gender differences in depressive symptoms in a large prospective community survey.

Greater household and childrearing responsibilities that women bear may be particularly likely to contribute to depression for women who work outside the home. According to the role conflict perspective (e.g., Allen, Herst, Bruck, & Sutton, 2000; Hochschild, 1989), a heavy workload outside the home can interfere with a person's obligations in their marital role and can alter marital behavior by reducing the amount of energy and time that they are able to spend as a marital partner. To the extent that this theory is true, then compared to men, greater demands on women in terms of family responsibilities may leave them particularly vulnerable to depression because of the stressors inherent in balancing the demands in family and work domains. Therefore, it appears that women are exposed to greater levels of domestic stress (e.g., household and emotional work) than are men, which therefore could result in higher levels of depression for women.

In addition to household and childrearing stress, there may be gender differences in exposure to other kinds of stressful relationship events, including traumatic life events. For example, Christian-Herman, O'Leary, and Avery-Leaf (2001) evaluated the impact of severe negative marital events on incidence of major depression among women with no history of depression. They found that the rate of major depression among women who had experienced a severe event (most commonly involving separation or divorce, extramarital affairs, and physical aggression) was higher than reported incidence rates in representative samples of women. Similarly, Cano and O'Leary (2000) studied the impact of negative relationship-oriented events that devalue the individual in relation to the self or others (i.e., humiliating events; Brown, Harris, & Hepworth, 1995), such as discovery of infidelity, separation or divorce, or separation or divorce due to infidelity or violence, and reported that the occurrence of these events was associated with increased risk of major depression. To the extent that relationship stressors are predictive of depression, then gender differences in exposure to these stressors could result in gender differences in depression. Indeed, it does appear that women are more likely than men to be exposed to such stressors. For example, research from epidemiological data in the United States has found that men are approximately 1.5 times more likely than women to report lifetime history of extramarital affairs (Laumann, Gagnon, Michael, & Michaels, 1994) and that men are more likely than women to be the perpetrator of severe physical aggression (Straus & Gelles, 1986). Therefore, to the extent that women may be more likely than men to experience household stressors and relationship traumas, then women would be more likely than men to be at risk for the emotional sequela of these events, including elevated risk for depression.

In summary, research has shown that compared to men, women are more likely to report greater marital distress and are more likely to experience relationship stressors such as household stress and relationship traumas. As these stressors have been associated with increased risk for depression, then gender differences in exposure to marital distress and relationship stressors could contribute to gender differences in depression. Taken with the findings regarding the limited gender differences in the association between marital distress and depression, the results from existing studies suggest that gender differences in depression are more likely to be the result of gender differences in exposure to adverse relationship events than the result of gender differences in response to adverse relationship events.

SPOUSES OF DEPRESSED INDIVIDUALS

In addition to examining associations between depression and marital status and functioning within a given individual, researchers have expanded their focus to include an assessment of the spouses of depressed persons. In particular, researchers have suggested that spouses' response to depression may impact the course (e.g., maintenance, recovery, and relapse rate) of depression. If there is a differential spouse response to partner's depression, dependent upon whether the depressed person is female or male, the response of spouses may also contribute to an overall gender difference in depression. In this section, we provide an overview of the research focused on the cognitive and behavioral responses of spouses to depressed individuals. As most of this research has not directly examined gender differences, we conclude this section with a discussion of a need for research evaluating whether spousal response to depression varies as a function of gender.

Cognitive and Behavioral Responses of Spouses to Depressed Individuals

There is mounting evidence that spouses of depressed persons are likely to report negative appraisals of their partners. For example, compared to spouses of nondepressed women, spouses of depressed women have been shown to rate their wives more negatively on both depression-related and depression-neutral traits, suggesting a generalized negativity toward the depressed person (Sacco, Dumont, & Dow, 1993). In the same study, spouses of depressed women made more dispositional attributions for negative events involving the patient when compared to clinical controls. Spouses of depressed women have also been shown to rate their wives more negatively on measures of dependence, directiveness, and detachment when compared to spouses of nondepressed women (Birtchnell, 1991). Finally, there is evidence to suggest that negative appraisals of depressed persons may persist even in the face of remitted depression. In a study that compared spouses of remitted depressed individuals to spouses of clinical controls, Levkovitz, Lamy, Ternachiano, Treves, and Fenning (2003) reported that the spouses of remitted individuals ranked their partners lower on positive and higher on negative qualities.

In addition to studying the cognitions of spouses of depressed individuals, investigators have also studied the behaviors of couples in which one person is depressed. Behavioral observation of couples engaged in problem-solving interactions indicates that, compared to couples in which neither person is depressed, couples in which one person is depressed exhibit more negative behaviors (e.g., Johnson & Jacob, 1997; Kahn, Coyne, & Margolin, 1985; Kowalik & Gotlib, 1987; Ruscher & Gotlib, 1988; Sher & Baucom, 1993). Couples in which one person is depressed also perceive their interactions to be more negative in comparison to couples in which neither person is depressed (e.g., Kowalik & Gotlib, 1987; McCabe & Gotlib, 1993). Given the previously discussed association between depression and marital distress, investigators have attempted to isolate behaviors that are unique to depression, independent of their shared association with marital distress. To date, it appears that the most unique characteristic of the interactions of couples with a depressed individual are depressive behaviors (e.g., physical or psychological complaints, self-derogatory statements) exhibited by the depressed person (e.g., Schmaling & Jacobson, 1990).

Turning our focus from research on the behavior of couples with a depressed member in general to a more specific focus on the behavior of spouses of depressed individuals, Levkovitz et al. (2003) found that spouses of both acute and remitted depressed persons reported high levels of criticism toward their partners. Furthermore, spouses living with a depressed person have reported fewer expressions of affection than spouses of community controls (Coyne, Thompson, & Palmer, 2002). There is additional evidence that the behavior exhibited by the spouses of depressed persons is associated with the course of depression in the affected individual. For example, Hooley and Teasdale (1989) found that marital distress, spouse criticism, and perceived criticism predicted 9-month relapse rates among unipolar depressed individuals. Most relevant to the present discussion, people that were married to spouses who were critical of them, as measured in terms of objective ratings of criticism as well as subjective ratings made by the depressed person, were more likely to relapse during the following 9 months.

Although the majority of studies described above do not directly address gender effects, there are a few studies that have begun to evaluate gender differences in the cognitive and behavioral responses of spouses to depressed individuals. For example, there is some research suggesting less positive communication among couples with a depressed wife as compared to couples with a depressed husband (Johnson & Jacob, 1997). In comparison, a separate study found that marital problem-solving interactions are more likely to result in significantly decreased positivity and increased negativity among wives of depressed men, without comparable effects for husbands of depressed women (Johnson & Jacob, 2000). Still other studies have failed to find gender differences in perceptions of problem-solving interactions for depressed individuals or their spouses (Gotlib & Whiffen, 1989). Given the limited and mixed nature of these findings, as well as the small sample sizes used in prior studies (which result in low statistical power for detecting gender differences), future research on gender differences in cognitive and behavioral responses of spouses of depressed individuals, and gender differences in the magnitude of the associations between spouse's responses and the course of depression, is clearly warranted.

Taken together, the findings reported in this section suggest that spouses of depressed individuals tend to be critical of the depressed person, as evidenced in their cognitions and behaviors. Moreover, such criticism appears to be associated with the course of depression. Most of the studies conducted to date have been based upon small samples, or samples consisting only of depressed men or women. Thus, there have been few analyses of gender differences in the responses of spouses to depressed people, although there is some evidence from marital interaction studies to suggest that spouse behavior is dependent upon the gender of the depressed individual. If spouse behavior is associated with the course of depression, and if there are gender differences in the spouses' behavior, then such differences could contribute to the gender difference in depression. This clearly is an area that warrants additional research.

Characteristics of Depression Likely to Contribute to Spouse Response

There are several explanations that could be offered to account for the negative cognitions and behaviors exhibited by spouses of depressed individuals. For example, it may be that these reactions develop in response to perceived problems in the relationship. In fact, Coyne et al. (2002) reported that, compared to husbands whose wives were not depressed, husbands of depressed women were more likely to report frequent arguments, differing social needs, and being blamed for things going wrong. In addition, spouses of current and remitted depressed women have reported greater use of coercive problem-solving tactics than spouses of controls (Hammen & Brennan, 2002). There is further evidence to suggest that spouses' perceived marital difficulties persist when patient depression remits. When compared to clinical controls, spouses of people with remitted depressive or bipolar disorders have reported lower levels of emotional and practical support, as well as poorer consensus, unity, and expressions of affection (Levkovitz, Fenning, Horesch, Barak, & Treves, 2000).

In addition to complaints concerning the marriage, it has been proposed that one specific negative outcome of living with a depressed person is an increased burden of care for the spouse. Specifically, Coyne et al. (1987) described *objective* burdens as being those related to disruptions in family routine, such as financial strain or missing work because of a partner's depression. As a complement, *subjective* burdens referred to attitudes and emotional responses of spouses to a depressed partner. When compared to individuals living with a remitted depressed person, those living with a currently depressed person reported greater overall burden. Sources of burden receiving the highest ratings included partner's lack of energy, emotional strain, and possibility of future depression. Moreover, subjective burden mediated the association between depression and spousal adjustment, suggesting that increases in burden may, in fact, explain psychological distress in spouses. Similarly, Benazon and Coyne (2000) found that partner's depression was significantly associated with subjective burdens reported by spouses and that increased burden explained the association between wives' and husbands' level of depression.

If negative cognitions and behaviors on the part of the spouse result from the burdens the spouse experiences, then gender differences in burdens could contribute to gender differences in spousal cognitive and behavioral responses to depression. As reported by Benazon and Coyne (2000), significant differences in specific burdens were reported by wives and husbands. For example, husbands of depressed women were more disturbed by their partner's crying, whereas wives of depressed men complained more of emotional and physical strain. Teichman, Bar-El, Shor, and Elizure (2003) found that wives of depressed men reported greater household responsibilities than husbands of depressed women, resulting in them feeling particularly overwhelmed with these responsibilities. Future research is needed to determine if relationship problems and burdens are associated with negative spousal attitudes and behaviors and if there are gender differences in these associations.

To explain the development of negative spousal appraisals, a focus has recently turned to the so-called *self-propagatory* processes in depression (Joiner, 2000). One such process, excessive reassurance-seeking, serves the purpose of eliciting support and indications of self-worth from others in the face of negative affect. However, according to this model (Coyne, 1976; Joiner, 2000), people with depression often doubt the sincerity of assurances received and are therefore prompted to engage in additional reassurance-seeking, resulting in frustration and hostility among interaction partners. One similar process, self-verification, suggests that people are driven by a need for cognitive consistency to seek feedback from others that is in line with one's own self-view (Joiner, 2000). As such, people with depression may demonstrate a tendency to specifically engage in negative feedback-seeking, which, in turn, may promote rejection and criticism from others. Such rejection would then serve to maintain depressotypic views of the self and the depression itself. There has been some preliminary evidence for certain aspects of each of these models within the context of intimate relationships. For example, Benazon and Coyne (2000) found that excessive reassurance-seeking was significantly associated with negative spouse appraisals, and Weinstock and Whisman (2004) found that negative feedback-seeking was significantly associated with depressive symptoms in a sample of dating couples.

Within the available literature, there has been little theoretical or empirical development of these models vis-à-vis gender differences in depression. Yet given aspects of life that are particularly salient to women, such as relationship function and partnership (Dudek et al., 2001), one might hypothesize that depressed women may be more likely to engage in excessive reassurance-seeking or negative feedback-seeking, which may in turn increase risk for partner rejection and maintenance of depression. However, this remains an empirical question. Additional studies that address possible gender effects and examine the full models, as theorized, within the context of intimate relationships, would help clarify role of the self-propagatory processes in an understanding of gender differences as they relate to depression in marriage.

In summary, existing research suggests that spouses of depressed individuals have negative appraisals of and are critical of their depressed partners and that criticism in turn is associated with relapse rates of depression. There are several potential models that could contribute to the cognitions and behaviors of spouses of depressed persons, including perceived marital problems, increased burden of care, excessive reassurance-seeking, and negative feedback-seeking. Although gender differences in these areas have generally not been explored, their evaluation in future research would help clarify the role of spouse response in an overall model of gender difference in depression, especially within the context of marriage.

CONCLUSION

In this chapter, we have reviewed the literature on marriage and depression, with a focus on marital status, marital functioning, and spouses of depressed individuals. In reviewing this literature, we have highlighted theoretical and empirical evidence for the perspective that gender differences in depression could be due, in part, to gender differences in marriage. Specifically, available evidence suggests that being married is associated with lower risks of depression in comparison to being separated or divorced – the results for gender differences in this association are inconclusive. Second, the available data suggest that marital distress (a) is associated with depression, (b) may be more strongly associated with depression for women than for men, and (c) may be a stronger predictor of longitudinal change in depression for women than for men. Moreover, there is some evidence to suggest that compared to men, women appear to report greater levels of marital distress and may be at heightened risk for other negative marital outcomes, such as marital stressors. Finally, there is emerging evidence that spouses of depressed individuals tend to be critical of the depressed person and that this criticism may affect the course of depression. Gender differences in spouse's cognitions and behavior, and in the processes that may give rise to such cognitions and behavior, have rarely been evaluated. In many cases, conclusions regarding gender differences in marital functioning and depression should be considered tentative, as many studies that have evaluated the association between marital functioning and depression have not evaluated gender differences or have yielded inconsistent findings. Clearly, evaluating gender differences in future studies, especially in terms of spouses of depressed individuals, is an important topic for future research.

Insofar as marital functioning may contribute to the onset or maintenance of depression, the findings reviewed in this chapter suggest that couple therapy may be effective in the treatment of depression in women. Indeed, there are several clinical trials that have found that couple therapy is an effective treatment for depression (for a review, see Baucom, Shoham, Mueser, Daiuto, & Stickle, 1998). To date, there is no research evaluating whether there are gender differences in response to couple therapy for depression. Furthermore, the studies reviewed in this chapter also suggest that preventing relationship distress and divorce should help to prevent the occurrence of depression. To date, there are no studies that have evaluated the impact of relationship distress prevention programs (e.g., premarital counseling) on the prevention of depression, although they have been shown to be effective in preventing relationship distress and divorce (for a review, see Baucom et al., 1998). Evaluating gender differences in response to couple therapy as a treatment for depression and premarital and relationship enhancement counseling as a prevention of depression could contribute to a greater understanding of gender differences in marital functioning and depression.

References

Allen, T. D., Herst, D. E. L., Bruck, C. S., & Sutton, M. (2000). Consequences associated with work-to-family conflict: A review and agenda for future research. *Journal of Occupational Health Psychology, 5,* 278–308.

Amato, P. R. (2000). The consequences of divorce for adults and children. *Journal of Marriage and the Family, 62,* 1269–1287.

Aneshensel, C. S., Frerichs, R. R., & Clark, V. A. (1981). Family roles and sex differences in depression. *Journal of Health and Social Behavior, 22,* 379–393.

Aseltine, R. H., Jr., & Kessler, R. C. (1993). Marital disruption and depression in a community sample. *Journal of Health and Social Behavior, 34,* 237–251.

Barrett, A. E. (2000). Marital trajectories and mental health. *Journal of Health and Social Behavior, 41,* 451–464.

Barrett, J., Oxman, T. E., & Gerber, P. (1987). Prevalence of depression and its correlates in a general medical practice. *Journal of Affective Disorders, 12,* 167–174.

Baucom, D. H., Shoham, V., Mueser, K. T., Daiuto, A. D., & Stickle, T. R. (1998). Empirically supported couple and family interventions for marital distress and adult mental health problems. *Journal of Consulting and Clinical Psychology, 66,* 53–88.

Beach, S. R., & O'Leary, K. D. (1993). Marital discord and dysphoria: For whom does the marital relationship predict depressive symptomatology? *Journal of Social and Personal Relationships, 10,* 405–420.

Beach, S. R. H., Davey, A., & Fincham, F. D. (1999). The time has come to talk of many things: A commentary on Kurdek (1998) and the emerging field of marital processes in depression. *Journal of Family Psychology, 13,* 663–668.

Beach, S. R. H., Katz, J., Kim, S., & Brody, G. H. (2003). Prospective effects of marital satisfaction on depressive symptoms in established marriages: A dyadic model. *Journal of Social and Personal Relationships, 20,* 355–371.

Beach, S. R. H., Sandeen, E. E., & O'Leary, K. D. (1990). *Depression in marriage: A model for etiology and treatment.* New York: Guilford Press.

Benazon, N. R., & Coyne, J. C. (2000). Living with a depressed spouse. *Journal of Family Psychology, 14,* 71–79.

Bernard, J. (1975). *The future of marriage.* New York: Bantam.

Bianchi, S. M., Subaiya, L., & Kahn, J. R. (1999). The gender gap in the economic well-being of nonresident fathers and custodial mothers. *Demography, 36,* 195–203.

Birtchnell, J. (1991). Negative modes of relating, marital quality and depression. *British Journal of Psychiatry, 158,* 648–657.

Bolger, N., DeLongis, A., Kessler, R. C., & Schilling, E. A. (1989). Effects of daily stress on negative mood. *Journal of Personality and Social Psychology, 57,* 808–818.

Bracke, P. (2000). The three-year persistence of depressive symptoms in men and women. *Social Science and Medicine, 51,* 51–64.

Brown, G. W., Harris, T. O., & Hepworth, C. (1995). Loss, humiliation and entrapment among women developing depression: A patient and nonpatient comparison. *Psychological Medicine, 25,* 7–21.

Bruce, M. L. (1998). Divorce and psychopathology. In B. P. Dohrenwend (Ed.), *Adversity, stress, and psychopathology* (pp. 219–232). London: Oxford University Press.

Bruce, M. L., & Kim, K. M. (1992). Differences in the effects of divorce on major depression in men and women. *American Journal of Psychiatry, 149,* 914–917.

Cano, A., & O'Leary, K. D. (2000). Infidelity and separations precipitate major depressive episodes and symptoms of nonspecific depression and anxiety. *Journal of Consulting and Clinical Psychology, 68,* 774–781.

Christian-Herman, J. L., O'Leary, K. D., & Avery-Leaf, S. (2001). The impact of severe negative events in marriage on depression. *Journal of Social and Clinical Psychology, 20,* 25–44.

Cohen, J. (1988). *Statistical power analysis for the behavioral sciences* (2nd ed.). Hillsdale, NJ: Lawrence Erlbaum Associates.

Coryell, W., Endicott, J., & Keller, M. (1992). Major depression in a nonclinical sample: Demographic and clinical risk factors for first onset. *Archives of General Psychiatry, 49,* 117–125.

Cowan, C. P., & Cowan, P. A. (1988). Who does what when partners become parents: Implications for men, women, and marriage. *Marriage and Family Review, 13,* 1–12.

Coyne, J. C. (1976). Depression and the response of others. *Journal of Abnormal Psychology, 85,* 186–193.

Coyne, J. C., Kessler, R. C., Tal, M., Turnbull, J., Wortman, C. B., & Greden, J. F. (1987). Living with a depressed person. *Journal of Consulting and Clinical Psychology, 55,* 347–352.

Coyne, J. C., Thompson, R., & Palmer, S. C. (2002). Marital quality, coping with conflict, marital complaints, and affection in couples with a depressed wife. *Journal of Family Psychology, 16,* 26–37.

Craig, T. J., & Van Natta, P. A. (1979). Influence of demographic characteristics on two measures of depressive symptoms. *Archives of General Psychiatry, 36,* 149–154.

Cross, S. E., & Madson, L. (1997). Models of the self: Self-construals and gender. *Psychological Bulletin, 122,* 5–37.

Davila, J., Karney, B. R., Hall, T. W., & Bradbury, T. N. (2003). Depressive symptoms and marital satisfaction: Within-subject associations and the moderating effects of gender and neuroticism. *Journal of Family Psychology, 17,* 557–570.

Dehle, C., & Weiss, R. (1998). Sex differences in prospective associations between marital quality and depressed mood. *Journal of Marriage and the Family, 60,* 1002–1011.

Doherty, W. J., Su, S., & Needle, R. (1989). Marital disruption and psychological well-being: A panel study. *Journal of Family Issues, 10,* 72–85.

Dudek, D., Zieba, A., Jawor, M., Szymaczek, M., Opila, J., & Dattilio, F. M. (2001). The impact of depressive illness on spouses of depressed patients. *Journal of Cognitive Psychotherapy, 15,* 49–57.

Ensel, W. M. (1982). The role of age in the relationship of gender and marital status to depression. *Journal of Nervous and Mental Disease, 170,* 536–543.

Fincham, F. D., Beach, S. R. H., Harold, G. T., & Osborne, L. N. (1997). Marital satisfaction and depression: Different causal relationships for men and women? *Psychological Science, 8,* 351–357.

Fincham, F. D., & Bradbury, T. N. (1993). Marital satisfaction, depression, and attributions: A longitudinal analysis. *Journal of Personality and Social Psychology, 64,* 442–452.

Fowers, B. J. (1991). His and her marriage: A multivariate study of gender and marital satisfaction. *Sex Roles, 24,* 209–221.

Glass, J., & Fujimoto, T. (1994). Housework, paid work, and depression among husbands and wives. *Journal of Health and Social Behavior, 35,* 179–191.

Gore, S., & Mangione, T. W. (1983). Social roles, sex roles and psychological distress: Additive and interactive models of sex differences. *Journal of Health and Social Behavior, 24*, 300–312.

Gotlib, I. H., & Whiffen, V. E. (1989). Depression and marital functioning: An examination of specificity and gender differences. *Journal of Abnormal Psychology, 98*, 23–30.

Gutierrez-Lobos, K., Woelfl, G., & Scherer, M. (2000). The gender gap in depression reconsidered: The influence of marital and employment status on the female/male ratio of treated incidence rates. *Social Psychiatry and Psychiatric Epidemiology, 35*, 202–210.

Hafner, J. (1993). *The end of marriage: Why monogamy isn't working*. London: Century.

Hammen, C., & Brennan, P. A. (2002). Interpersonal dysfunction in depressed women: Impairments independent of depressive symptoms. *Journal of Affective Disorders, 72*, 145–156.

Hjemboe, S., & Butcher, J. N. (1991). Couples in marital distress: A study of personality factors as measured by the MMPI-2. *Journal of Personality Assessment, 57*, 216–237.

Hochschild, A. (1989). *The second shift: Working parents and the revolution at home*. New York: Viking.

Hooley, J. M., & Teasdale, J. D. (1989). Predictors of relapse in unipolar depressives: Expressed emotion, marital distress, and perceived criticism. *Journal of Abnormal Psychology, 98*, 229–235.

Horwitz, A. V., & White, H. R. (1991). Becoming married, depression, and alcohol problems among young adults. *Journal of Health and Social Behavior, 32*, 221–237.

Horwitz, A. V., White, H. R., & Howell-White, S. (1996). The use of multiple outcomes in stress research: A case study of gender differences in response to marital dissolution. *Journal of Health and Social Behavior, 37*, 278–291.

Johnson, S. L., & Jacob, T. (1997). Marital interactions of depressed men and women. *Journal of Consulting and Clinical Psychology, 65*, 15–23.

Johnson, S. L., & Jacob, T. (2000). Sequential interactions in the marital communication of depressed men and women. *Journal of Consulting and Clinical Psychology, 68*, 4–12.

Joiner, T. E., Jr. (2000). Depression's vicious scree: Self-propagating and erosive processes in depression chronicity. *Clinical Psychology: Science and Practice, 7*, 203–218.

Jordon, J. V., Kaplan, A. G., Miller, J. B., Stiver, I. P., & Surrey, J. L. (1991). *Women's growth in connection*. New York: Guilford Press.

Kahn, J., Coyne, J. C., & Margolin, G. (1985). Depression and marital disagreement: The social construction of despair. *Journal of Social and Personal Relationships, 2*, 447–461.

Kendler, K. S., Thornton, L. M., & Prescott, C. A. (2001). Gender differences in the rates of exposure to stressful life events and sensitivity to their depressogenic effects. *American Journal of Psychiatry, 158*, 587–593.

Kessler, R. C., Berglund, P., Demler, O., Jin, R., Koretz, D., Merikangas, K. R., et al. (2003). The epidemiology of major depressive disorder: Results from the National Comorbidity Survey Replication (NCS-R). *JAMA, 289*, 3095–3105.

Kessler, R. C., & Essex, M. (1982). Marital status and depression: The importance of coping resources. *Social Forces, 61*, 484–507.

Kowalik, D. L., & Gotlib, I. H. (1987). Depression and marital interaction: Concordance between intent and perception of communication. *Journal of Abnormal Psychology, 96*, 127–134.

Kreider, R. M., & Fields, J. M. (2001). *Number, timing, and duration of marriages and divorces: Fall 1996.* Current Population Reports, P70–80. Washington, DC: U.S. Census Bureau.

Kurdek, L. A. (1998). The nature and predictors of the trajectory of change in marital quality over the first 4 years of marriage for first-married husbands and wives. *Journal of Family Psychology, 12*, 494–510.

Laumann, E. O., Gagnon, J. H., Michael, R. T., & Michaels, S. (1994). *The social organization of sexuality: Sexual practices in the United States.* Chicago: University of Chicago Press.

Lehtinen, V., & Joukamaa, M. (1994). Epidemiology of depression: Prevalence, risk factors and treatment situation. *Acta Psychiatrica Scandinavica, 377*, 7–10.

Levkovitz, Y., Fenning, S., Horesch, N., Barak, Y., & Treves, I. (2000). Perception of ill spouse and dyadic relationship in couples with affective disorder and those without. *Journal of Affective Disorders, 58*, 237–240.

Levkovitz, Y., Lamy, D., Ternachiano, P., Treves, I., & Fenning, S. (2003). Perceptions of dyadic relationship and emotional states in patients with affective disorder. *Journal of Affective Disorders, 75*, 19–28.

Maciejewski, P. K., Prigerson, H. G., & Mazure, C. M. (2001). Sex differences in event-related risk for major depression. *Psychological Medicine, 31*, 593–604.

McCabe, S. B., & Gotlib, I. H. (1993). Interaction of couples with and without a depressed spouse: Self-report and observations of problem-solving situations. *Journal of Social and Personal Relationships, 10*, 589–599.

McGrath, E., Keita, G. P., Strickland, B. R., & Russo, N. F. (Eds.). (1990). *Women and depression: Risk factors and treatment issues.* Washington, DC: American Psychological Association.

Menaghan, E. G., & Lieberman, M. A. (1986). Changes in depression following divorce: A panel study. *Journal of Marriage and the Family, 48*, 319–328.

Mirowsky, J. (1996). Age and the gender gap in depression. *Journal of Health and Social Behavior, 37*, 362–380.

Nolen-Hoeksema, S. (1987). Sex differences in unipolar depression: Evidence and theory. *Psychological Bulletin, 101*, 259–282.

Nolen-Hoeksema, S., Larson, J., & Grayson, C. (1999). Explaining the gender difference in depressive symptoms. *Journal of Personality and Social Psychology, 77*, 1061–1072.

Roberts, B. W., & Robins, R. W. (2000). Broad dispositions, broad aspirations: The intersection of personality traits and major life goals. *Personality and Social Psychology Bulletin, 26*, 1284–1296.

Romanoski, A. J., Folstein, M. F., Nestadt, G., Chahal, R., Merchant, A., Brown, C. H., et al. (1992). The epidemiology of psychiatrist-ascertained depression and DSM-III depressive disorders: Results from the Eastern Baltimore Mental Health Survey clinical reappraisal. *Psychological Medicine, 22*, 629–655.

Ruscher, S. M., & Gotlib, I. H. (1988). Marital interaction patterns of couples with and without a depressed partner. *Behavior Therapy, 19*, 455–470.

Sacco, W. P., Dumont, C. P., & Dow, M. G. (1993). Attributional, perceptual, and affective responses to depressed and nondepressed marital partners. *Journal of Consulting and Clinical Psychology, 61,* 1076–1082.

Schmaling, K. B., & Jacobson, N. S. (1990). Marital interaction and depression. *Journal of Abnormal Psychology, 99,* 229–236.

Schumm, W. R., Webb, F. J., & Bollman, S. R. (1998). Gender and marital satisfaction: Data from the National Survey of Families and Households. *Psychological Reports, 83,* 319–327.

Seltzer, J. A. (2000). Families formed outside of marriage. *Journal of Marriage and the Family, 62,* 1247–1268.

Sher, T. G., & Baucom, D. H. (1993). Marital communication: Differences among maritally distressed, depressed, and nondistressed-nondepressed couples. *Journal of Family Psychology, 7,* 148–153.

Simon, R. W., & Marcussen, K. (1999). Marital transitions, marital beliefs and mental health. *Journal of Health and Social Behavior, 40,* 111–125.

Straus, M. A., & Gelles, R. J. (1986). Societal change and change in family violence from 1975 to 1985 as revealed by two national surveys. *Journal of Marriage and the Family, 48,* 465–479.

Strazdins, L., & Broom, D. H. (2004). Acts of love (and work): Gender imbalance in emotional work and women's psychological distress. *Journal of Family Issues, 25,* 356–378.

Teichman, Y., Bar-El, Z., Shor, H., & Elizur, A. (2003). Cognitive, interpersonal, and behavioral predictors of patients' and spouses' depression. *Journal of Affective Disorders, 74,* 247–256.

Turner, R. J., Wheaton, B., & Lloyd, D. A. (1995). The epidemiology of social stress. *American Sociological Review, 60,* 104–125.

Uebelacker, L. A., & Whisman, M. A. (in press). Moderators of the association between relationship distress and depression in a national population-based sample. *Journal of Family Psychology.*

Umberson, D., Wortman, C. B., & Kessler, R. C. (1992). Widowhood and depression: Explaining long-term gender differences in vulnerability. *Journal of Health and Social Behavior, 33,* 10–24.

Vega, B. R., Bayon, C., Franco, B., Canas, F., Graell, M., & Salvador, M. (1993). Parental rearing and intimate relations in women's depression. *Acta Psychiatrica Scandinavica, 88,* 193–197.

Weinstock, L. M., & Whisman, M. A. (2004). The self-verification model of depression and interpersonal rejection in heterosexual dating relationships. *Journal of Social and Clinical Psychology, 23,* 240–259.

Weissman, M. M. (1987). Advances in psychiatric epidemiology: Rates and risks for major depression. *American Journal of Public Health, 77,* 445–451.

Weissman, M. M., Bland, R., Joyce, P. R., Newman, S., Wells, J. E., & Wittchen, H. (1993). Sex differences in rates of depression: Cross-national perspectives. *Journal of Affective Disorders, 29,* 77–84.

Weissman, M. M., Bruce, M., Leaf, P., Florio, L., & Holzer, C. (1991). Affective disorders. In L. Robins & E. Regier (Eds.), *Psychiatric disorders in America* (pp. 53–80). New York: Free Press.

Wheaton, B. (1990). Life transitions, role histories, and mental health. *American Sociological Review, 55,* 209–223.

Whisman, M. A. (1999). Marital dissatisfaction and psychiatric disorders: Results from the National Comorbidity Survey. *Journal of Abnormal Psychology, 108,* 701–706.

Whisman, M. A. (2001). The association between marital dissatisfaction and depression. In S. R. H. Beach (Ed.), *Marital and family processes in depression: A scientific foundation for clinical practice* (pp. 3–24). Washington, DC: American Psychological Association.

Whisman, M. A., & Bruce, M. L. (1999). Marital distress and incidence of major depressive episode in a community sample. *Journal of Abnormal Psychology, 108,* 674–678.

Whisman, M. A., Sheldon, C. T., & Goering, P. (2000). Psychiatric disorders and dissatisfaction with social relationships: Does type of relationship matter? *Journal of Abnormal Psychology, 109,* 803–808.

11

Depression in Women Who Are Mothers

An Integrative Model of Risk for the Development of Psychopathology in Their Sons and Daughters

Sherryl H. Goodman and Erin Tully

Driven both by knowledge of higher rates of depression in women than men and by assumptions about incompatibilities between depression and good quality parenting, research on children with depressed mothers has soared over the past few decades. In an attempt to account for the many factors involved in the transmission of risk for psychopathology from depressed mothers to their offspring, Goodman and Gotlib (1999) developed an integrative and developmentally sensitive model for understanding mechanisms whereby children of depressed parents might be at risk for the development of psychopathology. The model is consistent with the premise for this book in that it takes a bio-psycho-social perspective on mechanisms and, therefore, we use it as the framework for this chapter. We move beyond previous reviews by examining the current empirical status of each of the proposed mechanisms within this framework. Further, given this book's focus on women and depression, we explore the extent to which daughters may be more affected by their mothers' depression than sons, that is, gender specificity in the intergenerational transmission of depression. Finally, we draw implications from the current status of the model overall and of gender specificity, in particular, for further research as well as potential steps for prevention, treatment, and social policy.

This chapter begins by briefly describing the reasons for concern about children with depressed mothers, including the nature and extent of associations between depression in mothers and adverse outcomes in children, across the full age range from infancy to adolescence. Then, we examine evidence for each of the individual mechanisms proposed in the Goodman and Gotlib model to explain adverse outcomes in the children. Next, we examine evidence for gender specificity in the transmission of risk for the development of psychopathology in children with depressed mothers. Finally, we elaborate on implications for prevention, treatment, and policy from the conclusions we draw.

NATURE AND EXTENT OF RISKS: ADVERSE OUTCOMES IN CHILDREN
ASSOCIATED WITH MATERNAL DEPRESSION

Why Be Concerned about Children with Depressed Mothers?

Depression Is Common. Although estimates vary, probably between one
fourth and one third of women experience a clinically significant major
depressive episode at some point during their lives (Kendler & Prescott,
1999). Rates are about twice as high in women as in men (Kessler, 2000),
with the rates for women increasing during adolescence, when hormonal
changes and other major psychosocial changes likely interact with underlying predispositions to precipitate the depression (Angold, Costello, &
Erkanli, 1999; Frank & Young, 2000). Depression is also highly likely to
recur, with one's risk for a second episode being about 50% and for a third
episode as high as 80% (Angst, 1988; Piccinelli & Wilkinson, 1994). Over
50% relapse within 2 years of recovery and later episodes require lower
levels of precipitants compared to initial levels (Keller, Shapiro, Lavori,
& Wolfe, 1982; Kendler, Thornton, & Gardner, 2000). Given the increase in
rates of depression among girls during adolescence and the high likelihood
of recurrence, it should not be a surprise that depression rates are especially
high among women of childbearing ages (Blazer, Kessler, McGonagle, &
Swartz, 1994). Of particular concern, depression occurs in about 10–15% of
women during pregnancy and the postpartum (O'Hara, Zekoski, Philipps,
& Wright, 1990) and the highest rates of depression are in women of childbearing age (Kessler, McGonagle, Swartz, Blazer, & Nelson, 1993). Thus,
children are very likely to be exposed to depression in their mothers very
early in life and, moreover, to experience repeated exposures throughout
their infancy, childhood, and adolescence.

Defining Depression. Researchers define clinically significant depression using two primary approaches, categorical and dimensional, as is
described in other chapters in this handbook. The important point here is
that women who are identified as depressed based on having high scores on
dimensional scales, such as the commonly used Beck Depression Inventory, now in its second edition (BDI-II; Beck, Steer, & Brown, 1997), and
the Centers for Epidemiological Studies-Depression Scale (CES-D; Radloff,
1977), may differ in important ways from those who are diagnosed with
a depressive disorder. High scores on these scales may reflect general distress rather than depression as a specific disorder and may reveal more
transient depression than would meet diagnostic criteria (Coyne, 1994).
Only a small percentage of individuals in general population samples who
score high on depression rating scales meet diagnostic criteria for depression (Gotlib, Lewinsohn, & Seeley, 1995). Nonetheless, individuals who

score high on such scales have been found to have diminished parenting abilities (Harnish, Dodge, Valente, & Group, 1995) and, thus, they are still of concern.

Impairment Associated with Depression. Depression is known to be an impairing condition. As defined by the *Diagnotic and Statistical Manual of Mental Disorders (DSM-IV)*, a depressive episode must be accompanied by "clinically significant distress or impairment in social, occupational, or other important areas of functioning" (American Psychiatric Association, 1994, p. 356). Although other chapters in this handbook address impairment in terms of costs to the individual and society, broadly defined, of particular concern for this chapter is how depression interferes with quality of parenting. Several researchers have directly observed depressed mothers interacting with their children across ages ranging from infancy through adolescence and found impairments in functioning that raise serious concerns. As has been reviewed recently (Gotlib & Goodman, 1999; Lovejoy, Graczyk, O'Hare, & Neuman, 2000), compared to both nondepressed psychiatric and nonpsychiatric controls, depressed mothers display less positive affect and more sad and irritable affect; engage in more negative, intrusive, and hostile behaviors; and in discipline episodes tend to alternate between harsh, punitive discipline and lax and permissive parenting. Thus, there are ample reasons to be concerned about children with depressed mothers. Next, we turn to the literature on the nature and extent of associations between depression in mothers and adverse outcomes in children, across the full age range from infancy to adolescence.

Adverse Child Outcomes Associated with Maternal Depression: Potential Precursors as well as Signs of Psychopathology

A large body of research has accumulated showing that maternal depression is associated with adverse outcomes for children. These outcomes include not only evidence of psychopathology in the children, but also difficulties with or deficits in aspects of social, cognitive, or behavioral functioning, which might indicate vulnerabilities for the development of psychopathology or actual early signs of depression (Goodman & Gotlib, 1999). Because these wide-ranging outcomes vary widely with the age of the children under study, the findings will be reviewed separately for children in different developmental stages. Because some researchers have investigated effects of depression during pregnancy on fetal development, we will begin there.

Fetal Functioning. Researchers are beginning to recognize the need to be concerned about depression during pregnancy. Depressed mood has been found to be at least as high during pregnancy as during the postpartum (Evans et al., 2001; Green & Murray, 1994) and is of particular concern given the potential effects on fetal development. Knowledge of biobehavioral processes in fetal development, especially the role of corticotrophin releasing factor (CRF) in both fetal development and in models of depression in adults, underscores the importance of studying infants born to mothers who are depressed during pregnancy. In particular, hypersecretion of CRF during pregnancy raises concern for potential dysfunctions in the newborn's neuroregulatory systems, that is, frontal brain activity and hypothalamic-pituitary-adrenocortical (HPA) activity, in children who were exposed prenatally to their mothers' depression (Wadhwa et al., 2002). Because of the relative neglect of prenatal depression in the literature, few studies of infants with depressed mothers began during pregnancy and, thus, we know little about the consequences to infants of fetal exposure to maternal depression.

Of the few studies of maternal depression that began in pregnancy, one found that mothers' prenatal cortisol predicted newborns' cortisol levels (Lundy et al., 1999). Elevated cortisol in newborns is a concern because it reflects increased sensitivity to stress, which is implicated in most leading etiological models of depression (Sanchez, Ladd, & Plotsky, 2001). In another study that included data from pregnancy, number of prenatal months of exposure to maternal depression marginally predicted left frontal lobe activation from electroencephalogram (EEG; Ashman & Dawson, 2002; Dawson, Frey, Panagiotides, Osterling, & Hessl, 1997). This pattern of EEG asymmetry is associated with the experience of withdrawal emotions in children and characterized depressed adults and adolescents (Davidson, Ekman, Saron, Senulis, & Friesen, 1990; Dawson et al., 1992, 1999; Finman, Davidson, Colton, Straus, & Kagan, 1989). The research may have been limited by the fact that depression during pregnancy was measured retrospectively when the mothers were 13–15 months postpartum and included women who were in partial remission or subthreshold. If this finding is corroborated, it suggests the need to examine mechanisms such as genetics and intrauterine factors for the association between maternal depression and frontal brain activity in prenatally exposed infants. Finally, a study is underway in our lab to test associations between various indices of fetal exposure to maternal depression (diagnostic interviews, self-report questionnaires, and cortisol) and infants' neuroregulatory systems. We are studying women who met diagnostic criteria for depression at least once in the past and obtain multiple measures prospectively, one each month beginning typically in the third month of pregnancy. We expect to find that prenatal exposure to maternal depression and its correlates of stress and, possibly, hypersecretion of cortisol will predict infants' neuroregulatory dysfunctions. It is the latter that we expect would impose on these high-risk children a vulnerability to the development of psychopathology.

Infants. Compared to prenatal depression, much more is known about effects of postnatal depression on infants with respect to emotional, behavioral, and biological functioning. Given stage-salient needs of infants, limits on depressed mothers' abilities to provide the sensitive, responsive care needed for the development of healthy attachment relationships (Egeland & Farber, 1984) and emotional self-regulation have been of particular concern (Tronick & Gianino, 1986). As predicted, maternal depression is associated with infants' showing more negative affect, more self-directed regulatory behaviors, and less secure attachment relationships (Gotlib & Goodman, 1999; Tronick & Gianino, 1986; van IJzendoorn, Goldberg, Kroonenberg, & Frenkel, 1992). Infants of depressed mothers have also been shown to have higher cortisol levels, especially following interaction with their depressed mothers (Field, 1992); the harsh parenting that is sometimes associated with maternal depression has also been linked to higher cortisol levels in the children (Hertsgaard, Gunnar, Erickson, & Nachmias, 1995). Both findings suggest an association between children's HPA axis functioning and the depressed mothers' failure to provide sensitive, responsive care.

There is also documentation of frontal EEG asymmetries in infants of depressed mothers. Field and her colleagues report greater relative right frontal EEG asymmetry relative to controls in infants of depressed mothers as early as 1 week old, as well as in 1-month-old infants, which is correlated with infants' EEG at 3 months and 3 years (Jones, Field, Davalos, & Pickens, 1997), suggestive of continuity of this effect. Dawson and her colleagues show similar patterns in 18 month olds (Dawson et al., 1997). Thus, infants exposed to their mothers' postpartum depression are showing longer-term disturbances in aspects of functioning that are known vulnerabilities to the development of depression.

Toddlers and Preschool-Aged Children (Early Childhood). Unlike the prenatal and infant research, most of the studies of older children with depressed mothers are cross-sectional. Researchers typically sample children in a particular age range and test for associations with concurrent maternal depression. Thus, the studies are limited because researchers cannot determine if the prior exposures may better explain the children's functioning than the mother's concurrent depression. For example, children who had been exposed to maternal depression during fetal development are expected to show various manifestations of behavioral and neurobiological sequelae of such exposure later in development. Studies of toddlers or older children typically are unable to determine the role of prior exposures. In our lab, we are using a longitudinal design to study the consequences on later development of fetal exposure to maternal depression. If consistent with stress responses in animals exposed to early adverse experiences (e.g., maternal separation) and with knowledge of early human neurodevelopment (Graham, Heim, Goodman, Miller, & Nemeroff, 1999; Matthews, 2002), we expect that fetal exposure to maternal depression will be associated in preschool-age children, for example, with increased sensitivity to stress, a predilection to experience negative or withdrawal emotions, and with early signs of vulnerability to the development of stress-related psychopathology. That is, depression during pregnancy might set the biological and cognitive stage for children's later development of depression. Thus, it is important to interpret these findings with the caveat in mind that what is usually unknown or untested is the contribution of any earlier exposures to maternal depression, having had the initial exposure at an earlier stage, and the effects of repeated or continuous exposure.

Within these constraints, we find accumulating evidence that maternal depression is associated with toddlers' and preschoolers' greater resistance to mothers' control efforts, more comforting behavior when the mother is distressed, and more early signs of depression and anxiety (Gotlib & Goodman, 1999). For example, toddlers of depressed mothers show more dysregulated aggression and heightened emotionality and have more externalizing problems at age 5 (Zahn-Waxler, Cummings, Iannotti, & Radke-Yarrow, 1984; Zahn-Waxler, Iannotti, Cummings, & Denham, 1990). Maternal depression was associated with higher rates of internalizing problems (e.g., depression or anxiety) in 4-year-old boys and girls, but with externalizing problems (e.g., conduct disorder or attention deficit disorder) only in girls (Marchand & Hock, 1998). Mothers who reported high levels of depression symptoms reported higher levels of behavior problems in their 5-year-old children, with even stronger associations when whose symptoms were severe, chronic, and recent (Brennan et al., 2000). In addition to these outcomes, there is evidence for the continuing association between maternal depression and abnormal EEG patterns, as was noted in infants with depressed mothers. Dawson and colleagues recently showed abnormal, albeit a more generalized abnormal, brain electrical activity pattern, at age $3^1/_2$ in the form of reduced activation from both right and left frontal and parietal regions (Dawson et al., 2003). Overall, these findings are of particular concern because emotional and behavioral problems identified during early childhood are predictive of ongoing problems in later years (Campbell, 1995).

School-Aged Children. Beginning when children enter school, it is possible to reliably measure psychopathology using diagnostic interviews and other instruments with the children, their parents, and teachers. From several studies, we know that rates of depression are higher in children with depressed mothers relative to a variety of controls (Beardslee et al., 1988; Billings & Moos, 1985; Goodman, Adamson, Riniti, & Cole, 1994; Lee & Gotlib, 1989; Malcarne, Hamilton, Ingram, & Taylor, 2000;

Orvaschel, Walsh-Allis, & Ye, 1988; Weissman et al., 1984; Welner, Welner, McCrary, & Leonard, 1977). Rates of depression in the school-aged children of depressed mothers vary from 20 to 41%, depending on the severity or impairment of the parent's depression, whether both parents are depressed, and a number of other sociodemographic variables (Gotlib & Goodman, 1999). Psychopathology in children with depressed mothers is not limited to depression. Researchers have also found higher rates of social phobia, separation anxiety and other anxiety disorders, attention deficit disorders, disruptive behavior disorders, overall levels of behavior problems, and poorer social functioning relative to controls (Biederman et al., 2001; Luoma et al., 2001; Weissman et al., 1984). In addition, children with depressed mothers, in comparison to children whose mothers are medically ill or have other psychiatric disorders, also have poorer academic performance and other behavioral problems in school (Anderson & Hammen, 1993).

Similarly, adolescent offspring of depressed parents have been found to have higher rates of depression (Beardslee et al., 1988; Beardslee, Schultz, & Selman, 1987; Hammen et al., 1987; Hirsch, Moos, & Reischl, 1985) as well as higher rates of other disorders (Orvaschel et al., 1988; Weissman et al., 1984) relative to controls. Moreover, depression in offspring of depressed parents, compared with same-age offspring of nondepressed parents, has an earlier age of onset, longer duration, and is associated with greater functional impairment and higher likelihood of recurrence (Hammen & Brennan, 2001; Hammen, 1990; Hammen, Shih, Altman, & Brennan, 2003; Keller et al., 1986; Lieb, Isensee, Hofler, Pfister, & Wittchen, 2002; Warner, Weissman, Fendrich, Wickramaratne, & Moreau, 1992). In addition, levels of internalizing and externalizing behavior problems are higher in children with depressed mothers relative to controls (Brennan, Hammen, Katz, & Le Brocque, 2002; Fergusson, Lynskey, & Horwood, 1993; Forehand, 1988). Thus, depression in offspring of depressed parents appears to be pernicious.

Adolescence is a particularly important stage of development for studying the effects of maternal depression, because rates of depression increase in girls and the gender gap between girls and boys emerges during this period. Thus, it is intriguing that one study found an association between maternal depression and rates of depressive symptoms during middle childhood and adolescence for girls, but not boys (Fergusson, 1995), although others do not find differences between sons and daughters in the association between maternal depression and adolescent behavior problems (Fowler, 2002). Gender-specific findings are discussed in a later section of this chapter. In addition to studies of depression and other disorders in adolescent offspring, researchers have also studied aspects of functioning that might reveal vulnerabilities to the later development of depression. For example, adolescents with depressed mothers, and the daughters in particular, display more dysphoric and less happy affect (Hops, Sherman, & Biglan, 1990) and show early signs of cognitive vulnerability to depression, such as being more likely than other adolescents to blame themselves for negative outcomes and less likely to recall positive self-descriptive adjectives (Hammen & Brennan, 2001; Jaenicke et al., 1987). Evidence for interpersonal vulnerabilities comes from studies showing that adolescent offspring of depressed mothers have poorer peer relationships and less adequate social skills than teens of nondepressed control mothers (Beardslee et al., 1987; Billings & Moos, 1985; Forehand & McCombs, 1988; Hammen & Brennan, 2003).

In sum, there is strong evidence at all ages, even beginning prenatally, that offspring of depressed mothers are at risk for the development of psychopathology as well as delays and deficiencies in emotional, social, and cognitive development. Next, we examine evidence for each of the individual mechanisms proposed in the Goodman and Gotlib model to explain adverse outcomes in the children of depressed mothers.

ETIOLOGICAL MECHANISMS LINKING MATERNAL DEPRESSION AND ADVERSE CHILD OUTCOMES: THE GOODMAN AND GOTLIB MODEL

Goodman and Gotlib (1999) proposed an integrative, developmentally sensitive model for understanding how maternal depression impairs child development. Consistent with the premise of this book, the model is integrative in the sense that it cuts across lines that have traditionally separated studies of genetic, neuroendocrine, and other biological factors from research in cognitive or interpersonal factors in depression. The model also makes essential the consideration of principles of developmental psychopathology, such as the normative tasks that children are undertaking at each developmental stage, the parenting and other supports that are needed to facilitate that development, and how those tasks and the necessary supports might be negatively affected by different manifestations of maternal depression.

As shown in Figure 11.1, the model begins with the mother. The model progresses from the depressed mother to a set of mediators and moderators of the association between maternal depression and adverse child outcomes. A major motivation for proposing the model was to guide the identification of those variables that might mediate and moderate the association between maternal depression and the development of psychopathology in the children (Baron & Kenny, 1986). Empirically established mediators would reveal the mechanisms through which maternal depression comes to be associated with adverse child outcomes. Potential mediators include any of the ways in which children or families with a depressed mother differ from those without a depressed mother.

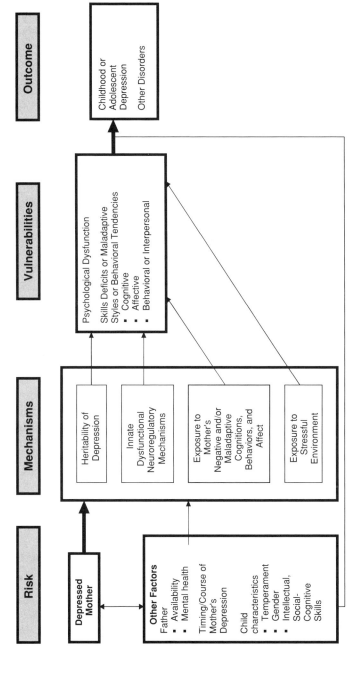

FIGURE 11.1. Integrative model for the transmission of risk to children of depressed mothers (Goodman & Gotlib, 1999).

The theory- and research-based mediators proposed in the model include (a) heritability of depression; (b) innate dysfunctional neuroregulatory mechanisms; (c) exposure to negative maternal cognitions, behaviors, and affect; and (d) the stressful context of the children's lives. Research on mediators holds promise for revealing the pernicious processes for which targeted interventions might be developed to prevent the development of psychopathology in the children.

Also important are potential moderators, defined as variables that affect the strength or direction of the association between maternal depression and the development of child psychopathology. Among the moderators proposed in the model are children's age and gender. Research on moderators is important for its potential to clarify how it is that some children with depressed mothers develop normally, whereas others do not. With knowledge of which children are most at risk, predictive efficacy is improved, as is our understanding of the circumstances under which maternal depression is more or less strongly associated with negative outcomes in the children.

Also essential to a developmental psychopathology perspective is a consideration of the developmental pathways to disorder. Depression and other disorders for which these children are at risk are known to unfold over time (Cicchetti & Toth, 1998). Thus, we propose a set of vulnerability factors, characteristics, or tendencies of the children, including behavioral, affective, and psychobiological functioning, which may be risk factors or early signs of a disorder that has not yet emerged. Each of the four proposed mechanisms in the model might lead to one or more of the vulnerabilities, including physiological (especially the HPA axis), cognitive (e.g., dysfunctional cognitions, low self-esteem, helplessness or hopelessness, negatively biased attention or memory functions), and behavioral or interpersonal (e.g., inadequate social and social-cognitive skills, over- or underregulated behavior, problems with concentration, low mastery motivation).

The strength of the model lies in its potential to guide research that will facilitate an understanding of the complexities of the associations between mothers' depression and children's outcomes. Although each aspect of the model is important, the proposed mechanisms are the component that holds the most promise for furthering our understanding of the development of psychopathology in children with depressed mothers, and they provide the foundation for studying the other components of the model, such as how gender and age of the child affect the associations. Thus, the mediators are an appropriate focus for this chapter, especially given that this component has been the subject of much of the research in this area of study that has been published since the model was proposed. In contrast to earlier research that sought to determine the extent to which children of depressed mothers are at risk for the development of psychopathology,

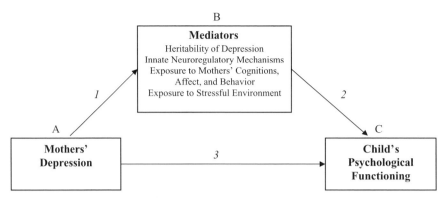

FIGURE 11.2. A mediated effects model for the associations between depression in mothers and children's psychological functioning.

much of the recent work has attempted to reveal what it is about being the child of a depressed mother that places one at risk. Accordingly, in the next section of this chapter, we examine empirical evidence for each of the four mechanisms proposed to explain adverse outcomes in the children in the Goodman and Gotlib model.

We preface the review of the support for the mechanisms with a brief discussion of the nature of evidence needed to demonstrate that a variable functions as a mediator. Figure 11.2 shows a mediated effects model for maternal depression and child functioning. As clarified by Holmbeck, a mediator "specifies how (or the mechanism by which) a given effect occurs" (Holmbeck, 1997, p. 599). Inherent in that definition is the idea that tests of mediation are based on the assumption that the independent or predictor variable (A), in our case depression in mothers, is significantly associated with the dependent variable (C), which in our case is any aspect of child psychological functioning that is of concern. It is that now well-established association for which we seek the mediator. Then, three further significant associations must be found for mediation to be established (Baron & Kenny, 1986). First, maternal depression must be significantly associated with the proposed mediator (B). Second, the proposed mediator must be significantly associated with the child outcome measure. And third, the association between the predictor (maternal depression) and the child outcome measure must be significantly less after controlling for the mediator, which indicates that a significant proportion of that association is accounted for by the proposed mediator.

Mechanism 1: Genetic Influences

The criteria for support of genetic mediation, then, are that (a) depression in women of childbearing age is highly heritable; (b) heritability is

also associated with depression in children, as well as with the other disorders and the vulnerabilities that have been associated with maternal
depression; and (c) controlling for heritability of depression, the association between maternal depression and children's depression and other
outcomes decreases significantly.

Both family studies and quantitative genetic studies (those using twin
or adoption designs) provide evidence for heritability of depression in
adults (Kendler et al., 1995; Kendler, Myers, & Prescott, 2000; Sullivan,
Neale, & Kendler, 2000). Heritability of major depression broadly defined
(using *DSM-III-R* criteria, which did not require impairment) is significantly greater in women than in men, but when impairment is required as
a criterion, heritability contributes to depression about equally in men as
in women (Kendler, Gardner, Neale, & Prescott, 2001; Kendler & Prescott,
1999). Also, postpartum depression itself has been found to be heritable
(O'Hara, 1986). Finally, recent molecular genetics studies also provide supporting evidence for heritability of depression. Recently published findings
implicate a region of chromosome 2 containing the CREB1gene as a susceptibility gene for unipolar depression in women in particular (Levinson
et al., 2002; Zubenko, Hughes, Stiffler, Zubenko, & Kaplan, 2002; Zubenko
et al., 2003). Kendler and colleagues' findings (Kendler et al., 2001; Kendler
& Prescott, 1999) are consistent with the possibility that at least some
genes may be related to risk for depression differently in men compared
to women. Also promising is the serotonin transporter (5-HTTLPR) gene,
which is particularly interesting given the accumulating evidence for the
role of the serotonin system in major depression and in the effectiveness of
serotonin reuptake-inhibitor drugs in the treatment of depression (Ressler
& Nemeroff, 2000). In a birth cohort sample of young adults, a functional
polymorphism of this gene moderated the prospective association between
stressful life events and depression (Caspi et al., 2003). Thus, the first step
to establish mediation is satisfied, given the strong support for heritability
of depression in adults.

The second step to establish heritability as a mediator is to show that
heritability is also associated with depression in children, as well as with
the other disorders and the vulnerabilities that have been associated with
maternal depression. Due to the methodological complexities of measuring
depression in children, evidence for the strength of heritability of depression in children is less clear than in adults (Happonen et al., 2002; Rice,
Harold, & Thapar, 2002). Harrington and colleagues reviewed the evidence from family studies and found that depression in children, especially when strictly defined, is associated with increased rates of depression in relatives (Harrington, 1996; Harrington et al., 1997). Twin studies
find partial support for heritability of depression in children, with heritability estimates higher for levels of symptoms that were below clinical
cutoffs (Rende, Plomin, Reiss, & Hetherington, 1993), for depression in

adolescents rather than in children (Eley & Stevenson, 1999; K. T. Murray & Sines, 1996; Scourfield et al., 2003; Silberg et al., 2001; Thapar & McGuffin, 1994), for parent-reported symptoms relative to child-reported symptoms (Eaves et al., 1997), and for girls more than boys, especially based on parent report (K. T. Murray & Sines, 1996; Scourfield et al., 2003). Thus, although there is no single estimate of heritability of depression in children, heritability likely plays an important role in at least a subset of children with depression. Moreover, because heritability of depression may not be specific to depression, other disorders found in higher rates in children with depressed mothers compared to controls may be explained by the mechanism of heritability (Moldin, 1999; Tsuang & Faraone, 1990). Finally, behavior genetics studies have also found that heritability is high for several of the proposed vulnerabilities in the Goodman and Gotlib model, lending further support to the idea that children with depressed mothers inherit vulnerabilities to depression, rather than inheriting depression as a disorder (Goodman & Gotlib, 1999). For example, behavioral inhibition and shyness (Cherny, Fulker, Corley, Plomin, & DeFries, 1994), low self-esteem (Loehlin & Nichols, 1976), neuroticism (Tellegen et al., 1988), sociability (Plomin et al., 1993), subjective well-being (Lykken & Tellegen, 1996), and expression of negative emotion (Plomin et al., 1993) are all highly heritable.

In sum, there is strong evidence for genetics as one mechanism to explain the association between depression in mothers and at least the vulnerabilities that have been implicated in the pathway to depression in children, and, with several qualifiers, to depression itself, the first and second steps in establishing mediation. As for the third step, to determine if the association between maternal depression and child outcome decreases after controlling for heritability, genetic mechanisms for the transmission of depression that could serve as the mediator need to be identified. As reviewed earlier in this section, the potential of soon being able to identify susceptibility genes for major depression would lead to studies of the biological mechanisms regulated by these genes, revealing how it is that children who inherit these gene variations are at greater risk for the development of depression relative to others.

Mechanism 2: Innate Dysfunctional Neuroregulatory Mechanisms

The second mechanism refers to neurobiological processes that are specific to infants born to women depressed during pregnancy. The criteria to establish support for mediation through innate dysfunctional neuroregulatory mechanisms are that (a) depression in women of childbearing age is associated with adverse influences on fetal development and with implications for the offspring's neuroregulatory processes; (b) neuroregulatory processes are associated with depression in children, as well as with the other disorders and the vulnerabilities that have been associated with maternal depression; and (c) controlling for measures of neuroregulation, the association between maternal depression and children's depression and other outcomes decreases significantly.

The first criterion has been supported by a few studies in which aspects of the fetal environment have been associated with depression in pregnant women. Levels of plasma and urinary cortisol, norepinephrine, beta-endorphin, and corticotrophin releasing hormone (CRH) are positively associated with depressed mood during pregnancy (Field, 2002; Handley, Dunn, Waldron, & Baker, 1980; Smith et al., 1990). Of those, none is known to be associated with fetal levels other than cortisol; maternal levels of cortisol account for 50% of the variance in the fetus' levels of cortisol (Glover, Teixeira, Gitau, & Risk, 1998). Further, mothers' prenatal levels of cortisol predict newborns' cortisol levels (Lundy et al., 1999) and number of months depressed during pregnancy predict preschool-aged children's baseline cortisol levels (Dawson, 1999, as cited in Ashman & Dawson, 2002). These associations between mothers' and children's cortisol levels, and indices of HPA axis functioning, may explain the role of early stress in the development of depression (Graham et al., 1999). Ongoing research in our lab may reveal the extent to which, among women with histories of major depressive episodes, cortisol levels during pregnancy are associated with infants' dysregulated emotions.

Depression during pregnancy may also be related to adverse income outcomes due to depressed women being less likely than other women to obtain adequate prenatal care, to report healthy eating and sleeping patterns, and to avoid smoking (Milberger, Biederman, Faraone, Chen, & Jones, 1996). Although an association between the use of antidepressants by depressed women and abnormal fetal development has been suggested, the concern is probably unfounded. Recent evidence suggests that women's use of antidepressants during pregnancy does not influence the offspring's neuroregulatory abilities (Nulman et al., 1997). In sum, a few studies show that maternal depression during pregnancy increases the level of fetal exposure to cortisol and is associated with other likely risks to fetal development, each of which is likely to compromise the offspring's neuroregulatory processes. Replications of these findings are needed. Also needed are findings from studies such as our current one that will examine the associations between fetal exposure to cortisol, as well as inadequate prenatal care, and infants' own neuroregulatory response to stress.[1]

---

[1] Mechanism 2 specifies that dysfunctional neuroregulatory abilities are innate, that is, present at birth, whether inherited or acquired during fetal development. Evidence is also accumulating that postnatal exposure to maternal depression is associated with children's cortisol levels and abnormal EEG patterns, which could also be mechanisms through which maternal depression is associated with children's adverse outcomes (Ashman & Dawson, 2002). Although important, these later acquired dysfunctions are not the focus of the

Are neuroregulatory processes associated with depression in children, as well as with the other disorders and the vulnerabilities in children that have been associated with maternal depression? Neuroregulatory systems, including both frontal brain activity (abnormal EEG patterns) and HPA functioning (abnormal cortisol responses to stress) have both been associated with depression in adults (Davidson et al., 1990; Ressler & Nemeroff, 2000). In children, supportive evidence is accumulating. Elevated CRH levels in pregnant women has been associated with impaired neurodevelopment in the fetus (Wadhwa et al., 2002), which may be the first indication of stress-related physiological dysregulation in offspring. Elevated cortisol in children increases sensitivity to stress (Sanchez et al., 2001), which increases children's vulnerability to develop depression (Graham et al., 1999). Similarly, abnormal EEG patterns and, particularly, a pattern of relative right frontal activation are associated with the tendency to experience withdrawal emotions such as sadness or fear (Davidson & Fox, 1989), making the children vulnerable to depression (Ashman & Dawson, 2002). Thus, support for this criterion is limited to evidence for the role of the mediator in the emergence of vulnerabilities to depression and not yet for depression or other disorders.

There is mounting evidence that the association between maternal depression and children's depression and other outcomes decreases significantly when controlling for measures of neuroregulation (EEG patterns or cortisol responses to stress). Dawson and colleagues' have found that 3-year-olds' frontal brain activation mediates the relation between maternal depression and child behavior problems, supporting the idea that maternal depression affects children's patterns of brain activity, whether through genetics or fetal or early postnatal experiences (Dawson et al., 2003). Indeed, the frontal lobe develops rapidly the first 2 years of life, suggesting that it may be sensitive to early adverse experiences (Chugani & Phelps, 1986). Dawson et al.'s (2003) findings are also consistent with increasing knowledge of the role of the frontal lobe in the regulation of emotion and affect and, thus, in increasing children's vulnerability to the development of psychopathology. In particular, the pattern of right frontal EEG asymmetry (left frontal activation), which characterizes depressed adults and adolescents, has been found to be associated with the experience of withdrawal emotions in children (Davidson et al., 1990; Dawson et al., 1999; Dawson, Grofer Klinger, Panagiotides, Hill, & Spieker, 1992; Finman et al., 1989).

dysfunctional neuroregulatory abilities proposed as a mechanism in the Goodman and Gotlib (1999) model. They are discussed later in that there is evidence that depressed mothers' insensitive or unresponsive behavior (Mechanism 3) mediates the relation between maternal depression and children's emotion regulation abilities, increasing children's vulnerability factor for the later development of depression.

Mechanism 3: Exposure to Mother's Negative or Maladaptive Cognitions, Behaviors, and Affect

The third mechanism refers to two processes: (a) the effects of adverse parenting on children's developing psychobiological systems and (b) the prospect that children may model or, through other social learning processes, acquire their depressed mothers' negative or maladaptive cognitions, behaviors, and affect. The criteria to establish support for mediation through these mechanisms are that (a) depression in mothers is associated with adverse parenting and to children's exposure to the cognitive, behavioral, and affective manifestations of the mother's depression; (b) these exposures are associated with the outcomes in children that have been associated with maternal depression; and (c) controlling for measures of maternal parenting and cognition, behavior, and affect, the association between maternal depression and children's depression and other outcomes decreases significantly.

The first criterion is well supported and extensively described in several reviews. In a recent meta-analysis of 46 observational studies, a strong effect size was found for the association between maternal depression and negative (hostile/coercive) parenting behavior, a moderate effect size for the association with disengagement from the child, and a small effect size for the association with (less) positive behavior (Lovejoy et al., 2000). The effect size for the association between maternal depression and positive behavior was moderated by socioeconomic status and age of the child such that the effect size was moderate for economically disadvantaged women and for women of infants and significantly smaller for nondisadvantaged women and women with children between the ages of 1 and 5 years.

Researchers have also found that the negatively biased cognitions that are consistently found to be associated with depression in adults (Gotlib, Gilboa, & Sommerfeld, 2000) extend to their parenting. Depressed mothers, relative to those with no depression, endorse more negative views of themselves as parents (Gelfand & Teti, 1990; Goodman, Sewell, Cooley, & Leavitt, 1993) and report less confidence in being able to positively influence their children (Kochanska, Radke-Yarrow, Kuczynski, & Friedman, 1987).

There is a strong evidence base supporting associations between these aspects of parenting and the development of depression and other problems in children and adolescents (Cicchetti & Toth, 1998; Cole & Rehm, 1986; Garber & Martin, 2002; Hetherington & Martin, 1986; Rubin & Mills, 1991; Sheeber, Hops, & Davis, 2001; Zahn-Waxler, Usher, & Goldman, 2003). For example, depressed mothers have more negative, critical perceptions of their children relative to other mothers, and this pattern has been associated with children's tendency to engage in self-blaming attributions for negative events and with lower perceived self-worth (Goodman et al., 1994; Jaenicke et al., 1987; Kochanska et al., 1987; Radke-Yarrow, Belmont, Nottelman, & Bottomly, 1990). These findings suggest that children are internalizing their parents' negative views into their own self-perceptions of being unworthy, which increases their risk for the development of depression (Beck & Young, 1985). Researchers also find that family environments of depressed children, and of children with high levels of internalizing problems, relative to controls, are characterized by aversive social exchange, parents' use of authoritative discipline and guilt-induction, less warmth, and more criticism (Cicchetti & Toth, 1998; Cole & Rehm, 1986; Davis, Sheeber, & Hops, 2002; Donenberg & Weisz, 1998; Gallimore & Kurdek, 1992; Hamilton, Asarnow, & Tompson, 1999; Hetherington & Martin, 1986; Hops, Lewinsohn, Andrews, & Roberts, 1990; Rubin & Mills, 1991; Stark, Humphrey, Laurent, Livingston, & Christopher, 1993; Zahn-Waxler et al., 2003).

The third criterion, that controlling for measures of maternal parenting and depression-related cognition, behavior, and affect, the association between maternal depression and children's depression and other outcomes decreases significantly, has been submitted to empirical test more often than has this criterion for the other mechanisms. Most of the studies reported here relied on structural equation modeling to determine support for mediation. Table 11.1 summarizes the findings. In contrast to support for these mediators, mothers' ability to maintain stability and routines in their children's activities failed to mediate the association between maternal depression and either mother- or teacher-reported child adjustment (Roderick, 2002). Similarly, family coercion failed to mediate the association between maternal depression during the early school years and boys' and girls' later antisocial behavior or depression, suggesting that other processes are more relevant (Compton, Snyder, Schrepferman, Bank, & Shortt, 2003). However, consistent with support for mediation, in one of our studies we found the negative attitudes mothers expressed toward their children moderated the association between maternal depression and children's lower global self-worth (Goodman et al., 1994).

Intervention studies are another potential source of support for mediation. Essentially, interventions are experimental tests of the mechanisms or mediators. If interventions are successful in reducing the aspects of parenting that are of concern in women with depression and the intervention is associated with improvement in child functioning, they offer clear support for the mechanism. This is essentially what Lyons-Ruth, Connell, Grunebaum, and Botein (1990) tested, but she and her colleagues failed to find support for the mechanism in an intervention targeting women at high social risk (experiencing a combination of maternal depression, poverty, and inadequate caretaking). Although the treatment (a home-visiting service with the goals of helping families to meet basic needs and for the mother to interact positively and appropriately with her infant) was

TABLE 11.1. *Empirically Supported Mediators of the Association between Maternal Depression and Outcomes in Children*

Citation	Children's Age	Mediator	Outcome
Lundy, 2002	6 months	Synchrony in mother–infant interactions	Quality of mother–infant attachment
Dawson et al., 2003	14 months	Insensitivity	Reduced left frontal brain activity
Snyder, 1991	4 to 5 years	Harsh, inconsistent, or ineffective discipline	Conduct problems
Ghodsian, Zayicek, & Wolkind, 1984	Infant through age 4	Frequency of physical punishment	Child behavior problems
L.C. Murray & Kempton, 1993	2 months old	Maternal speech characterized by negative affect and not infant focused	Cognitive development at 18 months
Kam, 2001	2nd grade	Warmth	Emotion regulation skills
Harnish et al., 1995	1st grade	Mother–child interaction quality	Externalizing problems (but not in African Americans)
McCarty & McMahon, 2003	11–12 years old	Cold, hostile, or difficult mother–child relationship	Disruptive behavior problems
McCarty & McMahon, 2003	11–12 years old	Less maternal social support	Internalizing problems
Hammen, Burge, & Stansbury, 1990	8–16 years old	Dysfunctional communication	Negative self-concept
Nelson, 2001	15 years old	Critical attitudes toward the child	Externalizing problems
Hilsman, 2001	12–14 years old	Parental criticism	Depressive disorder
Prinstein & LaGreca, 1999	5–6 years old	Maternal social skills	Social competence (girls only)

associated with infants' significant improvement in mental and physical development and in attachment security, these changes were not associated with changes in mothers' increased involvement or decreased hostility and intrusiveness with their infants. In contrast, other intervention studies yielded results that are consistent with the mechanisms in the model. As reviewed recently, interventions designed to improve the quality of parenting and parent–child relationships are demonstrating effectiveness in reducing levels of depression in children, from infancy through adolescence (Gladstone & Beardslee, 2002).

Mechanism 4: Exposure to Stressful Environments.

The fourth mechanism refers to the role of stressors in the association between maternal depression and adverse child outcomes. The criteria to establish support for mediation through exposure to stressful environments are that (a) depression in mothers is associated with children's exposure to stressful environments; (b) exposure to stressful environments is associated with depression in children, as well as with the other disorders and the vulnerabilities that have been associated with maternal depression; and (c) controlling for measures of stressful environments, the association between maternal depression and children's depression and other outcomes decreases significantly.

Support is strong for the first criterion, that depression in mothers is associated with children's exposure to stressful environments. As reviewed recently, maternal depression increases children's exposure to stress in several ways (Hammen, 2002). First, the mothers' depression itself, by its symptoms, impairment, and episodic course, is stressful to children (Compas, Langrock, Keller, Merchant, & Copeland, 2002). Second, depression in adults typically occurs in the context of chronic and episodic stressors, including economic, work-related, and marital and other family conflicts. Third, depression has been found to increase the likelihood of further stressful events. Thus, there is strong support for this criterion, making it all the more compelling to determine how these stressors relate to the risk for the development of psychopathology in children with depressed mothers.

In turn, exposure to stressful environments is associated with depression in children, as well as with the other disorders and the vulnerabilities that have been associated with maternal depression. In the classic stress-diathesis model, stress increases the likelihood that children will develop depression and symptoms of other disorders in association with a range of vulnerabilities, including genetic, neurobiological, cognitive, interpersonal, and contextual. Qualitative and quantitative reviews of the literature reveal that, despite complications associated with conceptualization and measurement of the key constructs, both chronic adversities and episodic stressors are associated with depression symptoms and disorders and other internalizing and externalizing disorders in children and adolescents (Compas, Grant, & Ey, 1994; Grant et al., 2003). Stress is also implicated in the development of many of the proposed vulnerabilities, as in Dawson's premise that the stressfulness of infants' interaction with an insensitive depressed mother is associated with the development of the infants' pattern of brain activity characterized by reduced activity in the left frontal region (Dawson et al., 2003).

Studies designed to test the third criterion are emerging, with some support and some mixed findings. For example, a high-risk environment mediated the association between maternal depression and child behavior problems, although not the association with toddlers' insecure attachment (Cicchetti, Rogosch, & Toth, 1998). In other studies, chronic family stress was found to mediate the relationship between self-reported depression symptoms in a community sample of mothers and adolescents' depression symptoms (Brennan et al., 2002). However, measures of social disadvantage did not explain the association between postnatal depression and 11-year-old children's IQ scores (Hay & Pawlby, 2001). In other studies, maternal depression was found to mediate the association between poverty in single-mother families and 4- to 9-year-old children's socioemotional problems (Eamon & Zuehl, 2001). Correspondingly, maternal depression symptoms mediated the relation between parents' problematic drinking and 6- to 12-year-old children's internalizing problems (El-Sheikh & Flanagan, 2001).

Given the common co-occurrence of women's depression and marital discord, and the known stressfulness for children of marital discord itself, it is important to understand whether the effect of one is explained by the effect of the other in predicting the development of psychopathology in children. Several researchers have addressed this question. In one such study, family discord, defined as marital discord, low family intimacy, and parenting impairments, was found to be a significant mediator of the association between histories of maternal depression symptoms and both conduct problems and depressive symptoms of children in middle adolescence, but only in girls (Davies & Windle, 1997). However, in a later study using a longitudinal design, Davies and Dumenci (1999) found that although marital distress mediated the association between maternal depressive symptoms and externalizing symptoms in adolescents, maternal depressive symptoms mediated the effects of marital distress on adolescents' depressive symptoms, supporting a pathway from marital distress through maternal depression to adolescents' depression. Others find that the specific nature of the marital conflict is what matters. For example, parents' use of observable conflict strategies involving depressive behaviors such as physical distress, withdrawal, sadness, and fear were found to mediate the association between depressive symptoms in mothers and their children's internalizing symptoms whereas destructive and constructive marital conflict strategies did not serve as mediators (Du Rocher Schudlich & Cummings, 2003). These latter studies suggest the need to further consider the roles of maternal depression and marital discord in relation to each other and the risk for the development of psychopathology in children. It will be important to also consider the role of heritability in the model that best explains associations among maternal depression, stressors, and the emergence of depression in children in that genetic factors are significant in influencing individual differences in susceptibility to environmentally mediated risk as well as risk for depression (Silberg & Rutter, 2001).

GENDER SPECIFICITY

Of particular relevance to addressing prevention, treatment, and social policy of depression in women who are mothers are mechanisms through which depression is maintained in families through intergenerational transmission of risk from depressed mothers to daughters, who may themselves become depressed mothers. Adolescent girls are at particularly high risk for depression. Although there are no consistent gender differences in rates of depression in preadolescent children (Angold, 1988), females are about twice as likely as males to be depressed during adolescence (Nolen-Hoeksema & Girgus, 1994). Gender differences in depressive symptoms appear to emerge around age 13 at which time girls' depressive symptoms continue to increase and boys' symptom levels remain relatively stable (Ge, Conger, & Elder, 2001; Ge, Lorenz, Conger, Elder, & Simons, 1994). Although girls' relatively elevated rates of depression do not emerge until adolescence, gender differences in vulnerabilities for the development of psychopathology and early signs of depression likely develop earlier in life (Zahn-Waxler et al., 2003). With respect to Goodman and Gotlib's model, it is important to explore gender of the child as a moderating variable that affects the associations between mothers' depression and children's depression and other disorders and vulnerabilities.

Although researchers have suggested that the greatest risk for depression posed by maternal depression may be to adolescent girls, relatively little research exists to support gender differences in the risk for psychopathology among offspring of depressed mothers (Cummings & Davies, 1994; Hops, 1992). The limited research on greater risk for girls relative to boys has shown that adolescent daughters of depressed mothers have more dysphoric and less happy affect than adolescent daughters of nondepressed mothers, though no significant difference in affect was found for adolescent boys of depressed mothers compared to sons of nondepressed mothers (Hops, Sherman et al., 1990). Additionally, depressed mothers and their daughters, but not depressed mothers and their sons, have been shown to have synchronous bouts of depression (Radke-Yarrow,

Nottelman, Belmont, & Welsch, 1993). Moreover, in community samples, mothers' depressive symptoms are associated with adolescent daughters', but not sons', depressive symptoms (Davies & Windle, 1997; Fergusson, Horwood, & Lynskey, 1995; Hops, 1992, 1996). However, although maternal depression seems to affect mood in girls more than in boys, maternal depression appears to affect intellectual functioning more in boys than in girls (Hay & Pawlby, 2001).

Several mechanisms have been proposed to help explain the gender difference in the prevalence of depression in general samples of adolescents who are not necessarily children of depressed mothers. These mechanisms include gender-related changes in biological and hormonal functioning, cognitive styles and coping patterns, amount and type of stress encountered, and socialization practices (Nolen-Hoeksema & Girgus, 1994; Peterson, Sarigiani, & Kennedy, 1991). The role of girls' changing body morphology in the development of depression during puberty has been supported in many studies over the past two decades (e.g., Angold & Rutter, 1992; Ge, Conger, & Elder, 1996). Recently, however, Angold and colleagues (1999) have found that associations between levels of testosterone and estradiol and diagnoses of depression in girls eliminate the effect of changes in body morphology on the probability of becoming depressed (Angold et al., 1999). Thus, increased levels of depression in adolescent girls appear to be related to hormonal changes, rather than the morphological changes in puberty. However, this study did not address whether these associations between hormone levels and depression are different for girls and boys, and thus do not necessarily support hormone levels as a mediator responsible for gender differences in rates of depression. Somewhat more, though still limited, evidence supports the etiological role of socialization of affective and cognitive vulnerabilities in girls (Hankin & Abramson, 1999). For example, studies provide empirical evidence that girls' identification with feminine stereotypes (e.g., needing to appear thin and becoming dissatisfied with body shape and appearance) increase during adolescence and gender-role identification is associated with increases in depression (Nolen-Hoeksema & Girgus, 1994). This socialization pattern may be related to greater rumination, more negative inferential styles, and lower self-worth as well as with greater perceived impact of negative events in girls relative to boys (Ge et al., 1994; Hankin & Abramson, 1999).

Furthermore, Sheeber, Davis, and Hops (2002) suggest several familial socialization processes that indirectly increase the risk of depression in girls. For example, girls' expressions of depressive-like behaviors are often normalized and even encouraged, and parents may provide differential reinforcement for gender-typic behaviors related to risk for depression, such as less instrumental and more relationship-focused behaviors. Empirical evidence to support Sheeber's hypothesized socialization processes comes largely from observational studies earlier in development. It is important to keep in mind that the studies do not provide insight into the direction of influence (i.e., whether differential socialization occurs before or subsequent to gender differences in behavior). Parents have been observed to respond to emotional displays' in girls through feeling-focused, passive interventions and to boys' through problem-focused, active interventions (Dino, Barnett, & Howard, 1984; Eisenberg, Cumberland, & Spinrad, 1998; Nolen-Hoeksema, 1998). Moreover, girls are encouraged to display dependency and affectionate behavior (Hops, Sherman et al., 1990; Maccoby, 1990) and nurturing and caring behavior (Radke-Yarrow, Zahn-Waxler, Richardson, & Susman, 1994; Zahn-Waxler, Cole, & Barrett, 1991; Zahn-Waxler & Robinson, 1995; Zahn-Waxler & Smith, 1992). Girls also receive more encouragement for being concerned about the feelings of others and having a sense of responsibility for others (Hops, 1996; Keenan & Shaw, 1997; Maccoby, 1990). Later, during adolescence, when there is greater pressure to conform to sex-role stereotypes (Hill & Lynch, 1983; Nolen-Hoeksema & Girgus, 1994), girls' engagement in feminine-type activities is related to self-reported depression (Nolen-Hoeksema & Girgus, 1994). Given the lack of teaching and encouragement to use problem-solving strategies earlier in life, adolescent girls are less equipped to effectively deal with the challenges of adolescence (Allgood-Merten, Lewinsohn, & Hops, 1990).

In terms of mechanisms explaining gender specificity, some researchers report that heritability estimates for depression are higher for girls than boys (Eley & Stevenson, 1999; Murray & Sines, 1996). Regarding the third mechanism, girls may be particularly vulnerable to modeling their mothers' depressive affect, cognitions, and behaviors given the greater salience of the same gender parent. Depressed mothers display depressive interpersonal behaviors and poor problem-solving and coping skills to deal with depression, which increase susceptibility to depression (Compas, Malcarne, & Fondacaro, 1988; Garber, Braafladt, & Zeman, 1991; Kashani, Burbach, & Rosenberg, 1988). Children of depressed mothers have been shown to demonstrate less active strategies for responding to negative affect (Garber et al., 1991; Nolen-Hoeksema, Wolfson, Mumme, & Guskin, 1995), and modeling of these behavioral vulnerabilities may be particularly salient to daughters of depressed mothers (Bandura, 1986). Additionally, maternal depression is associated with poor parenting, which may be a specific risk to girls through 1) impaired mother-daughter relationships, which are already characterized by high levels of conflict during adolescence (Hill, Holmbeck, Marlow, Green, & Lynch, 1985; Steinberg, 1987, 1988), and (2) adolescent girls developing a sense of responsibility for the family and taking on the role of parenting other children and caring for the depressed mother, a burden that increases risk for depression (McGrath, Keita, Strickland, & Russo, 1990; Stephens, Hokanson, & Welker, 1987). Finally, concerning the fourth mechanism, girls may be more sensitive than boys to heightened levels of stress, and maternal depression is a nonspecific stressor that may exacerbate the risk already posed to girls. Girls have been found to respond to stressors, such as marital discord and economic hardship, with internalizing symptoms (Conger, Conger, Matthews, & Elder, 1999; Cummings & Davies, 1994) and likely may respond to the stress of having a depressed mother with similar internalizing problems (Compas et al., 2002). Therefore, accumulating evidence supports the moderating role of gender in three of the four mechanisms proposed by Goodman and Gotlib's model, suggesting the importance of considering gender of the children when studying the transmission of risk from depressed mothers to their children.

CONCLUSION

Although researchers are beginning to design studies that test the processes by which maternal depression is associated with the development of psychopathology in the children, this line of research is new and many questions remain unanswered. The Goodman and Gotlib model provides theoretical bases for the design of prevention and treatment of disorders in the offspring of mothers with depression as well as for social policy. With this chapter, we took the further steps of summarizing the published findings in support of empirical tests of mediation and providing guides to the directions needed for future research to further test mediation. The longer-term goals of these lines of research are to be able to provide empirical support for the design of prevention and treatment studies and for policy recommendations. In particular, support for any of the four mechanisms provides information essential to identify targets of change for intervention and treatment. These suggestions are made in light of awareness that knowledge of risk for psychopathology in children of depressed mothers has not, in itself, been sufficiently compelling to drive a public health campaign to identify and treat depression in women who are mothers. Thus even though it may be that treating women's depression may minimize risk for psychopathology in the children (see Goodman et al., in press), research on mechanisms becomes necessary in that our best hope for preventing depression from being passed onto other generations is to learn to prevent it by intervening in the mechanisms.

Among the research findings, the research on genetic mechanisms suggests the need for interventions designed to minimize children's exposure to high-risk environments to reduce the role of environmental risks that are known to interact with genetic risk (Silberg & Rutter, 2002). Further, if research emerges for neuroregulatory mechanisms as mediators of the association between maternal depression and the development of psychopathology in children, then it may be important to test age-appropriate interventions designed to improve children's stress reactivity and other aspects of emotion regulation and their coping skills (Clarke et al., 1995; Compas et al., 2002).

Research on the effects of adverse parenting on children's developing psychobiological systems and other processes (e.g., socioemotional regulation) through which children are adversely affected by depressed mothers' affect, cognitions, and behavior supports what may have seemed to be an obvious and compelling next step: to intervene to improve the quality of parenting provided by mothers suffering from depression. Research is needed to determine whether the best approach is early intervention to treat the depression, with the implication that the poor qualities of parenting that have been found to be associated with depression in mothers would improve simultaneously or, alternatively, to target the intervention directly to improve qualities of parenting. In one recent small-scale study, we found preliminary support for the contention that parenting would improve as a secondary benefit of treatment for postpartum depression (Goodman, Broth, Hall, & Stowe, 2003). It is also important to consider that depression is known to be vastly underrecognized and untreated, partly due to attitudinal and other barriers to seeking treatment (Goodman & Tully, 2003). Research is needed to determine if women with depression are more receptive to interventions designed to improve quality of parenting relative to treatment for depression. Related questions include whether improving the quality of parenting itself can be effective in the context of maternal depression. Others have found that depression and stress in mothers interferes with the effectiveness of parenting interventions designed to treat children with aggressive and antisocial behavior (Kazdin & Whitley, 2003). Alternatively, researchers may find that effective parenting interventions have the secondary benefit of alleviating at least mild symptoms of depression as women begin to experience more behavioral successes and more positive affect in their interactions with their children and alter their cognitions to perceive themselves as effective in their parenting.

Researchers also need to determine the role of timing of interventions that would effectively interrupt the pathway from maternal depression to the development of psychopathology in the children. In particular, we are concerned about children who are exposed prenatally who may or may not be subject to later exposures, either in the postpartum period or later. It is essential to test the extent to which prenatal depression predicts not only infant outcomes but also outcomes at later, critical points in development such as at age 4 (e.g., development of emotion-regulation abilities and social development as their social worlds become more complex with peers and teachers). In some of our current and planned studies on this question, we extend findings from animal models, which show that prenatal challenges influence reactivity of the offspring's stress response system later in life (Newport, Stowe, & Nemeroff, 2002; Sanchez et al., 2001). Overall, the interventions need to be shown to provide children with the environmental supports that are known to be needed to facilitate healthy development, such as appropriate levels of stimulation for infants (Ashman & Dawson, 2002; Field, 2002), responsive and warm caregiving (Lyons-Ruth, Lyubchik, Wolfe, & Bronfman, 2002), and models of healthy affective, behavioral, and cognitive functioning (Garber & Martin, 2002). These interventions should be sensitive to the children's developmental level as well as to their gender, and targets of the interventions may vary accordingly.

Research on the fourth mechanism, exposure to stressful environments, shows clear support for the role of stressors in a model of the associations between maternal depression and the development of psychopathology in the children. However, the exact role of stressors is still unclear, leaving it up to future research to determine whether interventions would more effectively target the stressors or the depression. Research on the associations between marital distress and depression is informative on this matter: models that best account for the research findings are bidirectional, including the effects of marital distress and depression on each other, and take into account that the direction of influence may differ for men and women (Fincham, Beach, Harold, & Osborne, 1997). As we reviewed recently (Goodman & Tully, 2003), for women more than men, marital dissatisfaction is likely to lead to depression, and marital therapy has been found to be effective in treating depression (Beach, Fincham, & Katz, 1998; Beach & Jones, 2002). Other research indicates that stress itself interferes with effective parenting (Lyons-Ruth et al., 2002; Radke-Yarrow & Klimes-Dougan, 2002), thereby lending further support for interventions to improve quality of parenting. Whether those interventions would be more effective if they target the quality of parenting or aim to reduce the level of stress itself needs to be the subject of future research.

The accumulated evidence of gender as a moderator of the proposed mechanisms for risk of psychopathology in children of depressed mothers suggests that it is very important to consider gender of the offspring when developing prevention and intervention strategies for depression and other disorders in children of depressed mothers. Future research endeavors should focus on developing models of risk for depression specific to daughters of depressed mothers who not only have heightened risk for depression themselves but also pose the threat of continuing the intergenerational pattern of transmission of psychopathology from mother to child, passing the risk to their offspring.

References

Allgood-Merten, B., Lewinsohn, P., & Hops, H. (1990). Sex differences and adolescent depression. *Journal of Abnormal Psychology, 99,* 55–63.
American Psychiatric Association. (1994). *Diagnostic and statistical manual of mental disorders* (4th ed.). Washington, DC: American Psychiatric Association.

Anderson, C. A., & Hammen, C. L. (1993). Psychosocial outcomes of children of unipolar depressed, bipolar, medically ill, and normal women: A longitudinal study. *Journal of Consulting & Clinical Psychology, 61*, 448–454.

Angold, A. (1988). Childhood and adolescent depression: II. Research in clinical populations. *British Journal of Psychiatry, 153*, 476–492.

Angold, A., Costello, E. J., & Erkanli, A. (1999). Pubertal changes and depression in girls. *Psychological Medicine, 29*, 1043–1053.

Angold, A., & Rutter, M. (1992). The effects of age and pubertal status on depression in a large clinical sample. *Development & Psychopathology, 4*, 5–28.

Angst, J. C. (1988). Clinical course of affective disorders. In R. J. Daly (Ed.), *Depression illness: Prediction of course and outcome* (pp. 1–47). Berlin: Springer-Verlag.

Ashman, S. B., & Dawson, G. (2002). Maternal depression, infant psychobiological development, and risk for depression. In S. H. Goodman & I. H. Gotlib (Eds.), *Children of depressed parents: Mechanisms of risk and implications for treatment* (pp. 37–58). Washington, DC: American Psychological Association Books.

Bandura, A. (1986). *Social foundations of thought and action: A social cognitive theory.* Englewood Cliffs, NJ: Prentice-Hall.

Baron, R. M., & Kenny, D. A. (1986). The moderator-mediator variable distinction in social psychological research: Conceptual, strategic, and statistical considerations. *Journal of Personality and Social Psychology, 51*, 1173–1182.

Beach, S. R. H., Fincham, F. D., & Katz, J. (1998). Marital therapy in the treatment of depression: Toward a third generation of therapy and research. *Clinical Psychology Review, 18*, 635–661.

Beach, S. R. H., & Jones, D. J. (2002). Marital and family therapy for depression in adults. In I. H. Gotlib & C. L. Hammen (Eds.), *Handbook of depression*. New York: Guilford.

Beardslee, W. R., Keller, M. B., Lavori, P. W., Klerman, G. K., Dorer, D. J., & Samuelson, H. (1988). Psychiatric disorder in adolescent offspring of parents with affective disorder in a non-referred sample. *Journal of Affective Disorders, 15*, 313–322.

Beardslee, W. R., Schultz, L. H., & Selman, R. L. (1987). Level of social-cognitive development, adaptive functioning, and DSM-III diagnoses in adolescent offspring of parents with affective disorders: Implications for the development of the capacity for mutuality. *Developmental Psychology, 23*, 807–815.

Beck, A. T., Steer, R. A., & Brown, G. K. (1997). *Beck Depression Inventory* (2nd ed.). San Antonio, TX: The Psychological Corporation.

Beck, A. T., & Young, J. E. (1985). Depression. In D. H. Barlow (Ed.), *Clinical handbook of psychological disorders: A step-by-step treatment manual* (pp. 206–244). New York: Guilford.

Biederman, J., Faraone, S. V., Hirshfeld-Becker, D. R., Friedman, D., Robin, J. A., Rosenbaum, J. F. (2001). Patterns of psychopathology and dysfunction in high-risk children of parents with panic disorder and major depression. *American Journal of Psychiatry, 158*, 49–57.

Billings, A. G., & Moos, R. H. (1985). Children of parents with unipolar depression: A controlled 1-year follow-up. *Journal of Abnormal Child Psychology, 14*, 149–166.

Blazer, D. G., Kessler, R. C., McGonagle, K. A., & Swartz, M. S. (1994). The presence and distribution of major depression in a national community sample:

The National Comorbidity Survey. *American Journal of Psychiatry, 151,* 979–986.

Brennan, P. A., Hammen, C., Katz, A. R., & Le Brocque, R. M. (2002). Maternal depression, paternal psychopathology, and adolescent diagnostic outcomes. *Journal of Consulting & Clinical Psychology, 70,* 1075–1085.

Brennan, P. A., Hammen, C., Anderson, M. J., Bor, W., Najman, J. M., Williams, G. M. (2000). *Chronicity, severity, and timing of maternal depressive symptoms: Relationships with child outcomes at age five. Developmental Psychology, 36,* 759–766.

Campbell, S. B. (1995). Behavior problems in preschool children: A review of recent research. *Journal of Child Psychology and Psychiatry, 36,* 113–149.

Caspi, A., Sugden, K., Moffitt, T. E., Taylor, A., Craig, I. W., Harrington, H., et al. (2003). Influence of life stress on depression: Moderation by a polymorphism in the 5-HTT gene. *Science, 301,* 386–389.

Cherny, S. S., Fulker, D. W., Corley, R. P., Plomin, R., & DeFries, J. C. (1994). Continuity and change in infant shyness from 14 to 20 months. *Behavior Genetics, 24,* 365–379.

Chugani, H. T., & Phelps, M. E. (1986). Maturational changes in cerebral function in infants determined by 18FDG positron emission tomography. *Science, 231,* 840–843.

Cicchetti, D., Rogosch, F. A., & Toth, S. L. (1998). Maternal depressive disorder and contextual risk: Contributions to the development of attachment insecurity and behavior problems in toddlerhood. *Development & Psychopathology, 10,* 283–300.

Cicchetti, D., & Toth, S. (1998). The development of depression in children and adolescents. *American Psychologist, 53,* 221–241.

Clarke, G. N., Hawkins, W., Murphy, M., Sheeber, L., Lewinsohn, P., & Seeley, J. (1995). Targeted prevention of unipolar depressive disorder in an at-risk sample of high school adolescents: A randomized trial of a group cognitive intervention. *Journal of the American Academy of Child & Adolescent Psychiatry, 34,* 312–321.

Cole, D. A., & Rehm, L. P. (1986). Family interaction patterns and childhood depression. *Journal of Abnormal Child Psychology, 14,* 297–314.

Compas, B. E., Grant, K., & Ey, S. (1994). Psychosocial stress and child/adolescent depression: Can we be more specific? In H. F. Johnston (Ed.), *Handbook of depression in children and adolescents.* New York: Plenum Press.

Compas, B. E., Langrock, A. M., Keller, G., Merchant, M. J., & Copeland, M. E. (2002). Children coping with parental depression: Processes of adaptation to family stress. In I. H. E. Gotlib (Ed.), *Children of depressed parents: Mechanisms of risk and implications for treatment* (pp. 227–252). Washington, DC: American Psychological Association.

Compas, B. E., Malcarne, V. L., & Fondacaro, K. M. (1988). Coping with stressful events in older children and young adolescents. *Journal of Consulting and Clinical Psychology, 56,* 405–411.

Compton, K., Snyder, J., Schrepferman, L., Bank, L., & Shortt, J. W. (2003). The contribution of parents and siblings to antisocial and depressive behavior in adolescents: A double jeopardy coercion model. *Development & Psychopathology, 15,* 163–182.

Conger, R. D., Conger, K. D., Matthews, L. S., & Elder, G. H., Jr. (1999). Pathways of economic influence on adolescent adjustment. *American Journal of Community Psychology, 27,* 519–541.

Coyne, J. C. (1994). Self-reported distress: Analogues or ersatz depression? *Psychological Bulletin, 116*, 29–45.

Cummings, E. M., & Davies, P. T. (1994). Maternal depression and child development. *Journal of Child Psychology and Psychiatry, 35*, 73–112.

Davidson, R. J., Ekman, P., Saron, C., Senulis, R., & Friesen, W. V. (1990). Approachwithdrawal and cerebral asymmetry: Emotional expression and brain physiology I. *Journal of Personality and Social Psychology, 58*, 330–341.

Davidson, R. J., & Fox, N. A. (1989). Frontal brain asymmetry predicts infants' response to maternal separation. *Journal of Abnormal Psychology, 98*, 127–131.

Davies, P. T., & Dumenci, L. (1999). The interplay between maternal depressive symptoms and marital distress in the prediction of adolescent adjustment. *Journal of Marriage and the Family, 61*, 238–254.

Davies, P. T., & Windle, M. (1997). Gender-specific pathways between maternal depressive symptoms, family discord, and adolescent adjustment. *Developmental Psychology, 33*, 657–668.

Davis, B., Sheeber, L., & Hops, H. (2002). Coercive family processes and adolescent depression. In J. Snyder (Ed.), *The Oregon model of antisocial behavior: Research and interventions* (pp. 173–194). Washington, DC: APA Press.

Dawson, G. (1999, August). *The effects of maternal depression on children's emotional and psychobiological development*. Paper presented at the National Institute of Mental Health Conference on Parenting, Bethesda, MD.

Dawson, G., Ashman, S. B., Panagiotides, H., Hessl, D., Self, J., Yamada, E., et al. (2003). Preschool outcomes of children of depressed others: Role of maternal behavior, contextual risk, and children's brain activity. *Child Development, 74*, 1158–1175.

Dawson, G., Frey, K., Panagiotides, H., Osterling, J., & Hessl, D. (1997). Infants of depressed mothers exhibit atypical frontal brain activity: A replication and extension of previous findings. *Journal of Child Psychology and Psychiatry, 38*, 179–186.

Dawson, G., Frey, K., Self, J., Panagiotides, H., Hessl, D., Yamada, E., et al. (1999). Frontal brain electrical activity in infants of depressed and nondepressed mothers: Relation to variations in infant behavior. *Development and Psychopathology, 11*, 589–605.

Dawson, G., Grofer Klinger, L., Panagiotides, H., Hill, D., & Spieker, S. (1992). Frontal lobe activity and affective behavior of infants of mothers with depressive symptoms. *Child Development, 63*, 725–737.

Dino, G. A., Barnett, M. A., & Howard, J. A. (1984). Children's expectations of sex differences in parents' responses to sons and daughters encountering interpersonal problems. *Sex Roles, 11*, 709–717.

Donenberg, G. R., & Weisz, J. R. (1998). Guilt and abnormal aspects of parent-child interaction. In J. Bybee (Ed.), *Guilt and children* (pp. 245–267) San Diego, CA: Academic Press, Inc.

Du Rocher Schudlich, T. D., & Cummings, E. M. (2003). Parental dysphoria and children's internalizing symptoms: Marital conflict styles as mediators of risk. *Child Development, 74*, 1663–1681.

Eamon, M. K., & Zuehl, R. M. (2001). Maternal depression and physical punishment as mediators of the effect of poverty on socioemotional problems of children in single-mother families. *American Journal of Orthopsychiatry, 71,* 218–226.

Eaves, L. J., Silberg, J. L., Meyer, J. M., Maes, H. H., Simonoff, E., Pickles, A., et al. (1997). Genetics and developmental psychopathology: II. The main effects of genes and environment on behavioral problems in the Virginia Twin Study of Adolescent Behavioral Development. *Journal of Child Psychology and Psychiatry, 38,* 965–980.

Egeland, B., & Farber, E. (1984). Infant-mother attachment: Factors related to its development and changes over time. *Child Development, 55,* 753–771

Eisenberg, N., Cumberland, A., & Spinrad, T. L. (1998). Parental socialization of emotion. *Psychological Inquiry, 9,* 241–273.

Eley, T. C., & Stevenson, J. (1999). Exploring the covariation between anxiety and depression symptoms: A genetic analysis of the effect of age and sex. *Journal of Child Psychology and Psychiatry, 40,* 1273–1284.

El-Sheikh, M., & Flanagan, E. (2001). Parental problem drinking and children's adjustment: Family conflict and parental depression as mediators and moderators of risk. *Journal of Abnormal Child Psychology, 29,* 417–432.

Evans, J., Heron, J., Francomb, H., Oke, S., Golding, J., & Colleagues. (2001). Cohort study of depressed mood during pregnancy and after childbirth. *British Medical Journal, 323,* 257–260.

Fergusson, D. M., Horwood, L. J., & Lynskey, M. T. (1995). Maternal depressive symptoms and depressive symptoms in adolescents. *Journal of Child Psychology and Psychiatry, 36,* 1161–1178.

Fergusson, D. M., Lynskey, M. T., & Horwood, L. J. (1993). The effect of maternal depression on maternal ratings of child behavior. *Journal of Abnormal Child Psychology, 21,* 245–269.

Field, T. (1992). Infants of depressed mothers. *Development and Psychopathology, 4,* 49–66.

Field, T. (2002). Prenatal effects of maternal depression. In S. H. Goodman & I. H. Gotlib (Eds.), *Children of depressed parents: mechanisms of risk and implications for treatment* (pp. 59–88). Washington, DC: APA Books.

Fincham, F. D., Beach, S. R. H., Harold, G. T., & Osborne, L. N. (1997). Marital satisfaction and depression: Different causal relationships for men and women? *Psychological Science, 8,* 351–357.

Finman, R., Davidson, R. J., Colton, M. B., Straus, A. M., & Kagan, J. (1989). Psychophysiological correlates of inhibition to the unfamiliar in children [Abstract]. *Psychophysiology, 26*(26), S24.

Forehand, R., Brody, G., Slotkin, J., Fauber, R., McCombs, A. (1988). Young Adolescent & maternal depression: Assessment, interrelations, and family predictors. *Journal of Consulting and Clinical Psychology, 56,* 422–426.

Forehand, R., & McCombs, A. (1988). Unraveling the antecedent-consequence conditions in maternal depression and adolescent functioning. *Behaviour Research & Therapy, 26*(5), 399–405.

Fowler, E. P. (2002). *Longitudinal reciprocal relations between maternal depressive symptoms and adolescent internalizing and externalizing symptoms.* Unpublished Doctoral Dissertation, Vanderbilt University, Nashville.

Frank, E., & Young, E. (2000). Pubertal challenges and adolescent challenges. In E. Frank (Ed.), *Gender and its effects on psychopathology* (pp. 85–102). Washington, DC: American Psychiatric Press.

Gallimore, M., & Kurdek, L. A. (1992). Parent depression and parent authoritative discipline as correlates of young adolescents' depression. *Journal of Early Adolescence, 12,* 187–196.

Garber, J., Braafladt, N., & Zeman, J. (1991). The regulation of sad affect: An information-processing perspective. In K. A. Dodge (Ed.), *The development of emotion regulation and dysregulation* (pp. 208–242). Cambridge, England: Cambridge University Press.

Garber, J., & Martin, N. C. (2002). Negative cognitions in offspring of depressed parents: Mechanisms of risk. In S. H. Goodman & I. H. Gotlib (Eds.), *Children of depressed parents: Mechanisms of risk and implications for treatment* (pp. 121–154). Washington, DC: American Psychological Association.

Ge, X., Conger, R. D., & Elder, G. H. (1996). Coming of age too early: Pubertal influences on girls' vulnerability to psychological distress. *Child Development, 67,* 3386–3400.

Ge, X., Conger, R. D., & Elder, G. H. J. (2001). The relation between puberty and psychological distress in adolescent boys. *Journal of Research on Adolescence, 11,* 49–70.

Ge, X., Lorenz, F. O., Conger, R. D., Elder, G. H., & Simons, R. L. (1994). Trajectories of stressful life events and depressive symptoms during adolescence. *Developmental Psychology, 30,* 467–483.

Gelfand, D. M., & Teti, D. M. (1990). The effects of maternal depression on children. *Clinical Psychology Review, 10,* 329–353.

Ghodsian, M., Zayicek, E., & Wolkind, S. (1984). A longitudinal study of maternal depression and child behavior problems. *Journal of Child Psychology and Psychiatry, 25,* 91–109.

Gladstone, T. R. G., & Beardslee, W. R. (2002). Treatment, intervention, and prevention with children of depressed parents: A developmental perspective. In S. H. Goodman & I. H. Gotlib (Eds.), *Children of Depressed Parents: Mechanisms of Risk and Implications for Treatment* (pp. 277–305). Washington, DC: American Psychological Association.

Glover, V., Teixeira, J., Gitau, R., & Risk, N. (1998, April). *Links between antenatal maternal anxiety and the fetus.* Paper presented at the International Conference on Infant Studies, Atlanta, GA.

Goodman, S. H., Adamson, L. B., Riniti, J., & Cole, S. (1994). Mothers' expressed attitudes: Associations with maternal depression and children's self-esteem and psychopathology. *Journal of the American Academy of Child and Adolescent Psychiatry, 33,* 1265–1274.

Goodman, S. H., Broth, M., Hall, C. M., & Stowe, Z. N. (2003). *Treatment of postpartum depression in mothers: Secondary benefits to the infants.* Paper presented at the Society for Research in Child Development, Tampa.

Goodman, S. H., Broth, M. R., Hall, C. M., & Stowe, Z. N. (in press). Treatment of postpartum depression in mothers: Secondary benefits to the infants. *Infant Mental Health Journal: Special Issue on Perinatol Mood Disorders.*

Goodman, S. H., & Gotlib, I. H. (1999). Risk for psychopathology in the children of depressed mothers: A developmental model for understanding mechanisms of transmission. *Psychological Review, 106*, 458–490.

Goodman, S. H., Sewell, D. R., Cooley, E. L., & Leavitt, N. (1993). Assessing levels of adaptive functioning: The Role Functioning Scale. *Community Mental Health Journal, 29*, 119–131.

Goodman, S. H., & Tully, E. (2003). *Depression in spouses and parents: Problems and solutions.* Paper presented at the Sex, Marriage, and Family & the Religions of the Book: Modern Problems, Enduring Solutions.

Gotlib, I. H., Gilboa, E., & Sommerfeld, B. K. (2000). Cognitive functioning in depression: Nature and origins. In R. J. Davidson (Ed.), *Wisconsin Symposium on Emotion* (Vol. 1). New York: Oxford University Press.

Gotlib, I. H., & Goodman, S. H. (1999). Children of parents with depression. In W. K. Silverman & T. H. Ollendick (Eds.), *Developmental issues in the clinical treatment of children and adolescents* (pp. 415–432). New York: Allyn & Bacon.

Gotlib, I. H., Lewinsohn, P. M., & Seeley, J. R. (1995). Symptoms versus a diagnosis of depression: Differences in psychosocial functioning. *Journal of Consulting and Clinical Psychology, 63*, 90–100.

Graham, Y. P., Heim, C., Goodman, S. H., Miller, A. H., & Nemeroff, C. B. (1999). The effects of neonatal stress on brain development: Implications for psychopathology. *Development and Psychopathology, 11*, 545–565.

Grant, K. E., Compas, B. E., Stuhlmacher, A. F., Thurm, A. E., McMahon, S. D., & Halpert, J. A. (2003). Stressors and child and adolescent psychopathology: Moving from markers to mechanisms of risk. *Psychological Bulletin, 129*, 447–466.

Green, J. M., & Murray, D. (1994). The use of the Edinburgh Postnatal Depression Scale in research to explore the relationship between antenatal and postnatal dysphoria. In J. Cox & J. Holden (Eds.), *Perinatal psychiatry: Use and misuse of the Edinburgh Postnatal Depression Scale* (pp. 180–198). London: Gaskell.

Hamilton, E. B., Asarnow, J. R., & Tompson, M. C. (1999). Family interaction styles of children with depressive disorders, schizophrenia-spectrum disorders, and normal controls. *Family Process, 38*, 463–476.

Hammen, C., & Brennan, P. A. (2001). Depressed adolescents of depressed and nondepressed mothers: Tests of an interpersonal impairment hypothesis. *Journal of Consulting and Clinical Psychology, 69*, 284–294.

Hammen, C., & Brennan, P. A. (2003). Severity, chronicity, and timing of maternal depression and risk for adolescent offspring diagnoses in a community sample. *Archives of General Psychiatry, 60*, 253–258.

Hammen, C., Burge, D., Burney, E., & Adrian, C. (1990). Longitudinal study if diagnoses in children of women with unipolar and bipolar affective disorder. *Archives of General Psychiatry, 47*, 1112–1117.

Hammen, C., Gordon, D., Burge, D., Adrian, C., Jaenicke, C., & Hiroto, D. (1987). Maternal affective disorders, illness, and stress: Risk for children's psychopathology. *American Journal of Psychiatry, 144*, 736–741.

Hammen, C. L. (2002). Context of stress in families of children with depressed parents. In S. H. Goodman & I. H. Gotlib (Eds.), *Children of depressed parents: Mechanisms of risk and implications for treatment* (pp. 175–202). Washington, DC: American Psychological Association.

Hammen, C. L., Burge, D., & Stansbury, K. (1990). Relationship of mother and child variables to child outcomes in a high risk sample: A causal modeling analysis. *Developmental Psychology, 26*, 24–30.

Hammen, C. L., Shih, J., Altman, T., & Brennan, P. A. (2003). Interpersonal impairment and the prediction of depressive symptoms in adolescent children of depressed and nondepressed mothers. *Journal of the American Academy of Child & Adolescent Psychiatry, 42*, 571–577.

Handley, S. L., Dunn, T. L., Waldron, G., & Baker, J. M. (1980). Tryptophan, cortisol and puerperal mood. *British Journal of Psychiatry, 136*, 498–508.

Hankin, B. L., & Abramson, L. Y. (1999). Development of gender differences in depression: Description of possible explanations. *Annals of Medicine, 31*, 372–379.

Happonen, M., Pulkkinen, L., Kaprio, J., Van der Meere, J., Viken, R. J., & Rose, R. J. (2002). The heritability of depressive symptoms: Multiple informants and multiple measures. *Journal of Child Psychology and Psychiatry and Allied Disciplines, 43*, 471–480.

Harnish, J. D., Dodge, K. A., Valente, E., & Colleagues.(1995). Mother-child interaction quality as a partial mediator of the roles of maternal depressive symptomatology and socioeconomic status in the development of child behavior problems. *Child Development, 66*, 739–753.

Harrington, R. (1996). Family-genetic findings in child and adolescent depressive disorders. *International Review of Psychiatry, 8*, 355–368.

Harrington, R., Rutter, M., Weissman, M. M., Fudge, H., Groothues, C., Bredenkamp, D., et al. (1997). Psychiatric disorders in the relatives of depressed probands: I. Comparison of prepubertal, adolescent and early adult onset cases. *Journal of Affective Disorders, 42*, 9–22.

Hay, D. F., & Pawlby, S. (2001). Intellectual problems shown by 11-year-old children whose mothers had postnatal depression. *Journal of Child Psychology and Psychiatry and Allied Disciplines, 42*, 871–889.

Hertsgaard, L., Gunnar, M., Erickson, M., & Nachmias, M. (1995). Adrenocortical response to the strange situation in infants with disorganized/disoriented attachment relationships. *Child Development, 66*, 1100–1106.

Hetherington, E. M., & Martin, B. (1986). Family factors and psychopathology in children. In J. S. Werry (Ed.), *Psychological disorders of childhood* (3rd ed.). New York: Wiley.

Hill, J. P., Holmbeck, G. N., Marlow, L., Green, T. M., & Lynch, M. E. (1985). Menarcheal status and parent-child relations in families of seventh-grade girls. *Journal of Youth and Adolescence, 14*, 301–316.

Hill, J. P., & Lynch, M. E. (1983). The intensification of gender-related role expectations during early adolescence. In A. Peterson (Ed.), *Girls at puberty: Biological and psychological perspectives* (pp. 201–208). New York: Plenum Press.

Hilsman, R. (2001). *Maternal expressed emotion, children's self-cognitions, and psychopathology in children of depressed and nondepressed mothers.* Vanderbilt University, Nashville.

Hirsch, B. J., Moos, R. H., & Reischl, T. M. (1985). Psychosocial adjustment of adolescent children of a depressed, arthritic, or normal parent. *Journal of Abnormal Psychology, 94*, 154–164.

Holmbeck, G. (1997). Toward terminological, conceptual, and statistical clarity in the study of mediators and moderators: Examples from the child-clinical and pediatric psychology literatures. *Journal of Consulting and Clinical Psychology, 65,* 599–610.

Hops, H. (1992). Parental depression and child behaviour problems: Implications for behavioural family intervention. *Behaviour Change, 9,* 126–138.

Hops, H. (1996). Intergenerational transmission of depressive symptoms: Gender and developmental considerations. In P. Fiedler (Ed.), *Interpersonal factors in the origin and course of affective disorders.* London: Gaskell/Royal College of Psychiatrists.

Hops, H., Lewinsohn, P., Andrews, J. A., & Roberts, R. E. (1990). Psychosocial correlates of depressive symptomatology among high school students. *Journal of Clinical Child Psychology, 19,* 211–220.

Hops, H., Sherman, L., & Biglan, A. (1990). Maternal depression, marital discord, and children's behavior: A developmental perspective. In G. R. Patterson (Ed.), *Depression and aggression in family interaction* (pp. 185–208). Hillsdale, NJ: Erlbaum.

Jaenicke, C., Hammen, C. L., Zupan, B., Hiroto, D., Gordon, D., Adrian, C., et al. (1987). Cognitive vulnerability in children at risk for depression. *Journal of Abnormal Child Psychology, 15,* 559–572.

Jones, N. A., Field, T., Davalos, M., & Pickens, J. (1997). EEG stability in infants/children of depressed mothers. *Child Psychiatry & Human Development, 28,* 59–70.

Kam, C.-M. (2001). *Effects of maternal depression on children's social and emotional functioning during early school year.* Pennsylvania State University, University Park.

Kashani, J. H., Burbach, D. J., & Rosenberg, T. K. (1988). Perception of family conflict resolution of depressive symptomatology in adolescents. *Journal of the American Academy of Child & Adolescent Psychiatry, 27,* 42–48.

Kazdin, A. E., & Whitley, M. K. (2003). Treatment of parental stress to enhance therapeutic change among children referred for aggressive and antisocial behavior. *Journal of Consulting & Clinical Psychology, 71,* 504–515.

Keenan, K., & Shaw, D. (1997). Developmental and social influences on young girls' early problem behavior. *Psychological Bulletin, 121,* 95–113.

Keller, M. B., Beardslee, W. R., Dorer, D. J., Lavori, P. W., Samuelson, H., & Klerman, G. R. (1986). Impact of severity and chronicity of parental affective illness on adaptive functioning and psychopathology in children. *Archives of General Psychiatry, 43,* 930–937.

Keller, M. B., Shapiro, R. W., Lavori, P. W., & Wolfe, N. (1982). Recovery in major depressive disorder: Analysis with the life table and regression models. *Archives of General Psychiatry, 39,* 905–910.

Kendler, K. S., Gardner, C. O., Neale, M. C., & Prescott, C. A. (2001). Genetic risk factors for major depression in men and women: similar or different heritabilities and same or partly distinct genes? *Psychological Medicine, 31,* 605–616.

Kendler, K. S., Kessler, R. C., Walters, E. E., MacLean, C., Neale, M. C., Heath, A. C., et al. (1995). Stressful life events, genetic liability, and onset of an episode of major depression in women. *American Journal of Psychiatry, 152,* 833–842.

Kendler, K. S., Myers, J., & Prescott, C. A. (2000). Parenting and adult mood, anxiety, and substance use disorders in women: An epidemiological, multiinformant, retrospective study. *Psychological Medicine, 30*, 281–294.

Kendler, K. S., & Prescott, C. A. (1999). A population-based twin study of lifetime major depression in men and women. *Archives of General Psychiatry, 56*, 39–44.

Kendler, K. S., Thornton, L. M., & Gardner, C. O. (2000). Stressful life events and prior episodes in the aetiology of major depression in women: An evaluation of the "kindling" hypothesis. *American Journal of Psychiatry, 157*, 1243–1251.

Kessler, R. C. (2000). Gender differences in major depression: Epidemiological findings. In E. Frank (Ed.), *Gender and its effects on psychopathology* (pp. 61–84). Washington, DC: American Psychiatric Press.

Kessler, R. C., McGonagle, K. A., Swartz, M., Blazer, D. G., & Nelson, C. B. (1993). Sex and depression in the National Comorbidity Survey I: Lifetime prevalence, chronicity and recurrence. *Journal of Affective Disorders, 29*, 85–96.

Kochanska, G., Radke-Yarrow, M., Kuczynski, L., & Friedman, S. (1987). Normal and affectively ill mothers' beliefs about their children. *American Journal of Orthopsychiatry, 57*, 345–350.

Lee, C. M., & Gotlib, I. H. (1989). Clinical status and emotional adjustment of children of depressed. *American Journal of Psychiatry, 146*, 478–483.

Levinson, D. F., Zubenko, G. S., Crowe, R. R., DePaulo, J. R., Scheftner, W. S., et al. (2003). Genetics of recurrent early-onset depression (Gen RED): Design and preliminary clinical characteristics of a repository sample for genetic linkage studies. *American Journal of Medical Genetics (Neuropsychiatric Genetics), 119B*, 118–130.

Lieb, R., Isensee, B., Hofler, M., Pfister, H., & Wittchen, H.-U. (2002). Parental major depression and the risk of depression and other mental disorders in offspring. *Archives of General Psychiatry, 59*, 365–374.

Loehlin, J. C., & Nichols, R. C. (1976). *Heredity, environment, and personality*. Austin: University of Texas Press.

Lovejoy, M. C., Graczyk, P. A., O'Hare, E., &, & Neuman, G. (2000). Maternal depression and parenting behavior: A meta-analytic review. *Clinical Psychology Review, 20*, 561–592.

Lundy, B., Jones, N., Field, T., Pietro, P., Nearing, G., Davalos, M., et al. (1999). Prenatal depression effects on neonates. *Infant Behavior and Development, 22*, 119–129.

Lundy, B. L. (2002). Paternal socio-psychological factors and infant attachment: The mediating role of synchrony in father-infant interactions. *Infant Behavior and Development, 25*, 221–236.

Luoma, I., Tamminen, T., Kaukonen, P., Laippala, P., Puura, K., Salmelin, R., et al. (2001). Longitudinal study of maternal depressive symptoms and child well-being. *Journal of the American Academy of Child & Adolescent Psychiatry, 40*, 1367–1374.

Lykken, D. T., & Tellegen, A. (1996). Happiness is a stochatic phenomenon. *Psychological Science, 7*, 186–189.

Lyons-Ruth, K., Connell, D. B., Grunebaum, H. U., & Botein, S. (1990). Infants at social risk: Maternal depression and family support as mediators of infant development and security of attachment. *Child Development, 61*, 85–98.

Lyons-Ruth, K., Lyubchik, A., Wolfe, R., & Bronfman, E. (2002). Parental depression and child attachment: Hostile and helpless profiles of parent and child behavior among families at risk. In S. H. Goodman & I. H. Gotlib (Eds.), *Children of depressed parents: Mechanisms of risk and implications for treatment* (pp. 89–120). Washington, DC: American Psychological Association.

Maccoby, E. E. (1990). Gender and relationships: A developmental account. *American Psychologist, 45*, 513–520.

Malcarne, V. L., Hamilton, N. A., Ingram, R. E., & Taylor, L. (2000). Correlates of distress in children at risk for affective disorder: Exploring predictors in the offspring of depressed and nondepressed mothers. *Journal of Affective Disorders, 59*, 243–251.

Marchand, J. F., & Hock, E. (1998). The relation of problem behaviors in preschool children to depressive symptoms in mothers and fathers. *The Journal of Genetic Psychology, 159*, 353–366.

Matthews, S. G. (2002). Early programming of the hypothalamo-pituitary-adrenal axis. *Trends in Endocrinology & Metabolism, 13*, 373–380.

McCarty, C. A., & McMahon, R. J. (2003). Mediators of the relation between maternal depressive symptoms and child internalizing and disruptive behavior disorders. *Journal of Family Psychology, 17*, 545–556.

McGrath, E., Keita, G. P., Strickland, B. R., & Russo, N. F. (1990). *Women and depression: Risk factors and treatment issues*. Washington, DC: American Psychological Association.

Milberger, S., Biederman, J., Faraone, S. V., Chen, L., & Jones, J. (1996). Is maternal smoking during pregnancy a risk factor for attention deficit hyperactivity disorder in children? *American Journal of Psychiatry, 153*, 1138–1142.

Moldin, S. O. (1999). Report of the NIMH's Genetic Workgroups: Summary of Research. *Biological Psychiatry, 45*, 559–602.

Murray, K. T., & Sines, J. O. (1996). Parsing the genetic and nongenetic variance in children's depressive behavior. *Journal of Affective Disorders, 38*, 23–34.

Murray, L. C., & Kempton, C., Woolgar, M., & Hooper, R.(1993). Depressed mothers' speech to their infants and its relation to infant gender and cognitive development. *Journal of Child Psychology and Psychiatry and Allied Disciplines, 34*, 1083–1101.

Nelson, D. R. (2001). *Children of depressed mothers: The role of expressed emotion*. Unpublished Doctoral Dissertation, University of California at Los Angeles, Los Angeles.

Newport, D. J., Stowe, A. N., & Nemeroff, C. B. (2002). Parental depression: Animal models of an adverse life event. *The American Journal of Psychiatry, 159*, 1265–1283.

Nolen-Hoeksema, S. (1998). Ruminative coping with depression. In C. S. Dweck (Ed.), *Motivation and self-regulation across the life span* (pp. 237–256). New York: Cambridge University Press.

Nolen-Hoeksema, S., & Girgus, J. (1994). The emergence of gender differences in depression during adolescence. *Psychological Bulletin, 115*, 424–443.

Nolen-Hoeksema, S., Wolfson, A., Mumme, D., & Guskin, K. (1995). Helplessness in children of depressed and nondepressed mothers. *Developmental Psychology, 31*, 377–387.

Nulman, I., Rovet, J., Stewart, D. E., Wolpin, J., Gardner, H. A., Teis, J. G., et al. (1997). Neurodevelopment of children exposed in utero to antidepressant drugs. *New England Journal of Medicine, 336,* 258–262.

O'Hara, M. W. (1986). Social support, life events, and depression during pregnancy and the puerperium. *Archives of General Psychiatry, 43,* 569–573.

O'Hara, M. W., Zekoski, E. M., Philipps, L. H., & Wright, E. J. (1990). A controlled, prospective study of postpartum mood disorders: Comparison of childbearing and non-childbearing women. *Journal of Abnormal Psychology, 99,* 3–15.

Orvaschel, H., Walsh-Allis, G., & Ye, W. (1988). Psychopathology in children of parents with recurrent depression. *Journal of Abnormal Child Psychology, 16,* 17–28.

Peterson, A. C., Sarigiani, P. A., & Kennedy, R. E. (1991). Adolescent depression: Why more girls? *Journal of Youth and Adolescence, 20,* 247–271.

Piccinelli, M., & Wilkinson, G. (1994). Outcome of depression in psychiatric settings. *British Journal of Psychiatry, 164,* 297–304.

Plomin, R., Emde, R. N., Braungart, J. M., Campos, J., Corley, R. P., Fulker, D. W., et al. (1993). Genetic change and continuity from fourteen to twenty months: The MacArthur Longitudinal Twin Study. *Child Development, 64,* 1354–1376.

Prinstein, M. J., & LaGreca, A. M. (1999). Links between mothers' and children's social competence and associations with maternal adjustment. *Journal of Clinical Child Psychology, 28,* 197–210.

Radke-Yarrow, M., Belmont, B., Nottelman, E., & Bottomly, L. (1990). Young children's self-conceptions: Origins in the natural discourse of depressed and normal mothers and their children. In M. Beeghly (Ed.), *The self in transition: Infancy to childhood* (pp. 345–361). Chicago: University of Chicago Press.

Radke-Yarrow, M., & Klimes-Dougan, B. (2002). Parental depression and offspring disorders: A developmental perspective. In S. H. Goodman & I. H. Gotlib (Eds.), *Children of depressed parents: Mechanisms of risk and implications for treatment* (pp. 155–174). Washington, DC: American Psychological Association.

Radke-Yarrow, M., Nottelman, E., Belmont, B., & Welsch, J. D. (1993). Affective interactions of depressed and nondepressed mothers and their children. *Journal of Abnormal Child Psychology, 21,* 683–695.

Radke-Yarrow, M., Zahn-Waxler, C., Richardson, D., & Susman, A. (1994). Caring behavior in children of clinically depressed and well mothers. *Child Development, 65,* 1405–1414.

Radloff, L. S. (1977). The CES-D Scale: A self-report depression scale for research in the general population. *Applied Psychological Measurement, 1,* 385–401.

Rende, R. D., Plomin, R., Reiss, D., & Hetherington, E. M. (1993). Genetic and environmental influences on depressive symptomatology in adolescence: Individual differences and extreme scores. *Journal of Child Psychology and Psychiatry, 34,* 1387–1398.

Ressler, K. J., & Nemeroff, C. B. (2000). Role of serotonergic and noradrenergic systems in the pathophysiology of depression and anxiety disorders. *Depression & Anxiety, 12* (Suppl 1), 2–19.

Rice, F., Harold, G. T., & Thapar, A. (2002). The genetic aetiology of childhood depression: A review. *Journal of Child Psychology and Psychiatry and Allied Disciplines, 43,* 65–79.

Roderick, H. A. (2002). *Family stability as a mediator of the relationship between maternal attributes and child psychosocial adjustment.* Unpublished Doctoral Dissertation, State University of New York at Albany.

Rubin, K. H., & Mills, R. S. (1991). Conceptualizing developmental pathways to internalizing disorders in childhood. *Canadian Journal of Behavioural Science, 23,* 300–317.

Sanchez, M. M., Ladd, C. O., & Plotsky, P. M. (2001). Early adverse experience as a developmental risk factor for later psychopathology: Evidence from rodent and primate models. *Development and Psychopathology, 13,* 419–449.

Scourfield, J., Rice, F., Thapar, A., Harold, G. T., Martin, N. C., & McGuffin, P. (2003). Depressive symptoms in children and adolescents: Changing aetiological influences with development. *Journal of Child Psychology and Psychiatry and Allied Disciplines, 44,* 968–976.

Sheeber, L., Davis, B., & Hops, H. (2002). Gender-specific vulnerability to depression in children of depressed mothers. In S. H. Goodman & I. H. Gotlib (Eds.), *Children of depressed parents: Mechanisms of risk and implications for treatment* (pp. 253–274). Washington, DC: American Psychological Association.

Sheeber, L., Hops, H., & Davis, B. (2001). Family processes in adolescent depression. *Clinical Child and Family Psychology Review, 4,* 19–35.

Silberg, J., Pickles, A., Rutter, M., Hewitt, J., Simonoff, E., Maes, H., et al. (2001). The influence of genetic factors and life stress on depression among adolescent girls. *Archives of General Psychiatry, 56,* 225–232.

Silberg, J., & Rutter, M. (2001). Genetic moderation of environmental risk for depression and anxiety in adolescent girls. *British Journal of Psychiatry, 179,* 116–121.

Silberg, J., & Rutter, M. (2002). Nature-nurture interplay in the risks associated with parental depression. In S. H. Goodman & I. H. Gotlib (Eds.), *Children of depressed parents: Mechanisms of risk and implications for treatment* (pp. 13–36). Washington, DC: American Psychological Association.

Smith, R., Cubis, J., Brinsmead, M., Lewin, T., Singh, B., Owens, P., Eng-Cheng, C., et al. (1990). Mood changes, obstetrical experience and alterations in plasma cortisol, beta-endorphin and corticotrophin releasing hormone during pregnancy and the puerperium. *Journal of Psychosomatic Research, 34,* 53–69.

Snyder, J. (1991). Discipline as a mediator of the impact of maternal stress and mood on child conduct problems. *Development & Psychopathology, 3,* 263–276.

Stark, K. D., Humphrey, L. L., Laurent, J., Livingston, R., & Christopher, J. (1993). Cognitive, behavioral, and family factors in the differentiation of depressive and anxiety disorders during childhood. *Journal of Consulting and Clinical Psychology, 61,* 878–886.

Steinberg, L. (1987). Recent research on the family at adolescence: The extent and nature of sex differences. *Journal of Youth and Adolescence, 16,* 191–197.

Steinberg, L. (1988). Reciprocal relation between parent-child distance and pubertal maturation. *Developmental Psychology, 24,* 122–128.

Stephens, R. S., Hokanson, J. E., & Welker, R. (1987). Responses to depressed interpersonal behavior: Mixed reactions in a helping role. *Journal of Personality and Social Psychology, 52,* 1274–1282.

Sullivan, P. F., Neale, M. C., & Kendler, K. S. (2000). Genetic epidemiology of major depression: Review and meta-analysis. *American Journal of Psychiatry, 157,* 1552–1562.

Tellegen, A., Lykken, D. T., Bouchard, T. J., Wilcox, K. J., Segal, N. L., & Rich, S. (1988). Personality similarity in twins reared apart and together. *Journal of Personality and Social Psychology, 54,* 1031–1039.

Thapar, A., & McGuffin, P. (1994). A twin study of depressive symptoms in childhood. *British Journal of Psychiatry, 165,* 259–265.

Tronick, E. Z., & Gianino, A. F. (1986). The transmission of maternal disturbance to the infant. In T. Field (Ed.), *Maternal depression and infant disturbance* (pp. 5–12). San Francisco: Jossey-Bass.

Tsuang, M. T., & Faraone, S. V. (1990). *The genetics of mood disorders.* Baltimore: Johns Hopkins University Press.

van IJzendoorn, M. H., Goldberg, S., Kroonenberg, P. M., & Frenkel, O. J. (1992). The relative effects of maternal and child problems on the quality of attachment: A meta-analysis of attachment in clinical samples. *Child Development, 63,* 840–858.

Wadhwa, P. D., Glynn, L., Hobel, C. J., Garite, T. J., Porto, M., Chicz-DeMet, A., et al. (2002). Behavioral perinatology: Biobehavioral processes in human fetal development. *Regulatory Peptides, 108,* 149–157.

Warner, V., Weissman, M. M., Fendrich, M., Wickramaratne, P., & Moreau, D. (1992). The course of major depression in the offspring of depressed parents: Incidence, recurrence, and recovery. *Archives of General Psychiatry, 49,* 795–801.

Weissman, M. M., Prusoff, B., Gammon, G. D., Merikangagas, K. R., Leckman, J. F., Kidd, K. K. (1984). Psychopathology in the children (ages 6–18) of depressed and normal parents. *Journal of the American Academy, 23,* 78–84.

Welner, Z., Welner, A., McCrary, M., & Leonard, M. A. (1977). Psychopathology in children of inpatients with depression: A controlled study. *Journal of Nervous and Mental Disease, 164,* 408–413.

Zahn-Waxler, C., Cole, P. M., & Barrett, K. C. (1991). Guilt and empathy: Sex differences and implications for the development of depression. In K. A. Dodge (Ed.), *The development of emotion regulation and dysregulation.* Cambridge, UK: Cambridge University Press.

Zahn-Waxler, C., Cummings, E. M., Iannotti, R. J., & Radke-Yarrow, M. (1984). Young children of depressed parents: A population at risk for affective problems. In D. Cicchetti (Ed.), *Childhood depression. New directions for child development* (Vol. 26, pp. 81–105). San Francisco: Jossey-Bass.

Zahn-Waxler, C., Iannotti, R. J., Cummings, E. M., & Denham, S. (1990). Antecedents of problem behaviors in children of depressed mothers. *Development & Psychopathology, 2,* 271–291.

Zahn-Waxler, C., & Robinson, J. (1995). Empathy and guilt: Early origins of feelings of responsibility. In K. W. Fischer (Ed.), *Self-conscious emotions: The psychology of shame, guilt, embarrassment, and pride* (pp. 143–173). New York: Guilford Press.

Zahn-Waxler, C., & Smith, K. D. (1992). The development of prosocial behavior. In M. Hersen (Ed.), *Handbook of social development: A lifespan perspective.* New York: Plenum Press.

Zahn-Waxler, C., Usher, B. A., & Goldman, J. (2003). *Intropunitive, introspective, and interpersonal risk factors for depression*. Paper presented at the Society for Research in Child Development, Tampa, FL.

Zubenko, G. S., Hughes, H. B., Stiffler, J. S., Zubenko, W. N., & Kaplan, B. B. (2002). Genome survey for susceptibility loci for recurrent, early-onset major depression: Results at 10cM. *American Journal of Medical Genetics (Neuropsychiatric Genetics)*, *114*, 413–422.

Zubenko, G. S., Hughes, H. B., Maher, B. H., Stiffler, J. S., Zubenko, W. N., & Marazita, M. L. (2003). Genetic linkage of region containing the CREP1gene to depressive disorders in women from families with recurrent, early-onset, major depression. *American Journal of Medical Genetics (Neuropsychiatric Genetics)*, *114*, 980–987.

PART IV

SOCIAL, POLITICAL, AND ECONOMIC MODELS OF RISK

Social Suffering, Gender, and Women's Depression

Jeanne Marecek

> There is nothing the matter with one but a temporary nervous depression – a slight hysterical tendency – what is one to do? (p. 10)
>
> I get unreasonably angry with John sometimes. I'm sure I never used to be so sensitive. I think it is due to this nervous condition.
>
> But John says that if I feel so I shall neglect proper self-control; so I take pains to control myself – before him, at least, and that makes me very tired. (p. 11)
>
> But these nervous troubles are dreadfully depressing.
>
> John does not know how much I really suffer. He knows there is no reason to suffer, and that satisfies him. (p. 14)
>
> I don't feel as if it was worth while to turn my hand over for anything, and I'm getting dreadfully fretful and querulous.
>
> I cry at nothing, and cry most of the time. (p. 19)
>
> Charlotte Perkins Gilman, 1899/1973

These lines are from *The Yellow Wallpaper*, a short story composed by Charlotte Perkins Gilman at the turn of the 19th century. Gilman was no stranger to social suffering. The story was, as Elaine Hedges puts it, "wrenched out of her own life" (1973, p. 37). Gilman was an outspoken leader and theorist of the feminist movement of the late 1800s, regarded by some as its leading American intellectual. Gilman's focus was primarily on women's economic dependency in marriage, which she regarded as the linchpin of female subordination (Gilman, 1898). Despite her political convictions, Gilman herself was trapped by the prevailing conventions of marriage and motherhood. Shortly after the birth of her first child, she fell into grinding misery. She described her condition as a "growing melancholia," consisting of "every painful mental sensation, shame, fear, remorse, a blind oppressive confusion, utter weakness." (Gilman, 1935, p. 90). She experienced "a constant dragging weariness miles below zero. Absolute incapacity. Absolute misery." She would often "crawl into closets and under beds to hide from the grinding pressure of that profound distress" (Gilman, 1935, p. 91).

Gilman's words bespeak what we today would recognize as profound depression. She speaks of exhaustion, melancholy, misery, guilt, constant crying, and weakness. Yet, in keeping with the diagnostic conventions of her times, she was diagnosed not with depression but rather with neurasthenia and "nervous prostration" with a "slight hysterical tendency." Dr. Silas Weir Mitchell, a leading psychiatrist of the day, treated her. Mitchell was known throughout North America and Europe as an expert in the treatment of women with ailments like Gilman's. Indeed, Sigmund Freud is said to have expressed a keen interest in Mitchell's treatment methods because he himself sought to help women in similar distress.

Mitchell's prescription was a rest cure that enforced continual bed rest and constant feeding with rich, heavy food such as cream and sweets. It was common for his female patients to gain up to 40 pounds on this regimen and such weight gain was considered curative. Mitchell demanded that patients refrain not only from physical activity but also from intellectual endeavors and social stimulation. Gilman's fictional heroine, for instance, is chastised for jotting brief notes in a small notebook in violation of her doctor's orders. The regimen was both authoritarian and infantilizing, requiring female patients to submit without question to isolation, passivity, and medical control over their lives.

Throughout the novella, the heroine of *The Yellow Wallpaper* voices tentative critiques of bourgeois marriage, of the stifling norms of wifely and motherly behavior by which she must abide, of gendered power relations, and of Dr. Mitchell's rest cure. In the end, however, she is unable to sustain those critiques and plunges into the depths of madness, perhaps succumbing to suicide. Charlotte Perkins Gilman's own story has a better ending, though not a triumphant one. At the end of a month in Mitchell's sanitarium, she fled. Her condition, she proclaimed, was worse than when she had entered. She also fled her marriage because she had come to believe that she could not regain her mental equilibrium unless she lived independently. Although she married a second time, she did not live with her second husband nor depend on him for financial support for herself and her child. Instead she eked out a precarious living through her writing and speaking, sometimes taking in lodgers to make ends meet.

Why begin a discussion of women and depression with a lengthy description of a woman who lived 100 years ago and was never diagnosed as depressed? Charlotte Perkins Gilman's suffering has several lessons for those of us who are interested in the connections between gender and psychosocial suffering, and, more broadly, in psychopathology as an object of scholarly inquiry. I am not concerned to debate whether Gilman was really depressed or not. For sure, she suffered mightily. Gilman's story warns us against the intellectual mistake of reifying diagnostic categories such as depression. Like any psychiatric diagnosis, depression is a cultural category arising in a particular time and place. In Gilman's time, other categories named her suffering. Moreover, the diagnostic category of depression encompasses just a small fraction of the total field of depressive suffering.

Gilman's situation holds lessons about the gender politics of psychological suffering and of its diagnosis and treatment. *The Yellow Wallpaper* can be read as a meditation on the gap between expert knowledge and lived experience. Dominant discourses of her time, such as that of Dr. Silas Weir Mitchell, put forward an etiology of Gilman's condition that was grounded in neural physiology. (For Mitchell and other psychiatrists of his time, the term *nervous exhaustion* had the literal meaning of a nervous system overtaxed to the point of depletion.) Gilman and her heroine give voice to a different etiology, one grounded in social relations, gendered structures of power, and the confines of conventional domesticity. Gilman lived and wrote at a time (not unlike the latter decades of the 20th century) when some women were asserting new rights and demanding changes in gendered power relations in public and private life. It is hardly surprising that she and several other prominent feminists and female leaders seemed particularly prone to the conditions then called neurasthenia, nervous prostration, nervous exhaustion, and hysteria. Living in a time of social flux, they flouted the norms of proper feminine decorum. Often they were vilified or ridiculed by their social circle or by the press. No doubt they sometimes experienced self-doubts and internal conflict as they found themselves embroiled in situations fraught with contradiction.

The theme of this chapter is that cultural narratives organize, provide significance for, and influence the form, frequency, and social relations of women's depression. My orientation therefore diverges sharply from most of the other contributors, as well as the editors. I take categories of disorder, such as depression, as cultural artifacts shaped in response to prevailing concerns at different periods of history. I do not assume that such categories are universal, fixed natural entities awaiting scientific investigation. Gilman's tale points toward two contrasting narratives of women's suffering. The medical establishment of her day put forward explanatory narratives grounded in physiology and notions of inherent feminine weakness. Such explanations direct our gaze toward the individual in isolation, set apart from culture and social context. In contrast, Gilman put forward a narrative concerned with gendered power dynamics, thwarted self-expression, and culturally imposed limitations. She pointed not to inherent deficiency but to structural inequities of power. Accounts like Gilman's insist that depression and other forms of disorder are *social* suffering; they situate suffering in social, cultural, and political contexts.

The bifurcation between biologically based accounts of psychological distress and social accounts has proven remarkably durable in North American psychology. Biological foundationalism too has persisted in

North American intellectual life, particularly in psychology, in spite of a long line of trenchant critiques (e.g., Haraway, 1981). Even today, discussions of female psychology typically are initiated by first considering bodily difference (such as genes, hormones, brain differences, and neurochemistry). In that way, biology is made to appear as if it were the grounding for social experience.

In North America, the dominant accounts of depression and other categories of suffering have focused on the individual as the locus of pathology, risk, and resilience. Like Gilman, many feminist researchers have endeavored to shift attention to social relations and cultural context. Their accounts of women's suffering have emphasized social structures, cultural institutions, and social relations rather than individual deficiency or pathology. However, such models of suffering typically have been marginalized by psychological and psychiatric researchers and in the psychotherapeutic professions. The research technologies typically espoused by North American psychologists channel inquiry away from societal conditions and into the exploration of individual psyches removed from culture and history (Henriques, Hollway, Urwin, Venn, & Walkerdine, 1984). Even when psychologists set out to study the societal bases of suffering, the constraints of the discipline – its structures of thought, its measurement technologies, its preference for laboratory experiments, and its language practices – often stealthily transform projects into investigations centered on individuals lifted out from culture, society, and history. In the specialty journals in clinical psychology, the individual often appears as a carrier of weakness or deficiency, whether because of biological inheritance, faulty childhood socialization, or a life history of traumatic events. Surveying research on women several years ago, Mary Crawford and I coined the phrase "woman as problem" to describe this mode of producing knowledge (Crawford & Marecek, 1989). It remains a prominent approach to conceptualizing psychopathology (Marecek, 2001).

DEPRESSION: IS IT A "WOMAN'S PROBLEM"?

For at least three decades, therapists and researchers have identified depression as a "woman's problem." Depression was one of three high-prevalence diagnoses identified in the Conference on Women and Mental Health jointly convened by the National Institute of Mental Health (NIMH) and the American Psychological Association (APA) in 1979 (Brodsky & Hare-Mustin, 1980). Roughly 10 years later, "women and depression" was the subject of a Task Force sponsored by the Women's Program Office of the APA (McGrath, Keita, Strickland, & Russo, 1990). Ten years later (in 2000), the APA sponsored a "Summit on Women and Depression" (Mazure, Keita, & Blehar, 2002).

The claim that depression is a woman's problem rests mainly on rates and counts of people with depression. Depression is considered a woman's problem because greater numbers of women than men experience depression and because depression is especially common among women. Throughout the latter part of the 20th century, researchers in several Western European and North American societies noted a preponderance of women among those who experience depression (Weissman & Klerman, 1977). Women in present-day North America and Europe appear to have rates of clinical depression that are between two and three times higher than men's (Bebbington, 1996; Kessler, McGonagle, Swartz, Blazer, & Nelson, 1993; Weissman et al., 1993). Moreover, in community surveys, women usually report more or more severe symptoms of depression than do men. This gender difference appears to emerge sometime in adolescence and to persist thereafter throughout the lifespan (Mirowsky, 1996; Nolen-Hoeksema & Girgus, 1994).

Overall, the risk of depression appears to have expanded to younger and younger age cohorts during the latter part of the 20th century. Prior to World War II, it appears that depression occurred mainly to people beyond the age of 40. By the 1970s, however, experts began to diagnose depression among people in their 20s and 30s. The idea of that children and adolescents too could suffer from depression took hold in the late 1970s and was linked to the rise of cognitive theories of depression. Prior to that, psychoanalytic formulations of childhood ego development held that children lacked the capacity for depression. (As we shall see later, this idea that depression requires a degree of psychic maturity was also applied to so-called primitive peoples by colonial psychiatrists.) By the 1990s, experts began alerting the general public to the problem of widespread depression among teenagers. Today even children between 7 and 11 years of age are said to be at risk for depression. Are the rates of depression actually changing? Or have successive generations of diagnosticians and therapists become more and more astute at uncovering depression? Or have our understandings of depression and its symptom criteria shifted to encompass broader range of emotional experience and thus to envelop more and more people? We need to be careful not to essentialize depression or its symptom criteria as monolithic and unchanging entities that exist in isolation apart from our descriptions of them. Historians of medicine have shown how various diagnoses and diagnostic categories (among them, schizophrenia, depression, psychopathy, nymphomania, chlorosis, posttraumatic stress disorder (PTSD), multiple personality disorder, psychogenic amnesia, and eating disorders) and their symptom criteria have shifted in accord with cultural and political trends and, in some cases, disappeared altogether. We must also avoid the unwarranted assumption that scientific knowledge is necessarily progressive, always arriving at closer approximations of the truth.

Before we declare depression to be a woman's problem, let us step back to consider that claim more closely. A number of issues bear consideration:

Is the Gender Gap Universal?

Evidence that women have higher rates of depression than men is limited to certain locales. The gender gap seems to be specific to Western Europe and North America. Agricultural societies show little or no gender difference in rates of depression (McCarthy, 1990). Data from non-Western countries are skimpy, but reports from India, Thailand, Rhodesia, and Sri Lanka report no excess of depression among women. Thus, we must modify the blanket statement that depression is a women's problem to a more cautious one, that it may be a Western European and North American woman's problem. We should not be surprised that the gender gap is culture specific. The incidence, patterning, and epidemiology of many disorders differs considerably from one cultural setting to another (Demyttenaere et al., 2004; World Heath Organization, 2000). As we will see, what we call depression appears to be nearly absent in some parts of the world. Although simple cross-national comparisons of rates of depressive symptoms tell us little about culture, they suggest that the culture gap in depression far exceeds the gender gap. What is surprising is how seldom culture has been a focus of serious inquiry by psychiatrists or clinical psychologists.

Has the Gender Gap Been Constant over Time?

Comparisons across historical eras are tenuous because definitions of depression are not fixed. Nonetheless, it seems unlikely that the current male–female difference found in Euro-American societies has been constant over time. One can find mention of depression-like thoughts, emotions, and somatic complaints in medical and religious writings from Greek and Roman times onward. However, historical writings do not mention a specific female proclivity toward depression. Turning to modern times, Elizabeth Lunbeck's (1992) landmark study of the records of the Boston Psychiatric Hospital in the early 20th century offers a glimpse into the diagnosis and treatment of women and girls in that era. In Lunbeck's analysis, depression and melancholia were not diagnoses of high prevalence for women and girls treated at the Boston Psychiatric Hospital; indeed, these categories were only rarely applied to women. The historical evidence is too scanty and too fragmentary to conclude either that the gender gap has been constant across history or that it is historically contingent. However, in the absence of historical evidence for a gender gap in depression, it seems unwise to presume that such a gap has always prevailed.

Is Depression a Culture-Bound Syndrome?

Brute comparisons of rates of depression across national boundaries are not particularly meaningful. Expressions of suffering and demoralization take different forms in different cultures, within and across national boundaries. The configuration of emotion, thought, and action that we call depression is an enactment of suffering that is specific to contemporary Western societies. It typically involves negative affect, such as sadness and the inability to experience pleasure, a profound loss of interest in activities, and diminished zest for life. It also includes profound pessimism and negative beliefs about the self, as well as bodily symptoms of fatigue, reduced activity, trouble eating and sleeping, and loss of interest in sex (cf. Hollon, Thase, & Markowitz, 2002). These experiences have been codified into symptom criteria in the *Diagnostic and Statistical Manual of Mental Disorders* (*DSM-IV*; American Psychiatric Association, 1994) and the *International Classification of Diseases, Tenth Revision* (World Health Organization, 2002).

The *DSM-IV* criteria of depression include only a small fraction of depressive phenomena. Moreover the affective and cognitive elaboration of suffering that we call depression is particular to present-day, Western, industrialized societies. These societies, of course, comprise only a small fraction of the world's population. In global perspective, it is the Western expression of depression that is atypical (Jadhav, 1996). For most of the world's population, there are other modes of registering and communicating social suffering and demoralization. In South, Southeast, and East Asia, bodily complaints are prominent. In China, for example, suffering individuals report the sensation of chest pains, as if the heart is being squeezed (Kleinman & Good, 1985). A substantial proportion of depressed Sri Lankans describe burning sensations in the body, frequently on the soles of their feet. They do not register core depressive symptoms, either in their spontaneous accounts or when completing symptom inventories such as the General Health Questionnaire (Kuruppuarachchi & Williams, 2001). Indians (both men and women) frequently report semen loss and accompanying malaise. In Nepal, researchers from a psychosocial organization administered a standard inventory of depression (the Center for Epidemiologic Studies Depression Scale [CES-D]) to roughly 100 women whose circumstances placed them at high risk of depression. Only 3% of the women scored in the depressed range on the full scale; however, 18% exceeded the cutoff for depression when only the somatic subscale was considered, a six-fold difference (Eller & Mahat, 2003). In Nigeria, suffering individuals report feeling like ants are crawling inside their heads. "Soul loss" is a culture-specific enactment of depressive suffering among the Hmong people of Laos. In short, depressive suffering takes varied, culture-specific forms. These culture-specific forms are overlooked, even in diagnostic systems designed for international use.

The meanings and moral valuation of depressive suffering and the prescribed channels for managing suffering are also culture specific. Present-day Western accounts of depressive suffering (such as the account embodied in the *DSM-IV*) characterize such suffering as an aberrant experience and as a secular and medical (perhaps even neurochemical) problem. Such characterizations are not universal. Indeed, as Obeyesekere (1985) points out, the idea that such painful affects and cognitions as despair, worthlessness, pessimism, self-abnegation, and withdrawal constitute a problem is far from universal. Among devout Buddhists in Sri Lanka, these painful affects and experiences arise from the recognition of the existential human condition. Theravada Buddhist teachings counsel followers to cultivate willful dysphoria through prescribed meditation exercises. These include ruminating on death, feces, and bodily decay, as well as meditations on the impermanence of material goods, social relationships, and worldly pleasures. Many other spiritual traditions (including Islam and some forms of Christianity) have urged believers to cultivate grief, self-abnegation, and resignation to suffering as part of their spiritual practice. In a number of religious traditions, the embrace of grief and suffering is a mark of maturity, full moral personhood, and wisdom. My point is not that medicalized and secular views of depressive suffering are false or unhelpful. Rather, I only remind the reader that our views of depression are socially fashioned and culturally located. As a framework for theorizing and research, they shape the questions we ask, the phenomena we observe, and the interpretations we offer.

The range of depressive phenomena that we experience, their meanings (including the moral meanings), and the channels for relief are culturally located. We should therefore anticipate that gender differences in depression too will be bound up with culture. From a cultural psychologist's vantage point, the question is not whether more women than men meet the symptom criteria for depression in one country or another. Rather we should ask what strategies for expressing and managing suffering are available to women and men in particular settings.

Does the Diagnostic Category of Depression Have a History?

Whether seen in the long historical view or in a much shorter one, ideas of what constitutes depression are not fixed. An intriguing study by Jackson (1986) describes two idioms of dysphoria in Western history: melancholy and acedia. The genealogy of each of these terms reflects the religious, moral, medical, and psychological meanings that are intertwined in our ideas about depressive suffering. Over the centuries, the meanings of melancholy have shifted, variously emphasizing somatic, psychological, and moral aspects. Acedia also underwent shifts in meaning before disappearing entirely. At one time, the construct acedia referred to a state of both indolence and sorrow. Thus, acedia had overtones of moral failing as well as personal distress.

Focusing more narrowly on the past century, we still observe flux in the diagnosis of depression. You will remember that Charlotte Perkins Gilman described her mental state in terms that closely resemble current symptom criteria for major depressive illness. Yet although Gilman received three diagnoses from leading psychiatrists, none of them alluded to depression. Perhaps clinical depression was understood differently in Gilman's day or perhaps it was eclipsed by other diagnostic categories that were believed to be the special province of women from Gilman's background. Contemporary researchers have expended considerable effort to craft explicit, objective, reliable criteria for diagnosing depression. Nonetheless, the meanings of depressive suffering ebb and flow in response to cultural currents beyond researchers' control. Psychiatric diagnoses depend largely on people's verbal reports of their subjective experiences. Hence, they are always entangled in linguistic practices and cultural meaning systems.

Two current cultural trends seem pertinent to North Americans' subjective experiences of depressive suffering. For the past several years, massive government-supported and commercial campaigns have been explicitly directed toward remaking popular beliefs about depression. The goal of these campaigns has been to inform people how vulnerable to depression they and their loved ones are, to give them tools to self-diagnose depression, and to promote antidepressant drugs. The government-sponsored campaigns have successively targeted younger and younger age groups (high school and college students) and, most recently, men (Kluger, 2003). Pharmaceutical companies that sell antidepressant drugs have sponsored even more insistent marketing campaigns. Whatever benevolent motives they might have, such mass media campaigns are potent cultural interventions. They promulgate a particular vocabulary for communicating one's suffering and for understanding emotional life. They accentuate psychologized discourses of suffering, selfhood, and social life, displacing philosophical, sociopolitical, or spiritual discourses (Rose, 1996). They also direct sufferers to remedies (notably antidepressant medication) aimed exclusively at symptomatic relief. Societal or sociopolitical changes are not mentioned as a means of alleviating depressive suffering.

The second trend pertains to the culture of psychotherapy. Recent decades have seen the rise to prominence of the concept of trauma as a primary explanation for women's suffering. In this context, trauma refers to experiences of gender-linked victimization, such as rape, sexual abuse, and many forms of intimate violence and intimidation. For feminists, the trauma idiom has offered a compelling means of narrating women's suffering. It has become a staple in the booming marketplace of popular psychology and women's self-help. The trauma idiom, along with the diagnostic category of PTSD, is in wide usage among feminist therapists and others who claim special expertise in helping women. For instance, when Diane Kravetz and I interviewed 100 self-identified feminist therapists in the mid-1990s, several told us that trauma or PTSD was the diagnosis of choice for women. Taking the trauma history was an essential part of a clinical assessment. As many of these therapists understood it, the benefit of the label PTSD was that it had no implication of psychopathology. Instead, it signified to the client that she was having a normal response to a traumatic situation (Marecek, 1999). In contrast to the practitioners' embrace of PTSD, only a handful made mention of depressive disorders. Moreover, not one of these 100 experts on therapy for women identified the treatment of depression as one of her or his specialties. For the time being, the attractiveness of the diagnostic category of PTSD – a portmanteau that embraces a grab bag of symptoms and dysfunctions – may be eclipsing the category of depression, at least among feminist therapists, a group that constitutes an important subset of therapists who work with women. I hope that this volume succeeds in calling attention to depressive suffering and in sparking interest in developing psychotherapies that address the needs of depressed women.

This is not the place to elaborate the pros and cons of the trauma movement and the emergent field of traumatology. (For critical feminist readings of trauma, PTSD, and related issues, see Burstow, 2003; Haaken, 1998; and Lamb, 1999.) The point is that cultural and professional discourses about women's suffering bring forward alternate constructions of suffering, alternate names for it, and alternate enactments of it. In the real world of psychotherapy practice, diagnosing is less a matter of scientific accuracy than a negotiation with clients (and often third-party payers) to find a useful and acceptable way of framing the clients' problems. Such cultural practices and discourses confound researchers' efforts to produce true accounts of women and men who are depressed. In recent years, alternate discourses have often signaled feminist contestation over women's diagnoses. Indeed, feminist therapists' embrace of the categories of trauma and PTSD can be read as an effort to contest the biomedical perspective that has come to dominate the mental health professions. Ironically, however, PTSD has gained credence in the psychiatric mainstream by conforming more and more closely to the biomedical framework. Indeed, a prominent agenda for those who call themselves traumatologists has been to establish its neurobiological basis.

Truth in Tests? Measuring Depression

The discovery of antidepressant medications in the 1960s opened the way to a new kind of research on treatment efficacy. This research paradigm required methods of diagnosing clinical depression that were more reliable than clinical judgment and that yielded finer calibrations of depression.

Thus, rating scales and inventories replaced clinicians' judgments. Such rating scales can be seen as means of narrating depressive suffering. One of the most common measures of depression is the Beck Depression Inventory (BDI; Beck, Ward, Mendelson, Mock, & Erbaugh, 1961). The BDI emphasizes cognitive and affective features that are germane to Beck's theories about the cognitive antecedents of depression. In fact, the BDI was originally designed to serve as an instrument for charting a client's week-to-week progress in Beck's cognitive behavior therapy, not a means of diagnostic classification. Inactivity, apathy, lassitude, weakness, and the wide variety of somatic complaints (such as burning feet and semen loss) are not assessed in detail.

Questions about how depression is measured bear directly on claims of a gender gap in depression. Is the gender gap a gap in depressive suffering or in the ways we measure it? Do men and boys express, experience, and enact such suffering differently than women and girls? If so, we must ask whether depression inventories index masculine and feminine enactments of depression equally. In our time and place, the emotion practices of men and boys are very different than those of women and girls (Shields, 2002). Stapley and Haviland (1989), for instance, reported that one of the strongest gender differences in self-reports of emotions was boys' tendency to deny experiences of negative emotion. Also, boys were less likely than girls to give elaborated reports of emotion connected to affiliative situations or relationships. Boys elaborated on emotions connected to activity, aggression, and achievement. If these childhood emotion practices persist in adulthood, men's experiences of depressive suffering may not be fully tapped by depression inventories. Depression inventories for adults typically do not include items concerned with disturbances in activities (e.g., sports) or feelings related to aggression (such as contempt and anger). Instead they accentuate disturbances in relationships and feelings related to sadness and despair.

Norms of masculinity in many subgroups of Euro-American societies put pressure on males to be cool, detached, and tough (McLean, Carey, & White, 1996; Oransky, 2002). An extreme example can be seen in an interview that Oransky had with a 15-year-old boy a few months after the World Trade Center attack in New York City. On the day of the attacks, the boy confided, he had concealed his fear that his stockbroker father had died in the attack because he believed his peers would ridicule him if he appeared upset. This boy and his peers, all from affluent white families, ridiculed emotional expression as "girly" and "faggy." Manly men, in these boys' eyes, deal with adverse events by stoically "sucking it up" and maintaining an unperturbed front. Do such norms shape men's enactment of depressive suffering? A study of clinically depressed men suggests that they do (Vrendenburg, Krames, & Flett, 1986). The men's self-descriptions emphasized work-related problems (e.g., inability to perform

adequately; difficulty making decisions) and somatic concerns (e.g., physical complaints; concerns about their general health). The men did not
acknowledge crying, sadness, or dejection. Depression inventories may
inadvertently accentuate emotion practices that men and boys feel obliged
to repudiate because they are culturally coded as feminine. Inventories
may also omit emotion practices consonant with norms of masculinity,
for example, consuming alcohol or drugs. In settings in which masculine
norms of emotional suppression and toughness hold sway, the extent of
men's depressive suffering may be underrepresented.

The Epistemology of Epidemiology

The statistical scene setting that epidemiological data entail serves the
rhetorical function of implying that unambiguous facts about depression
are being provided (Reekie, 1994). Rates and counts transform depressive suffering into a bounded category, although in reality, the set of
experiences we call depression is not a fixed entity, but an emergent set
of social practices. The reliance on statistical counts also reproduces a
binary logic of sorting depressed women from normal women. It leads
researchers to focus on searching for psychopathological characteristics or
social experiences that distinguish depressed women from their "normal"
counterparts. Yet, depression is both an ordinary mood state and a clinical
condition; the line between ordinary and pathological is under continual
renegotiation.

Whether or not more women than men are depressed, it is important
to study depressive suffering as a gendered phenomenon. It arises in the
context of gendered social relations and gendered social institutions. The
enactment, expression, and management of depressive suffering are necessarily bound up with cultural configurations of masculinity and femininity.
Indeed, the characterization of depression as a woman's problem emerged
alongside the emergence of the second wave of feminism in the United
States. Various ways of figuring women's depression and understanding
its origins closely parallel the ways that late 20th century women came to
understand themselves. It is to this history that I now turn.

WOMEN'S DEPRESSION AND WOMEN'S LIBERATION

Middle Class Women and the Problem with No Name

Like other progressive movements of the 1960s, the women's liberation
movement viewed psychotherapy as a suspect cultural institution. Feminists viewed therapists as complicit in perpetuating women's oppression. Movement members unleashed relentless critique of psychoanaly-
sis, which was the prevailing theoretical framework of the mental health

professions. Psychoanalytic theories of female depression viewed normal femininity as incorporating such qualities as masochism, low self-esteem, dependency, disappointment, and inhibited hostility. These traits inevitably developed once a girl recognized that she lacked a penis. Freud described this realization as a permanent "wound to her narcissism" 1925/1974), one for which she blamed her mother. A "second reproach" directed to her mother followed: "It is that her mother did not give her enough milk, did not suckle her long enough" (Freud, 1931/1974).

For feminists of the 1960s, this kind of theorizing was intolerable. For them, women's depression was not a result of inferior anatomy but a social problem that demanded societal solutions. Naomi Weisstein's *Kinder, Küche, Kirche: Psychology Constructs the Female,* first printed in 1968 and reprinted 30 times thereafter, issued a scathing critique: (1) what is advanced as scientific dogma about women merely recycles cultural stereotypes and (2) psychology's claims about female nature grossly underrate the influence of social context. A few years later, Phyllis Chesler's *Women and Madness* (1972) charged therapists with putting women in a double bind. Norms of femininity required certain behaviors (such as emotional expressiveness and dependence) that were simultaneously regarded as psychiatric symptoms. Femininity was thus rendered pathological. At the same time, nonconformity too was judged pathological.

In keeping with the progressive political ethos of the 1960s and 1970s, women's liberationists put forward social models of women's depression. They sought its causes in the conditions of women's lives under patriarchy: Important among these were subordination in marriage; constricted economic, social, and political roles and opportunities; and adverse life events and circumstances. *The Feminine Mystique,* published in 1963 by Betty Friedan, painted a searing portrait of educated, affluent, suburban wives weeping into the kitchen sink on long, empty afternoons. Although we might suppose these women were depressed, Friedan called their demoralization "the problem with no name." The book launched a wide public discussion about middle-class marriage. Friedan's portrayal of middle class family life as a psychological prison for women resonated with many female readers and served to galvanize middle-class White women's participation in the fledgling women's movement.

Sociological research on the relation between marital status and psychological distress offered support for Friedan's charges. Walter Gove and his colleagues (e.g., Gove & Tudor, 1973) examined the rates of mental illness (not specifically depression) among men and women with various marital statuses. Using data from the 1960s, Gove repeatedly produced evidence that single, divorced, and widowed women had lower rates of mental illness than comparable men; married women, in contrast, had higher rates. In a similar study, Laurie Radloff (1975) focused specifically on depression, examining symptom inventories gathered in community surveys. Radloff too reported that married women were at elevated risk for depression in comparison to other women.

Other sociologists investigated married life more closely. Lopata (1971) argued that the housewife role was so unstructured and diffuse that women who were housewives could obtain little sense of efficacy or accomplishment. Moreover, caring for small children in the confines of a nuclear household produced social isolation. Others noted that cultural expectations (bolstered by the opinions of mental health professionals) held mothers responsible for their children's lifelong mental health, happiness, and success (Caplan, 1989). Such impossible standards generated guilt, blame, and a sense of failure. Further, some theorists proclaimed that older women suffered a depressive empty nest syndrome when their last child departed from the home, leaving them without a role and sense of purpose.

The 1970s were a time when dramatic changes were initiated in gender arrangements in North America, particular among middle-class individuals. Legislative changes cleared the way for women's participation in higher education, professional occupations, and public life. Women gained a new degree of control over their reproductive choices. They entered the workforce in vast numbers. Progressive couples began to experiment with new childcare arrangements from day care to shared parenting to communal child rearing. Ultimately, few of the reforms went as far as feminists had hoped and many subgroups of women did not benefit from them. Many early claims and reforms pertained mostly to privileged women. For example, the possibility of personal fulfillment through paid work could only be realized by women who had access to rewarding jobs. For less advantaged women, low-wage jobs such as domestic service, assembly-line labor, and service jobs could well be tantamount to drudgery and economic exploitation. Nonetheless, the ethos of the times fostered scientific interest in social models of depression and other psychological difficulties. Popular culture, public institutions, and social researchers were open to the idea that social arrangements, institutional structures, and cultural values were implicated in women's depression.

Adverse Life Experiences

The idea that bouts of depression can be triggered by adversity has a long history in social research and robust support. Freud drew the connection between depression (or melancholia, as he termed it) and bereavement in 1917. The landmark studies of Brown and Harris (1978) identified a number of adverse life circumstances that raised the risk of depression among the women they studied in Camberwell, a working-class district of London. A more recent study found that more than 80% of people with major depression reported that adverse life events had taken place shortly before they became depressed (Mazure, Keita, and Blehar, 2002).

The connection between adversity and depression (and many other forms of psychological distress) is a robust one. But can it explain the gender gap? Are women subject to a greater number of adverse experiences than men? Are there certain experiences that are unique to women and precursors of depression? Feminist researchers and practitioners have focused on gender-based violence and its effects on women's psychic life. They insisted that it was a common occurrence and often a devastating one for women. They also insisted that it was linked to gendered power relations and cultural sanctions for male dominance, sexual aggression, and physical violence. Some depression researchers argued that sexual violation and intimate violence – adverse experiences that predominantly affect women and girls – might account for the gender gap in depression (e.g., Cutler & Nolen-Hoeksema, 1991). Rape, sexual abuse, and intimate partner violence, however, lead to any of a large array of psychological and behavioral difficulties (Koss, Koss, & Woodruff, 1991). It is difficult to make the case that they are specifically connected to depression. Moreover, the response of any particular woman to such violation depends on a complex of factors. These include the circumstances surrounding the victimization, its subjective meaning to the victimized woman, the aftermath (including experiences with medical personnel, police, and legal system), the victim's psychological resources, and the social supports available to her. Although the connection between adverse events and depressive suffering seems unassailable, it does not seem particularly useful to search for a single objectively defined class of events that will account for women's depression.

THE RISE OF PERSONOLOGICAL THEORIES

Social models of women's depression resonated with feminist critiques of society and the mood of progressive social change that marked the 1970s. The 1980s, however, witnessed a sea change in American public life. The 1980 presidential election marked a dramatic swing to social and political conservatism, which included the rise of the religious right, a cultural backlash against feminism, the remedicalization of psychiatry, and the corporatization of medicine. Following in the culture's footsteps, scientific psychology shifted to the right as well. For example, the focus of attention shifted to genetics, the evolutionary basis of gendered behavior, and the neurochemistry of psychological disorders. Many in the mainstream of psychology and psychiatry saw this shift as a corrective to the misguided and unscientific emphasis on the sociopolitical and societal contributions to psychological disorder.

In a less overt way, the study of gender in psychology shifted in the conservative direction as well. For many psychologists, the focus of interest shifted from social forces to the psychic interior. The theorists who captured the stage were psychotherapists and personality theorists who proposed personological explanations of women's depression. These explanations emphasized such factors as personality traits (e.g., lack of assertion, acquiescence, dependence on others, suppression of anger, and low self-esteem), patterns of thinking, modes of interacting, or self-structures that were presumed to be distinctive to women. Although theorists regarded these qualities as originating in early social conditioning, they conceptualized them as qualities within the person, not as relational practices arising from and maintained by ongoing social life.

One example of a personological theory is response style theory (Nolen-Hoeksema, 1991). According to this theory, when women confront negative emotions, they engage in rumination. That is, they focus passively on their feelings of distress and on the possible causes and consequences of these feelings. Men, in contrast, engage in distraction, pushing away bad feelings by diversionary activities such as sports, drinking, and watching TV. Nolen-Hoeksema argued that rumination intensified depressed affect and hopelessness and thus could precipitate a full-blown clinical depression; distraction was a more effective coping strategy. Nolen-Hoeksema's claim of a global male–female difference in response style mirrored a prevalent gender stereotype. However, the idea that women ruminate and men distract themselves was not substantiated in further studies. People's reports of their coping strategies show more complex patterns and less gender differentiation. Both men and women sometimes ruminate and sometimes distract themselves (Strauss, Muday, McNall, & Wong, 1997). Other studies showed that the choice of strategy is domain specific, that is, it depends on what the negative feeling is about. Nolen-Hoeksema herself eventually abandoned her idea that rumination and distraction were gender-linked traits (Nolen-Hoeksema & Jackson. 2001).

More generally, explanations for women's depression based on presumed gender differences in personality risk oversimplification and overgeneralization. The search for gender difference falsely assumes that women are a homogenous group with uniform experiences; this conceals the considerable diversity among women (Hare-Mustin & Marecek, 1994). The idea that depressive symptoms reflect overconformity to societal norms of femininity does not take into account the multiple forms that depressive suffering takes. Nor does it adequately account for the multiple and sometimes contradictory norms of femininity. It also ignores the substantial numbers of men who are diagnosed with depression, most of whom were presumably not socialized to be feminine.

Personological explanations of women's depression suited the conservative times in which they rose to prominence. Although they view femininity as a product of social conditioning, they do not develop a close analysis of when, where, and how such conditioning occurs. Nor do they explain how it is that women can be recruited to projects of femininity that exact such a price in emotional pain and dysfunction and why they remain mired in them. They risk portraying women as docile victims of culture, who have no means of resistance. Furthermore, personological explanations place the onus of change on individual women. They advocate individualistic technologies of change (psychotherapy, assertiveness training, and cognitive retraining), while leaving social structures untouched.

CULTURAL FEMINISM AND WOMEN'S DEPRESSION

The especial genius of Woman I believe to be electrical in movement, intuitive in function, spiritual in tendency. [M]ale and female represent the two sides of the great radical dualism. Margaret Fuller, 1845/1976

The 1980s ushered in a revival of the notion of a great radical gender dualism, which caught the fancy of a broad swath of North American women. A subset of feminist thinkers extolled women's gentle heroism, harmony with nature, pacifism, superior morality, and ethic of caring for others. This body of thought became known as cultural feminism. In psychology, writers such as Carol Gilligan and the Stone Center group claimed that women were endowed with a unique psychology that encompassed qualities of intuition, empathy, beneficence, and the capacity to nurture others. From these claims sprung another model of women's depression.

The crux of the model was a radical dualism reminiscent of Margaret Fuller's, but framed within psychodynamic theory. In this model, girls and boys develop distinctively different selves and different capacities for relationships during the earliest months of life. In the mother–infant dyad, girls develop a self that is more permeable, less bounded, and more attuned to others. They develop both a capacity and a need for intimate emotional relationships. Nancy Chodorow (1978), who formulated the original version of this theory, was careful to locate her developmental account in the context of the gender and family arrangements specific to late-20th-century, Euro-American, postindustrial capitalism. However, as psychologists and psychotherapists took up Chodorow's ideas, her meticulous sociological groundwork slipped away. What remained was the idea that "the" mother–daughter relationship (now constituted as universal, not as a contingent social arrangement) produced a specifically feminine developmental trajectory and personality structure. All women were said to be oriented to caring for others and to have a special empathic attunement to others' needs. At the same time, they required emotional intimacy and empathic connection in their personal relationships. When intimacy and connection were unavailable, women could not develop and grow psychologically (Miller & Stiver, 1997).

Relational theorists have drawn on these ideas to develop accounts of women's depression (e.g., Jack, 1991; Kaplan, 1991; Miller & Stiver, 1997). Dana Jack (1991) has put forward the most substantial account. Jack's model of women's depression involves what she has called "Silencing the Self," which is a constellation of self-abnegating beliefs and practices. For example, some key beliefs are that caring demands self-sacrifice and that one should inhibit one's actions and speech to avoid displeasing others. Some key practices include judging oneself according to standards held by others and presenting a compliant and agreeable facade despite inner feelings of anger and resentment. In Jack's model, women have a special fear of emotional abandonment that emerges from their needs for emotional intimacy. Therefore, they are prone to silencing the self, that is, suppressing their needs and feelings for fear of losing relationships. Silencing the self in this way is ultimately self-defeating, however. It leads to a loss of self and it thwarts any chance for genuine intimacy and connection. This impasse places women at risk for depression.

This is not the place to recount in detail the intricate bodies of writing concerning women's relational identity or the controversies they evoked (Becker, 2005; Bohan, 1993; Hare-Mustin & Marecek, 1990; Westkott, 1997). One of the strengths of this body of theory is its careful attention to the minute details of women's daily lives and especially to what women themselves say. The nuanced accounts of women's sorrows, joys, ambivalences, and frustrations often evoke in female readers a sense of deep familiarity. Nonetheless, efforts to validate systematically the gender dualism on which these theories are grounded have not been successful. For example, a meta-analysis of 113 studies found only a small difference between men and women in the "ethic of care" proposed by Carol Gilligan (Jaffee & Hyde, 2000). Moreover, two studies, both using a scale that Jack herself has designed, have found that, contrary to her claim, men had higher scores than women on a Silencing the Self Scale (Gratch, Bassett, & Attra, 1995; Jack & Dill, 1992).

Claims about the emotional experiences and identity development of infants cannot easily be assessed. We do know, however, that even within the United States, childcare practices vary considerably depending on cultural background, family structure, residence patterns, and economic situations. The idea that the affective relationship between mothers and their baby daughters takes a single universal form seems implausible. Moreover, much of the effort to identify a feminine intrapsychic makeup that puts women at risk for depression rests on observations of women who are already depressed. Kaplan, Miller, and Stiver, for example, drew upon case studies of women in their psychotherapy practices. Jack studied a group of clinically depressed women. What these theorists described may aptly characterize women once they are already depressed. But that does not suffice as an account of how they *became* depressed (Barnett & Gotlib, 1988).

BEYOND THE MEDICAL MODEL: CULTURAL AND CONSTRUCTIONIST STUDIES OF DEPRESSION

Scientific knowledge is shaped in accord with the social and cultural circumstances of those who produce it. As Ludmilla Jordanova has said,

It is a mistake to separate the knowledge claims of medicine from its practices, institutions, and so on. All are socially fashioned, and so it may ultimately be more helpful to think in terms of mentalities, modes of thought, and medical culture than in terms of "knowledge," which implies the exclusion of what is inadmissible. (1995, p. 362)

Many parts of the knowledge-producing process – deciding what should be construed as significant facts, choosing which ones are relevant and important to the question at hand, and crafting an interpretation – are rooted in a researcher's epistemological commitments and cultural location. Whether implicit or explicit, whether consciously acknowledged or not, researchers hold some vision of how the world works and that vision guides our investigations. At best, we can hope to become conscious of that vision and cognizant of how our place in the world contributes to it.

Depression became identified as a woman's problem in the context of the second wave of the North American women's movement some 35 years ago. Indeed, women's depression became an object of scientific curiosity because a group of early feminist psychologists refused to take women's unhappiness as normative. Their critiques challenged the normal versus pathological distinction. They also challenged the medical model and insisted instead that women's subordination was the soil from which depression sprouted. Later epochs of second-wave feminism put forward different images of depressed women, each with a distinctive moral geography and a distinctive understanding of what constituted *woman* and *femininity*. Today, most researchers on women and depression operate on far less edgy ground than those early feminist critics. They hew to conventional positivist epistemologies and seem to regard canonical research and diagnostic technologies as unproblematic. Among other contributions, their work has made women's health (if not gender and health) a subject of legitimate inquiry. Moreover, their work has served to insinuate some sociocultural issues into traditional views of psychopathology (cf. Marecek, 1993). But all epistemologies and technologies are accompanied by characteristic blind spots, deforming their objects of study in characteristic ways. For example, work on women and depression has seldom looked beyond the notion of gender as qualities located inside individual men and women. Much of the research remains lodged in a reductive framework in which internal qualities such as low self-esteem and suppressed anger are identified as the vital forces behind depressive suffering.

There are few studies concerned with gender and depression that have looked beyond the medical metaphor of depression, which holds it to be akin to a physical disease. Medicalized constructs of depression reduce a complex and socially embedded experience to a single identity. They discount a depressed woman's account of her experience, no matter how complex, astute, and rich it might be. Her words are meaningful only insofar as they meet or fail to meet some symptom criterion. (See Stoppard & McMullen, 2003, for an alternative approach.) Moreover, medical models of depression seem to obviate the need for investigating the constitutive nature of society, culture, and history. As Szekely has commented, "Once a phenomenon has been constructed as a disease, the sociocultural can only be viewed as a factor that further undermines the weak personality of the individual" (1989, p. 176). In this final section, I consider some areas of inquiry about women and depressive suffering that come to mind once we heed Jordanova's words and think about mentalities and modes of thought, rather than knowledge.

The Cultural Politics of Diagnosis

The demarcation of depression as a category of disorder is not a neutral scientific accomplishment, but an endeavor imbued with cultural values and political interests. Indeed, of all psychiatric categories, depression raises the most questions of cross-cultural validity (Jadhav, Weiss, & Littlewood, 2001). Transforming certain emotions and practices into disorders pathologizes them and those who engage in them.

Psychiatric diagnoses inform cultural notions of normality and abnormality, mental health and illness. Historians of medicine have elucidated many examples of psychiatric diagnoses that were used to control behavior and stigmatize those who did not conform to prevailing standards of conduct. For example, scholars have detailed various diagnoses used for women who violated standards of sexual, maternal, or wifely behavior. Postcolonial theorists have detailed how diagnoses of actors as "mad" and actions as "madness" justified the coercive power of the colonizers (e.g., Mills, 2000).

Suman Fernando (2003) recently recounted the reactions of colonial psychiatrists when they realized that depression was comparatively rare among non-Western colonial subjects in some parts of Africa, South Asia, and Southeast Asia. One might think such apparent immunity from depression would be regarded as a psychological strength. However, the absence of depression among colonial subjects was deemed to result from moral deficiency (such as "irresponsibility"), psychological immaturity ("psychic underdevelopment"), and primitive and childlike natures. This might lead us to ponder the meanings and morality of depression in present-day North American men and women. Depression seems to carry a certain social stigma, whether the sufferer is a man or a woman. Beyond that, is the lower prevalence of depression among men regarded as a sign of moral or psychological immaturity? Is women's depressive suffering taken as an indicator of advanced development, greater maturity, moral superiority, or psychological strength? This seems unlikely. It seems more likely that a diagnosis of depression may convey connotations of weakness, lack of agency, and perhaps self-pity. An important topic for further investigation is whether a diagnosis of depression changes its meaning and moral value according to the social position of the sufferer.

Antidepressants and the Construction of Subjectivity

At present, in the United States, most people (about 75%) who are treated for depression receive antidepressant medication (Kluger, 2003). Antidepressants account for nearly 8% of retail drug sales in the United States (Pomerantz, 2003). In North America, drugs that modulate other aspects of psychic life are plentiful as well. This everyday presence of psychotropic drugs has been incorporated into ordinary people's folk psychology. For example, some people now explain depression as a chemical imbalance; some even refer to chemical depression. Such simplistic biomedical constructions of self, identity, and disorder may contribute to a mechanistic view of human life. Moreover, by obscuring how the *social* is implicated, they discourage a sociocultural analysis of depressive suffering. The literary critic Jonathan Metzl (2002) has explored a genre of literature that he calls the Prozac novel (e.g., *Black Swans*, *Prozac Highway*, *Prozac Nation*, all written by women). Metzl considers how sufferers understand the transformations of personality and self wrought by antidepressants. This kind of study need not be left to literary critics and it need not be limited to fictional texts.

Producing Gender through Depressive Suffering

Bringing a constructionist viewpoint to the study of women's depression opens some novel questions. For social constructionists, gender is not something people are, but what they do and say to produce themselves as (particular kinds of) men and women (Marecek, Crawford, & Popp, 2004; West & Zimmerman, 1987). Similarly, depression is not something people have, but a set of practices authorized by the culture through which people express to others that they are suffering. How have enactments of depressive suffering changed over historical time? Are different enactments available to people who occupy different social positions? How do those enactments reproduce and reaffirm those social positions? What does women's depressive suffering accomplish in the social worlds they inhabit? Does the enactment of suffering serve as a means to effect changes in contentious social roles and relationships? Is it a means to renegotiate mutual rights and responsibilities (cf. Hunt, 2000)? Rather than asking if adverse events cause depression, a constructionist inquiry might construe adversity as a problem that prompts a woman to initiate a line of action. Depression in this framework is one of many possible lines of action, not a passive giving up.

Many scientific stories can be told about women's depressive suffering. Each is constructed from a particular theoretical vantage point and each reflects the gender politics of a particular cultural moment. All are partially true. Our task is not to sort through them in search of the single correct one. Nor is it to find a rubric that would bring them all together into a comprehensive master account. That is impossible because the premises behind the stories and the disciplinary commitments they entail are incommensurate. In my view, we serve our scholarly purposes well by stepping back to examine these stories as culture artifacts. We will not only situate women's depressive suffering in social and historical perspective, but situate ourselves in that perspective as well. That way, we can glimpse how psychology and culture – like discourses in a mirrored room – mutually constitute each other.

References

American Psychiatric Association. (1994). *Diagnostic and statistical manual of mental disorders* (4th ed.). Washington, DC: American Psychiatric Association.

Barnett, P. A., & Gotlib, I. H. (1988). Psychosocial functioning and depression: Distinguishing among antecedent, concomitants, and consequences. *Psychological Bulletin, 104,* 97–126.

Bebbington, P. (1996). The origins of sex-differences in depressive disorder–bridging the gap. *International Review of Psychiatry, 6,* 295–232.

Beck, A. T., Ward, C. H., Mendelson, M., Mock, J., & Erbaugh, J. (1961). An inventory for measuring depression. *Archives of General Psychiatry, 4,* 561–571.

Becker, D. (2005). *The myth of empowerment.* New York: New York University Press.

Bohan, J. (1993). Regarding gender: Essentialism, constructionism, and feminist psychology. *Psychology of Women Quarterly, 17,* 5–22.

Brodsky, A. M., & Hare-Mustin, R. T. (Eds.) (1980). *Women and psychotherapy.* New York: Guilford.

Brown, G. W., & Harris, T. O. (1978). *Social origins of depression.* London: Tavistock.

Burstow, B. (2003). Toward a radical understanding of trauma and trauma work. *Violence against Women, 9,* 1293–1317.

Caplan, P. J. (1989). *Don't blame mother.* New York: Harper & Row.

Chesler, P. (1972). *Women and madness.* New York: Doubleday.

Chodorow, N. (1978). *The reproduction of mothering.* Berkeley: University of California Press.

Crawford, M., & Marecek, J. (1989). Psychology reconstructs the female, 1968–1988. *Psychology of Women Quarterly, 13*, 147–166.

Cutler, S. E., & Nolen-Hoeksema, S (1991). Accounting for sex differences in depression through female victimization: Childhood sexual abuse. *Sex Roles: A Journal of Research, 24*, 425–438.

Demyttenaere, K., Bruffaerts, R., Posada-Villa, J., Gasquet, I., Kovess, V., Lepine, J. P., et al. (2004). Prevalence, severity, and unmet need for treatment of mental disorders in the World Health Organization World Mental Health Surveys. *Journal of the American Medical Association, 291*, 2581–2590.

Eller, L.-S., & Mahat, G. (2003). Psychological factors in Nepali former commercial sex workers with HIV. *Journal of Nursing Scholarship, 35*, 53–60.

Fernando, S. (2003). *Cultural diversity, mental health, and psychiatry: The struggle against racism.* New York: Palgrave-Macmillan.

Freud, S. (1974). Some psychical consequences of the anatomical distinction between the sexes. In J. Strouse (Ed.), *Women and analysis.* New York: Grossman. (Original work published 1925).

Freud, S. (1974). Female sexuality. Reprinted in J. Strouse (Ed.), *Women and analysis.* New York: Grossman. (Original work published 1931).

Friedan, B. (1963). *The feminine mystique.* New York: Dell.

Fuller, M. (1976). Woman in the nineteenth century. In B. Chevigny (Ed.), *The woman and the myth: Margaret Fuller's essential writings* (pp. 239–279). Old Westbury, NY: The Feminist Press. (Original work published 1845).

Gilman, C. P. (1898). *Women and economics.* Boston: Small, Maynard, & Co.

Gilman, C. P. (1899/1973). *The yellow wallpaper.* Brooklyn, NY: The Feminist Press at the City University of New York.

Gilman, C. P. (1935/1991). *The living of Charlotte Perkins Gilman.* Madison, WI: University of Wisconsin Press.

Gove, W. R., & Tudor, J. F. (1973). Adult sex roles and mental illness. *American Journal of Sociology, 78*, 812–835.

Gratch, L. V., Bassett, M. E., & Attra, S. L. (1995). The effects of gender and ethnicity on self-silencing and depression. *Psychology of Women Quarterly, 19*, 509–515.

Haaken, J. (1998). *Pillar of salt.* New Brunswick, NJ: Rutgers University Press.

Haraway, D. J. (1981). In the beginning was the word: The genesis of biological theory. *Signs, 6*, 469–481.

Hare-Mustin, R. T., & Marecek, J. (1990). *Making a difference: Psychology and the construction of gender.* New Haven, CT: Yale University Press.

Hare-Mustin, R. T., & Marecek, J. (1994). Asking the right questions: Feminist psychology and sex differences. *Feminism & Psychology, 4*, 531–537.

Hedges, E. R. (1973). Afterword. In *The yellow wallpaper.* Brooklyn, NY: The Feminist Press at the City University of New York.

Henriques, J., Hollway, W., Urwin, C., Venn, C., & Walkerdine, V. (1984). *Changing the Subject: Psychology, Social Regulation, and Subjectivity.* London: Routledge.

Hollon, S., Thase, M. E., & Markowitz, J. C. (2002). Treatment and prevention of depression. *Psychological Science in the Public Interest, 3*(2), 41–77.

Hunt, L. M. (2000). Strategic suffering: Illness narratives as social empowerment among Mexican cancer patients. In C. Mattingly and L. C. Garro (Eds.), *Narrative and the cultural construction of illness and healing* (pp. 88–107). Berkeley: University of California Press.

Jack, D. C. (1991). *Silencing the self: Women and depression.* Cambridge, MA: Harvard University Press.

Jack, D.C., & Dill, D. (1992). The Silencing the Self Scale: Schemas of intimacy associated with depression in women. *Psychology of Women Quarterly, 16,* 97–106.

Jackson, S. W. (1986). *Melancholia and depression: From Hippocrates' time to modern times.* New Haven, CT: Yale University.

Jadhav, S. (1996). The cultural origins of western depression. *International Journal of Social Psychiatry, 42,* 269–286.

Jadhav, S., Weiss, M. G., & Littlewood, R. (2001). Cultural experience of depression among White Britons in London. *Anthropology and Medicine, 8,* 47–70.

Jaffee, S., & Hyde, J. S. (2000). Gender differences in moral orientation: A meta-analysis. *Psychologicaly Bulletin, 126,* 703–726.

Jordanova, L. (1995). The social construction of medical knowledge. *Social History of Medicine, 8,* 362–372.

Kaplan, A. G. (1991). The "self-in-relation": Implications for depression in women. In J. V. Jordan, A. G. Kaplan, J. B. Miler, I. P. Stiver, & J. L. Surrey (Eds.), *Women's growth in connection* (pp. 206–222). New York: Guilford.

Kessler, R. C., McGonagle, K. A., Swartz, M., Blazer, D. G., & Nelson, C. B. (1993). Sex and depression in the National Comorbidity Survey I: Lifetime prevalence, chronicity, and recurrence. *Journal of Affective Disorders, 29,* 85–96.

Kleinman, A., & Good, B. (1985). Introduction: Culture and depression. In A. Kleinman and B. Good (Eds.), *Culture and depression.* Berkeley: University of California Press.

Kluger, J. (2003, September 22). Real men get the blues. *Time,* pp. 48–49.

Koss, M. P., Koss, P. G., & Woodruff, W. J. (1991). Deleterious effects of criminal victimization on women's health and medical utilization. *Archives of Internal Medicine, 151,* 342–348.

Kuruppuarachchi, K. A. L, A. & Williams, S. S. (2001). Cross-cultural psychiatric interviews and research instruments. *British Journal of Psychiatry, 179,* 461–462.

Lamb, S. (1999). New versions of victims: Feminists struggle with the concept. New York: New York University Press.

Lopata, H. Z. (1971). *Occupation housewife.* New York: Oxford University Press.

Lunbeck, E. (1992). *The psychiatric persuasion.* Princeton, NJ: Princeton University Press.

Marecek, J. (1993). Disappearances, silences, and anxious rhetoric: Gender in abnormal psychology textbooks. *Journal of Theoretical and Philosophical Psychology, 13,* 114–123.

Marecek, J. (1999). Trauma talk in feminist clinical practice. In S. Lamb (Ed.), *New versions of victims: Feminists struggle with the concept* (pp. 158–182). New York: New York University Press.

Marecek, J. (2001). Disorderly constructs: Feminist frameworks for clinical psychology. In R. K. Unger (Ed.), *The handbook of the psychology of sex and gender* (pp. 303–316). New York: John Wiley.

Marecek, J., Crawford, M., & Popp, D. (2004). On constructing gender, sex, and sexuality. In A. Eagly, R. Sternberg, & A. Beall (Eds.), *The psychology of gender* (pp. 192–216). New York: Guilford.

Mazure, C., Keita, G. P., & Blehar, M. (Eds.) (2002). *Summit on women and depression: Proceedings and recommendations*. Washington, DC: American Psychological Association. (Available online at www.apa.org.pi/wpo/women&depression.pdf.)

McCarthy, M. (1990). The thin ideal and eating disorders in women. *Behavior Research and Therapy, 28*, 205–214.

McGrath, E., Keita, G. P., Strickland, B. R., & Russo, N. F. (Eds.). (1990). *Women and depression: Risk factors and treatment issues*. Washington, DC: American Psychological Association.

McLean, C., Carey, M., & White, C. (1996). *Men's ways of being*. Boulder, CO: Westview Press.

Metzl, J. M. (2002). Prozac and the pharmacokinetics of narrative form. *Signs, 27*, 347–380.

Miller, J. B., & Stiver, I. P. (1997). *The healing connection: How women form relationships in therapy and life*. Boston: Beacon Press.

Mills, J. H. (2000). *Madness, cannabis, and colonialism*. London: Macmillan.

Mirowsky, J. (1996). Age and the gender gap in depression. *Journal of Health and Social Behavior, 37*, 362–380.

Nolen-Hoeksema, S. (1991). Responses to depression and their effects on the duration of depressive episodes. *Journal of Abnormal Psychology, 100*, 569–582.

Nolen-Hoeksema, S., & Girgus, J. (1994). The emergence of gender differences in depression in adolescence. *Psychological Bulletin, 115*, 424–443.

Nolen-Hoeksema, S., & Jackson, B. (2001). Mediators of the gender difference in rumination. *Psychology of Women Quarterly, 25*, 37–47.

Obeyesekere, G. (1985). Depression, Buddhism, and the work of culture. In A. Kleinman & B. Good (Eds.), *Culture and depression*. Berkeley, CA: University of California Press.

Oransky, M. (2002). *Doing boy*. Unpublished manuscript, Swarthmore College.

Pomerantz, J. M. (2003). Dr. Pomerantz responds. *Drug Benefit Trends, 15*, 8–17.

Radloff, L. (1975). Sex differences in depression: The effects of occupation and marital status. *Sex Roles, 1*, 249–265.

Reekie, G. (1994). Reading the problem family: Poststructuralism and the analysis of social problems. *Drug and Alcohol Review, 13*, 457–465.

Rose, N. (1996). *Inventing ourselves: Psychology, power, and personality*. Cambridge, UK: Cambridge University Press.

Scattolon, Y., & Stoppard, J. M. (1999) "Getting on with life": Women's experiences and ways of coping with depression. *Canadian Psychology, 40*, 205–219.

Shields, S. A. (2002). *Speaking from the heart: Gender and the social meaning of emotion*. Cambridge, UK: Cambridge University Press.

Stapley, J. C., & Haviland, J. M. (1989). Beyond depression: Gender differences in normal adolescents' emotional experiences. *Sex Roles, 20*, 295–308.

Stoppard, J., & McMullen, L. (2003). *Situating sadness*. New York: New York University Press.

Strauss, J., Muday, T., McNall, K., & Wong, M. (1997). Response Style Theory revisited: Gender differences and stereotypes in rumination and distraction. *Sex Roles, 36*, 771–792.

Szekely, E. A. (1989). From eating disorders to women's situations: Extending the boundaries of psychological inquiry. *Counseling Psychology Quarterly, 2*, 167–184.

Vrendenburg, K., Krames, L., & Flett, G. L. (1986). Sex differences in the clinical expression of depression. *Sex Roles, 14*, 37–49.

Weissman, M. M., Bland, R., Joyce, P. R., Newman, S., Wells, J. E., & Wittchen, H. (1993). Sex differences in the rates of depression: Cross-national perspectives. *Journal of Affective Disorders, 29*, 77–84.

Weissman, M. M., & Klerman, G. L. (1977). Sex differences and the epidemiology of depression. *Archives of General Psychiatry, 34*, 98–111.

Weisstein, N. (1968). *Kinder, küche, kirche as scientific law: Psychology constructs the female*. Boston: New England Free Press.

West, C., & Zimmerman, D. H. (1987). Doing gender. *Gender and Society, 1*, 125–151.

Westkott, M. C. (1997). On the new psychology of women: A cautionary view. In M. R. Walsh (Ed.), *Women, men and gender: Ongoing debates* (pp. 362–372). New Haven, CT: Yale University Press.

World Health Organization International Consortium in Psychiatric Epidemiology. (2000). Cross-national comparisons of the prevalence and correlates of mental disorders. *Bulletin of the World Health Organization, 78*, 413–426.

World Health Organization. (2002). *International classification of diseases, Tenth Revision*. Geneva, Switzerland: World Health Organization, 2003.

13

Women, Work, and Depression

Conceptual and Policy Issues

Mary Clare Lennon

Social scientists have been concerned about the relationship between work and mental health among women for several decades. Since the 1960s, there have been large-scale changes in women's employment, social roles, social theory, and social policy, all of which have shaped the scholarly literature. This chapter summarizes this literature and its evolution in the context of the social changes that have shaped the study of women, work, and depression. It gives a broad overview of the literature, with particular focus on research questions and debates that have emerged in the past decade. Implications for future research and social policy are highlighted, as well.

EARLY RESEARCH ON WOMEN, WORK, AND DEPRESSION

Social science research on employment and women's well-being reached its peak in the 1980s and 1990s (see Klumb & Lampert, 2004, for a review). As the number of women in the labor force increased, researchers became concerned about the impact of employment on women's psychological well-being. A number of studies had found that employed wives exhibited fewer symptoms of psychological distress and depressive symptomatology than did nonemployed wives (e.g., Pearlin, 1975; Radloff, 1975; Rosenfield, 1980). Explanations for this finding generally focused on the importance of the prevailing female sex role of housewife and mother. It was argued that tasks involved in women's traditional domestic roles were unskilled, repetitive, and isolating and thus apt to be psychologically distressing (Gove & Tudor, 1973). Further evidence for this perspective was found in the higher rates of depressive symptoms among mothers compared to women without children (Gore & Mangione, 1983; McLanahan & Adams, 1987; Radloff, 1975) and among wives compared to husbands (Gove & Geerken, 1977; Radloff, 1975).

Over time, the study of depression and work among women has become far more nuanced and complex. Changes in sociological, economic, and psychological theories; changes in the reality of women's lives; and changes in research methodology have generated new perspectives on the relationship between work and depression. Perhaps the most striking change both in social theory and in the social world is the present-day anachronism of the term *traditional female sex role*. Even when women's roles centered primarily around marriage, motherhood, and housework, the term referred to a historically specific phenomenon, one that was rapidly to become out of date. The use of the term *sex roles* has been criticized by sociologists for reifying a set of relationships (e.g., between women and men or husbands and wives) by characterizing them in terms of fixed, and theoretically complementary, roles structured by gender (e.g., Lopata & Thorne, 1978). The relevance of the concept of a traditional sex role also became questionable because of the increasing participation of women, and especially mothers, in the workforce. In 1965, 39.3% of women aged 16 and older were in the labor force; by 2004 this had increased to 59.2%. The increase is even more striking among women with children under the age of 6, whose labor force participation more than doubled between 1965 and 2004, from 25.3 to 62.2% (U.S. Department of Labor, 2005).

Employment rates of married mothers increased over time as well, reaching the same level as that for single mothers in the mid-1990s, when approximately two of every three mothers were employed. After that time, however, rates for married mothers remained fairly stable; but, starting in 1994, those for single mothers climbed rapidly, reaching around 80% in 2002 (Burtless, 2004). The booming economy during the late 1990s, and various policy changes, such as the liberalization of the Earned Income Tax Credit (EITC), increases in subsidized child care, and the implementation of welfare reform, have all contributed to the increased employment of single mothers. As described later, these changes in women's employment, and especially the increased employment of single mothers, have altered the perspectives brought by social scientists to the study of the relation of work to women's mental health.

WORK AND DEPRESSION

The association between work and depression, as currently portrayed in the literature, is complex. I will simplify this complexity by presenting an overview of four broad themes that may be found in the literature. As described below, evidence exists to support each of these four themes.

- Depression as a consequence of unemployment
- Depression as a consequence of job characteristics
- Depression as a consequence of work and family stress
- Depression as a barrier to employment

Depression as a Consequence of Unemployment

The detrimental effects of unemployment on general psychological well-being are well established (see Banks & Jackson, 1982; Dew, Penkower, & Bromet, 1991; Horwitz, 1984; Warr & Jackson, 1987; see also Brenner, 1973; Dooley, Catalano, & Rook, 1988: for ecological studies). Analyses of the National Comorbidity Survey by Marcotte, Wilcox-Gok, and Redmon (1999) indicate that the 12-month prevalence of major depressive disorder[1] among unemployed women is twice as high as that found among employed women (21 vs. 10%). Although these results are cross-sectional, prospective studies indicate that job loss plays a causal role in the onset of depression and other psychological problems. Unemployment is found to increase symptoms of anxiety and depression whereas reemployment reduces such symptoms (Dooley, Catalano, & Wilson, 1994; Warr & Jackson, 1985). In a 1-year prospective study, Dooley and colleagues (1994) find that unemployment is associated with a twofold increase in the risk for clinical depression. Although the literature reports similar effects of unemployment for women and men (Dew et al., 1991), much of the longitudinal research has been conducted on samples of men.

A number of investigators examine the social and economic processes through which unemployment may cause depression and other psychopathologies. As one example, Pearlin and colleagues (1981) view unemployment as a primary stressor that may set into motion disruptive secondary stressors, such as economic deprivation (see also Conger & Elder, 1994; Price, Choi, & Vinokur, 2002). As another example, Jahoda (1982) asserts that unemployment is stressful because it involves the loss of connection to important social institutions and associations.

Since preunemployment economic circumstances of women (especially single mothers) are generally more precarious than men's (Bianchi, 1995; Dew et al., 1991), it is likely that the economic effects of unemployment should be more problematic for women. On the other hand, women generally have stronger social networks and social support, especially from kin, than men (Bleiszner & Adams, 1992; DiLeonardo, 1987). Thus men may experience more ill effects of unemployment that are due to the reduction of social contacts at work (Gerstel, Reissman, & Rosenfield, 1985). Should these expectations be supported by empirical data, it would suggest that alternative intervention strategies may be needed for women and men to reduce the negative impact of unemployment.

---

[1] The term *major depressive disorder* and *clinical depression* are used throughout this chapter to refer to disorders that meet criteria set forth in the *Diagnostic and Statistical Manual of Mental Disorder* of the American Psychological Association (1980, 1994). The term *depressive symptoms* is used when clinical criteria are not assessed but reports of the numbers of symptoms of depression are, such as with the Center for Epidemiologic Studies Depression Scale (CES-D; Radloff, 1977).

Recent work by Dooley (2003) and colleagues (Dooley, Prause, & Ham-Rowbottom, 2000; Prause & Dooley, 2001) proposes a broader view of employment circumstances, one that incorporates underemployment. They argue that a focus on unemployment alone does not capture many of the employment-related problems that exist in the current economy, such as those that derive from increased income inequality and limited wage growth for low-income families. Dooley and colleagues focus on the concept of economically inadequate work, defined as involuntary part-time work and work that pays poverty-level wages (Dooley, 2003; see also Ehrenreich, 2001; Kalleberg, 2000). Because women make up a large share of low-wage workers (Kim, 2000), the concept of economically inadequate work and its relation to women's mental health is a promising area for further study (Dooley, et al., 2000; Prause & Dooley, 2001).

Proposals to expand the scope of research on unemployment to include suboptimal employment draw attention to the quality of jobs, a field with a long history of research. Since women, and particularly low-income women, tend to work in jobs characterized by poorer wages and working conditions than men, job conditions may be especially salient for understanding women's mental health. Research in this area is considered in the next section of this chapter.

Depression as a Consequence of Job Characteristics

A large body of literature investigates the association between job conditions and psychological distress or depressive symptomatology. Fewer studies examine the linkages of job conditions with major depressive disorder. The literature examines various aspects of jobs, including job insecurity (Hellgren & Sveike, 2003; Matthews & Power, 2002), physical demands (Mausner-Dorsch & Eaton, 2000), shift work (Bildt & Michelsen, 2002; Kandolin, 1993), social support (Bildt & Michelsen, 2002), and, as described below, levels of work control and demands (e.g. Ganster, 1989; Muntaner, Eaton, Diala, Kessler, & Sorlie, 1998; Spector, 1986; Syme, 1988). Because control over work appears to be the key dimension of work associated with depression and its symptoms, this aspect is considered in detail.

Definitions of work control vary from study to study. A prominent theoretical framework (Kohn, Naoi, Schoenbach, Schooler, & Slomczynski, 1990; Kohn & Schooler, 1982, 1983) attempts to explain social class differences in psychological functioning by examining class differences in occupational self-direction. Occupations that permit such self-directed work generally involve substantively complex work with data, people, or things; work that is not closely supervised; and work that is not routine (Kohn & Schooler, 1982). These structural imperatives of work facilitate the use of initiative and personal judgment on the part of the worker. Self-directed work, which occurs more frequently in high socioeconomic status

(SES) occupations (and in predominantly male occupations), promotes the worker's intellectual flexibility and enhances the perception of self as competent and in control of one's own fate.

Kohn and his colleagues document the mental health consequences of substantively complex work across a variety of activities (occupations, housework, leisure time activities) and across a range of cultures (Kohn & Schooler, 1982, 1983; Kohn et al., 1990; Miller, Schooler, Kohn, & Miller, 1979; Schooler, Kohn, Miller, & Miller, 1983). Support for this model with regard to women's occupations, however, is cross-sectional (Miller et al., 1979; see also Roxburgh, 1996; Lennon, 1987).

Research by Link and colleagues (1993) focuses on the related concept of direction, control, and planning (DCP) of others' work activities. Like Kohn and colleagues, Link et al. argue that DCP may protect against the development of depression and psychological distress by contributing to a sense of mastery over the environment. Their empirical results support this hypothesis in that exposure to occupations that permit direction, control and planning accounts for the associations of SES with depressive disorder and psychological distress. Moreover, the association between DCP and depressive disorder is mediated by mastery. In other words, individuals in high DCP occupations feel a greater sense of control over their lives, and this sense of control reduces the likelihood of becoming depressed. These relationships cannot fully be accounted for by social selection processes, as assessed by family history of mental disorder, remote life-threatening illness or injury, and being raised by a mother alone. Importantly, the effects of DCP are the same for women and men, indicating that this form of control appears protective against depression regardless of gender.

A related framework, proposed by Karasek and colleagues (Karasek, Gardell, & Windell, 1987; Karasek et al., 1988; Karasek & Theorell, 1990), asserts that control over work is related to health outcomes primarily among individuals who experience high levels of psychological demands on the job – high-strain jobs are those in which the worker experiences excessive demands but little control. Although research documents an association between job strain and coronary heart disease (Karasek, 1979; Karasek et al., 1988; Karasek & Theorell, 1990; Theorell et al., 1985), empirical evidence for an interaction of job control and demands in predicting psychological distress is mixed. Although, many studies demonstrate direct effects of control and demands on depressive symptoms (Bromet, Dew, Parkinson, & Schulberg, 1988; Niedhammer, Goldberg, Leclerc, Bugel, & David, 1998), few document their joint effect. An exception is the investigation by Mausner-Dorsch and Eaton (2000), which finds that women in high-strain jobs are significantly more likely than women employed in other types of jobs to have a major depressive disorder.

Karasek's framework has been expanded to incorporate the role of social support in the workplace (Barnett & Marshall, 1991; Johnson,

Hall, & Theorell, 1989; Viswesvaran, Sanchez, & Fisher, 1999). Jobs that combine high demands, low support, and low control are associated with earlier onset of cardiovascular disease (Johnson et al., 1989) and greater declines in functional status among women (Cheng, Kawachi, Coakley, Schwartz, & Colditz, 2000). Whether this expanded model also predicts depression remains to be demonstrated. To date, research shows only independent effects – and not the interactive effects postulated – of control, demands, and support on depression (e.g., Niedhammer et al., 1998).

In her studies of family demands, Rosenfield (1989) suggests an alternative way of conceptualizing the role of workplace demands. She hypothesizes that excessive household demands among employed wives (such as having a number of young children at home or sole responsibility for housework) may undermine their sense of control, leading to higher levels of depressive symptoms. Thus, excessive demands have a similar psychological effect as low levels of control: to reduce personal sense of control or mastery. Future research is needed to determine whether these processes generalize to the workplace, with job demands affecting depression by diminishing a sense of mastery.

Depression as a Consequence of Work and Family Stress

Like Rosenfield (1989; Lennon & Rosenfield, 1992), a number of investigators emphasize the importance of considering home responsibilities, as well as job conditions, in studying women's mental health. Social expectations for women's lives have changed dramatically over the past four decades – at least with respect to employment. As more and more women have entered the workplace, general attitudes toward the employment of women and of mothers have become more positive (Simon & Landis, 1989; Brewster & Padavic, 2000). In a national opinion poll conducted in the mid-1970s, the majority of respondents agreed that preschool children suffer when their mothers work outside the home; by 1996, agreement with this statement declined by over 20 percentage points and only a minority of respondents expressed disapproval (Brewster & Padavic, 2000). As noted earlier, employment for mothers of young children is now the statistical norm.

Women's family responsibilities, however, have been less subject to change than have their employment prospects. In dual-earner families women perform about two-thirds of household chores (Lennon & Rosenfield, 1994; Shelton & John, 1996). Studies indicate that husbands of employed wives do not contribute more to housework than do husbands of nonemployed wives (Thompson & Walker, 1989). Of course, women who are single parents and employed are at an even greater disadvantage, because there is often no other adult in the home to do even one-third of the household chores.

Most studies that examine issues of work and family stress are cross-sectional. Also, most focus on married, not single, women. Results of these suggest that combining paid work with family responsibilities may be stressful in some circumstances. For example, the number of children at home is associated with increased depressive symptoms when women hold jobs low on control (Lennon & Rosenfield, 1992). Barnett and Marshall (1989) and Repetti (1988) show similar interactions between job conditions and family responsibilities. Using a different approach to this issue and directly measuring perceived work–family conflict, Frone and colleagues (1997) find that conflict between work and family in employed parents is associated longitudinally with elevated levels of depressive symptomatology. This result is also documented in cross-sectional investigations.

An important component of work and family stress involves child care. Difficulties in arranging, managing, and paying for child care are each associated cross-sectionally with higher levels of depressive symptoms (Lennon, Wasserman, & Allen, 1991; Ross & Mirowsky, 1988). These difficulties are likely to be exaggerated when women hold jobs with little flexibility or control, as well as among low-income parents, who invest proportionately more of their earnings in child care than do middle- or upper-income working parents.

Another way of looking at work and family issues is by considering family work as productive activity that may be characterized along similar dimensions as paid work (Bird & Ross, 1993; Griffith, Fuhrer, Stansfeld, & Marmot, 2002; Lennon, 1994; Schooler et al., 1983). In a comparison of the working conditions of employed women to those of full-time homemakers, homemakers report more control over their work activities and fewer time pressures, but they experience more interruptions, more routinized work, and greater physical demands (Lennon, 1994). In a recent study, Griffith and colleagues (2002) find that control in each domain independently predicts women's depressive symptoms in a large sample of British civil servants. The relation of home control to symptoms is most pronounced among those in lower social classes, indicating the importance of understanding the risks faced by women in lower SES groups.

Depression as a Barrier to Employment

The view that depression interferes with obtaining and retaining jobs derives from a general social selection argument. According to this perspective, the association between unemployment or job instability and mental illness is explained by the selection of the mentally ill out of the labor force. Thus, women who are depressed may have difficulty searching for, obtaining, or retaining jobs. In this view, the greater depressive symptomatology found among nonemployed women may be attributed to the fact that depression, with its concomitant sense of hopelessness and low self-esteem, inhibits the initiative needed for a job search. Even if employment is found, depressed women may exhibit job instability and poor job performance due to their illness. Research indicates that depressed individuals have a much greater likelihood than those without depression of taking disability days at work (Broadhead Blazer, George, & Tse, 1990; Kessler et al., 1999).

Many studies conducted prior to the 1990s on the relation of work to depression generally assumed a social causation perspective, although much of this work is cross-sectional (see Klumb & Lampert, 2004, for a review). The social selection perspective has gained greater currency in the current welfare reform environment. The Personal Responsibility and Work Opportunity Reconciliation Act (PRWORA) of 1996 made employment for women on welfare mandatory after 2 years of welfare receipt and imposed a 5-year lifetime limit for receipt of federal welfare payments[2]. Partly as a consequence of this legislation, there has been a dramatic decline in caseloads – nationally the number of recipients has been reduced by over 60% since 1993 (U.S. Department of Health and Human Services 2004).

Although the majority of women who left welfare obtained employment in the early years after welfare reform (Corcoran, Danziger, Kalil, & Seefeldt, 2000), the employment rate for women leaving welfare has declined since 2000 (Fremstad, 2004). Concern has been expressed about women who either remain on the rolls or leave due to sanctions because of mental health and other problems. Generally, research finds high rates of depression among women on welfare, with 12-month prevalence of major depressive disorder ranging from 15 to 26.7% (as measured by the Composite International Diagnostic Interview [CIDI]) and high levels of depressive symptoms (as measured by the Center for Epidemiologic Studies Depression Scale [CES-D]) ranging between 35 and 56.7% (see reviews by Lennon, Blome, & English, 2001, 2002).

Longitudinal data on the association between welfare receipt and depression are scarce, especially in the post-PRWORA era. Having current data is important because the incentives for employment have changed with welfare reform. Employment has become mandatory whereas it was voluntary under the previous welfare program, Aid to Families with Dependent Children (AFDC). Evidence from one recent study in Michigan indicates that women on welfare who were depressed in 1997 were less likely to be employed 5 years after welfare reform than were nondepressed welfare recipients (Corcoran, Danziger, & Tolman, 2004). However, other studies (Michalopoulos, Schwartz, & Adams-Ciardullo, 2000) find that depression is unrelated to employment among welfare recipients. These different findings may be due to differences in samples, in local labor

---

[2] However, some states have modified these requirements and the federal government permits a minority of recipients to be exempt from the lifetime limit).

markets, in local welfare policies, and in the measurement of depression (e.g., Michalopoulos, Schwartz, & Adams-Ciardullo use an indicator of depressive symptoms whereas Corcoran and colleagues use a diagnostic measure).

An alternative interpretation of the association between employment and depression is given by the social causation explanation. This asserts that aspects of the social environment in general (and employment in this instance) may cause depression. Much of the social science literature described in the first three sections is characterized by this approach. As outlined above, researchers have examined various dimensions of employment – becoming unemployed, being in a stressful job, and juggling work and family responsibilities – in relation to depression.[3]

CONCEPTUAL, METHODOLOGICAL, AND RESEARCH CHALLENGES

The study of depression and the workplace in women faces considerable conceptual and methodological challenges. Conceptual challenges are perhaps the most daunting, although these are not unrelated to methodological issues. In this section, I consider just two research areas in need of further conceptual and methodological development. These are the selection and causation debate and the role of social context.

Perhaps most problematic conceptually are assumptions related to social selection and social causation. Most studies on work and depression in women take one of these explanations as a starting point, and conduct research accordingly. Others treat selection and causation as competing explanations and test one against the other empirically. Although a number of investigators recognize that both processes are probably operative, much work is needed to specify the conditions under which this is likely to be the case.

Various scenarios are plausible. For example, social causation and social selection may characterize a woman's experience at varying points in the life cycle. Early childhood poverty may predispose a girl to depression; this depression, in turn, may interfere with her educational and occupational attainment, leading to unstable employment in poor quality jobs (e.g., those allowing little control). These suboptimal jobs may place her at risk for a recurrence of depression. Alternatively, one or the other process may work alone in a given person. For example, a child who is genetically predisposed to a mood disorder may develop clinical depression early in life that impairs social functioning, leading to poor occupational performance later: a social selection effect. Social causation processes may be primary

---

[3] Given differing views of causal direction posited by the social selection and social causation arguments, implications for intervention, treatment, and policy vary widely. These will be outlined later in this chapter.

in another woman who becomes depressed after losing a valued job due to downsizing or a recession. This effect might primarily be manisfested later in life, after a career has been established, suggesting the importance of life stage as context.

One challenge to researchers is to conceptualize the theoretically plausible scenarios and, as described below, specify the circumstances under which they may unfold. A related challenge is to obtain longitudinal data at the appropriate points in the life cycle on theoretically relevant constructs. Defining these time points and measuring these constructs accordingly are critical. Statistical procedures to test feedback processes will be important in this effort.

Another important conceptual challenge facing researchers involves understanding the role that social context may play in shaping the association between work and depression. For example, under strong economic conditions with very low levels of unemployment, the need for employees in a tight labor market may improve employment prospects for individuals with depression. In these circumstances, employers may accommodate workers' needs, for example, by allowing time off for medical emergencies and perhaps even providing health benefits to finance these visits. On the other hand, in an economy with high unemployment, depressed individuals are likely to have greater trouble finding and keeping jobs. In these circumstances, employers may demand more of their workers (Tausig & Fenwick, 1999), cut back on pay and benefits, and refuse to accommodate special needs.

Another contextual issue that is likely to affect the relation of work and depression among women is the gendered distribution of work opportunities. A glass ceiling that restricts upward mobility potentially limits women's exposure to higher level jobs. Should access to higher level managerial positions be protective against depression, as the work of Muntaner and colleagues (1998) suggests, then studies of women and work at this particular historical juncture may not contain sufficient variation to detect the hypothesized effect of managerial position.

An additional contextual factor that requires study consists of the social policies that affect women's options regarding employment. As previously noted, welfare policies – and their mandates concerning employment – play a critical role for low-income women. The EITC is another policy initiative that has increased employment and reduced poverty. The EITC refunds federal taxes to families below an income threshold, resulting in substantial income increases for some families. It has been estimated that almost 5 million people were removed from poverty in 2002 due to the EITC (Llobrera & Zahradnik, 2004), creating an important incentive for poor women to obtain jobs. Women's employment situations are also affected by policies regarding family leave, child care, health insurance, marriage, minimum wage, and immigration. Some of these policies may directly affect mental health (e.g., by moving women out of poverty or reducing job–family conflict)[4].

Taken together, understanding the complexities of social causation and social selection dynamics, as well as delineating the role of contextual factors, form an important future research agenda. The critical first step involves specifying the conceptual issues to derive theory-based hypotheses that may be tested empirically. Although existing research results are suggestive about future research directions, productive research in the future will arise from more than prior empirical results. The next steps in understanding the relation of work and depression require careful conceptualization and measurement of concepts, the identification of critical contextual factors, and the development of strategic research designs to examine selection and causation processes. The importance of theory in guiding this process cannot be overstated (see Klumb & Lampert, 2004).

IMPLICATIONS FOR TREATMENT, INTERVENTIONS, AND POLICY

To the extent that socially causal factors operate in childhood, a strong case may be made for primary prevention strategies in childhood and adolescence. Meta-analyses indicate the effectiveness of a variety of strategies for the prevention of depression and other mental health problems in children and youth (Durlak & Wells, 1997). Although many successful strategies are focused on mental health interventions, some effective approaches target educational attainment. For example, the prevention of school dropout has been shown to reduce depression in the experimental group (Felner et al., 1993) and has the additional benefit of enhancing occupational choices during adulthood. To the extent that better educational achievement increases employability and the quality of jobs attained, these interventions may prevent the development or recurrence of work-related depression in adulthood.

In addition, various adult-focused interventions have been shown to be effective in reducing depressive symptoms among low-income individuals. One promising approach (Vinokur, Sshul, Vuori, & Price, 2000) incorporates attention to mental health problems into job programs for the unemployed. Results from randomized trials indicate that those who receive an early intervention to prevent the harmful effects of job loss on mental and physical health show improvements in both occupational and mental health outcomes.

---

[4] For a general discussion of the importance of context to understanding the effects of risk factors on illness, see Schwartz and Carpenter (1999); for a discussion of the role of historical context, see Hunt (2002).

Interventions focused on employed individuals have also shown effectiveness. Treatment of depression increases the likelihood of continued employment (Simon et al., 2000; Wells et al., 2000) in randomized trials. The investigation by Simon and colleagues (2000) also shows that those with greater clinical improvement report missing fewer days from work for illness or health care visits.

Another promising approach, not geared toward improving mental health per se, involves ensuring that employed women obtain an income sufficient to raise their families out of poverty. A randomized experimental study in Minnesota that provided earnings supplements to women who left welfare for work had positive effects on employment and earnings. More important in the present context, women who received supplements showed lower levels of depressive symptoms (Knox, Miller, & Gennetian, 2000). Additional policy initiatives, such as refundable state EITCs, increasing eligibility for the federal EITC, and subsidized child care, housing, and transportation improve the material well-being of poor families and quite possibly enhance their psychological functioning and well-being.

A crucial mental health policy issue has to do with reductions in health insurance coverage nationally. Three trends contribute to this problem: cutbacks in employer-based health insurance coverage (Kaiser Commission on Medicaid and the Uninsured, 2003) the declines in Medicaid enrollment among women who leave welfare (Families USA, 1999), and the contraction of mental health coverage under managed care (HayGroup, 1999). Policy interventions to ensure universal health coverage and parity for mental health treatment are critical for reducing the prevalence of depression in women.

Another important policy issue derives from disparities in treatment for depression. Although not specifically a workplace issue, given the importance of occupational position for SES, these disparities will be greater among individuals in lower status occupational groups, precisely those at risk for developing depression. Such disparities in treatment of mental illness have been well documented over time (e.g., Hollingshead & Redlich, 1958), with more recent data from the National Comorbidity Survey showing that individuals with low income are far less likely to seek treatment for a mental health problem than are individuals with higher incomes (Katz, Kessler, Frank, Leaf, & Lin, 1997). Moreover, the poor in the United States are notably less likely than others to receive appropriate medication management for depression (Katz, Kessler, Lin, & Wells, 1998; see also Melfi, Crogan, & Hanna, 1999). Treatment disparities are apparent, as well, by race, with Blacks less likely to receive any medication for depression or, when medication is prescribed, more likely to be given older forms of antidepressants, which are less expensive but have more side effects, than the newer antidepressants (i.e., selective serotonin reuptake inhibitors such as Prozac; Melfi, Crogan, & Hanna, 1999; Melfi et al., 2000).

Clearly, these results require further investigation to understand the mechanisms through which these disparities arise, whether they be individual preference (patient or provider) or systematic factors, such as policies or practices related to insurance reimbursement. Although research is needed, as well, to understand the treatment experiences of all women with depression, treatment issues in poor women and African American women require special attention.

The broad agenda for the future with regard to treatment and intervention in women needs to focus on learning what interventions are effective in preventing depression and what, possibly job-based, interventions might prevent recurrence. In addition to preventive interventions in childhood, adult-focused interventions that increase or supplement earnings and move women out of poverty hold particular promise for reducing onset and recurrence of depression in poor women. Such interventions could be developed in conjunction with prevention strategies and direct treatment for women whose depression is not reduced by income-enhancing interventions.

In summary, research on the relation of employment to depression among women suggests several perspectives for understanding how women's mental health relates to employment status. Recent investigations highlight issues of causal direction, job quality, and the home environment. This review suggests that further consideration of contextual factors, including the macroeconomic and policy environments, is critically important for understanding factors that shape the relation of work to depression in women.

References

American Psychiatric Association. (1980). *Diagnostic and statistical manual of mental disorders* (3rd ed.). Washington, DC: American Psychiatric Association.

American Psychiatric Association. (1994). *Diagnostic and statistical manual of mental disorders*. (4th ed.). Washington, DC: American Psychiatric Association.

Banks, M. H., & Jackson, P. R. (1982). Unemployment and risk of minor psychiatric disorder in young people: Cross-sectional and longitudinal evidence. *Psychological Medicine 12*, pp. 789–798.

Barnett, R. C., & Marshall, N. L. (1989). *Multiple roles, spillover effects, and psychological distress* (Working Paper No. 200). Wellesley College Center for Research on Women.

Barnett R. C., & Marshall, N. L. (1991). The relationship between women's work and family roles on subjective well-being and psychological distress. In M. Frankenhaeuser, U. Lundberg, & M. Chesney, (Eds.), *Women, work and health: Stress and opportunities*. (pp. 111–136). New York: Plenum.

Bianchi, S. (1995). Changing economic roles of women and men. In R. Farley (Ed.), *State of the Union: America in the 1990s, Vol. 1. Economic trends* (pp. 107–154). New York: Russell Sage Foundation.

Bildt, C., & Michelsen, H. (2002). Gender differences in the effects from working conditions on mental health: A 4-year follow-up. *International Archives of Occupational and Environmental Health, 75*, 252–258.

Bird, C. E., & Ross, C. E. (1993). Houseworkers and paid workers: Qualities of the work and effects on personal control. *Journal of Marriage and the Family, 55*, 913–925.

Blieszner, R., & Adams, R. (1992). *Adult friendship*. Beverly Hills: Sage.

Brenner, M. H. (1973). *Mental illness and the economy*. Cambridge, MA: Harvard University Press.

Brewster, K. L., & Padavic, I. (2000). Change in gender-ideology, 1977–1996: The contributions of intracohort change and population turnover. *Journal of Marriage and the Family, 62*, 477–487.

Broadhead, W. E., Blazer, D. G., George L. K., & Tse, C. K. (1990). Depression, disability days, and days lost from work in a prospective epidemiologic survey, *Journal of the American Medical Association, 264*, 2524–2528.

Bromet, E. J., Dew, M. A., Parkinson, D. K., & Schulberg, H. C. (1988). Predictive effects of occupational and marital stress on mental health of a male workforce, *Journal of Organizational Behavior 9*, 1–13.

Burtless, G. (2004). *The labor force status of mothers who are most likely to receive welfare: Changes following reform*. Washington, DC: The Brookings Institution. Retrieved May 14, 2004, from http://www.brookings.edu/views/op-ed/burtless/20040330.htm.

Cheng, Y., Kawachi, I., Coakley, Jr. H., Schwartz, J., & Colditz, G. (2000). Association between psychosocial work characteristics and health functioning in American women: Prospective study. *British Medical Journal, 320*, 1432–1436.

Conger, R. D., & Elder, Jr. G. H. (1994). *Families in troubled times: adapting to change in rural America*. New York: Aldine de Gruyter.

Corcoran, M., Danziger, S. K., Kalil, A., & Seefeldt, K. S. (2000). How welfare reform is affecting women's work, *Annual Review of Sociology, 26*, 241–269.

Corcoran, M., Danziger, S., & Tolman, R. (2004). Long term employment of African-American and white welfare recipients and the role of persistent health and mental health problems. *Women & Health, 39*, 21–40.

Dew, M. A., Penkower, L., & Bromet, E. J. (1991). Effects of unemployment on mental health in the contemporary family. *Behavior Modification, 15*, 501–544.

DiLeonardo, M. (1987). The female world of cards and holidays: Women, families and the work of kinship. *Signs, 12*, 440–452.

Dooley, D. (2003). Unemployment, underemployment, and mental health: Conceptualizing employment status as a continuum. *American Journal of Community Psychology, 32*, 9–20.

Dooley, D., Catalano, R., & Rook, K. S. (1988). Personal and aggregate unemployment and psychological symptoms. *Journal of Social Issues, 44*, 107–123.

Dooley, D., Catalano, R., & Wilson, G. (1994). Depression and unemployment: Panel findings from the Epidemiologic Catchment Area study. *American Journal of Community Psychology, 22*, 745–765.

Dooley, D., Prause, J., & Ham-Rowbottom, K. A. (2000). Underemployment and depression: Longitudinal relationships. *Journal of Health & Social Behavior, 41*, 421–436.

Durlak, J. A., & Wells, M. (1997). Primary prevention mental health programs for children and adolescents: A meta-analytic review. *American Journal of Community Psychology, 25,* 115–152.

Ehrenreich, B. (2001). *Nickel and dimed: On (not) getting by in boom-time America.* New York: Metropolitan Books.

Families USA. (1999). *Losing health insurance: Unintended consequences of welfare reform.* Retrieved August 20, 2000, from http://www.familiesusa.org/unintend.pdf.

Felner, R. D., Brand, S., Adan, A. M., Mulhall, P. F., Flowers, N., Sartain, B., et al. (1993). Restructuring the ecology of the school as an approach to prevention during school transitions: Longitudinal follow-ups and extensions of the School Transitional Environment Project (STEP). *Prevention in Human Services, 10,* 103–136.

Fremstad, S. (2004). *Recent welfare reform research findings: Implications for TANF reauthorization and state TANF policies.* Washington DC: Center on Budget and Policy Priorities. Retrieved May 14, 2004, from http://www.cbpp.org/1-30-04wel.pdf.

Frone, M. R., Russell, M., & Cooler, M. L. (1997). Relation of work-family conflict to health outcomes: A four-year longitudinal study of employed parents. *Journal of Occupational & Organizational Psychology, 70,* 325–335.

Ganster, D.C. (1989). Work control and well-being: A review of research in the workplace. In S. L. Sauter, J. J. Hurrell, & C. L. Cooper (Eds). *Job control and worker health* (pp. 2–23). New York: Wiley.

Gerstel, N. R., Reissman, C. K., & Rosenfield, S. (1985). Explaining the symptomatology of separated and divorced women and men: The role of material conditions and social networks. *Social Forces, 64,* 84–101.

Gore, S., & Mangione, T. W. (1983). Social roles and psychological distress: Additive and interactive models of sex differences. *Journal of Health and Social Behavior, 24,* 300–312.

Gove, W. R., & Geerken, M. R. (1977). The effects of children and employment on the mental health of married men and women. *Social Forces, 56,* 66–76.

Gove, W., & Tudor, J. (1973). Adult sex roles and mental illness. *American Journal of Sociology, 78,* 812–835.

Griffith, J. M., Fuhrer, R., Stansfeld, S. A., & Marmot, M. (2002). The importance of low control at work and home on depression and anxiety: Do these effects vary by gender and social class? *Social Science & Medicine, 54,* 783–798.

HayGroup. (1999). *Health care plan design and cost trends–1988 through 1998.* Hay Group: Arlington, VA.

Hellgren, J., & Sverke, M. (2003). Does job insecurity lead to impaired well-being or vice versa? Estimation of cross-lagged effects using latent variable modeling. *Journal of Organizational Behavior, 24,* 215–236.

Hollingshead, A. & Redlich, F. C. (1958). *Social class and mental illness: A community study.* New York: Wiley.

Horwitz, A. V. (1984). The economy and social pathology. *Annual Review of Sociology, 10,* 95–119.

Hunt, K. (2002). A generation apart? Gender-related experiences and health in women in early and late mid-life. *Social Science & Medicine, 54,* 663–676.

Jahoda, M. (1982). Employment and unemployment: A social psychological analysis. New York: Cambridge University Press.

Johnson, J. V., Hall, E. M., & Theorell, T. (1989). Combined effects of job strain and social isolation on cardiovascular disease morbidity and mortality in a random sample of the Swedish male working population, *Scandinavian Journal of Work, Environment & Health, 15*, 271–279.

Kaiser Commission on Medicaid and the Uninsured. (2003). *Health insurance coverage in America: 2002 data update.* Retrieved May 15, 2004, from http://www.kff.org/content/archive/1407/ Uninsured%20in%20America.pdf.

Kalleberg, A. L. (2000). Nonstandard employment relations: Part-time, temporary and contract work, *Annual Review of Sociology, 26*, 341–365.

Kandolin, I. (1993). Burnout of female and male nurses in shift work. *Ergonomics, 36*, 141–147.

Karasek, R. (1979). Job demands, job decision latitude, and mental strain: Implications for job redesign. *Administrative Science Quarterly, 24*, 285–308.

Karasek, R., Theorell, T., Schwartz, J., Schnall, P., Pieper, C. & Michela, J. (1988). Job characteristics in relation to the prevalence of myocardial Infarction in the U.S. HES and HANES. *American Journal of Public Health, 78*, 910–918.

Karasek, R., Gardell, B., & Windell, J. (1987). Work and nonwork correlates of illness and behaviour in male and female Swedish white-collar workers. *Journal of Occupational Behavior, 8*, 87–207.

Karasek, R., & Theorell, T. (1990). *Healthy work: Stress, productivity and the reconstruction of working life.* New York: Basic.

Katz, S. J., Kessler, R. C., Frank, R. G., Leaf, P., & Lin, E. (1997). Mental health care use, morbidity, and socioeconomic status in the United States and Ontario. *Inquiry, 34*, 38–49.

Katz, S. J., Kessler, R. C., Lin, E., & Wells, K. B. (1998). Medication management of depression in the United States and Ontario. *Journal of General Internal Medicine, 13*, 77–85.

Kessler, R. C., Barber, C., Birnbaum, H. G., Frank, R. G., Greenberg, P. E., Rose, R. M., et al. (1999). Depression in the workplace: effects on short-term disability. Health Affairs, *18*, 163–171.

Kim, M. (2000). Women paid low wages: Who they are and where they work. *Monthly Labor Review Online, 123* (9), 26–30.

Klumb, P. L., & Lampert, T. (2004). Women, work, and well-being 1950–2000: a review and methodological critique. *Social Science & Medicine, 58*, 1007–1024.

Knox, V., Miller, C., & Gennetian, L. A. (2000). *Reforming welfare and rewarding work: A summary of the final report on the Minnesota Family Investment Program.* New York: Manpower Demonstration Research Corporation.

Kohn, M. L., Naoi, A., Schoenbach, C., Schooler, C., & Slomczynski, K. M. (1990). Position in the class structure and psychological functioning in the United States, Japan, and Poland. *American Journal of Sociology, 95*, 964–1008.

Kohn, M. L., & Schooler, C. (1982). Job conditions and personality: A longitudinal assessment of their reciprocal effects. *American Journal of Sociology, 87*, 1257–1286.

Kohn M. L., & Schooler, C. (1983). *Work and personality: An inquiry into the impact of social stratification.* Norwood, NJ: Ablex.

Lennon, M. C. (1987). Sex differences in distress: The impact of gender and work roles. *Journal of Health and Social Behavior, 28*, 290–305.

Lennon, M. C. (1994). Women, work, and well-being: The importance of work conditions. *Journal of Health and Social Behavior, 35*, 235–247.

Lennon, M. C., Blome, J., & English, K. (2001). *Depression and low-income women: Challenges in an era of devolution*. New York: The Research Forum, Retrieved May 10, 2004, from http://www.researchforum.org.

Lennon, M. C., Blome, J., & English, K. (2002). Depression among women on welfare: A review of the literature, *Journal of the American Medical Womens Association, 57*, 27–31, 40.

Lennon, M. C., & Rosenfield, S. (1992). Women and distress: The contribution of job and family conditions. *Journal of Health and Social Behavior, 33*, 316–327.

Lennon, M. C., & Rosenfield, S. (1994). Relative fairness and the division of family work: The importance of options. *American Journal of Sociology, 100*, 506–531.

Lennon, M. C., Wasserman, G. A., & Allen, R. (1991). Husbands' involvement in child care and depressive symptoms among mothers of infants. *Women and Health, 17*, 1–23.

Link, B. G., Lennon, M. C., & Dohrenwend, B. P. (1993). Socioeconomic status and depression: The role of occupations involving direction, control and planning. *American Journal of Sociology, 98*, 1351–1387.

Llobrera, J., & Zahradnik, B. (2004). *A hand up: How state earned income tax credits help working families escape poverty in 2004*. Washington DC: Center on Budget and Policy Priorities. Retrieved May 14, 2004, from http://www.cbpp.org/5-14-04sfp.htm.

Lopata, H. Z., & Thorne, B. (1978). On the term sex roles. *Signs, 3*, 718–721.

Marcotte, D. E., Wilcox-Gok, V., & Redmon, D. P. (1999). Prevalence and patterns of major depressive disorder in the United States labor force. *Journal of Mental Health Policy and Economics, 2*, 123–131.

Matthews, S., & Power, C. (2002). Socio-economic gradients in psychological distress: A focus on women, social roles and work-home characteristics. *Social Science & Medicine, 54*, 799–810.

Mausner-Dorsch, H., & Eaton, W. W. (2000). Psychosocial work environment and depression: Epidemiologic assessment of the demand-control model. *American Journal of Public Health, 90*, 1765–1770.

McLanahan, S. S., & Adams, J. (1987). Parenthood and psychological well-being, *Annual Review of Sociology, 13*, 237–257.

Melfi, C. A., Crogan, T. W., & Hanna, M. (1999). Access to treatment for depression in a Medicaid population. *Journal of Health Care for the Poor and Underserved, 10*, 201–215.

Melfi, C. A., Croghan, T. W., Hanna, M. P., & Robinson, R. L. (2000). Racial variation in antidepressant treatment in a Medicaid population. *Journal of Clinical Psychiatry, 61*, 16–21.

Michalopoulos, C., Schwartz, C., & Adams-Ciardullo, D. (2000). *What Works Best for Whom: Impacts of 20 Welfare-to-Work Programs by Subgroup: The National Evaluation of Welfare-to-Work Strategies*. New York: Manpower Demonstration Research Corporation.

Miller, J., Schooler, C., Kohn, M. L., & Miller, K. A. (1979). Women and work: The psychological effects of occupational conditions. *American Journal of Sociology, 85,* 66–94.

Muntaner, C., Eaton, W. W., Diala, C. L., Kessler, R. C., & Sorlie, P. D. (1998). Social class, assets, organizational control and the prevalence of common groups of psychiatric disorders. *Social Science & Medicine, 47,* 2043–2053.

Niedhammer, I., Goldberg, M., Leclerc, A., Bugel, I., & David, S. (1998). Psychosocial factors at work and subsequent depressive symptoms in the Gazel cohort. *Scandinavian Journal of Work, Environment & Health, 24,* 197–205.

Pearlin, L. I. (1975). Sex roles and depression. In N. Datan & L. H. Ginsberg (Eds.), *Life-span developmental psychology: normative live crises.* New York: Academic.

Pearlin, L. I., Menaghan, E. G., Lieberman, M. A., & Mullan, J. T. (1981). The stress process. *Journal of Health and Social Behavior, 22,* 337–356.

Prause, J., & Dooley, D. (2001) Favourable employment status change and psychological depression: A two-year follow-up analysis of the National Longitudinal Survey of Youth. *Applied Psychology, 50,* 282–304.

Price R. H., Choi, J. N., & Vinokur, A. D. (2002). Links in the chain of adversity following job loss: how financial strain and loss of personal control lead to depression, impaired functioning, and poor health, *Journal of Occupational Health Psychology, 7,* 302–12.

Radloff, L. S. (1975). Sex differences in depression: The effects of occupation and marital status. *Sex Roles, 1,* 243–265.

Radloff, L. S. (1977). The CES-D scale: A self-report depression scale for research in the general population. *Applied Psychological Measurement, 1,* 385–401.

Repetti, R. L. (1988). Family and occupational roles and women's mental health. In R. M. Schwartz (Ed.), *Women at work* (pp. 97–129). Los Angeles: Institute of Industrial Relations Publications, University of California.

Rosenfield, S. (1980). Sex differences in depression: Do women always have higher rates? *Journal of Health and Social Behavior, 21,* 33–42.

Rosenfield, S. (1989). The effects of women's employment: Personal control and sex differences in mental health. *Journal of Health and Social Behavior, 30,* 77–91.

Ross, C. E., & Mirowsky, J. (1988). Child care and emotional adjustment to wives' employment. *Journal of Health and Social Behavior, 29,* 127–138.

Roxburgh, S. (1996). Gender differences in work and well-being: Effects of exposure and vulnerability. *Journal of Health and Social Behavior, 37,* 265–277.

Schooler, C., Kohn, M. L., Miller, K. A., & Miller, J. (1983). Housework as work. In M. L. Kohn & C. Schooler (Eds.), *Work and personality: An inquiry into the impact of social stratification* (pp. 242–260). Norwood, NJ: Ablex.

Schwartz, S., & Carpenter, K. (1999). The right answer for the wrong question: Consequences of type III error for public health research. *American Journal of Public Health, 89,* 1175–1180.

Shelton, B. A., & John, D. (1996). The division of household labor. *Annual Review of Sociology, 22,* 299–322.

Simon, G. E., Revicki, D., Heiligenstein, J., Grothaus, L., VonKorff, M., Katon, W., et al. (2000). Recovery from depression, work productivity, and health care costs among primary care patients. *General Hospital Psychiatry, 22,* 153–162.

Simon, R. J., & Landis, J. M. (1989). Women's and men's attitudes about a women's place and role, *Public Opinion Quarterly, 53*, 256–276.

Spector, P. E. (1986). Perceived control by employees: A meta-analysis of studies concerning autonomy and participation at work. *Human Relations, 39*, 1005–1016.

Syme, S. L. (1988). *Social epidemiology and the work environment. International Journal of Health Services, 18*, 635–645.

Tausig, M., & Fenwick, R. (1999). Recession and well-being. *Journal of Health & Social Behavior, 40*, 1–16.

Theorell, T., Knox, S., Svensson, J., & Waller, D. (1985). Blood pressure variations during a working day at age 28: Effects of different types of work and blood pressure level at age 18. *Journal of Human Stress, 11*, 36–41.

Thompson, L., & Walker, A. J. (1989). Gender in families: Women and men in marriage, work, and parenthood. *Journal of Marriage and the Family, 51*, 845–871.

U.S. Department of Health and Human Services (2004). *U.S. welfare caseload information*. Retrieved on May 14, 2004, from http://www.acf.dhhs.gov/news/stats/newstat2.shtml.

U.S. Department of Labor. (2005). *Labor force statistics from the Current Population Survey*. Retrieved May 9, 2005, from http://www.data.bls.gov.

Vinokur, A. D., Sshul, Y., Vuori, J., & Price, R. H. (2000). Two years after a job loss: Long-term impact of the JOBS program on reemployment and mental health. *Journal of Occupational Health Psychology, 5*, 32–47.

Viswesvaran, C., Sanchez, J. I., & Fisher, J. (1999). The role of social support in the process of work stress: A meta-analysis. *Journal of Vocational Behavior, 54*, 314–334.

Warr, P., & Jackson, P. (1985). Factors influencing the psychological impact of prolonged unemployment and of re-employment. *Psychological Medicine, 15*, 795–807.

Warr, P., & Jackson, P. (1987). Adapting to the unemployed role: A longitudinal investigation, *Social Science & Medicine, 25*, 1219–1224.

Wells, K. B., Sherbourne, C., Schoenbaum, M., Duan, N., Meredith, L., Unutzer, J., et al. (2000). Impact if disseminating quality improvement programs for depression in managed primary care: A randomized controlled trial. *JAMA, 283*, 212–20.

14

Culture, Race/Ethnicity, and Depression

Pamela Braboy Jackson and
David R. Williams

INTRODUCTION

Mental health problems affect approximately 26% of U.S. adults in a given year and 50% at some point in their life time (Demyttenaere et al., 2004; U.S. Department of Health and Human Services [DHHS], 1999). The leading causes of disability in the United States, in fact, include four mental health conditions: major depression, bipolar disorder, schizophrenia, and obsessive-compulsive disorder (OCD; Murray & Lopez, 1996). Depressive disorders, however, are more common than schizophrenia and OCD's. For example, in the Epidemiological Catchment Area (ECA) study, about 9.5% of adults over the age of 18 were diagnosed with a depressive disorder in a given year compared to 1.1% for schizophrenia and 2.3% for OCD (Regier et al., 1993). It is estimated that 16% of American adults meet criteria for major depressive disorder at some time in their life (Kessler et al., 2003) and the prevalence of depressive disorders is markedly increasing in younger cohorts compared to older ones (Kasen, Cohen, Chen, & Castille, 2003).

However, racial differences in the risk of depression is unclear. African Americans appear to have equivalent or lower rates of major depression than Whites (Kessler, McGonagle, Schwartz, Glazer, & Nelson, 1993; Kessler et al., 2003; Robins & Regier, 1991) but higher rates than Whites of depressive symptoms (Vega & Rumbaut, 1991; Williams & Harris-Reid, 1999). The limited available data for other racial groups suggest that the prevalence of mental health problems is similar to those of Whites but the need for more research on this topic has been explicitly stated by a growing contingent of mental health scholars (U.S. DHHS, 2001). In fact, there is consistent evidence that many physical health outcomes conform to the hypothesized inequalities suffered by ethnic minorities (see Williams, Yu, Jackson, & Anderson, 1997). Despite the vast literature demonstrating that mental health outcomes are similarly graded by socioeconomic status (SES;

Turner, Wheaton, & Lloyd, 1995), there is less support in the mental health literature of the racial inequality hypothesis.

In this chapter, we discuss some of the cultural characteristics of racial/ethnic groups that may correlate with the risk of experiencing, and resilience to, depression among women. We begin by briefly exploring the changing definitions of culture and race/ethnicity. We then pay particular attention to the literature on gender differences in depression and move on to some of the ways in which culture might help explain patterns of depression. Our empirical analysis begins with an examination of the structure of depression in a nationally representative sample of adult women. From there we examine the extent to which five American racial/ethnic groups differ in reports of depressive symptoms, after taking into consideration a wide range of cultural factors. The chapter concludes with a discussion of the challenges of understanding the influence of culture on symptoms of depression.

Culture. The word *culture* finds its roots in the Latin word *colere*, which means to cultivate or foster growth. This definition was extended during the 16th century from a referral of crops to include human beings, especially in terms of the development of the mind or intellect. By the 19th century, the concept of culture nad expanded further to include entire societies so that not every society was considered "cultured." To become cultured, according to imperialist thought originating from Britain, one had to possess knowledge of the arts, letters, and sciences of the Western world. The 20th century witnessed the most extensive alteration in the definition of culture. Culture has come to represent the way of life of any group of peoples (Linton, 1945). Many definitions can be found in current dictionaries (see Merriam-Webster, 2004). The definition we feel captures the relevance of culture for better understanding gender differences in depression is offered by Murphy (1986, p. 14), who states,

> Culture means the total body of tradition borne by a society and transmitted from generation to generation. It thus refers to the norms, values, standards by which people act, and it includes the ways distinctive in each society of ordering the world and rendering it intelligible. Culture is . . . a set of mechanisms for survival, but it provides us also with a definition of reality. It is the matrix into which we are born, it is the anvil upon which our persons and destinies are forged.

Culture, therefore, consists of learned concepts and behaviors, a worldview, and the resulting products such as customs. Of course, there is a great deal of ethnic diversity within many nation states. This diversity has implications for the ubiquity of a single, unified notion of culture. Recent conflicts in Bosnia, for example, illustrate this problem. Within this country, bitter divisions remain between the country's Serb, Croat, and Bosniac (or Muslim) communities. As nations continue to embrace a vast array of immigrants from around the world, the inherently contradictory nature of multiculturalism will continue to challenge the study of culture and mental health. Culture as "contained, controlled, and homogenized by the national state" (Joppke, 1996, p. 450) is no longer an option in a world system characterized by migration flows. Nonetheless, cultural communities are formed within the boundaries of the larger nation state and still expect their citizens to adhere to basic civic rules. The issue relevant for this study is the extent to which cultural elements are associated with symptoms of depression among an ethnically diverse group of women who coexist within the boundaries of the larger state system. A broad body of sociological research suggests that certain components of culture contribute to the patterning of depression among women. We believe that a focus on culture can highlight potential vulnerabilities among women who belong to certain ethnic minority populations. Before proceeding, we briefly define race/ethnicity and depression as used throughout this chapter.

Race. Race is a social construct (American Association of Physical Anthropology, 1996; Omi & Winant, 1986; Owens & King, 1999; Williams, 1997). The first U.S. census enumerated three groups based on race: Whites, Blacks (as three-fifths of a person), and those Indians who paid taxes. Some race scholars contend that the racial composition of the United States reflects the conscious design of U.S. immigration and naturalization laws (Lopez, 2001). Guidelines laid out by the U.S. Office of Management and Budget (OMB; 1978) for categorizing race and ethnicity initially stipulated four racial categories – (1) White, (2) Black or African American, (3) American Indian or Alaska Native, (4) Asian or Pacific Islander – and one ethnic category – (5) Hispanic. Included among Hispanics are Mexicans, Puerto Ricans, Cubans, Dominicans, and Peruvians. A revision of these guidelines by OMB in 1997 alters these categories by placing Asians in a separate category and creating a new category for Native Hawaiians and other Pacific Islanders (Grieco & Cassidy, 2001).

Although we acknowledge that there is considerable diversity within each of these broader categories, the data utilized in this study only allow us to consider five racial/ethnic groups: Whites, African Americans and Asians and two subgroups within the Hispanic category, Mexican Americans and Puerto Ricans. These groupings do not capture any biological distinctions between groups but reflect certain aspects of ethnicity such as common geographic origins, family patterns, language, values, cultural norms, and traditions. Race, however, as it is popularly understood, does predict variations in depressive symptoms among some groups in the United States (Gallo, Cooper-Patrick, & Lesikar, 1998; Vega & Rumbaut, 1991) and in Canada (Wu, Noh, Kaspar, & Schimmele, 2003). These differences in rates of symptoms of depression have come to be understood in the context of the social meaning of race (Omi & Winant, 1994; Williams, 1997). That is, to the extent that racial/ethnic groups have differential access to power and other valued resources in society (including deference, respect, and access to safe neighborhoods), race has meaning as a social category.

LITERATURE REVIEW

Gender and Depression

The prevalence of major depression is strongly related to gender. For example, in the ECA study, women had rates of major depression that were more than twice as high as those of men (Weissman, Bruce, Leaf, Florio, & Holzer, 1991). Similarly, in the National Comorbidity Survey (NCS), the prevalence rate of ever experiencing an episode of major depression and dysthymia is almost twice as high for women (21.3) as for men (12.7) (Kessler & Zhao, 1999).

Most studies of depressive symptoms find that women are more depressed than men (Ross & Mirowsky, 1995), with this gender gap emerging as early as adolescence (Avison & McAlpine, 1992; Gallo et al., 1998; Joyner & Udry, 2000; Nolen-Hoeksema, 1994). In fact, many studies show that prior to adolescence, there is no difference in rates of depressive symptoms between boys and girls (Brooks-Gunn & Petersen, 1991) or that boys exhibit higher levels of depression than girls (Angold, Costello, & Worthman, 1998; Nolen-Hoeksema, Girgus, & Seligman, 1991). Gender differences in rates of first onset of major depressive episodes appear to begin between ages 13 and 15 (Weissman & Olfson, 1995) and the gender disparity in major depression tapers off around age 50 (Kessler et al., 1993).

Similar to approaches in psychiatric epidemiology, we are especially interested in the patterns of endorsement of symptoms of depression across racial/ethnic groups (Gallo et al., 1998; Takeuchi, Kuo, Kim, & Leaf, 1989). Our analyses will examine differences among women in reported symptoms of depression using the Center for Epidemiologic Studies Depression Scale (CES-D), one of the most widely used measures of depressive symptoms. For a more detailed discussion of the measurement of depression, see Chapter 1 of this volume by Wilhelm.

Race/Ethnicity, Gender, and Depression

There is some research on gender differences in the prevalence of depressive symptoms among ethnic minorities (see Salgado de Snyder, Cervantes, & Padilla, 1990; Vega & Rumbaut, 1991, for a review). Although the gender pattern described above persists across a variety of studies that focus on White adults, research on ethnically diverse samples do not find this consistent gender disparity in studies that utilize the CES-D. Studies find, for example, that African American women have higher symptom scores than African American men (Vega & Rumbaut, 1991). However, never-married and formerly married (separated and divorced) African American men have higher rates of major depression than their female counterparts (Williams, Takeuchi, & Adair, 1992). In terms of Asian American populations, research finds that Chinese and Filipino (Kuo, 1984; Ying, 1988) women report more symptoms of depression than men. On the other hand, Japanese males have higher depression scores than Japanese females (Kuo, 1984) and no significant gender differences have been found in levels of depression among Koreans (Hurh & Kim, 1988; Kuo, 1984). There is a growing body of research that focuses our attention on Latino mental health. This literature has found that Cuban (Narrow, Rae, Moscicki, Locke, & Regier, 1990) and Mexican American (Moscicki, Locke, Rae, & Boyd, 1989) females have higher depression scores than their male counterparts.

Despite this growing body of research, we know little about the extent to which racial/ethnic differences exist in symptom reporting and overall levels of depression between women (see Golding and Aneshensel, 1989). We know even less about how ethnic minority women compare to each other in reports of depressive symptomatology. This may be a more interesting comparison because many scholars contend that ethnic minorities experience similar barriers to optimal mental health, such as discrimination, poor housing options, and exposure to violence (Williams & Collins, 1995).

Theory

A review of the determinants of gender differences in depressive disorders concluded that these differences are neither artifactual nor primarily due to genetics and biological factors (Piccinelli & Wilkinson, 2000). Instead, they reflect differences in the social experiences of males and females. Specifically, gender differences in early exposures in childhood, sociocultural roles, response to adverse experiences, psychological attributes, and patterns of coping appear to play a critical role in the observed gender differences in depression. Although the full etiology of depression is not clear, the model of learned helplessness is a widely cited parsimonious explanation of why men and women differ in reports of depression (Seligman, 1974, 1991). Some believe that women have a greater tendency to assume that they are not capable of achieving a task after repeated experiences of failure at the task; thus a sense of helplessness is a learned and conditioned response to repeated failures. According to this view, learned helplessness creates three basic deficits: cognitive, emotional, and motivational. It is the emotional deficit that arguably leads to depression.

In general, women are taught to be more passive than men (Carli, 1989, 1990; Eder & Parker, 1987). Socialization processes have, in fact, been directly implicated in much of the work on gender and depression (Avison & McAlpine, 1992; Compas & Orosan, 1993; Kessler et al., 1993; Peterson, Sarigiani, & Kennedy, 1991; Sorenson, Rutter, & Aneshensel, 1991; Turner & Lloyd, 1995). The learned helplessness model, however, faced several criticisms and has been shown to be an insufficient explanation for gender differences in depression (Ross & Mirowsky, 1995). This model was soon replaced with a reformulated helplessness theory.

Helplessness theory focused on particular aspects of the attribution process (Abramson, Seligman, & Teasdale, 1978). For example, depressed individuals might feel that this problem is their fault (internal), that things cannot change (stability), and that this affects every aspect of their life (global). Problems with this model initiated further work leading to a more detailed hopelessness theory (Abramson, Alloy, & Metalsky, 1989). The more complex *hopelessness theory* contends that prior to becoming hopeless the person has (a) a negative cognitive or attribution style and (b) some unfortunate, stressful experience. Because both of these factors are involved, some people with depression-prone thinking do not become depressed (by avoiding traumatic experiences) and some people go through awful experiences without getting depressed (by avoiding negative thinking). Following is a description of aspects of the minority experience that may facilitate or protect women from feelings of depression.

Studies find that women and ethnic minorities report a lower sense of control over life circumstances (or mastery), higher levels of demoralization, a greater sense of hopelessness, and greater exposure to negative life events compared to their counterparts (Essed, 1991; Hughes & Demo, 1989; Metalsky & Joiner, 1992; Pearlin, 1989; Ross & Sastry, 1999; Rumbaut, 1989; Schieman & Turner, 2001). Furthermore, discrimination has been proposed as a major life stressor (Clark, Anderson, Clark, & Williams, 1999; Krieger, 1999; Williams, Neighbors, & Jackson, 2003) that plays a role in producing feelings of learned helplessness and subsequent depression among ethnic minorities (Dion & Giordano, 1990). A related stressor, acculturative stress, is also associated with depressive symptomatology among Asian and Hispanic women (Noh, Beiser, Kaspar, Hou, & Rummens, 1999; Saldana, 1995; Salgado de Snyder, 1987). Here, minorities are attempting to adjust to the host culture. To the extent that the host society is hostile and adopts discriminatory practices (including resistance to bilingual education programs), the impact of acculturative stress on psychological well-being may be compounded (Falcon & Campbell, 1991; Finch, Kolody, & Vega, 2000).

Heavily concentrated areas provide a certain level of comfort for many ethnic groups, especially where there are ethnic enclaves (Wilson & Portes, 1980). Ethnic enclaves are generally defined as "a concentration of ethnic firms in physical space – generally a metropolitan area – that employ a significant proportion of workers from the same minority" (Portes & Jensen, 1992, p. 418). Included within neighborhoods characterized by ethnic enclaves are dense networks of obligations and traditions that facilitate the culture of the minority group. Although recent census data document the growth in the Latino population throughout the United States, many studies find that the diversity represented in the United States is highly concentrated in a few states. Historically, Mexican Americans were heavily concentrated in southwestern and western states, and Puerto Ricans were clustered in New York and the northeast. On the other hand, the southern region has been the home for many African Americans. These regions can serve as buffer zones that act to preserve cultural difference as well as facilitate survival in a hostile environment. As such, we take into consideration region of the United States in our model of depression because this cultural factor has proven important in research on minority mental health (Finch et al., 2000).

Country of birth (i.e., whether women were born in the United States) is another cultural variable that we consider in this chapter. There is considerable variation in the rates of depression across countries. For example, the recent World Mental Health Study found that rates of any mood disorder were much lower in Mexico, China, and Japan (among other countries) than in the United States (Demyttenaere et al., 2004). Other data suggest that immigrants in the United States often have better mental health than their native-born counterparts (Vega & Rumbaut, 1991).

Another aspect of culture that may affect reports of mental health is the way in which people learn to present their symptoms. In many Asian societies, the self is conceptualized in terms of group harmony and interpersonal relationships. To maintain this harmony (and relationships), the expression of depressive affect is often discouraged. There is some evidence, in fact, that Asian American patients report their somatic symptoms, such as dizziness, while not reporting their emotional symptoms. When later questioned about symptom presentation, they acknowledge having emotional symptoms (Lin & Cheung 1999). In a study of older adults, however, Gallo and colleagues (1998) found no significant difference between Blacks and Whites in the report of somatic symptoms. These findings support the view that adults express or present symptoms in culturally acceptable ways (Kleinman, 1988). Although the CES-D has been used in cross-cultural research (see Garcia & Marks, 1989; Golding & Aneshensel, 1989; Roberts, 1980), this symptom checklist was not developed with specific minority patient populations. It remains unclear, therefore, how symptom presentation varies across ethnic groups. We take a closer look at the constellation of depressive symptoms. We examine whether there exists racial/ethnic differences in the type (and frequency) of symptoms women report on this scale.

A growing body of research suggests that religious involvement can affect mental health in multiple ways (Ellison, Boardman, Williams, & Jackson, 2001; Ellison & Levin, 1998; Pargament, Smith, Koenigh, & Perez,

1998). Religiously based friendship networks can provide emotional and instrumental support and religious beliefs and values can provide systems of meaning that can help individuals to appraise stressors in ways that can reduce their negative effects on psychological well-being. Religious involvement can also discourage negative health practices that can lead to reduced risk-taking behavior and exposure to stress. Research reveals that compared to the religiously affiliated, people with no religious affiliation and from certain affiliations have an elevated risk of depressive symptoms and disorders (McCullough & Larson, 1999). Similarly, a recent meta-analysis of 147 studies concluded that religious involvement was inversely related to symptoms of depression (Smith, McCullough, & Poll, 2003). Our analysis considers the possibility that differences in depressive symptoms vary according to religious affiliation.

Another aspect of culture that has been identified to explain gender differences in depression is the division of labor in society. Women's social roles are viewed as especially problematic for the following reasons. First, women experience more role conflict and overload as they attempt to meet home and work demands (Nolen-Hoeksema, 1994). Second, regardless of employment status, women are held responsible for much of the household chores, including child care (Ciscel, Sharp, & Heath, 2000; Stohs, 2000). Housework is not considered prestigious nor is it recognized as legitimate work. Given the customary roles played by women in conjunction with traditional role expectations, we take women's social roles into consideration in our model of depression. We also consider social role engagements here because there is some work indicating racial/ethnic group differences in the division of labor in the household (Ericksen, Yancey, & Ericksen, 1979; Hyde & Texidor, 1988; McAdoo, 1988; Ross, 1987).

In essence, more research is needed on symptom reporting across racial/ethnic groups (Neighbors & Lumpkin, 1990). One purpose of this chapter is to examine reports of symptoms across five American ethnic groups. Our goal is to build on existing research by more clearly describing the structure of depression among an ethnically diverse adult sample (see Weissman et al., 1996). Given the aspects of culture discussed in this chapter, we then consider a variety of culturally relevant factors that might be important in exploring such differences, including resources (e.g., education, self-esteem), religion, and social role involvements (e.g., marital status).

DATA AND METHODS

Study Design and Sample

This chapter presents original analyses of data from the National Survey of Families and Households (NSFH). The NSFH was conducted in 1987 and

1988 (Sweet, Bumpass, & Call, 1988). The survey was based on respondents from a multistage probability sample of households in the United States ($N = 13,017$). The main sample included 9,643 households. The survey also included an oversample of minorities, single-parent families, families with stepchildren, cohabiting couples, and recently married persons ($n = 3,374$), yielding a total sample of 13,017 respondents. The response rate for the study was 75%. The sample can be viewed as representative of noninstitutionalized persons in the United States. Two-hour structured interviews with respondents were conducted by trained interviewers in each respondent's home. The survey provides information on a variety of topics, including mental health.

We utilize information provided by women who self-identify as non-Hispanic White ($N = 5,545$), African American ($N = 1,519$), Mexican American ($N = 372$), Puerto Rican ($N = 125$), and Asian American ($N = 76$). These are the five largest racial/ethnic subsamples in the NSFH and, therefore, should provide enough statistical power to assess the relationships under investigation.

Depression. Depression is assessed using a 12-item version of the CES-D (Radloff, 1977; Weissman & Klerman, 1977). Respondents were asked on how many days during the past week (0–7) they

- Felt like not eating?
- Had trouble keeping your mind on what you were doing?
- Had trouble shaking off the blues, even with help from family and friends?
- Slept restlessly?
- Talked less than usual?
- Felt depressed?
- Felt lonely?
- Felt sad?
- Felt fearful?
- Felt that everything you did was an effort?
- Felt that you could not get going?
- Felt bothered by things that usually don't bother you.

This scale does not measure a clinical disorder, but it is a highly reliable and valid measure of depressive symptoms (see Ross, Mirowsky, & Huber, 1983). Previous research indicates good internal consistency (Roberts, 1980) and discriminatory power among clinical subsamples (Boyd & Weissman, 1982; Weissman & Klerman, 1977).

Control Variables. Age, education, household income, marital status, employment status, and parental status were included as controls because they have been identified as predictors of depression (Jackson, 1997,

2003). Age is coded in years. Education is measured by years completed. Household income is coded as one of eight categories, each having a $10,000 range except the last category, which includes those who report household incomes over $70,000.

Two dummy variables were created to assess marital status with the currently married serving as the comparison (and omitted) category. The first includes the never married and the second represents the formerly married. Employment status is also assessed by two dummy variables comparing those who are employed full-time (defined as working at least 35 hr/week) and those working part-time (5–34 hr/wk) to those who are not working for pay at this time (4 or fewer hr/wk). The employed are assigned a value of "1" on each dummy variable and all others are assigned a value of "0." In terms of parental status, respondents who have children are assigned a value of "1" and childless adults receive a value of "0."

A set of secondary social role involvements were also taken into consideration. A composite measure was created that included sons/daughters, siblings, friends, organizational members, and church goers. *Son/daughters* are those who see or communicate with a parent once a month or more. *Siblings* are identified as those who saw or talked with their brother or sister (including step and half siblings) several times or more last year. Respondents are involved in *friendships* if they report spending an evening with friends at least once a month. *Organizational* membership is defined as those who participate at least several times a year in clubs or organizations (excluding church). *Churchgoers* are defined as those who participate in church-affiliated groups at least once a month. Dummy variables were created for each of these secondary relationships with "0" indicating the absence of the role and "1" indicating the possession of the role. The number of secondary roles variable, therefore, increases by 1 with each possessed role. A separate dichotomous variable measuring whether women were *caregivers* was also included. Respondents who report providing care or assistance for someone (due to disability or chronic illness) during the past 12 months are considered caregivers.

Besides actual role obligations, women are often responsible for household chores. As such, we included a measure of household strain. Respondents were asked "How would you describe the work you do around the house? Would you say it is: interesting/boring, appreciated/ unappreciated, overwhelming/manageable, complicated/simple, lonely/ sociable, poorly done/well done?" Responses were recorded on a 7-point scale and recoded so that high values represent high strain (boring, unappreciated, overwhelming, complicated, lonely, poorly done).

Self-concept, which refers to positive or negative feelings of self-worth, was included as a control and is measured by three questions. Respondents were asked their level of agreement with the following: (1) "On the whole, I am satisfied with myself"; (2) "I am able to do things as well as other people"; and (3) I have always felt pretty sure my life would work out the way I wanted it to." Responses ranged from 1 = strongly agree to 5 = strongly disagree. Items were recoded so that high scores represent high self-esteem. These items are consistent with Rosenberg's (1979) self-esteem scale and demonstrate good reliability in this sample (alpha = .63).

Another set of cultural factors included in the analysis is religion, region, and nativity. Religion was coded using a series of dummy variables comparing the following religious denominations with those who claim to have no religion: (1) conservative Protestant groups (e.g., Southern Baptist, Independent Baptist, Primitive Baptist, Missionary Baptist,, Gospel Baptist, Church of God in Christ, Church of God, Church of Christ, Pentecostal, Adventist, Alliance, Assemblies of God, Holiness, Apostolic, Evangelical), (2) mainline Protestant groups (Methodists, Presbyterian, Episcopalian, Congregationalists, Reformed, and Lutheran churches), (3) Catholics, and (4) other denominations. Region of the country in which the respondent currently resides was measured by three dummy variables comparing those who report being from the Northeast, North Central, and Western part of the country (coded 1) to those who report being from the South (coded 0). Nativity is assessed with a dummy variable comparing those born in the United States with those born outside of the United States (1 = U.S. born). Finally, a sample weight was included as a control since we combine the groups in the analysis. This variable was calculated by researchers at the University of Wisconsin for person-level analysis of primary respondents in the NSFH. It is the product of the basic sampling weight screening nonresponse adjustment, interview nonresponse adjustment, and poststratification adjustment (see Sweet et al., 1988).

RESULTS

We begin with the distribution of mean scores for depressive symptoms across the five racial/ethnic groups. As shown in Table 14.1, the pattern of

TABLE 14.1. *Descriptive Statistics of CES-D Scale across Five Racial/Ethnic Groups of Women in the NSFH (1987)*

	Race/ethnicity				
	Non-Hispanic White	African American	Mexican American	Puerto Rican American	Asian American
Average CES-D score	28.12	31.34	31.71	32.23	25.78
S.D.	17.12	19.26	20.13	20.73	17.07
Median	22	25	24	25	19
N	5,250	1,401	338	102	74

average scores for depression is consistent with much previous research. We find that non-Hispanic Whites have lower levels of depression than African Americans, Mexican Americans, and Puerto Rican women. Also consistent with much work on minority mental health, Asian American women report the fewest symptoms of depression (Vega & Rumbaut, 1991; Wu et al., 2003).

Turning to a further consideration of differences (or similarities, or both) in the endorsement of symptoms of depression across groups, we present the mean distribution of each item on the CES-D. As shown in Table 14.2, there are both striking similarities and clear patterns of differences in reports of various symptoms. For example, we see some subgroup differences in the average number of days reported for feeling "sad." Asian Americans report the fewest number of days feeling sad (2.05), followed in succession by non-Hispanic Whites (2.45), African Americans (2.63), Mexican Americans (2.82), and Puerto Rican Americans (2.92).

Interestingly, when looking at the type and frequency of symptoms reported, there is a strong similarity found among African American and Mexican American women. In fact, there is no significant difference in the number of days that more than half of the items on the CES-D were experienced in the past week (poor appetite, trouble concentrating, the blues, slept restlessly, talked less than usual, could not get going, and bothered by things that usually do not bother you). Not surprisingly, then, the overall depression score is similar for these women.

Table 14.3 reorders the 12 symptoms according to the average number of days they were experienced by each group. When looking at the rankings of the mean scores, some similarities become glaringly obvious. Among non-Hispanic White and African American women, the same four items rank in the same order at the lower end of the scale (in terms of number of days experiencing a symptom): having a poor appetite, feeling blue, talking less than usual, and feeling fearful. Furthermore, three of these four items fall in the lowest tier for Mexican American women (feeling blue, talking less than usual, feeling fearful), all four among Puerto Rican women, and two for Asian American women (feeling blue, feeling fearful). In other words, there appears to be a constellation of depressive symptoms that are not experienced as frequently as are other symptoms. In fact, feeling blue is a prominent symptom for all groups (see below) but is among the items experienced least often within the week.

An empirical examination of the factor structure of the CES-D among women across race/ethnicity is presented in Table 14.4. In this analysis, depression is considered a latent (or unobserved) variable that can be explained by the association of a set of observed variables (Bartholomew, 1987). The factor analysis (or observed correlations in the observed variables) reveals the presence of this latent variable. We focus particular attention on those factor loadings that are greater than .80. These items are indicated by bold type.

TABLE 14.2. *Descriptive Statistics of CES-D across Five Racial/Ethnic Groups of Women, in the NSFH (1987) (Standard Deviations in Parentheses)*

	Race/ethnicity				
	Non-Hispanic White	African American	Mexican American	Puerto Rican American	Asian American
Average no. days in past week:					
Poor appetite	2.19 (1.83)	2.55 (2.03)[a,d,e]	2.57 (1.99)[a,e]	2.38 (2.00)[a]	2.27 (1.96)
Trouble concentrating	2.49 (1.95)	2.57 (2.11)[e]	2.59 (2.09)[a,e]	2.59 (2.13)[a]	2.07 (1.61)[a]
The blues	2.05 (1.81)	2.44 (2.07)[a,e]	2.47 (2.12)[a,e]	2.40 (1.96)[a,e]	1.87 (1.51)[a]
Slept restlessly	2.76 (2.14)	2.90 (2.28)[a,d,e]	2.82 (2.14)[d,e]	3.22 (2.46)[a,e]	2.28 (1.82)[a]
Talked less	1.90 (1.66)	2.21 (1.91)[a]	2.20 (1.94)[a]	2.29 (2.02)[a]	2.08 (1.76)[a]
Depressed	2.43 (1.93)	2.76 (2.15)[a,c,d,e]	2.92 (2.22)[a,e]	2.97 (2.29)[a,e]	2.11 (1.63)[a]
Lonely	2.35 (2.08)	2.64 (2.25)[a,c,e]	2.48 (2.19)[a,e]	2.65 (2.27)[a,e]	2.01 (1.71)[a]
Sad	2.45 (1.98)	2.63 (2.18)[a,c,d,e]	2.82 (2.17)[a,e]	2.92 (2.23)[a,e]	2.05 (1.61)[a]
Fearful	1.82 (1.66)	2.11 (1.89)[a,c,d,e]	2.38 (2.21)[a,e]	2.46 (2.18)[a,e]	1.84 (1.46)
Everything an effort	2.61 (2.08)	3.21 (2.46)[a,c,d,e]	2.95 (2.28)[a,e]	2.97 (2.40)[a]	2.55 (2.01)
Could not get going	2.65 (2.03)	2.80 (2.18)[a,d,e]	2.79 (2.22)[a,e]	2.60 (2.12)	2.25 (1.67)[a]
Bothered	2.60 (1.85)	2.85 (2.02)[a,d,e]	2.86 (2.10)[a,d,e]	2.55 (1.94)	2.31 (1.65)[a]

[a] Mean scores significantly different from non-Hispanic whites.

[b] Mean scores significantly different from African Americans.

[c] Mean scores significantly different from Mexican Americans.

[d] Mean scores significantly different from Puerto Rican Americans.

[e] Mean scores significantly different from Asian Americans.

Note: *p < .05.

TABLE 14.3. *Rank-Order of CES-D Items by Average Score*

			Race/ethnicity		
Non-Hispanic White	**African American**	**Mexican American**	**Puerto Rican American**	**Asian American**
1. Slept restlessly	Everything an effort	Everything an effort	Slept restlessly	Everything an effort
2. Could not get going	Slept restlessly	Depressed	Depressed	Bothered
3. Everything an effort	Bothered	Bothered	Everything an effort	Slept restlessly
4. Bothered	Could not get going	Sad	Sad	Poor appetite
5. Can't concentrate	Depressed	Slept restlessly	Lonely	Could not get going
6. Sad	Lonely	Could not get going	Could not get going	Depressed
7. Depressed	Sad	Can't concentrate	Can't concentrate	Talked less
8. Lonely	Can't concentrate	Poor appetite	Bothered	Can't concentrate
9. Poor appetite	Poor appetite	Lonely	Fearful	Sad
10. Blues	Blues	Blues	Blues	Lonely
11. Talked less	Talked less	Fearful	Poor appetite	Blues
12. Fearful	Fearful	Talked less	Talked less	Fearful

TABLE 14.4. *Factor Structure of CES-D Items Using Principal Component Analysis among Five Groups of Women in the NSFH (1987)*

	Non-Hispanic White (5,545)	African American (1,519)	Mexican American (372)	Puerto Rican American (125)	Asian American (76)
Poor appetite	.664	.599	.681	.716	.869
Trouble concentrating	.767	.732	.835	.791	.767
The blues	.819	.819	.832	.847	.848
Slept restlessly	.705	.667	.721	.789	.846
Talked less	.714	.754	.814	.801	.836
Depressed	.844	.860	.837	.811	.873
Lonely	.793	.769	.813	.716	.844
Sad	.861	.834	.858	.858	.899
Fearful	.717	.720	.792	.840	.882
Everything an effort	.670	.769	.703	.806	.728
Could not get going	.781	.752	.793	.850	.856
Bothered	.716	.675	.725	.704	.751
Extraction sum of squared loadings					
Total	6.876	6.734	7.410	7.600	8.367
% of Variance	57.301	56.113	61.753	63.335	69.726
α	.926	.930	.942	.946	.958

342

As shown in Table 14.4, there are several patterns worth noting. First, sadness appears to be a consistent item reported regardless of ethnicity. This item had the highest factor loading for this construct for every group with the exception of African Americans. Second, the same three items loaded for non-Hispanic White and African American women (feeling blue, depressed, and sad). These three items are often used as a key part of the initial screening for clinical depression in structured psychiatric instruments such as the Diagnostic Interview Schedule (DIS) used in the ECA study and the Composite International Diagnostic Interview (CIDI) used in the World Mental Health Study.

Another pattern that emerges from this table is the total number of items that load on the latent variable (i.e., depression) across racial/ethnic groups. Although only three items reach the designated .80 mark for non-Hispanic White and African American women, a significantly higher number of items weighed heavily on the latent construct depression for other women. For Mexican American women there were six items, for Puerto Rican women there were seven items, and for Asian American women, nine items loaded above .80 on the factor scale. Instructively, the percentage of variance explained was higher for Asians (70%), Puerto Ricans (63%), and Mexicans (62%) than for Whites (57%) and Blacks (56%). In general, however, a single factor was found for each of these subgroups of women and the CES-D demonstrates high reliability for all of the groups (ranging from .93 to .96).

Although sadness is an important indicator of the construct of depressive symptoms across all groups of women, it is worth noting that its bivariate correlations with other symptoms is different for Asian Americans than for the other groups of women. In analyses not shown in table form, "sadness" is correlated with feeling "depressed" and "lonely" for non-Hispanic White ($r = .74, r = .74$), African American ($r = .75, r = .75$), Mexican American ($r = .76, r = .74$), and Puerto Rican women ($r = .75, r = .73$). On the contrary, "sadness" is highly correlated with the item "depression" ($r = .91$) and weakly associated with "loneliness" ($r = .22$) among Asian American women. In the sample of Asian American women, the item that is similarly correlated with "depression" is feeling "fearful" ($r = .76$). The relationship between "sadness" and feeling "fearful" is somewhat lower among non-Hispanic White ($r = .57$), African American ($r = .59$), Mexican American ($r = .65$), and Puerto Rican women ($r = .68$).

We now examine whether differences in depression remain when cultural factors are taken into consideration. We take two approaches to the question of racial/ethnic differences in reports of depressive symptoms. First, we utilize ordinary least squares (OLS) regression to examine the association between each set of predictor variables and symptom reporting, paying particular attention to changes in the race/ethnicity coefficients and the amount of variation explained in the total depression score. The net R^2 in Table 14.5 is the incremental variance explained over the regression

TABLE 14.5. *OLS Regression Predicting Depressive Symptoms among Women (Unstandardized Regression Coefficients; N = 6,328)*

Independent variables	Adjusted for						All	
	Age	and	SES	Culture	Roles	Strain/S-E	B	s.e.
1. Age	-.61***		-1.41***	-.57***	-1.74***	-.24+	-.128***	.16
2. Race/Ethnicity (White)								
a. African American	2.31***		1.12*	2.36***	1.40*	3.44***	2.11***	.61
b. Mexican	1.01		-1.75*	.93	.91	1.69*	-.24	.94
c. Puerto Rican	.90		-1.65	1.42	-.55	1.50	.11	1.61
d. Asian	-2.92+		-2.53	-2.05	-2.74+	-1.55	-.29	1.75
3. Education			-.63***				-.51***	.08
4. HHIncome			-.62***				-.33***	.06
5. Religion (No religion)								
a. Conservative Prot.				-.04			.55	.85
b. Mainline Prot.				-2.80**			-.93	.89
c. Catholic				-.30			.86	.87
d. Other				-.83			.57	1.07
6. Region (South)								
a. Region1 – Northeast				.51			-.30	.61
b. Region2 – North Central				-.39			-1.10*	.53
c. Region3 – West				.62			-.01	.62
7. Nativity (1 = U.S. born)				1.78*			1.98*	.89

	(1)	(2)	(3)	(4)	(5)	(6)
8. Marital Status (Married)						
a. Never married				3.15***		1.27+ (.70)
b. Formerly married				4.77***		3.28*** (.54)
9. Employment Status (Not working)						
a. PT Employed				−2.67***		−1.90** (.63)
b. FT Employed				−3.84***		−2.32*** (.49)
10. Parent (1 = yes)				1.43**		−.51 (.56)
11. Secondary roles				−1.38***		−.54** (.20)
12. Caregiver (1 = yes)				3.73***		3.52*** (.58)
13. Household strain					.63***	.67*** (.03)
14. Self-esteem					−1.78***	−1.50*** (.11)
Intercept	32.14***	44.73***	30.96***	37.22***	38.87***	45.07*** (2.29)
R^2	.01	.05	.02	.05	.12	.16
Net R^2		.04	.01	.04	.11	.15

+$p < .10$; *$p < .05$; **$p < .01$; ***$p < .001$ (two-tailed tests); Omitted categories indicated in parentheses.

model that contained age and race/ethnicity variables. Second, we reestimate the depression model that includes all of the potential explanatory cultural and social factors simultaneously. The results from these models are presented in Table 14.5. All of the regression models are age adjusted.

As shown in the first column of Table 14.5, African American women report significantly higher levels of depressive symptomatology than non-Hispanic White women. On the contrary, there is no significant difference in levels of depression between Latinas (Mexican and Puerto Rican) and non-Hispanic Whites. Asian American women report slightly fewer symptoms (although marginally significant) than their White peers.

The next set of variables examined in the model was education and income. Interestingly, there is a change in the race/ethnicity coefficients. The coefficient comparing African American women to white women decreases significantly although Black women still report more symptoms than White women. In this model, we find that Mexican American women report fewer symptoms than their White peers. This change in the statistical significance of the coefficient suggests that these resources act as suppressor variables in our model. Once these factors are taken into account, then we see that there is a difference in the number of symptoms reported by Mexican American women compared to non-Hispanic White women. Similar to model 1, the coefficient for Puerto Rican women failed to reach statistical significance. The coefficient for Asian American women no longer reaches statistical significance. Not surprisingly, high education and income are associated with a decrease in the number of days depressive symptoms are reported.

The cultural variables assessing religion, region, and nativity constituted the next block of variables considered in the model predicting depressive symptoms. As shown in column 3, the race/ethnicity coefficient for African American women remains statistically significant. In this model, there are no significant differences in depressive symptomatology between the other minority groups and non-Hispanic White women. Two variables reached statistical significance: religion and nativity. More specifically, women who identify as mainline Protestant report fewer depressive symptoms than women who say they have no religion. Furthermore, women who were born in the United States report more symptoms of depression than their foreign-born counterparts.

Social roles may also explain racial/ethnic differences in symptom reporting. Turning to the fourth model in Table 14.5, we find that African American women continue to report more symptoms, even after taking into consideration social role involvements. Furthermore, the racial/ethnic coefficient for Asian American women reaches statistical significance in this model: Asian women report fewer symptoms than non-Hispanic White women. All of the social role variables are associated with depressive symptoms. The never- and formerly married report more symptoms than currently married women. Women who are employed either part-time or full-time report fewer symptoms than unemployed women. Women with children report more symptoms. Those with many secondary roles, however, report fewer symptoms. Consistent with much research, women who are caregivers report more depressive symptoms than those who are not engaged in this type of role relationship.

The final set of factors focuses on strain and self-concept. In column 5 of Table 14.5, we find that a consideration of these factors accounts for much of the variation in depressive symptoms (i.e., 11% increase in the variance explained). The changes in some of the racial/ethnic coefficients are also worth noting. First, African American women continue to report more symptoms of depression, but in this model the coefficient is much higher than in the other models. Second, Mexican American women report more symptoms than non-Hispanic White women when housework strain and self-concept are included in the model. This finding could indicate an increased burden of the traditional housewife role for many Mexican American women. There were no significant differences in depressive symptoms between Puerto Rican or Asian women, compared to non-Hispanic White women.

In sum, we find that African American women report more symptoms of depression than White women regardless of the cultural or other social factors taken into consideration. We also consistently find no significant difference in the reporting of symptoms between Puerto Rican women and non-Hispanic White women (across models). The change in the race/ethnicity coefficient for Mexican American women speaks to an important statistical phenomenon of suppressor effects. Some cultural indicators act as suppressor variables as indicated by the emergence of the coefficient as statistically significant in the models in which SES, self-esteem, and household strain are considered. When SES is considered, Mexican Americans have significantly lower depressive symptom scores but when self-esteem and household strain are entered into the regression equation, Mexican Americans have significantly higher levels of depressive symptoms. Finally, the pattern for other minority women is markedly different for Asian American women. That is, the coefficient for Asian American women remains negative across the models. This coefficient sometimes reaches statistical significance (although marginal), depending upon the set of cultural and social factors taken into consideration in the model.

The final regression model included each set of cultural factors. As shown in Table 14.5, African American women report significantly more symptoms of depression than non-Hispanic White women. None of the other race/ethnicity coefficients reached statistical significance. Women with high levels of education and household incomes also report fewer depressive symptoms. In this final model, we also find that region surfaces as an important predictor of depression. Women who reside in the north central region of the country report fewer symptoms than women in the south. Nativity is also associated with depression: U.S. born women report more symptoms. The social role variables continue to significantly impact symptomatology with parenting being the exception. Finally, women with much household strain report more depressive symptoms whereas those who have high self-esteem report fewer days feeling depressed.

As a final point of exploration, we reestimated this full model using African American women as the comparison category (data not shown). Here, we find that the direction of the coefficients across race/ethnicity implies that all other women report fewer depressive symptoms than African American women. The coefficients that reached statistical signifi-cance, however, are confined to non-Hispanic White ($\beta = -2.25, p < .001$), Mexican American ($\beta = -2.31, p < .05$), and Puerto Rican ($\beta = -2.72, p < .10$) women.

CONCLUSION

Culture has been defined as a set of rules or standards that produce the range of behaviors considered proper and acceptable (Haviland, 2000). Culture, then, includes a variety of learned concepts and behaviors such as gender roles, decision-making processes, and verbal and nonverbal com-munication (symptom expression). Culture serves as a template that shapes behavior and consciousness. These cultural templates are constantly being negotiated, revised, and reproduced (Miraglia, Law, & Collins, 1999). Although the power to participate in this process of negotiation has his-torically been divided along gendered lines, women are in a better posi-tion today to define the way in which culture impacts their mental health. Some research finds, in fact, that Brazilian women actively use the condi-tion of *nervos* to change other peoples' behavior in various social settings (Rebhun, 1993). And, many Korean adults feel that the onset of depression is subject to individual control (Pang, 1994).

In this chapter we have outlined several factors related to culture that may help explain racial/ethnic group differences in depression among women. The empirical analysis considered symptom reports across racial/ethnic groups. In other words, we ask if there are certain groups of women in American society who display higher levels of depression and display certain symptoms compared to others. This focus moves the literature beyond the mere gender difference in depression to a con-sideration of the underlying cultural or social factors that may shed light on this emotional phenomenon labeled as depression. We explored racial/ethnic differences in depression after taking into account a variety of factors.

The single symptom that weighed heavily in the construct of depression (high factor loading) across all five subgroups was sadness. In other words, this feeling seems to have the same meaning for women. On the other hand, Latino women report more days feeling sad than other women. Perhaps there are elements to Latino culture that support this form of expressivity. At the same time, the pattern for Asian women (they report the least number of days feeling sad) is consistent with prior reports of selectively presenting emotional symptoms (Lin & Cheung, 1999).

Although sadness stood out as a consistent symptom of depression for all of the subgroups, we must reiterate the fact that clinical depression is different from normal sadness or grief. The latter generally runs for a shorter period of time and is typically less pervasive. It is the role of the diagnostician to determine if a mood state is normal (given a certain set of circumstances) or pathologic. Some symptoms of clinical depression, in fact, are more common in women than in men, especially oversleeping, overeating, and weight gain (Nemeroff, 1992). On the other hand, sadness as a prominent feature of depression may be indicative of a lesser form of depression known as dysthymia. Dysthymia is more likely to become intertwined with a person's self-concept (Akiskal, 1981, 1985). Many believe, in fact, that due to its early onset in childhood or adolescence, it may affect personality development and ways of coping with stress. Dysthymia has been linked to passive avoidance and dependent traits (U.S. DHHS, 1999).

Culture also impacts the way in which people deal with everyday problems. Some Asian American groups, for example, do not dwell on negative thoughts. Many Asians believe that avoidance is a more appropriate response than the outward expression of negative emotion (Kleinman, 1977). Research indicates that this strategy can be useful when faced with discrimination (Noh et al., 1999). African American and non-Hispanic White adults, on the other hand, rely on a similar set of strategies for dealing with life's problems (Broman, 1996; Menaghan & Merves, 1984). It is not too surprising to find, then, that the constellation of depressive symptoms is so similar for Black and White women. There is, nonetheless, a need for more research on cultural differences in coping styles, especially since the change in the results for Asian American women (the coefficient assessing race/ethnicity for this group) seems to indicate that certain cultural factors are important in predicting depressive symptomatology (see McCarty et al., 1999, for this type of research).

There is some work on the meaning of depression across cultural groups (see Kleinman, 1980, 1986; Lutz, 1988; Manson, Shore, & Bloom, 1985). For example, the constellation of symptoms of depression is captured in three separate folk illnesses in Korean society, referred to as *Hwabyung*, *Han*, and *Shinggyongshaeyak* (Pang, 1994). East Asians are also more likely to have physical manifestations associated with mood disorders. Therefore, it becomes important to take into account complaints of physical health problems. Physical illness is a risk factor for mental health problems (U.S. DHHS, 1999). Similarly, included within the Brazilian problem labeled *nervos* is a vast configuration of symptoms such as crying, trembling, fatigue, headaches, and sadness (Rebhun, 1993). Many of the manifestations of this condition fall under American categories of depression and anxiety – two dimensions of psychological distress (Mirowsky & Ross, 1989). The emotional phenomenon or emotional state called *nerves* or *nervos* is both culturally widespread and viewed primarily as a feminine condition (Barnett, 1989; Davis, 1989; Low, 1985). Among a sample of elderly Korean adults, Pang (1994) found that many respondents identified sadness as an emotional manifestation of depression. Similarly, older African Americans appear to report sadness more than their White peers (Gallo et al., 1998). It is not clear from these studies, however, whether there are gender differences in reports of sadness.

Much of the work on emotional displays demonstrates the complexity of culturally prescribed ways of reacting to life conditions. For example, although women are more likely than men to cry (Ross & Mirowsky, 1984), it is acceptable for men to cry in certain contexts such as sporting events, relationship breakups, and during periods of grief. As such, there appears to be a gender difference in the expression of sadness rather than a difference in the actual experience of sadness (Fabes & Martin, 1991). By examining the individual items on the CES-D across groups, we found that there appears to be differences in the relative ranking of sadness among the list of symptoms for different groups of women.

Future research must also give greater attention to the conceptualization and measurement of acculturation and rigorous examination of the conditions under which acculturation can affect mental health. The available evidence suggests that the patterns may be complex. In the Los Angeles site of the ECA, rates of depression were lower for immigrant Mexicans than for U.S.-born Mexican Americans (Burnam, Karno, Hough, Escobar, & Telles, 1987). Puerto Ricans in New York seem to have higher levels of depressive symptoms than island Puerto Ricans (Vega & Rumbaut, 1991). In an analysis of 30 published studies on acculturation and mental health among Hispanics, Rogler, Cortes, & Malgady, (1991) found that 12 studies showed a negative relationship – that is, individuals who are low in acculturation (not fully integrated into the host society) are more prone to suffer from low self-esteem and lack of psychological well-being. At the same time, 13 studies reported a positive relationship between acculturation and mental health, with individuals who were more highly integrated into the new society experiencing greater psychological strain in the forms of self-depreciation or self-hatred and isolation from traditional support systems. In three studies, a curvilinear relationship was found in which good mental health was a product of the ability to obtain balance between traditional cultural norms and the host country's norms and values. As one moved away from this place of equilibrium toward either accultura- tion extreme, psychological distress increased (Rogler et al. 1991).

The findings from this study further suggest that African American women are the most vulnerable to symptoms of depression, even compared to other ethnic minority women. This pattern is consistent with research on learned helplessness, which indicates a high sense of hopelessness among this population. Many investigators have discussed this issue in the context of discrimination in employment, housing, and wage inequality. In other words, there are structural features of society (which are also part of the cultural fabric) that may facilitate a depressed mood. Perceptions of these structural contingencies in terms of unfair treatment (due to discrimination or acculturation stress) and limited opportunities for life happiness should continue to be assessed in studies of women's depression.

This chapter builds on previous research by focusing on symptom reports across a nationally representative sample of women (see Gallo et al., 1998). Nonetheless, there are several limitations that should be noted. First, the measure of religiosity used in this study, religious affiliation, is a very limited indicator of the multidimensional nature of religious participation (Williams, 1994). Future research must also seek to understand the specific aspects of religious involvement that may have consequences for mental health. For example, little research attention has been given to the health consequences of religious involvement among Latinas. This is surprising because religion is a fundamental social and spiritual resource influencing Latinos' worldview. Religion and culture are closely intertwined for this population (Valdez, 1996). Research is needed that attends to the ways in which specific aspects of religiosity and spirituality, including religiously based systems of meaning, can have positive or negative consequences for mental health.

Second, the category of race/ethnicity includes a homogenous group of people conveniently placed into the crude racial/ethnic categories pro- vided by the U.S. census. Although this approach is consistent with previ- ous research and therefore allows us to speak to this work, it becomes espe- cially problematic in this study for the group of Asian American women. There are over 43 groups represented within this ethnic category and their historical experiences and current resource base can be markedly different (U.S. DHHS, 2001). This crude measure of race/ethnicity for this group may help explain the lack of significant differences between Asian women and non-Hispanic White women in the final model predicting depressive symptoms.

Despite these limitations, we advocate further work on the particular aspects of culture that differentially explain depression among women. For example, perhaps a set of sociodemographic characteristics is more consistently associated with depression among certain ethnic minorities (e.g., nativity) rather than all groups. Similarly, psychosocial resources such as self-esteem may not be an important component of the self-concept among some Asian populations, given the emphasis on collective (rather than individualistic) identity. Close attention should be paid, then, to the specific mechanisms linking race/ethnicity to depression. Because we found racial/ethnic differences in the reporting of certain symptoms, the challenge for future work is to explore the influence of culture on the process by which these factors result in feelings of depression.

References

Abramson, L. Y., Alloy, L. B., & Metalsky, G. I. (1989). Helplessness depression: A theory-based subtype of depression. *Psychological Review, 46*, 358–372.

Abramson, L. Y., Seligman, M. E. P., & Teasdale, J. D. (1978). Learned helplessness in humans: Critique and reformulation. *Journal of Abnormal Psychology, 87*, 49–74.

Akiskal, H. S. (1981). Clinical overview of depressive disorders and their pharmacological managements. In G. C. Palmer (Ed.), *Neuropharmacology of central nervous system and behavior disorders* (pp. 38–72). New York: Academic Press.

Akiskal, H. S. (1985). Interaction of biologic and psychologic factors in the origin of depressive disorders. *Acta Psychiatrica Scandinavica, 319* (Suppl.), 131–139.

American Association of Physical Anthropology. (1996). AAPA statement on biological aspects of race. *American Journal of Physical Anthropology, 101*, 569–570.

Angold, A. E., Costello, J., & Worthman, C. M. (1998). Puberty and depression: The roles of age, pubertal status and pubertal timing. *Psychological Medicine, 28*, 51–61.

Avison, W. R., & McAlpine, D. D. (1992). Gender differences in symptoms and depression among adolescents. *Journal of Health and Social Behavior, 33*, 77–96.

Barnett, E. A. (1989). Notes on nervios: A disorder of menopause. In D. L. Davis & S. M. Low (Eds.), *Gender, health, and illness: The case of nerves* (pp. 67–78). New York: Hemisphere.

Bartholomew, D. J. (1987). *Latent variable models and factor analysis.* New York: Oxford University Press.

Boyd, J. H., & Weissman, M. M. (1982). Screening for depression in a community sample. *Archives of General Psychiatry, 39*, 1195–1200.

Broman, C. L. (1996). Coping with personal problems. In H. W. Neighbors & J. S. Jackson (Eds.), *Mental health in black America,* (pp. 117–129). Thousand Oaks, CA: Sage.

Brooks-Gunn, J., & Petersen, A. C. (1991). Studying the emergence of depression and depressive symptoms during adolescence. *Journal of Youth & Adolescence, 20*, 115–119.

Burnam, A., Karno, M., Hough, R. L., Escobar, J. I., & Telles, C. (1987). Acculturation and lifetime prevalence of psychiatric disorders among Mexican Americans in Los Angeles. *Journal of Health and Social Behavior, 28*, 89–92.

Carli, L. (1989). Gender differences in interaction style and influence. *Journal of Personality and Social Psychology, 56*, 565–576.

Carli, L. (1990). Gender, language, and influence. *Journal of Personality and Social Psychology, 59*, 941–951.

Ciscel, D., Sharp, D., & Heath, J. (2000). Family work trends and practices: 1971–1991. *Journal of Family and Economic Issues, 21*, 23–36.

Clark, J. (1991). Getting there: Women in political office. *Annals of the American Academy of Political and Social Science, 514*, 63–76.

Clark, R., Anderson, N. B., Clark, V. R., & Williams, D. R. (1999). Racism as a stressor for African Americans: A biopsychosocial model. *American Psychologist, 54*, 805–816.

Compass, B., & Orosan, P. G. (1993). Adolescent stress and coping: Implications for psychopathology during adolescence. *Journal of Adolescence, 16*, 331–349.

Davis, D. L. (1989) The variable character of nerves in a Newfoundland fishing village. *Medical Anthropology, 2*, 63–78.

Demyttenaere, K., Bruffaerts, R., Posada-Villa, J., Gasquet, I., Kovess, V., Lepine, J. P., et al. (2004). Prevalence, severity, and unmet need for treatment of mental disorders in the World Health Organization World Mental Health Surveys. *Journal of the American Medical Association, 291*, 2581–2590.

Dion, K., & Giordano, C. (1990). Ethnicity and sex as correlates of depression symptoms in a Canadian university sample. *International Journal of Social Psychiatry, 36*, 30–41.

Eder, D., & Parker, S. (1987). The central production and reproduction of gender: The effect of extracurricular activities on peer-group culture. *Sociology of Education, 60*, 200–213.

Ellison, C. G., Boardman, J. D., Williams, D. R., & Jackson, J. S. (2001). Religious involvement, stress, and mental health: Findings from the 1995 Detroit area study. *Social Forces, 80*, 215–249.

Ellison, C. G., & Levin, J. S. (1998). The religion-health connection: Evidence, theory, and future directions. *Health Education Behavior, 25*, 700–720.

Ericksen, J. A., Yancey, W. L., & Ericksen, E. P. (1979). The division of family roles. *Journal of Marriage and the Family, 41*, 301–313.

Essed, P. (1991). *Understanding everyday racism: The Black middle-class experience.* Boston: Beacon Press.

Fabes, R. A., & Martin, C. L. (1991). Gender and age stereotypes of emotionality. *Personality and Social Psychology Bulletin, 17*, 532–540.

Falcon, P., & Campbell, P. J. (1991). The politics of language and the Mexican American: The English only movement and bilingual education. In G. W. Shepherd, Jr., & D. Penna (Eds.), *Racism and the underclass: State policy and discrimination against minorities* (pp. 145–158). New York: Greenwood Press.

Finch, B. K., Kolody, B., & Vega, W. A. (2000). Perceived discrimination and depression among Mexican-origin adults in California. *Journal of Health and Social Behavior, 41*, 295–313.

Gallo, J., Cooper-Patrick, L., & Lesikar, S. (1998). Depressive symptoms of whites and African Americans aged 60 years and older. *Journal of Gerontology 53B(5)*, P277–286.

Garcia, M., & Marks, G. (1989). Depressive symptomatology among Mexican-American adults: An examination of the CES-D scale. *Psychiatry Research, 27*, 137–148.

Golding, J. M., & Aneshensel, C. S. (1989). Factor structure of the Center for Epidemiologic Studies Depression scale among Mexican Americans and nonHispanic whites. *Psychological Assessment, 1*, 163–168.

Grieco, E. M., & Cassidy, R. C. (2001). *Overview of race and Hispanic origin, 2000.* United States Census 2000, U.S. Department of Commerce, Economics and Statistics Administration.

Haviland, W. A. (2000). *Anthropology.* San Diego, CA: Harcourt Trade.

Healey, J. F. (1995). *Race, ethnicity, gender, and class.* Thousand Oaks, CA: Pine Forge Press.

Hughes, M., & Demo, D. H. (1989). Self-perceptions of Black Americans: Self-esteem and personal efficacy. *American Journal of Sociology, 95*, 132–159.

Hurh, W. M., & Kim, K. C. (1988). *Uprooting and adjustment: A sociological study of Korean immigrants' mental health. (Final report to the National Institute of Mental Health).* Macomb, IL: Western Illinois University, Department of Sociology and Anthropology.

Hyde, B. L., & Texidor, M. S. (1988). A description of the father experience among Black families. *Journal of Black Nurses Association, 2*, 67–78.

Jackson, P. B. (1997). Role occupancy and minority mental health. *Journal of Health and Social Behavior, 38*, 237–255.

Jackson, P. B. (2003). Sho' me the money: The relationship between social class and mental health among married women. In D. Brown and V. Keith (Eds.), In and Out of Our Right Minds (pp. 173–98). New York: Columbia University Press.

Joppke, C. (1996). Multiculturalism and immigration: A comparison of the United States, Germany, and Great Britain. *Theory and Society, 25*, 449–500.

Joyner, K., & Udry, J. R. (2000). You don't bring me anything but down: Adolescent romance and depression. *Journal of Health and Social Behavior, 41*, 369–391.

Kasen, S., Cohen, P., Chen, H., & Castille, D. (2003). Depression in adult women: Age changes and cohort effects. *American Journal of Public Health, 93*, 2061–2066.

Kessler, R. C., McGonagle, K. A., Schwartz, M., Glazer, D. G., & Nelson, C. B. (1993). Sex and depression in the National Comorbidity Survey I: Lifetime prevalence, chronicity, and recurrence. *Journal of Affective Disorders, 25*, 85–96.

Kessler, R. C., Berglund, P., Demler, O., Jin, R., Koretz, D., Merikangas, K. R., et al. (2003). National Comorbidity Survey Replication. The epidemiology of major depressive disorder: Results from the National Comorbidity Survey Replication (NCS-R). *Journal of the American Medical Association, 289*, 3095–3105.

Kessler, R. C., & Zhao, S. (1999). Overview of descriptive epidemiology of mental disorders. In C. S. Aneshensel & J. C. Phelan (Eds.), *The handbook of the sociology of mental health* (pp. 127–150). New York: Kluwer Academic/Plenum.

Kleinman, A. (1977). Depression, somatization and the new cross-cultural psychiatry. *Social Science and Medicine, 11*, 3–10.

Kleinman, A. (1980). *Patients and healers in the context of culture: An exploration of the borderland between anthropology, medicine, and psychiatry.* Los Angeles, CA: University of California Press.

Kleinman, A. (1986). *Social origins of distress and disease: Depression, neurasthenia, and pain in modern China.* New Haven, CT: Yale University Press.

Kleinman, A. (1988). *Rethinking psychiatry: From cultural category to personal experience.* New York: Macmillan.

Krieger, N. (1999). Embodying inequality: A review of concepts, measures, and methods for studying health consequences of discrimination. *International Journal of Health Services*, *29*, 295–352.

Kuo, W. H. (1984). Prevalence of depression among Asian-Americans. *Journal of Nervous Mental Disorders*, *172*, 449–457.

Lin, K. M., & Cheung, F. (1999). Mental health issues for Asian Americans. *Psychiatric Services*, *50*, 774–780.

Linton, R. (1945). Present world conditions in cultural perspective. In R. Linton (Ed.), *The science of man in world crisis* (pp. 201–221). New York: Columbia University Press.

Lopez, I. F. H. (2001). Racial restrictions in the law of citizenship. In J. A. Kromkowski (Ed.), *Race and ethnic relations* (pp. 12–16). Guilford, CO: McGraw-Hill/Dushkin.

Low, S. (1985). Culturally interpreted symptoms or culture-bound syndromes: A cross-cultural review of 'nerves.' *Social Science and Medicine*, *21*, 187–196.

Lutz, C. A. (1988). *Unnatural emotions: Everyday sentiments on a Micronesian atoll and their challenge to western theory*. Chicago: University of Chicago Press.

Manson, S. M., Shore, J. H., & Bloom, J. D. (1985). The depressive experience in American Indian communities: A challenge for psychiatric theory and diagnosis. In A. Kleinman & B. Good (Eds.), *Culture and depression: Studies in the Anthropology and cross-cultural psychiatry of affect and disorder* (pp. 331–368). Los Angeles, CA: University of California Press.

McAdoo, H. P. (1988). *Black families*. Newbury Park: Sage.

McCarty, C. A., Weisz, J. R., Wanitromanee, K., Eastman, K. L., Suwanlert, S., Chaiyasit, W., et al. (1999). Culture, coping, and context: Primary and secondary control among Thai and American youth. *Journal of Child Psychology and Psychiatry*, *40*, 809–818.

McCullough, M. E., & Larson, D. B. (1999). Religion and depression: A review of literature. *Twin Research*, *2*(2), 126–136.

Menaghan, E. G., & Merves, E. S. (1984). Coping with occupational problems: The limits of individual efforts. *Journal of Health and Social Behavior*, *25*, 406–423.

Merriam-Webster collegiate dictionary (11th ed.). (2004). San Jose, CA: Fogware.

Metalsky, G. I., & Joiner, Jr., T. E. (1992). Vulnerability to depressive symptomatology: A prospective test of the diathesis-stress and causal mediation components of the hopelessness theory of depression. *Journal of Personality and Social Psychology*, *63*, 667–675.

Miraglia, E., Law, R., & Collins, P. (1999). *What is culture?* Retrieved August 15, 2004, from Washington State University site: http://www.wsu.edu:8001/vcwsu/commons/topics/culture.

Mirowsky, J., & Ross, C. (1989). *Social causes of psychological distress*. Hawthorne, NY: Aldine de Gruyter.

Moscicki, E. K., Locke, B. Z., Rae, D. S., & Boyd, J. H. (1989). Depressive symptoms among Mexican Americans: The Hispanic health and nutrition examination survey. *Social Psychology Psychiatric Epidemiology*, *25*, 260–268.

Murphy, R. (1986). *Culture and social anthropology: An overture* (2nd ed.). Englewood Cliffs, NJ: Prentice Hall.

Murray, C. J. L., & Lopez, A. D. (1996). *Summary: The global burden of disease: a comprehensive assessment of mortality and disability from diseases, injuries, and risk factors in 1990 and projected to 2020*. Cambridge, MA: Published by the Harvard

School of Public Health on behalf of the World Health Organization and the World Bank, Harvard University Press.

Narrow, W. E., Rae, D. S., Moscicki, E. K., Locke, B. Z., & Regier, D. A. (1990). Depression among Cuban Americans: The Hispanic health and nutrition examination survey. *Social Psychology Psychiatric Epidemiology*, *25*, 260–268.

Neighbors, H. W., & Lumpkin, S. (1990). The epidemiology of mental disorder in the Black population. In D. S. Ruiz (Ed.), *Handbook of mental health and mental disorder among Black Americans* (pp. 55–70). Westport, CT: Greenwood Press.

Nemeroff, C. B. (1992). New vistas in neuropeptide research in neuropsychiatry: Focus on corticotropin-releasing factor. *Neuropsychopharmacology*, *6*, 69–75.

Noh, S., Beiser, M., Kaspar, V., Hou, F., & Rummens, J. (1999). Perceived racial discrimination, depression, and coping: A study of Southeast Asian refugees in Canada. *Journal of Health and Social Behavior*, *40*, 193–207.

Nolen-Hoeksema, S. (1994). *Sex differences in depression*. Stanford, CA: Stanford University Press.

Nolen-Hoeksema, J., Girgus, S., & Seligman, M. E. P. (1991). Sex differences in depression and explanatory style in children. *Journal of Youth and Adolescence*, *20*, 233–245.

Omi, M., & Winant, H. (1986). *Racial formation in the United States: From the 1960s to the 1980s*. New York: Routledge.

Omi, M., & Winant, H. (1994). *Racial formation in the United States: From the 1960s to the 1990s*. New York: Routledge.

Owens, K., & King, M. C. (1999). Genomic views of human history. *Science*, *286*, 451–453.

Pang, K. Y. C. (1994). Understanding depression among elderly Korean immigrants through their folk illnesses. *Medical Anthropology Quarterly*, *8*, 209–216.

Pargament, K. I., Smith, B. W., Koenigh, H. G., & Perez, L. (1998). Patterns of positive and negative religious coping with major life stressors, *Journal for the Scientific Study of Religion*, *37*, 711–725.

Pearlin, L. I. (1989). The sociological study of stress. *Journal of Health and Social Behavior*, *30*, 241–256.

Petersen, A. C., Sarigiani, P. A., & Kennedy, R. E. (1991). Adolescent depression: Why more girls? *Journal of Youth and Adolescence*, *20*, 247–271.

Piccinelli, M., & Wilkinson, G. (2000). Gender differences in depression. *British Journal of Psychiatry*, *177*, 486–492.

Portes, A., & Jensen, L. (1992). Disproving the enclave hypothesis: Reply (in Comments and Replies). *American Sociological Review*, *57*, 418–420.

Radloff, L. (1977). The CES-D scale: A self-report depression scale for research in the general population. *Applied Psychological Measurement*, *1*, 385–401.

Rebhun, L. A. (1993). Nerves and emotional play in Northeast Brazil. *Medical Anthropology Quarterly*, *7*, 131–151.

Regier, D. A., Narrow, W. E., Rae, D. S., Manderscheid, R. W., Locke, B. F., & Goodwin, F. K. (1993). The de facto U.S. mental and addictive disorders service system: Epidemiologic Catchment Area prospective 1-year prevalence rates of disorders and services. *Archives of General Psychiatry*, *50*, 85–94.

Roberts, R. E. (1980). Reliability of the CES-D scale in different ethnic contexts. *Psychiatry Research*, *2*, 125–134.

Robins, L. N., & Regier, D. A. (1991). *Psychiatric disorders in America: The Epidemiologic Catchment Area Study.* New York: Free Press.

Rogler, L. H., Cortes, D. E., & Malgady, R. G. (1991). Acculturation and mental health status among Hispanics. *American Psychologist, 46,* 585–597.

Rosenberg, M. (1979). *Conceiving the self.* New York: Basic Books.

Ross, C. (1987). The division of labor at home. *Social Forces, 65,* 816–833.

Ross, C., & Mirowsky, J. (1984). Men who cry. *Social Psychology Quarterly, 47,* 138–146.

Ross, C., & Mirowsky, J. (1995). Sex differences in distress: Real or artifact? *American Sociological Review, 60,* 449–468.

Ross, C., Mirowsky, J., & Huber, J. (1983). Dividing work, sharing work, and in-between: Marriage patterns and depression, *American Sociological Review, 48,* 809–823.

Ross, C., & Sastry, J. (1999). The sense of personal control: Social-structural causes and emotional consequences. In C. S. Aneshensel and J. C. Phelan (Eds.), *Handbook of the sociology of mental health* (pp. 369–394). New York: Kluwer.

Rumbaut, R. G. (1989). Portraits, patterns, and predictors of the refugee adaptation process. In D. W. Haines (Ed.), *Refugees as immigrants: Cambodians, Laotians and Vietnamese in America* (pp. 138–182). Totowa, NJ: Rowan and Littlefield.

Saldana. D. H. (1995). Acculturative Stress. In D. H. Saldana (Ed.), *Hispanic psychology: Critical issues in theory* (pp. 43–56). Beverly Hills, CA: Sage.

Salgado de Snyder, V. N. (1987). Factors associated with acculturative stress and depressive symptomatology among married Mexican immigrant women. *Psychology of Women Quarterly, 11,* 475–488.

Salgado de Snyder, V. N., Cervantes, R. C., & Padilla, A. M. (1990). Gender and ethnic differences in psychosocial stress and generalized distress among Hispanics. *Sex Roles: A Journal of Research, 22,* 441–455.

Schieman, S., & Turner, H. A. (2001). When feeling other people's pain hurts: The influence of psychosocial resources on the association between self-reported empathy and depressive symptoms. *Social Psychology Quarterly, 64,* 376–389.

Seligman, M. E. P. (1974). Depression and learned helplessness. In R. J. Friedman & M. M. Katz (Eds.), *The psychology of depression: Contemporary theory and research.* New York: Winston-Wiley.

Seligman, M. E. P. (1991). *Helplessness: On depression, development, and death* (2nd ed.). New York: W. H. Freeman.

Smith, T., McCullough, M. E., & Poll, J. (2003). Religiousness and depression: evidence for a main effect and the moderating influence of stressful life events, *Psychological Bulletin, 129,* 614–636.

Sorenson, S., Rutter, C., & Aneshensel, C. (1991). Depression in the community: An investigation into age of onset, *Journal of Consulting Clinical Psychology, 59,* 541–546.

Stohs, J. (2000). Multiculutural women's experience of household labor, conflicts, and equity. *Sex Roles, 42,* 339–361.

Sweet, J., Bumpass, L., & Call, V. (1988). *The design and content of the National Survey of Families and Households* (Working Paper NFSH-1). Madison: Center for Demography and Ecology, University of Wisconsin-Madison.

Takeuchi, D. T., Kuo, H.-S., Kim, K., & Leaf, P. J. (1989). Psychiatric symptom dimensions among Asian Americans and native Hawains: An analysis of the symptom checklist. *Journal of Community Psychology, 17*, 319–329.

Turner, R. J., & Lloyd, D. (1995). Lifetime traumas and mental health: The significance of cumulative adversity. *Journal of Health and Social Behaviors, 36*, 360–376.

Turner, R. J., Wheaton, B., & Lloyd, D. (1995). Epidemiology of social stress. *American Sociological Review, 60*, 104–125.

U.S. Department of Health and Human Services. (1999). *Mental health: A report of the surgeon general – Executive summary.* Rockville, MD: U.S. Department of Health and Human Services, Substance Abuse and Mental Health Services Administration, Center for Mental Health Services, National Institutes of Health, National Institute of Mental Health.

U.S. Department of Health and Human Services. (2001). *Mental health: Culture, race, and ethnicity – A supplement to Mental health: A report of the Surgeon General.* Rockville, MD: U.S. Department of Health and Human Services, Public Health Service, Office of the Surgeon General.

U.S. Office of Management and Budget (1978). *Directive No. 15: Race and ethnic standards for federal statistics and administrative reporting.* Washington, DC: Office of Federal Statistical Policy and Standards, U.S. Department of Commerce.

Valdez, E. O. (1996). Chicano families and urban poverty: Familial strategies of cultural retention. In R. M. DeAnda (Ed.), *Chicanas and Chicanos in contemporary society* (pp. 63–74). Boston: Allyn and Bacon.

Vega, W. A., & Rumbaut, R. G. (1991). Ethnic minorities and mental health. *Annual Review of Sociology, 17*, 351–383.

Weissman, M. M., & Klerman, G. (1977). Sex differences and epidemiology of depression. *Archives of General Psychiatry, 34*, 98–111.

Weissman, M. M., Bland, R. C., Carino, G. J., Faravelli, C., Greenwald, S., Hwu, H. G., et al. (1996). Cross-national epidemiology of major depression and bipolar disorder. *Journal of the American Medical Association, 276*, 293–299.

Weissman, M. M., & Olfson, M. (1995). Depression in women: Implications for health care research. *Science, 269*, 799–801.

Weissman, M. M., Bruce, M. L., Leaf, P. J., Florio, L. P., & Holzer III, C. (1991). Affective disorders. In L. N. Robins & D. A. Regier (Eds.), *Psychiatric disorders in America: The Epidemiologic Catchment Area Study* (pp. 53–80). New York: The Free Press.

Williams, D. R. (1994). The measurement of religion in epidemiologic studies: Problems and prospects. In J. S. Levin (Ed.), *Religious factors in aging and health: Theoretical foundations and methodological frontiers* (pp. 125–148). Thousand Oaks, CA: Sage.

Williams, D. R. (1997). Race and health: Basic questions, emerging directions. *Annual Epidemiology, 7*, 322–333.

Williams, D. R., & Collins, C. (1995). US socioeconomic and racial differences in health: Patterns and explanations. *Annual Review of Sociology, 21*, 349–386.

Williams, D. R., & Harris-Reid, M. (1999). Race and mental health: Emerging patterns and promising approaches. In A. V. Horwitz & T. L. Scheid (Eds.), *A handbook for the study of mental health: Social contexts, theories, and systems* (pp. 295–314). New York: Cambridge University Press.

Williams, D. R., Neighbors, H. W., & Jackson, J. S. (2003). Racial/ethnic discrimination and health: Findings from community studies. *American Journal of Public Health*, *93*, 200–208.

Williams, D. R., Takeuchi, D. T., & Adair, R. K. (1992). Marital status and psychiatric disorders among Blacks and Whites. *Journal of Health and Social Behavior*, *33*, 140–157.

Williams, D. R., Yu, Y., Jackson, J. S., & Anderson, N. (1997). Racial differences in physical and mental health: Socioeconomic status, stress, and discrimination. *Journal of Health Psychology*, *2*, 335–351.

Wilson, K., & Portes, A. (1980). Immigrant enclaves: An analysis of the labor market experience of Cubans in Miami. *American Journal of Sociology*, *86*, 295–319.

Wu, Z., Noh, S., Kaspar, V., & Schimmele, C. (2003). Race, ethnicity, and depression in Canadian society. *Journal of Health and Social Behavior*, *44*, 426–441.

Ying, Y. W. (1988). Depressive symptomatology among Chinese-Americans as measured by the CES-D. *Journal of Clinical Psychology*, *44*, 739–746.

15

Trauma and Depression

Kristin M. Penza, Christine Heim,
and Charles B. Nemeroff

Considerable evidence exists to suggest that traumatic events contribute toward vulnerability for major depression throughout the lifespan (Brown & Harris, 1993; Finlay-Jones & Brown, 1981; Kendler, Karkowski, & Prescott, 1999). However, the timing of the trauma may constitute an especially important variable in the development of long-term vulnerabilities toward depression and other psychiatric disorders including posttraumatic stress disorder (PTSD). Traumatic events occurring early in life appear to result in persistent alterations in neurobiological stress systems, increasing one's vulnerability to develop major depression. These early-life stress-induced changes include neuroendocrine, neurochemical, and neuroanatomical alterations. Increasing data derived from clinical and preclinical studies lend support to the view that these neurobiological changes associated with trauma experienced early in life occur after undue stress during particular critical developmental periods. These studies also support the important contribution of early-life traumas in the development of symptoms of depression and anxiety. In addition, this neurobiological vulnerability secondary to early-life trauma may permanently increase susceptibility to depression by rendering individuals more sensitive to stress throughout their adult life. Women are particularly vulnerable to depression; approximately twice as many women (12%) as men (7%) endure a depressive episode each year (Nolen-Hoeksema, 1987). Lifetime risk of depression is also higher for women; 21% of women and 13% of men in the United States will experience an episode of major depression in their lifetime (Kessler et al., 1994). Higher exposure of women to early life trauma might contribute to this gender-related risk (Weiss, Longhurst, & Mazure, 1999).

This work is in part supported by NIMH grants MH-42088 and the Emory Conte Center for the Neuroscience of Mental Disorders (MH-58922).

TYPES OF EARLY TRAUMA RELATED TO DEPRESSION

To define early trauma, two main criteria must be considered: the developmental age range that is subsumed under *early life* and the characteristics of the events that would be considered *traumatic*. As for the early-life criterion, childhood is often defined using an upper age limit, usually between 12 and 18 years. More appropriate approaches refer to the developmental stage by using the onset of sexual maturation, that is, menarche in females, to more precisely define the early-life period. As for the traumatic event criterion, according to the *Diagnostic and Statistical Manual for Mental Disorders*, 4th Edition (*DSM-IV*; American Psychiatric Association, 1994), a traumatic event is an experience that is life threatening to the self or someone close to the self, accompanied by intense fear, horror, or helplessness. Any traumatic event occurring during childhood can be defined as early trauma. The most salient forms of early trauma in humans are sexual, physical, and emotional maltreatment (abuse or neglect) as well as parental loss (death or separation). Other forms of early trauma include accidents, physical illness, surgeries, natural disasters, and war or terrorism-related events. It should be noted that there are forms of early adversity in humans that do not fulfill the *DSM-IV* criterion for traumatic events, but may well exert pronounced influences on the child's subsequent vulnerability to stress and psychopathology throughout its lifespan. Such experiences include unstable families, inadequate parental care due to mental or physical illness, dysfunctional relationships between parent and child, and poverty. It should be noted that early life stress (ELS) is often complex, inasmuch as various forms of ELS coexist or are associated with each other (see Heim, Meinlschmidt, & Nemeroff, 2003).

Several types of childhood abuse and maltreatment are associated with increased risk for developing major depression in adulthood. These include physical, sexual, and emotional abuse; severe neglect; family violence; inadequate care following parental loss; or parenting characterized by rejection, lack of warmth, insecure attachment, or inconsistent rearing attitudes (Bifulco, Brown, & Harris, 1987; Crook & Raskin, 1975; Handwerker, 1999; Juang & Silbereisen, 1999; Kessler & Magee, 1993, 1994; McCloskey, Figueredo, & Koss, 1995; Mullan & Orrell, 1996). In addition to increased risk for major depression, individuals exposed to early-life trauma have a greater likelihood of developing other psychiatric syndromes in adulthood such as PTSD, bipolar disorder, borderline personality disorder, substance or alcohol use disorders, and anxiety disorders (Agid et al. 1999; Anda et al., 2002; Bifulco, Harris, & Brown, 1992; Brown & Anderson, 1991; Grilo, Sanislow, Fehon, Martino, & McGlashan, 1999; Johnson, Pike, & Chard, 2001; Kaplan & Klinetob, 2000; Kendler, Sheth, Gardner, & Prescott, 2002; Safren, Gershuny, Marzol, Otto, & Pollack, 2002). Although the biological basis for this link between early-life trauma and later development of these

disorders is obscure, recent evidence suggests that – like the increased vulnerability to depression – it is mediated by permanently altered, or dam-
aged, neurobiological stress response systems (Heim & Nemeroff, 2001;
Putnam & Trickett, 1997).

Childhood Sexual Abuse

Childhood sexual abuse (CSA) is undoubtedly the most extensively studied form of childhood adversity as regards risk for major depression. Multiple clinical studies support an association of CSA with increased incidence
of major depression in adulthood for women. Rates of CSA are typically
higher for women than men (female rate, 13.5 to 22.9%; male rate, 2.5 to
5.4%; Cecil & Matson, 2001; Molnar, Buka, & Kessler, 2001; Nelson et al.,
2002). Childhood sexual abuse is associated with major depression comorbid with anxiety disorders such as PTSD, more so than either non-comorbid
major depression or non-comorbid anxiety disorders (Levitan, Rector,
Sheldon, & Goering, 2003), even after controlling for dysfunctional parental
behavior (e.g., parental psychopathology, parental verbal and physical
abuse, and parental substance disorders; Molnar et al., 2001). Results are
mixed whether female gender is associated with increased comorbidity
risk in response to CSA. For women, CSA increases the risk of a multiplicity of *DSM-IV* Axis I diagnoses including mood, anxiety, substance and
alcohol use disorders, and borderline personality disorder (Molnar et al.,
2001; Nelson et al., 2002; Zlotnick, Mattia, & Zimmerman, 2001). In comparing men and women, Molnar et al. (2001) found a larger number of
associations between CSA and later onset of disorders for women (14 disorders), compared to men (5 disorders). However, Levitan et al. (2003)
reported finding no sex by CSA interactions in predicting several diagnoses, suggesting that childhood trauma, particularly CSA, significantly
increases risk for major depression and anxiety disorders, regardless of
gender. A similar lack of gender effect was reported by Kessler, Davis, and
Kendler (1997).

Clinical and community studies reveal longer duration and increased
severity of depressive episodes associated with CSA. Among depressed
outpatients, CSA is associated with longer index depressive episodes,
compared to patients with major depression without a history of CSA
(Zlotnick et al., 2001). Cecil and Matson (2001) reported that among adolescent females in a community sample with a history of CSA, longer CSA
duration is associated with greater depression severity, a history of childhood physical and emotional abuse, and lower levels of self-esteem. Peritraumatic dissociation, which occurs immediately following abuse, is associated with severity of CSA (e.g., penile penetration vs. fondling, having a
relationship with the perpetrator vs. a stranger; Johnson et al., 2001) and

with severity of major depression and PTSD among adult women (Johnson et al., 2001).

Childhood sexual abuse may increase vulnerability to major depression in adulthood in part via an increased incidence of adulthood trauma. Women with CSA exhibit an increased risk of suicide attempts, rape after 18 years of age, and divorce (Nelson et al., 2002). Indeed, CSA is associated with greater odds of lifetime suicide attempts (Ullman & Brecklin, 2002). CSA predicts suicidal ideation and attempts later in life to such a degree that CSA is associated with borderline personality disorder, a disorder typically marked by repeated episodes of suicidal behavior. Fergusson, Beautrais, and Horwood (2003) reported that among young depressed patients, a majority had not developed suicidal ideation or behavior. However, history of CSA was related to suicidal ideation and behavior. Other variables contributing variance toward suicidal behavior were family history of suicide, neuroticism, low self-esteem, novelty seeking, poor peer affiliations, and poor school achievement (Fergusson et al., 2003).

A dysfunctional family environment can increase the risk for sexual abuse trauma in children. For example, childhood sexual abuse is more likely to occur in homes with parental alcohol-related disorders (Nelson et al., 2002). Indeed, it appears that the negative environment might have as important an impact on later development of psychopathology as the abuse itself. In a sample of 862 women, CSA and low maternal care each independently predicted the presence of affective symptoms in adulthood (Hill et al., 2000). Maternal care was measured by the Parental Bonding Instrument (Parker, Tupling, & Brown, 1979). Nelson et al. (2002) reported finding that among more than 1,900 same-sex twin pairs, adults who had not experienced CSA, but had twin siblings who had, also had increased risk for adversities in adulthood. It appears that the family environment in which the CSA occurs has a potent negative impact, regardless of exposure to CSA, and this type of environment is likely associated with other types of abuse as well.

Other Types of Childhood Abuse

In addition to CSA, childhood physical abuse (CPA) also increases risk of developing depression later in life. CPA is commonly examined either alone or in combination with other types of childhood trauma. CPA is associated with an increased incidence of both comorbid and non-comorbid depression (Levitan et al., 2003). Compared to women with no abuse, CPA is significantly predictive of lifetime major depression (MacMillan et al., 2001), increased risk of first onset major depressive episodes in young adults (aged 30 years or younger; Wainwright & Surtees, 2002), and development of other adult onset disorders including anxiety disorders, alcohol abuse or dependence, antisocial behavior, illicit drug abuse or dependence, and suicide attempts (Dube, Anda, et al., 2003; Dube, Felitti, et al., 2003b; MacMillan et al., 2001).

Childhood emotional or psychological abuse also increases the risk of psychopathology, including major depression in adulthood. Kaplan and Kinetob (2000) compared 20 antidepressant-resistant with 20 antidepressant-responsive adults with depression. The treatment-resistant group was more severely depressed and had higher rates of childhood emotional abuse as well as more comorbid anxiety disorders than the treatment-responsive depressed comparison group. These findings indicate that the effects of childhood emotional abuse lead to long-term neurobiological effects, which contribute toward treatment resistance of depressive episodes. Bifulco and colleagues (2002) used a semi-structured interview measure of abuse, the Childhood Experiences of Care and Abuse (described in Moran, Bifulco, Ball, Jacobs, & Benaim, 2002), in a high-risk, community sample of London women. They explored the relationship of childhood psychological abuse to major depression and suicidal behavior in adulthood, as well as to other childhood traumas including physical abuse, sexual abuse, parental antipathy, and neglect. Psychological abuse was related to all other childhood adversities, most robustly to antipathy, neglect, and sexual abuse (Bifulco et al., 2002). Psychological abuse and other childhood traumas were associated with adult depression and suicidal behavior, demonstrating a dose-response relationship of adversities to later disordered behavior (Bifulco et al., 2002), identical to findings of Dube, Anda, et al. (2003) and Dube, Felitti, et al. (2003) as regards suicide attempts. Spertus, Yehuda, Wong, Halligan, and Seremetis (2003) reported that childhood emotional abuse and neglect predicted adulthood depression and anxiety symptoms among women, even when controlling for other types of childhood abuse (e.g., physical and sexual abuse).

Early Parental Loss and Neglect

Literature on the impact of early parental loss on the development of adulthood major depression has largely focused on early parental death (EPD) and early parental separation (EPS). However, the findings have been inconsistent, perhaps stemming from methodological inconsistencies (e.g., combining EPD and EPS into one category and considering features of EPS including duration, permanency, and gender of the separated parent; Tennant 1988). With regard to the extant literature on EPD, typically measured before the ages of 15 to 17 years, some studies report an association with the incidence of adulthood depression (Lloyd, 1980; Mireault & Bond, 1992; Pfohl, Stangl, & Tsuang, 1983), and other studies fail to confirm such an association (Bifulco et al., 1987; Birtchnell, 1980; Canetti et al., 2000; Favarelli et al., 1986; Hallstrom, 1987; Kendler et al., 1992; Kendler et al.,

2002; Oakley-Browne, Joyce, Wells, Bushnell, & Hornblow, 1995a; Perris, Holmgren, Von Knorring, & Perris, 1986; Roy, 1983; Tennant, Bebbington, & Hurry, 1980). Characteristics of the EPD are apparently unrelated to vulnerability to adult major depression, including age of the subject at the time of the EPD, how the parent died, forewarning of the death, the sex of the deceased parent, the subject's and the surviving parent's sex, and their interaction, participation in mourning activities, and witnessing the death (Luecken, 2000; Roy, 1983).

Using an alternative approach that involves comparing subjects with and without EPD, Luecken (2000) studied 30 college students who had experienced EPD before age 16 years and 31 control participants. There was no main effect of parental death on depression severity scores, but an interaction effect of EPD on adult depression was significant among those who experienced poor quality family relationships characterized by low levels of affection and family support and by high levels of family conflict (Luecken, 2000). Qualities of family relationships that were protective against depression following EPD included supportiveness among family members, open expression of feelings, and low levels of conflict and anger (Luecken, 2000). This finding suggests that the family environment, potentially both before and after parental death, constitutes an essential variable in predicting a mourning child's long-term risk for adult psychopathology. Previously, others have contended that examination of parental loss without considering the child's relationship with the surviving parent is missing a critical variable (Bowlby, 1980; Tennant, 1988).

Some studies combine EPD and EPS into one "early parental loss" variable. Many of these studies found clear associations of early parental loss and adulthood depression. For example, in a survey examination of more than 8,000 respondents, Kessler et al. (1997) found early parental loss events to be associated with the onset of mood disorders, more so than anxiety disorders, even after controlling for comorbidities and other adversities (e.g., parental psychopathology, interpersonal traumas such as rape, and natural disasters). No sex differences were found in this association. However, these findings may largely be driven by the presence of the EPS variable, as the studies exploring the distinct effects of EPD and EPS often find a larger effect for EPS. Oakley-Browne et al. (1995b) reported that prolonged separation from both parents had a stronger association with current or lifetime depression in women than parental death, separation or divorce, and other types of loss. Kendler et al. (2002) studied the time course of increased risk for depression or alcohol dependence after early parental loss in more than 7,000 same-sex and opposite-sex twin pairs. Their findings suggested that both EPD and EPS were associated with increased risk for major depression; however, this effect was stronger for EPS. For subjects with EPD, increased risk returned to baseline within 12 years. However, in subjects with EPS, increased risk returned to baseline only within 30 years.

It has been suggested that EPS might more profoundly affect a child than EPD due to the perception of parental rejection and the subsequent development of cognitive appraisals about the self and others. EPS also occurs far more frequently than EPD, as divorce is fairly common in our society (Canetti et al., 2000). In contrast to their negative report on EPD, Canetti et al. (2000) reported that compared to controls, Israeli adolescents who experienced separation (for any 5-year period before age 16 years) from one or both parents demonstrated higher levels of major depression and other psychological symptoms. Kendler et al. (1992) reported that EPS, not EPD, before 17 years of age increased the risk for major depression as well as several anxiety disorders including generalized anxiety disorder and panic disorder. Even among healthy controls without major depression, early parental loss, including EPS, is associated with other factors that increase vulnerability to psychiatric and medical disorders. These factors include low income, medical illness burden, lifetime cigarette smoking, frequent divorce, and living alone (Agid et al., 1999).

As discussed above, many authors have noted the significant impact of family quality factors in mediating the effects of EPS on the likelihood of depression in adulthood. Specifically, neglectful parental care subsequent to EPS appears to increase susceptibility to later psychopathology including major depression (Bifulco et al., 1987, 1992; Breier et al., 1988; Harris, Brown, & Bifulco, 1986). Oakley-Browne et al. (1995a) studied depressed and never-depressed women and measured various childhood adversities including permanent parental separation (divorce, death), temporary parental separation (1 month or more), quality of parental relationships, and leaving school. After controlling for other adversity variables, only low maternal care predicted recent and lifetime major depression (odds ratio = 4.1). In addition, nonsupportive family relationships, feeling burdened by the parent's need for emotional support, and lack of parental bonding contribute to the deleterious effects of EPS on later psychiatric symptoms (Breier et al., 1988; Kendler et al., 1992). Protective family qualities after early parental loss include supportive family relationships and positive parental qualities of the remaining parent such as empathy, warmth, meeting childhood needs, and promoting autonomy in the child (Breier et al., 1988; Luecken, 2000; Saler & Skolnick, 1992).

NEUROBIOLOGICAL EFFECTS OF EARLY TRAUMA

A majority of the extant data on the effects of early-life trauma on the dysregulation of neurobiological systems and the development of pathological (e.g., depressogenic, anxious) behavior is derived from laboratory animal studies. Robust and long-lasting effects of early adverse experiences on behavior and neurobiology have been reported in rodents and nonhuman primates (e.g., Henry, 1992; Higley, Suomi, & Linnoila; 1992; Joseph, 1999; Kaufman et al., 1997; Putnam & Trickett, 1997; Stein, Yehuda, Koverola, Hanna, 1997). In many studies, early-life stress results in permanent dysregulation of the neuroendocrine stress response.

Laboratory Animals

A commonly used experimental paradigm for studying early life trauma is maternal separation (MS), in which infant animals are separated from their mother for defined periods per day early in life. Anxious and depressive behaviors are observed as a result of MS among nonhuman primates and rodents (Kraemer, Ebert, Lake, & McKinney, 1984; Ladd, Owens, & Nemeroff, 1996; Ladd et al., 2000; McKinney, 1984; Plotsky & Meaney, 1993; Suomi, Eisele, Grady, & Harlow, 1975). Nonhuman primates exposed to MS demonstrate agitated behavior followed by a depressive phase consisting of slouched, withdrawn postures, social isolation, and reduced motor activity (Kaufman & Stynes, 1978; Laudenslager, Held, Boccia, Reite, & Cohen, 1990). These behavioral changes are long-lasting (Higley et al., 1992). Increased anxiety, social subordinance, and decreased capacity for affiliative behaviors have been observed among bonnet macaques up to 2 years after exposure to MS in infancy, compared to normal controls (Andrews & Rosenblum, 1994; Rosenblum, Forger, Noland, Trost, & Coplan, 2001). In addition to anxious and anhedonic behaviors, adult rats exposed to MS during the postnatal period have been observed to display increased alcohol preference (e.g., Francis, Caldji, Champagne, Plotsky, & Meaney 1999; Huot, Thrivikraman, Meaney, & Plotsky, 2001; Ladd et al., 2000).

Many of the central nervous system (CNS) changes associated with MS stress in laboratory animals are similar to those reported in adults with major depression and PTSD. In particular, the hypothalamic-pituitary-adrenal (HPA) axis, the major neuroendocrine stress response system, exhibits dysregulation in both of these disorders (Nemeroff, 1998; Plotsky, Owens, & Nemeroff, 1998; Roy, 1988; Trestman et al., 1991). Among nonhuman primates, permanent dysregulation of cerebrospinal fluid (CSF) corticotropin-releasing factor (CRF) systems has been observed as a result of early-life social stress (Coplan et al., 1996, 2001; Higley et al., 1992; Mathew et al. 2003). Bonnet macaques reared by mothers exposed to stressful conditions (e.g., variable foraging demand [VFD] in which a mother monkey is presented with differing degrees of access to food for her and her infant) exhibit increased CSF CRF concentrations as juveniles, compared with controls, and this effect is sustained into early adulthood (Coplan et al., 2001). The young adult VFD-reared macaques also demonstrate alterations in somatostatin, serotonin, and dopamine metabolite systems

(Coplan et al., 1998) as well as abnormal cortisol and immune responses to social stress (Smith, Batuman, Trost, Coplan, & Rosenblum, 2002).

Rodent studies also provide evidence for MS-induced long-term changes in CNS stress response systems and the HPA axis. Duration of infant MS (e.g., 2–24 hr) in rat pups is correlated with increases in corticosteroid responses to injected adrenocorticotrophic hormone (ACTH; Levine, Huchton, Wiener, & Rosenfeld, 1991; Rosenfeld, Wetmore, & Levine, 1992). Rats exposed to MS in infancy demonstrate marked increases in ACTH responses to psychological stress, compared to controls (e.g., Plotsky & Meaney, 1993; Ladd et al., 1996). Hyper-responsiveness of the HPA axis to stress is associated with alterations in multiple neurocircuits known to mediate the stress responses, for example, increased number, activity, and sensitization of CRF neurons in hypothalamic and limbic regions, decreased glucocorticoid receptor density in the hippocampus and prefrontal cortex, increased mineralcorticoid receptors in the hippocampus, increased locus coeruleus noradrenergic activity, decreased $GABA_A$/central benzodiazepine receptor binding, and decreased neuropeptide Y concentrations in selected brain regions (reviewed in Francis et al., 1999; Heim & Nemeroff, 2001; Heim, Meinlschmidt, & Nemeroff, 2003; Ladd et al., 2000; Plotsky, Sánchez, & Levine, 2001; Newport, Stowe, & Nemeroff, 2002).

Human Studies

Similar to the preclinical findings on the effects of early-life stress, childhood trauma such as abuse, neglect, and parental loss appears to produce lasting neurobiological effects in humans (reviewed in Heim & Nemeroff, 2001; Heim, Plotsky, & Nemeroff, 2004). Evidence from clinical studies indicates that early-life stress produces an array of long-lasting changes in the CNS, including neuroendocrine (HPA axis) and neurochemical (norepinephrine, benzodiazepine, opiate, dopaminergic, and neuropeptide systems) system alterations; structural changes affecting the corpus callosum, prefrontal cortex, hippocampus, and amygdala; and functional changes in limbic structures and the cerebellar vermis (Bremner, 2003; Nemeroff, 1996; Teicher et al., 2003; Vythilingam et al., 2002). These manifold neurobiological changes produced by early-life trauma are associated with, and are thought to contribute toward, increased vulnerability to psychopathology later in life, particularly major depression and PTSD. In addition, these changes due to early life trauma may increase vulnerability to addictive disorders, including illicit substance and alcohol use disorders (DeBellis, 2002).

The effects of early life trauma on the HPA axis, which is the major neuroendocrine stress response system, have been the focus of substantial research in patients with major depression and PTSD. Patients with major depression demonstrate CRF hypersecretion, blunted ACTH responses to exogenous CRF stimulation, high baseline levels of cortisol, and cortisol nonsuppression in response to the administration of the synthetic glucocorticoid, dexamethasone. PTSD is also associated with increased CRF secretion; however, in contrast to the findings in major depression, PTSD is associated with decreased plasma cortisol concentrations and supersuppression of cortisol to low doses of dexamethasone (reviewed in Nemeroff, 1998; Arborelius, Owens, Plotsky, & Nemeroff, 1999; Heim & Nemeroff, 2001; Yehuda, 1997).

Heim et al. (2000) used a standardized psychosocial laboratory stress test (Trier Social Stress Test, TSST) to measure neuroendocrine responses among 49 women divided into four groups: women with no history of childhood abuse (physical or sexual) or psychiatric disorder (controls, $n = 12$), women with a diagnosis of current major depression with a history of childhood abuse ($n = 13$), women without a diagnosis of current major depression who were abused as children ($n = 14$), and women with current major depression and without history of childhood abuse ($n = 10$). The TSST consists of an anticipation/preparation phase and a subsequent public speaking and mental arithmetic task in front of an audience (Kirschbaum, Pirke, & Hellhammer, 1993). In agreement with preclinical findings, childhood abuse was associated with robust HPA axis abnormalities. Specifically, women with histories of childhood abuse (regardless of the presence or absence of current major depression) displayed elevated peak plasma ACTH concentrations compared to controls and depressed women without a history of childhood abuse. Also, depressed women with a history of abuse demonstrated elevated peak plasma cortisol concentrations compared to the three other groups and a clear tachycardia compared to the control group.

Provocative endocrine challenge tests have also been utilized to assess the HPA axis in this population (Heim, Newport, Bansall, Miller, & Nemeroff, 2001). In the standard CRF stimulation test, abused women without depression exhibited increased ACTH to exogenous CRF, whereas abused women with depression and depressed women without abuse experience exhibited blunted ACTH responses. A blunted ACTH response is typically associated with major depression and has also been reported for patients with combat-related PTSD (Holsboer, Gerken, Stalla, & Mueller, 1985, Maes et al., 1992, Smith et al., 1989). Blunted ACTH responses to CRF have also been observed in sexually abused girls, the majority of whom were dysthymic, compared to nonabused girls (De Bellis et al., 1994). In contrast, the opposite response to CRF challenge, an increased ACTH response, have been found among adult women with PTSD related to mixed types of trauma histories (Rasmusson et al., 2001) and children with depression living with ongoing, chronic abuse, many of whom had comorbid PTSD (Kaufman et al., 1997). This pattern is similar to the finding of exaggerated ACTH responses to CRF in abused women without major depression (Heim et al., 2001).

Similarly, disparate findings have been reported with respect to cortisol responses to ACTH challenge tests. In response to an ACTH stimulation test, Heim et al. (2001) reported that abused women without major depression exhibit smaller plasma cortisol responses compared to abused women with major depression and normal controls. Rasmusson et al. (2001) reported elevated cortisol responses to ACTH among PTSD subjects compared to controls.

Reconciliation of these discordant findings in the CRF and ACTH challenge tests has yet to be completed, though differences in the methods as well as in trauma and patient characteristics in the various studies are considerable. Pituitary and adrenocortical responses may differ depending on several variables such as timing and duration of stress exposure, Categorical Axis I and II diagnoses, symptom severity, genetic risk, and ongoing stress in adulthood. Heim et al. (2001) speculate that an initial sensitization of the neuroendocrine stress response due to early-life trauma might evolve into blunted corticotrophic responsiveness with ongoing stress. In addition, precise timing of early-life stress might shed light on differential HPA-axis dysregulation (Heim et al., 2003). Recently, Carpenter et al. (2003) using retrospectively reported pre-school stress ratings (occurring before age 6 years) found a significant correlation between elevated CSF CRF concentrations and the magnitude of such stressors. However, both perinatal stress and preteen life stress (occurring between ages 6 and 13 years) predicted reduced CSF CRF concentrations. Adult major depression did not predict elevated CSF CRF concentrations, suggesting that early-life stress may be the primary transducer in altering CRF neural circuits. Clearly, the timing of the early-life stress is crucial in predicting HPA-axis activity alterations. Similar timing effects of stress have been reported in the preclinical literature by Mathew et al. (2002). Among nonhuman primates, VFD stressful rearing predicted later HPA-axis dysregulation, which changed depending on the age of the infant at exposure to VFD rearing. VFD exposure of infants at 18 weeks of age was associated with lower, not higher, CSF CRF concentrations. In marked contrast, VFD exposure at 10–12 weeks was associated with increased CSF CRF concentrations.

Increased availability of CRF has been reported to exert toxic effects on hippocampal neurons (Brunson, Eghbal-Ahmadi, Bender, Chen, & Baram, 2001), like cortisol. Both of these phenomena have been implicated in the decreased hippocampal volume detected among patients with major depression as well as patients with PTSD (Bremner, Licino, et al., 1997a; Bremner, Randall, et al., 1997b; Bremner et al., 2000; Sheline, Wang, Gado, Csernansky, & Vannier, 1996; Stein et al. 1997). However, other studies have also found no association of hippocampal atrophy with stress-related psychopathology (Ashtari et al., 1999; Pantel et al., 1997).

An examination of childhood trauma experiences sheds some light on these discrepant findings. Vythilingam et al. (2002) compared the hippocampal volumes of women with and without histories of childhood abuse and with or without current adulthood major depression. Subjects with histories of physical and/or sexual abuse and current major depression had an 18% smaller mean left hippocampal volume compared to the nonabused, depressed subjects. Compared to healthy controls, abused women with major depression had a 15% smaller left hippocampal volume (Vythilingam et al., 2002). These findings parallel those reported in laboratory animals (Janssens, Helmond, & Wiegant, 1994; Johnson, Kamilaris, Carter, & Calogero, 1996; Ladd et al., 1996, 2000; Plotsky & Meaney, 1993).

Positron emission tomography (PET) studies have revealed functional changes of frontal cortical regions during mental imagery in abused women with PTSD versus abused women without PTSD (Bremner et al. 1999; Shin et al. 1999). These findings, however, do not inform on potential cortical functional changes due to abuse itself. In a recent study, Pruessner, Champagne, Meaney, & Dagher (2004) investigated dopamine release using [11C]raclopride during a psychosocial stress task in healthy college students with high vs. low self-reported parental care. The stressor caused a significant release of dopamine in the ventral striatum in the stress versus resting condition in subjects reporting low parental care. Interestingly, the magnitude of the salivary cortisol responses to the stressor was significantly correlated with the reduction in [11C]raclopride binding in the ventral striatum. Also using PET, we recently observed a strong neural processing bias for negative emotional stimuli after early-life trauma and depression. Compared to controls, women with ELS and major depression demonstrate striking activation of the left amygdala and hippocampus in response to negative and lesser activation of the right hippocampus in response to positive pictures.

Discussion

Recent clinical research has corroborated much of the preclinical literature on early-life trauma and its associated neurobiological and behavioral consequences. The effect of early-life trauma on the CNS, particularly the HPA axis, is well established in laboratory animals. Studies on women with early-life trauma are concordant with the preclinical findings. Early-life trauma can induce HPA-axis dysregulation, which may, through chronic hypersecretion of CRF, contribute toward hippocampal atrophy, either directly or by increased cortisol secretion (Heim et al., 2000; Vythilingam et al., 2002). Traumatic experiences in adulthood, which are associated with the onset of major depression and anxiety disorders (Brown & Harris, 1993; Finlay-Jones & Brown, 1981; Kendler et al., 1999) may further impair

HPA-axis functioning (Heim et al., 2003). How early stress interacts with genetic vulnerability in the development of major depression is an active avenue of investigation. Caspi et al. (2003) reported that a functional polymorphism in the promoter region of the serotonin transporter gene exerts a major influence on vulnerability to depression after stress exposure.

Intergenerational effects of trauma and the development of psychopathology represent an area of needed future research. Parental alcoholism contributes toward the development of major depression in adulthood. However, the number of childhood adversities is increased in alcoholic homes compared to nonalcoholic homes, and the number of those adversities (e.g., childhood abuse, domestic violence, parental drug abuse, mental illness and divorce or separation) can account for the effect of parental alcoholism on later major depression (Anda et al., 2002). Parental history of psychopathology including major depression, mania, or schizophrenia is associated with a two- to threefold increase in rates of physical and/or sexual abuse (Walsh, MacMillan, & Jamieson, 2002). Parental psychiatric disorders may represent the first adverse life events for young children (Newport et al., 2002). Parents with psychiatric disorders may at times provide compromised care to their children, including neglect and lack of support, the same variables that influence the effect of an early parental loss on a child (Breier et al., 1988; Oakley-Browne et al., 1995a; Kendler et al., 1992). Parental psychopathology, including alcohol use disorders, is important to consider in determining and assisting families that may incur increased risk of childhood maltreatment and subsequent psychopathology (Walsh et al., 2002).

Treatment implications for HPA dysregulation represents another area of needed future studies. Psychological interventions will likely focus on treating aspects of psychopathology related to the experienced traumas. Threat-processing difficulties stemming from traumatic experiences might alter behavioral anticipation, focus, and information appraisal and processing (Van der Kolk, 2003). Therapeutic interventions should target coping styles in an effort to develop effective response styles (Van der Kolk, 2003). Because peritraumatic dissociation at the time of CSA (during which patients report mentally leaving their bodies and observing what happens from a distance) is associated with increased risk of later major depression, it has been suggested that an important element of treatment might be exposure-based interventions to promote processing of the experiences without dissociating, which might decrease symptoms related to CSA (Johnson et al., 2001). In addition, future research will aid in determining the relative efficacy of pharmacological and psychotherapeutic interventions for early-life trauma and associated neurobiological changes. In particular, CRF-1 receptor antagonists might be efficient to prevent or reverse the adverse consequences of early-life trauma (Heim & Nemeroff, 2001; Holsboer 1999). Considering the range of neuroendocrine stress system changes associated with major depression and early-life trauma, coping strategies to deal with stress are likely impacted. Particular intervention strategies may be more beneficial to patients with coping and stress-reactivity challenges. Indeed, a recent study found that psychotherapy (Cognitive Behavioral Analysis System of Psychotherapy) was more effective than antidepressant monotherapy (nefazodone) in treating patients with chronic major depression ($\geq$ 2 years) with a history of childhood trauma before age 15 years, including EPS or EPD, physical abuse, sexual abuse, and neglect (Nemeroff et al., 2003). Whether these results can be replicated in patients treated with selective serotonin reuptake inhibitors remains an important unanswered question.

References

Agid, O., Shapira, B., Zislin, J., Ritsner, M., Hanin, B., Murad, H., et al. (1999). Environment and vulnerability to major psychiatric illness: A case control study of early parental loss in major depression, bipolar disorder and schizophrenia. *Molecular Psychiatry, 4*, 163–172.

American Psychiatric Association. (1994). *Diagnostic and statistical manual of mental disorders* (4th ed.). Washington, DC: American Psychiatric Association.

Anda, R. F., Whitfield, C. L., Felitti, V. J., Chapman, D., Edwards, V. J., Dube, S. R., et al. (2002). Adverse childhood experiences, alcoholic parents, and later risk of alcoholism and depression. *Psychiatric Services, 53*(80), 1001–1009.

Andrews, M. W., & Rosenblum, L. A. (1994). The development of affiliative and agonistic social patterns in differentially reared monkeys. *Child Development, 65*, 1398–1404.

Arborelius, L., Owens, M. J., Plotsky, P. M., & Nemeroff, C. B. (1999). The role of corticotropin-releasing factor in depression and anxiety disorders. *Journal of Endocrinology, 160*, 1–12.

Ashtari, M., Greenwald, B. S., Kramer-Ginsberg, E., Hu, J., Wu, H., Patel, M., et al. (1999). Hippocampal/amygdala volumes in geriatric depression. *Psychological Medicine, 29*, 629–638.

Bifulco, A. T., Brown, G. W., & Harris, T. O. (1987). Childhood loss of parent, lack of adequate parental care and adult depression: A replication. *Journal of Affective Disorders, 12*, 115–128.

Bifulco, A., Harris, T., & Brown, G. W. (1992). Mourning or early inadequate care: Reexamining the relationship of maternal loss in childhood with adult depression and anxiety. *Developmental Psychopathology, 4*, 433–449.

Bifulco, A., Moran, P. M., Baines, R., Bunn, A., & Stanford, K. (2002). Exploring psychological abuse in childhood: II. Association with other abuse and adult clinical depression. *Bulletin of the Menninger Clinic, 66*, 241–258.

Birtchnell, J. (1980). Women whose mothers died in childhood: An outcome study. *Psychological Medicine, 10*, 699–713.

Boulby, J. (1980) Loss: Sadness and depression. In *Attachment and Loss*, Volume III. New York: Basic Books.

Breier, A. B., Kelsoe, J. R., Kirwin, P. D., Beller, S. T., Wolkowitz, O. M., & Pickar, D. (1988). Early parental loss and development of adult psychopathology. *Archives of General Psychiatry, 45*, 987–993.

Bremner, J. D. (2003). Long-term effects of childhood abuse on brain and neurobiology. *Child and Adolescent Psychiatric Clinics North America, 12*, 271–292.

Bremner, J. D., Licinio, J., Darnell, A., Krystal, J. H., Owens, M. J., & Southwick, S. M. (1997). Elevated CSF corticotropin-releasing factor concentrations in PTSD. *American Journal of Psychiatry, 154*, 624–629.

Bremner, J. D., Narayan, M., Anderson, E. R., Staib, L. H., Miller, H. L., & Charney, D. S. (2000). Hippocampal volume reduction in major depression. *American Journal of Psychiatry, 157*, 115–118.

Bremner, J. D., Narayan, M., Staib, L. H., Southwick, S. M., McGlashan, T., & Charney, D. S. (1999). Neural correlates of memories of childhood sexual abuse in women with and without posttraumatic stress disorder. *American Journal of Psychiatry, 156*, 1787–1795.

Bremner, J. D., Randall, P., Vermetten, E., Staib, L., Bronen, R. A., Mazure, C., et al. (1997). Magnetic resonance imaging-based measurement of hippocampal volume in posttraumatic stress disorder related to childhood physical and sexual abuse. *Biological Psychiatry, 41*, 23–32.

Brown, G. R., & Anderson, B. (1991). Psychiatric morbidity in adult inpatients with childhood histories of sexual and physical abuse. *American Journal of Psychiatry, 148*, 55–61.

Brown, G. W., & Harris, T. O. (1993). Aetiology of anxiety and depressive disorders in an inner-city population. 1. Early adversity. *Psychological Medicine, 23*, 143–154.

Brunson, K. L., Eghbal-Ahmadi, M., Bender, R., Chen, Y., & Baram, T. Z. (2001). Long-term, progressive hippocampal cell loss and dysfunction induced by early-life administration of corticotropin-releasing hormone reproduces the effects of early life stress. *Proceedings of the National Academy of Sciences of the United States of America, 98*, 8856–8861.

Canetti, L., Bachar, E., Bonne, O., Agid, O., Lerer, B., D-Nour, A. K., et al. (2000). The impact of parental death versus separation from parents on the mental health of Israeli adolescents. *Comprehensive Psychiatry, 41*, 360–368.

Carpenter, L. L., Tyrka, A. R., McDougle, J., Malison, R. T., Owens, M. J., Nemeroff, C. B., et al. (2003). Cerebrospinal fluid Corticotropin-releasing Factor and perceived early-life stress in depressed patients and healthy control subjects. *Neuropsychopharmacology, 29*, 777–784.

Caspi, A., Sugden, K., Moffitt, T. E., Taylor, A., Craig, I. W., Harrington, H., McClay, J., Mill, J., Martin, J., Braithwaite, A., Poutton, R. (2003). Influence of life stress on depression: Moderation by a polymorphism in the 5-HTT gene. *Science, 301*, 386–389.

Cecil, H., & Matson, S. C. (2001). Psychological functioning and family discord among African-American adolescent females with and without a history of childhood sexual abuse. *Child Abuse and Neglect, 25*, 973–988.

Coplan, J. D., Andrews, M. W., Rosenblum, L. A., Owens, M. J., Friedman, S., Gorman, J. M., et al. (1996). Persistent elevations of cerebrospinal fluid concentrations of corticotropin-releasing factor in adult nonhuman primates exposed to early-life stressors: Implications for the pathophysiology of mood and anxiety disorders. *Proceedings of the National Academy of Sciences of the United States of America, 93*, 1619–1632.

Coplan, J. D., Smith, E. L., Altemus, M., Scharf, B. A., Owens, M. G., Nemeroff, C. B., et al. (2001). Variable foraging demand rearing: Sustained elevations in cisternal cerebrospinal fluid corticotropin-releasing factor concentrations in adult primates. *Biological Psychiatry, 50*, 200–204.

Coplan, J. D., Trost, R. C., Owens, M. J., Cooper, T. B., Gorman, J. M., Nemeroff, C. B., et al. (1998). Cerebrospinal fluid concentrations of somatostatin and biogenic amines in grown primates reared by mothers exposed to manipulated foraging conditions. *Archives of General Psychiatry, 55*, 473–477.

Crook, T., & Raskin, A. (1975). Association of childhood parental loss with attempted suicide and depression. *Journal of Consulting and Clinical Psychology, 43*, 277.

DeBellis, M. D. (2002). Developmental traumatology: a contributory mechanism for alcohol and substance abuse disorder. *Psychoneuroendocrinology, 27*, 155–170.

DeBellis, M. D., Chrousos, G. P., Dorn, L. D., Burke, L., Helmers, K., Kling, M. A., et al. (1994). Hypothalamic-pituitary-adrenal dysregulation in sexually abused girls. *Journal of Clinical Endocrinology and Metabolism, 78*, 249–255.

Dube, S. R., Anda, R. F., Felitti, V. J., Chapman, D. P., Williamson, D. F., & Giles, W. H. (2003). Childhood abuse, household dysfunction, and the risk of attempted suicide throughout the life span: Findings from the Adverse Childhood Experiences Study. *JAMA: The Journal of the American Medical Association, 286*, 3089–3096.

Dube, S. R., Felitti, V. J., Dong, M., Giles, W. H., & Anda, R. F. (2003). The impact of adverse childhood experiences on health problems: evidence from four birth cohorts dating back to 1900. *Preventative Medicine, 37*, 268–277.

Favarelli, C., Sacchetti, E., Ambonetti, A., Conte, G., Pallanti, S., & Vita, A. (1986). Early life events and affective disorder revisited. *British Journal of Psychiatry; The Journal of Mental Science. 148*, 288–295.

Fergusson, D. M., Beautrais, A. L., & Horwood, L. J. (2003). Vulnerability and resiliency to suicidal behaviors in young people. *Psychological Medicine, 33*, 61–73.

Finlay-Jones, R., & Brown, G. W. (1981). Types of stressful life event and the onset of anxiety and depressive disorders. *Psychological Medicine, 11*, 803–815.

Francis, D. D., Caldji, C., Champagne, F., Plotsky, P. M., & Meaney, M. J. (1999). The role of corticotropin-releasing factor–norepinephrine systems in mediating the effects of early experience on the development of behavioral and endocrine responses to stress. *Biological Psychiatry, 46*, 1153–1166.

Grilo, C. M., Sanislow, C., Fehon, D. C., Martino, S., & McGlashan, T. H. (1999). Psychological and behavioral functioning in adolescent psychiatric inpatients who report histories of childhood abuse. *American Journal of Psychiatry, 156*, 538–543.

Hallstrom, T. (1987). The relationships of childhood socio-demographic factors and early parental loss to major depression in adult life. *Acta Psychiatrica Scandinavica*, 75, 212–216.

Handwerker, W. P. (1999). Childhood origins of depression: Evidence from native and nonnative women in Alaska and the Russian Far East. *Journal of Women's Health*, 8, 87–94.

Harris, T., Brown, G. W., & Bifulco, A. (1986). Loss of parent in childhood and adult psychiatric disorder: the role of lack of adequate parental care. *Psychological Medicine*, 16, 641–659.

Heim, C., Meinlschmidt, G., & Nemeroff, C. B. (2003). Neurobiology of early life stress and its relationship to PTSD. *Psychiatric Annals*, 33, 1–10.

Heim, C., & Nemeroff, C. B. (2001). The role of childhood trauma in the neurobiology of mood and anxiety disorders: Preclinical and clinical studies. *Biological Psychiatry*, 49, 1023–1039.

Heim, C., Newport, D. J., Bonsall, R., Miller, A. H., & Nemeroff, C. B. (2001). Altered pituitary-adrenal axis responses to provocative challenge tests in adult survivors of childhood abuse. *American Journal of Psychiatry*, 158, 575–581.

Heim, C., Newport, D. J., Heit, S., Graham, Y. P., Wilcox, M., & Bonsall, R., et al. (2000). Pituitary-adrenal and autonomic responses to stress in women after sexual and physical abuse in childhood. *JAMA: The Journal of the American Medical Association*, 284, 592–597.

Heim, C., Plotsky, P. M., & Nemeroff, C. B. (2004). Importance of studying the contributions of early adverse experience to neurobiological findings in depression. *Neuropsychopharmacology*, 29, 641–648.

Henry, J. P. (1992). Biological basis of the stress response. *Integrative Physiological and Behavioral Science: The Official Journal of the Pavlovian Society*, 27, 66–83.

Higley, J. D., Suomi, S. J., & Linnoila, M. (1992). A longitudinal assessment of CSF monoamine metabolite and plasma cortisol concentrations in young rhesus monkeys. *Biological Psychiatry*, 32(3), 127–145.

Hill, J., Davis, R., Byatt, M., Burnside, E., Rollinson, L., & Fear, S. (2000). Childhood sexual abuse and affective symptoms in women: A general population study. *Psychological Medicine*, 30, 1283–1291.

Holsboer, F. (1999). The rationale for corticotropin-releasing hormone receptor (CRH-R) antagonists to treat depression and anxiety. *Journal of Psychiatric Research*, 33(3), 181–214.

Holsboer, F., Gerken, A., Stalla, G. K., & Mueller, O. A. (1985). ACTH, cortisol and corticosterone output after ovine corticotropin-releasing factor challenge during depression and after recovery. *Biological Psychiatry*, 20, 276–286.

Huot, R. L., Thrivikraman, K. V., Meaney, M. J., & Plotsky, P. M. (2001). Development of adult ethanol preference and anxiety as a consequence of neonatal maternal separation in Long Evans rats and reversal with antidepressant treatment. *Psychopharmacology*, 158, 366–373.

Janssens, C. J., Helmond, F. A., & Wiegant, V. M. (1994). Increased cortisol response to exogenous adrenocorticotropic hormone in chronically stressed pigs: Influence of housing conditions. *Journal of Animal Science*, 72, 1771–1777.

Johnson, D. M., Pike, J. L., & Chard, K. M. (2001). Factors predicting PTSD, depression, and dissociative severity in female treatment-seeking childhood sexual abuse survivors. *Child Abuse and Neglect, 25*, 179–198.

Johnson, E. O., Kamilaris, T. C., Carter, C. S., & Calogero, A. E. (1996). The biobehavioral consequences of psychogenic stress in a small, social primate (Callithrix jacchus jacchus). *Biological Psychiatry, 40*, 317–337.

Joseph, R. (1999). The neurology of traumatic "dissociative" amnesia: Commentary and literature review. *Child Abuse and Neglect, 23*, 715–727.

Juang, L. P., & Silbereisen, R. K. (1999). Supportive parenting and adolescent adjustment across time in former East and West Germany. *Journal of Adolescence, 22*, 719–736.

Kaplan, M. J., & Klinetob, N. A. (2000). Childhood emotional trauma and chronic posttraumatic stress disorder in adult outpatients with treatment-resistant depression. *Journal of Nervous and Mental Disease, 188*(9), 596–601.

Kaufman, I. C., & Stynes, A. J. (1978). Depression can be induced in a bonnet macaque infant. *Psychosomatic Medicine, 40*, 71–75.

Kaufman, J., Birmaher, B., Perel, J., Dahl, R. E., Moreci, P., Nelson, B. et al. (1997). The corticotropin-releasing hormone challenge in depressed abused, depressed nonabused, and normal control children. *Biological Psychiatry, 42*, 669–679.

Kendler, K. S., Karkowski, L. M., & Prescott, C. A. (1999). Causal relationship between stressful life events and the onset of major depression. *American Journal of Psychiatry, 156*, 837–848.

Kendler, K. S., Neale, M. C., Kessler, R. C., Heath, A. C., Eaves, L. J. (1992). Childhood parental loss and adult psychopathology in women. A twin study perspective. *Archives of General Psychiatry, 49*, 109–116.

Kendler, K. S., Sheth, K., Gardner, C. O., & Prescott, C. A. (2002). Childhood parental loss and risk for first-onset of major depression and alcohol dependence: The time-decay of risk and sex differences. *Psychological Medicine, 32*, 1187–1194.

Kessler, R. C., Davis, C. G., & Kendler, K. S. (1997). Childhood adversity and adult psychiatric disorder in the US National Comorbidity Survey. *Psychological Medicine, 27*, 1101–1119.

Kessler, R. C., & Magee, W. J. (1993). Childhood adversities and adult depression: Basic patterns of association in a US national survey. *Psychological Medicine, 23*, 679–690.

Kessler, R. C., & Magee, W. J. (1994). Childhood family violence and adult recurrent depression. *Journal of Health and Social Behavior, 35*, 13–27.

Kessler, R. C., McGonagle, K. A., Shanyang, Z., Nelson, C. B., Hughes, M., Eshleman, S., et al. (1994). Life time and 12-month prevalence of DSM-III-R psychiatric disorders in the United States. *Archives of General Psychiatry, 51*, 8–19.

Kirschbaum, C., Pirke, K. M., & Hellhammer, D. H. (1993). The "Trier Social Stress Test"–a tool for investigating psychobiological stress responses in a laboratory setting. *Neuropsychobiology, 28*, 76–81.

Kraemer, G. W., Ebert, M. H., Lake, C. R., & McKinney, W. T. (1984). Cerebrospinal fluid measures of neurotransmitter changes associated with pharmacological alteration of the despair response to social separation in rhesus monkeys. *Psychiatry Research, 11*, 303–315.

Ladd, C. O., Huot, R. L., Thrivikraman, K. V., Nemeroff, C. B., Meaney, M. J., & Plotsky, P. M. (2000). Long-term behavioral and neuroendocrine adaptations to adverse early experience. *Progress in Brain Research, 122,* 81–102.

Ladd, C. O., Owens, M. J., & Nemeroff, C. B. (1996). Persistent changes in corticotropin-releasing factor neuronal systems induced by maternal deprivation. *Endocrinology, 137,* 1212–1218.

Laudenslager, M. L., Held, P. E., Boccia, M. L., Reite, M. L., & Cohen, J. J. (1990). Behavioral and immunological consequences of brief mother-infant separation: A species comparison. *Developmental Psychobiology, 23,* 247–264.

Levine, S., Huchton, D. M., Wiener, S. G., & Rosenfeld, P. (1991). Time course of the effect of maternal deprivation on the hypothalamic-pituitary-adrenal axis in the infant rat. *Developmental Psychobiology, 24*(8), 547–558.

Levitan, R. D., Rector, N. A., Sheldon, T., & Goering, P. (2003). Childhood adversities associated with major depression and/or anxiety disorders in a community sample of Ontario: Issues of co-morbidity and specificity. *Depression and Anxiety, 17,* 34–42.

Lloyd, C. (1980). Life events and depressive disorder reviewed: I. Events as predisposing factors. *Archives of General Psychiatry, 37,* 529–535.

Luecken, L. J. (2000). Attachment and loss experiences during childhood are associated with adult hostility, depression, and social support. *Journal of Psychosomatic Research, 49,* 85–91.

MacMillan, H. L., Fleming, J. E., Streiner, D. L., Lin, E., Boyle, M. H., Jamieson, E., et al. (2001). Childhood abuse and lifetime psychopathology in a community sample. *American Journal of Psychiatry, 158,* 1878–1883.

Maes, M., Claes, M., Vandewoude, M., Schotte, C., Martin, M., Blockx, P., et al. (1992). Adrenocorticotropin hormone, beta-endorphin and cortisol responses to CRF in melancholic patients. *Psychological Medicine, 22,* 317–329.

Mathew, S. J., Coplan, J. D., Smith, E. L., Scharf, B. A., Owens, M. J., Nemeroff, C. B., et al. (2003). Cerebrospinal fluid concentrations of biogenic amines and corticotropin-releasing factor in adolescent non-human primates as a function of the timing of adverse early rearing. *Stress, 5*(3), 185–193.

McCloskey, L. A., Figueredo, A. J., & Koss, M. P. (1995). The effects of systemic family violence on children's mental health. *Child Development, 66,* 1239–1261.

McKinney, W. T. (1984). Animal models of depression: An overview. *Psychiatric Developments, 2*(2), 77–96.

Mireault, G. C., & Bond, L. A. (1992). Parental death in childhood: perceived vulnerability, and adult depression and anxiety. *American Journal of Orthopsychiatry, 62,* 517–524.

Molnar, B. E., Buka, S. L., & Kessler, R. C. (2001). Child sexual abuse and subsequent psychopathology: Results from the National Comorbidity Survey. *American Journal of Public Health, 91,* 753–760.

Moran, P. M., Bifulco, A., Ball, C., Jacobs, C., & Benaim, K. (2002). Exploring psychological abuse in childhood: I. Developing a new interview scale. *Bulletin of the Menninger Clinic, 66,* 213–240.

Mullan, E., & Orrell, M. (1996). Early life experience in elderly women with a history of depression: A pilot study using the Brief Parenting Interview. *Irish Journal of Psychological Medicine, 13,* 18–20.

Nelson, E. C., Heath, A. C., Madden, P. A., Cooper, M. L., Dinwiddie, S. H., Bucholz, K. K., et al. (2002). Association between self-reported childhood sexual abuse and adverse psychosocial outcomes: Results from a twin study. *Archives of General Psychiatry*, *59*(2), 139–145.

Nemeroff, C. B. (1996). The corticotropin-releasing factor hypothesis of depression: New findings and new directions. *Molecular Psychiatry*, *1*, 336–342.

Nemeroff, C. B. (1998). The neurobiology of depression. *Scientific American*, *278*(6), 42–49.

Nemeroff, C. B., Heim, C. M., Thase, M. E., Klein, D. N., Rush, A. J., Schatzberg, A. F., et al. (2003). Differential responses to psychotherapy versus pharmacotherapy in patients with chronic forms of major depression and childhood trauma. *Proceedings of the National Academy of Sciences of the United States of America*, *100*, 14293–14296.

Newport, D. J., Stowe, Z. N., & Nemeroff, C. B. (2002). Parental depression: Animal models of an adverse life event. *American Journal Psychiatry*, *159*, 1265–1280.

Nolen-Hoeksema, S. (1987). Sex differences in unipolar depression. Evidence and theory. *Psychological Bulletin*, *101*, 259–282.

Oakley-Browne, M. A., Joyce, P. R., Wells, J. E., Bushnell, J. A., & Hornblow, A. R. (1995a). Disruptions in childhood parental care as risk factors for major depression in adult women. *Australian and New Zealand Journal of Psychiatry*, *29*, 437–448.

Oakley-Browne, M. A., Joyce, P. R., Wells, J. E., Bushnell, J. A., & Hornblow, A. R. (1995b). Adverse parenting and other childhood experience as risk factors for depression in women aged 18–44 years. *Journal of Affective Disorders*, *34*, 13–23.

Pantel, J., Schroder, J., Essig, M., Popp, D., Dech, H., Knopp, M. V. et al. (1997). Quantitative magnetic resonance imaging in geriatric depression and primary degenerative dementia. *Journal of Affective Disorders*, *42*, 69–83.

Parker, G., Tupling, H., & Brown, L. B. (1979). A parental bonding instrument. *British Journal of Medical Psychology*, *52*, 1–10.

Perris, C., Holmgren, S., Von Knorring, L., & Perris, H. (1986). Parental loss by death in the early childhood of depressed patients and of their healthy siblings. *British Journal of Psychiatry*, *148*, 165–169.

Pfohl, B., Stangl, D., & Tsuang, M. T. (1983). The association between early parental loss and diagnosis in the Iowa 500. *Archives of General Psychiatry*, *40*, 965–967.

Plotsky, P. M., & Meaney, M. J. (1993). Early, postnatal experience alters hypothalamic corticotropin-releasing factor (CRF) mRNA, median eminence CRF content and stress-induced release in adult rats. *Molecular Brain Research*, *18*(3), 195–200.

Plotsky, P. M., Owens, M. J., & Nemeroff, C. B. (1998). Psychoneuroendocrinology of depression: Hypothalamic-pituitary-adrenal axis. *The Psychiatric Clinics of North America*, *21*, 293–307.

Plotsky, P. M., Sanchez, M. M., & Levine, S. (2001). Intrinsic and extrinsic factors modulating physiological coping systems during development. In D. M. Broom (Ed.), *Coping with challenge* (pp. 169–196). Berlin: Dahlem University Press.

Pruessner, J. C., Champagne, F., Meaney, M. J., & Dagher, A. (2004). Dopamine release in response to a psychological stress in humans and its relationship to early life maternal care: a positron emission tomography study using [11C]raclopride. *Journal of Neuroscience, 24,* 2825–2831.

Putnam, F. W., & Trickett, P. K. (1997). Psychobiological effects of sexual abuse. A longitudinal study. *Annals of the New York Academy of Sciences, 821,* 150–159.

Rasmusson, A., Lipschitz, D. S., Wang, S., Hu, S., Vojvoda, D., Bremner, J. D., et al. (2001). Increased pituitary and adrenal reactivity in premenopausal women with posttraumatic stress disorder. *Biological Psychiatry, 50,* 965–977.

Rosenblum, L. A., Forger, C., Noland, S., Trost, R. C., & Coplan, J. D. (2001). Response of adolescent bonnet macaques to an acute fear stimulus as a function of early rearing conditions. *Developmental Psychobiology, 39,* 40–45.

Rosenfeld, P., Wetmore J. B., & Levine, S. (1992). Effects of repeated maternal separations on the adrenocortical response to stress of preweanling rats. *Physiology & Behavior, 52,* 787–791.

Roy, A. (1983). Early parental death and adult depression. *Psychological Medicine, 13,* 861–865.

Roy, A. (1988). Cortisol nonsuppression in depression: Relationship to clinical variables. *Journal of Affective Disorders, 14,* 265–270.

Safren, S. A., Gershuny, B. S., Marzol, P., Otto, M. W., & Pollack, M. H. (2002). History of childhood abuse in panic disorder, social phobia, and generalized anxiety disorder. *Journal of Nervous and Mental Disease, 190*(7), 453–456.

Saler, L., & Skolnick, N. (1992). Childhood parental death and depression in adulthood: Roles of surviving parent and family environment. *American Journal of Orthopsychiatry, 62,* 504–516.

Sheline, Y. I., Wang, P. W., Gado, M. H., Csernansky, J. G., & Vannier M. W. (1996). Hippocampal atrophy in recurrent major depression. *Proceedings of the National Academy of Sciences of the United States of America, 93,* 3908–3913.

Shin, L. M., McNally, R. J., Kosslyn, S. M., Thompson, W. L., Rauch, S. L., Alpert, N. M. et al. (1999). Regional cerebral blood flow during script-driven imagery in childhood sexual abuse-related PTSD: A PET investigation. *American Journal of Psychiatry, 156,* 575–584.

Smith, E. L., Batuman, O. A., Trost, R. C., Coplan, J. D., & Rosenblum, L. A. (2002). Transforming growth factor-beta 1 and cortisol in differentially reared primates. *Brain, Behavior, and Immunity, 16,* 140–149.

Smith, M. A., Davidson, J., Ritchie, J. C., Kudler, H., Lipper, S., Chappell, P., et al. (1989). The corticotropin-releasing hormone test in patients with posttraumatic stress disorder. *Biological Psychiatry, 26,* 349–355.

Spertus, I. L., Yehuda, R., Wong, C. M., Halligan, S., & Seremetis, S. V. (2003). Childhood emotional abuse and neglect as predictors of psychological and physical symptoms in women presenting to a primary care practice. *Child Abuse & Neglect, 27,* 1247–1258.

Stein, M. B., Yehuda, R., Koverola, C., & Hanna, C. (1997). Enhanced dexamethasone suppression of plasma cortisol in adult women traumatized by childhood sexual abuse. *Biological Psychiatry, 42,* 680–686.

Suomi, S. J., Eisele, C. D., Grady, S. A., & Harlow, H. F. (1975). Depressive behavior in adult monkeys following separation from family environment. *Journal of Abnormal Psychology, 84,* 576–578.

Teicher, M. H., Andersen, S. L., Polcari, A., Anderson, C. M., Navalta, C. P., & Kim, D. M. (2003). The neurobiological consequences of early stress and childhood maltreatment. *Neuroscience and Biobehavioral Reviews, 27,* 33–44.

Tennant, C. (1988). Parental loss in childhood: Its effect in adult life. *Archives of General Psychiatry, 45,* 1045–1050.

Tennant, C., Bebbington, P., & Hurry, J. (1980). Parental death in childhood and risk of adult depressive disorders. *Psychological Medicine, 10,* 289–299.

Trestman, R. L., Coccaro, E. F., Bernstein, D., Lawrence, T., Gabriel, S. M., Horvath, T. B., et al. (1991). Cortisol responses to mental arithmetic in acute and remitted depression. *Biological Psychiatry, 29,* 1051–1054.

Ullman, S. E., & Brecklin, L. R. (2002). Sexual assault history and suicidal behavior in a national sample of women. *Suicide & Life-threatening Behavior, 32,* 117–130.

Van der Kolk, B. A. (2003). The neurobiology of childhood trauma and abuse. *Child and Adolescent Psychiatric Clinics of North America, 12,* 293–317.

Vythilingam, M., Heim, C., Newport, J., Miller, A. H., Anderson, E., Bronen, R. et al. (2002). Childhood trauma associated with smaller hippocampal volume in women with major depression. *American Journal of Psychiatry, 159,* 2072–2080.

Wainwright, N. W., & Surtees, P. G. (2002). Childhood adversity, gender and depression over the life-course. *Journal of Affective Disorders, 72,* 33–44.

Walsh, C., MacMillan, H., & Jamieson, E. (2002). The relationship between parental psychiatric disorder and child physical and sexual abuse: findings from the Ontario Health Supplement. *Child Abuse & Neglect, 26,* 11–22.

Weiss, E. L., Longhurst, J. G., & Mazure, C. M. (1999). Childhood sexual abuse as a risk factor for depression in women: Psychosocial and neurobiological correlates. *American Journal of Psychiatry, 156,* 816–828.

Yehuda, R. (1997). Sensitization of the hypothalamic-pituitary-adrenal axis in posttraumatic stress disorder. *Annals of the New York Academy of Sciences, 821,* 57–75.

Zlotnick, C., Mattia, J., & Zimmerman, M. (2001). Clinical features of survivors of sexual abuse with major depression. *Child Abuse & Neglect, 25,* 357–367.

16

Public Health Approach to Depression and Women

The Case of the Disadvantaged Inner-City Woman

Claire E. Sterk, Katherine P. Theall,
and Kirk W. Elifson

INTRODUCTION

Nobody knows what depression is, where it comes from or how I can make it go away. I am depressed, but so is my community. It is a community of trouble and struggle and there is little happiness.... Depression is having no hope, having no energy, having nothing to look forward to, always feeling pain ... drugs allow me to get away from it.

> [Quote from a 32-year-old female respondent who resides in a disadvantaged neighborhood and who uses cocaine daily]

The focus of this chapter is on inner-city disadvantaged women and depression. According to the World Health Organization's Global Burden of Disease study, depression among women worldwide has been identified as the leading cause of disability adjusted life years among adult women worldwide today (Murray & Lopez, 1996). A similar picture has been presented by others (Kessler et al., 1994; National Institute of Mental Health, 1999). Compared to men, women have higher rates of initial onset of depression than men. Consequently, the number of females who experience a recurrence of a depressive episode is higher than that among their male counterparts (Kessler et al., 1994).[??]

The burden of severe chronic mental disorders, including depression, is seen clearly among those at the greatest social and economic disadvantage in our societies (Kessler & Zhao, 1999; World Health Organization, 2000). In other words, a high prevalence of chronic mental disorders tends to be concentrated among a small proportion of the population (Kessler & Zhao, 1999). Inner-city disadvantaged women are among that small proportion of people in which severe profiles of mental illness are concentrated. They tend to experience high levels of environmental triggers to depression and to encounter numerous everyday stressors, including those due to poverty and sexism. In the case of racial and ethnic minority women, they also encounter racism. Furthermore, women residing in disadvantaged communities also are likely to encounter stress due to other social problems such as experiences with homelessness, drug abuse, violence and abuse, and health problems such as the HIV/AIDS epidemic. In addition, the women tend to reside in communities that lack social capital.

In this chapter we take an ecological public health approach, which allows us to move from individual, including genetics, to community-level characteristics. Women living in impoverished neighborhoods may have the same genetic/biological vulnerability to depression as other women, but their exposure to environmental triggers is such that they are more likely to experience depression and other mental disorders. We will focus on two of the six traditional activities in the public health approach (Turnock, 2004) – (1) defining the problem and (2) identifying its causes, consequences, and protective factors – to aid in development of intervention strategies among a population most in need – in this case, disadvantaged women. A greater effort to address the link between mental health problems and social inequalities will provide a better basis from which to develop and test hypotheses about causes, treatment, and course of the different disorders, as well as to inform and evaluate policies.

In recent decades, an increasing awareness of the complex nature of depression and its devastating impact on the quality of life among those with depression has emerged. The term *depression* has entered the everyday vocabulary in the United States and people who refer to themselves as depressed or who are labeled by others as such do not necessarily have a mental illness. Often the term is used to describe short-lived experiences and feelings of hopelessness that are linked to specific events or stressors. An occasional depressed mood or occasional sadness may be no more than a reaction to a major loss, adversity, or disappointment. However, such experiences and feelings differ significantly from a depression that represents an actual mental illness and that involves the impairment of psychological, somatic, and social functioning (Akiskal, 2000). The spectrum of depression is wide (Angst & Merikangas, 1997) and its etiology remains unclear. Research on depression is hindered by the lack of reliable biological markers or valid behavioral tests for depression (Ellason, Ross, & Fuchs, 1995). The general consensus appears to be that the etiology of depression has biological, psychological, and social dimensions, thereby suggesting the need for interdisciplinary inquiries and multifaceted efforts. By focusing on disadvantaged inner-city women we will be able to address several of these dimensions. We consider the women's individual characteristics, and we also take the larger context into consideration; for example, community characteristics and public health threats such as those related to drug abuse and HIV/AIDS. The health of a community is a composite of physical, psychological, social, and economic variables. Consequently, the responsibility for overall community or public health is also multifaceted, residing in a number of systems (e.g., family, education, health, work, social services).

Various theoretical paradigms have been applied to studies on depression. These are, for instance, psychosocial (Beck, 1991; Clark & Steer, 1994), social (Brown, Harris, & Eales, 1996; Hobfoll, Ritter, Lavin, Hulsizer, & Cameron, 1995), biological (Gershon et al., 1982; Paykel, 1991), and woman-centered (Gilligan, 1982; Stopppard, 2000). Although no single perspective can provide all the answers to the complex problem of mood disorders such as depression, diagnosing and treating depression requires a fundamental understanding of all processes that may lead to psychopathology. Significant advances have occurred in these areas but much work remains to be done, particularly in investigating the individual risk and protective factors that make certain individuals more vulnerable to mental disorders as well as in gaining a better understanding of environmental and contextual factors. We will focus primarily on social, psychosocial, and woman-centered approaches.

It is also important to point out that discussions about an appropriate measure of depression are ongoing, especially among social and behavioral science researchers. Depression often is assessed and conceptualized on a continuum or a multidimensional scale, with the assessments being based on a person's self-reported symptoms and experiences and recorded on a structured questionnaire. The Beck Depression Inventory (BDI; Beck, Ward, Mendelson, Mock, & Erbaugh, 1961) and the Center for Epidemiologic Studies – Depression scale (CES-D, Radloff, 1977) appear to be most commonly used in research studies. The BDI is a 21-item self-report rating inventory measuring characteristic attitudes and symptoms of depression (Beck et al., 1961). The items in the first BDI were based on clinicians' notions of the descriptions provided by patients of symptoms of their depression. The purpose of the CES-D is to measure symptoms associated with depression experienced in the past week (Radloff, 1977). It is derived from several previously validated depression scales. However, there is evidence that the scale may not be as optimal as a depression screening measure among low-income women (Thomas, Jones, Scarinci, Mehan, & Brantley, 2001). It is important to recognize the need for appropriate measures for women, including subgroups of women.

THE EPIDEMIOLOGY OF DEPRESSION AMONG WOMEN

Research conducted in the United States (Kessler, McGonagle, Nelson et al., 1994; Regier, Farmer, & Rae, 1990) and throughout the world (Weissman et al., 1996) has consistently shown depression to be more common among women than men. Prevalence estimates of major depression among women have ranged between one and one-half to three times that of men (Kessler,

2003). A female-to-male prevalence ratio of 2:1 for chronic depression or dysthymia consistently is reported (Angst & Merikangas, 1997; Kessler, 2003), whereas no gender differences in the prevalence of mania have been reported in epidemiological (Kessler, 2003) or clinical studies (Goodwin, 1990). The risk of an initial depressive episode is higher among women than men (Eaton et al., 1997; Kessler, McGonagle, Nelson et al., 1994). The risk of recurrent depressive episodes, however, is similar (Kessler, 2003).

The nature and incidence of depression among women varies across the lifespan. A variety of factors unique to women's lives are suspected to play a role in the development of depression, including premenstrual dysphoric disorder, postpartum depression, and perimenopausal depression. Depressive disorders occurring during the puerperium are thought to contribute to the higher rate of mood disorders among women than men in the United States (Hobfoll et al., 1995; Liewellyn, Stowe, & Nemeroff, 1997). The combined period prevalence of postpartum major and minor depressive disorders is estimated to be between 5 and 25% (Gotlib, Whiffen, Mount, Milne, & Cordy, 1989; Harding, 1989; Leopold & Zoschnick, 1997; O'Hara & Swain, 1996).

A substantial body of research on depression among women also has identified the co-occurrence of depression and substance use disorders (Brady & Sonne, 1995; Grant & Pickering, 1996; Hasin et al., 1996; Merikangas et al., 1998; Regier et al., 1990). Three major epidemiologic studies in the United States have been conducted to examine the prevalence of psychiatric and substance use disorders in community samples – the Epidemiologic Catchment Area (ECA) Study, sponsored by the National Institute of Mental Health and conducted in the early 1980s (Regier et al., 1990); the National Comorbidity Study (NCS) conducted in 1991 (Kessler, McGonagle, Zhao et al., 1994); and the National Longitudinal Alcohol Epidemiologic Study (NLAES) conducted in 1992 (Grant & Pickering, 1996). Forty-five percent of individuals with an alcohol-use disorder and 72% of those with a drug-use disorder had at least one co-occurring psychiatric disorder according to estimates from the ECA. Estimates from the NCS revealed that approximately 86% of alcohol-dependent women and 78% of alcohol-dependent men met lifetime criteria for another psychiatric disorder, including drug dependence (Kessler, Crum, & Warner, 1997). Individuals in the NCS with alcohol abuse or dependence diagnoses were 2.6 times more likely to be diagnosed as having major depressive disorder than those without the diagnosis. In the NLAES, respondents with a diagnosis of alcohol abuse or dependence were 3.7 times more likely to have major depressive disorder and 7.2 times more likely when related to drug abuse or dependence (Grant & Pickering, 1996).

Women are more likely than men to experience co-occurring mental health and substance abuse or addiction disorders (National Institute on Drug Abuse, 1998). For many women with depression and substance use disorders, depression is the primary disorder and women use substances to modify the effects of a negative mood (Brady & Randall, 1999; Sterk, 1999; Wakschlag, Pickett, Cook, Benowitz, & Leventhal, 2002). However, for some women depression may be a consequence of a primary substance-use disorder (Marshall, 2000), often due to consequences of their substance use such as relationship problems, work and school failure, and criminal involvement.

Depression also is common among individuals infected with HIV and research has shown a high prevalence of major depression among those receiving treatment for HIV (Bing, Burnam, & Longshore, 2001), with an increased risk associated with late-stage HIV disease (Lyketsos, Hoover et al., 1996). The prevalence of depression has been reported to be at least twice as high in women with HIV infection compared to HIV-infected men (Evans et al., 2002; Ickovics et al., 2001). Furthermore, HIV-infected women report significantly diminished feelings of well-being, a weaker sense of coherence, and less social contact and support than men (Gebo, Keruly, & Moore, 2003; Komiti et al., 2001; Morrison et al., 2002), which can impact the risk for or severity of depressive symptoms.

Among those with HIV, co-occurrence of mental health and substance-use disorders is also common. In the Health Care Services and Utilization Study (HCSUS), nearly one-half of the adults being treated for HIV screened positive for one of four psychiatric disorders; heavy drinking or illegal drug use increased the likelihood of a psychiatric disorder (Bing et al., 2001). During the previous year, over one-third of HCSUS patients screened positive for major depression and over one-fourth experienced symptoms of prolonged mild depression. In another population-based ($N = 378{,}710$) study using hospital discharge abstract data, persons with a mental illness were more likely to have HIV/AIDS, and women were at increased risk of being dually diagnosed (Stoskopf, Kim, & Glover, 2001). Treisman and colleagues (2001) found, in an outpatient clinic population, that more than half of their patients had a major psychiatric disorder other than a personality disorder or substance abuse and 20% had other cognitive impairments. The authors also assert that without appropriate access to care for mental illness and substance abuse conditions, life for many of these patients will remain chaotic.

An additional consistent finding is the known association between lower income and higher rates of mental health disorders and comorbid conditions. Poverty, residence in an impoverished community, low socioeconomic status, unemployment, and lower levels of education have all been shown to increase the vulnerability for depressive symptoms (Belle, 1990; Blazer et al., 1985; Holzer et al., 1986; Ross & Huber, 1985). Research has shown that individuals living in poverty (defined as lower strata of education, income, and occupation) are two to three times more likely than those not living in poverty to have a mental disorder (Muntaner, Eaton, Diala, Kessler, & Sorlie, 1998; Norquist, Regier, Burke, & Regier, 1996) and are more likely to have higher levels of psychiatric distress (Bovasso, Eaton, & Armenian, 1999). Studies have also shown that levels of depression among welfare recipients and low-income women are much higher than in the general population (Hobfoll et al., 1995; Ritchey, La Gory, Fitzpatrick, & Mullis, 1990). Unfortunately, women and children are disproportionately represented in the poorer strata in the United States (Dalaker, 2001; Spraggins, 2003). In a nationally representative stratified random sample of women, Kahn and colleagues (2000) found differences in depression according to annual household income (categorized into fifths; 0–$10,000, $10,000–19,999, $20,000–34,999, $35,000–49,999, >$50,000). Results showed that women in the lowest quintile of household income were more likely to report depressive symptoms (33% vs. 9%, $p < 0.001$) and fair or poor health (15% vs. 2%, $p < 0.001$) compared to women in the highest quintile (Kahn, Wise, Kennedy, & Kawachi, 2000). In an unmatched case-control study based on the National Comorbidity Survey, Bassuk and colleagues (1998) compared the prevalence of *DSM-III-R* disorders among homeless and low-income housed mothers with the prevalence of these disorders among all women in the survey. Homeless and housed mothers had similar rates of psychiatric and substance use disorders. Both groups of low-income mothers had similar rates of psychiatric and substance use disorders and both had higher lifetime and current rates of major depression and substance abuse than did women in the general population (Bassuk, Buckner, Perloff, & Bassuk, 1998). Hobfoll and colleagues (1995) found, in a sample of low-income postpartum women, a point prevalence of 23% for postpartum depression, including 7% for major depressive disorder and 16% for minor depressive disorder. These rates were about twice those published for postpartum women from a middle-class sample in the United States at the time (Hobfoll et al., 1995).

Race and ethnicity are significant factors related to both poverty and mental health (U.S. Department of Health and Human Services, 2001). Although White women constitute the largest number of women in poverty, the poverty rate is higher among African American and Hispanic women (Dalaker, 2001; Spraggins, 2003). African Americans account for approximately 25% of the mental health needs in this country, though they comprise 11–12% of the national population (Satcher, 1999). African American women are less likely to have depression (16%) than are Caucasian women (22%). However, of those suffering from depression, almost half (47%) are afflicted with severe depression (Satcher, 1999). The lower prevalence of depression among African American women also may be a result of diagnostic screening and cultural factors, as discussed later in the chapter.

Hispanic women have the highest lifetime prevalence of depression (24%) of all women (Satcher, 1999). American Indian/Alaska Native individuals appear to suffer disproportionately from depression and substance abuse. A significant proportion of Asian American women suffer from psychiatric disorders, particularly among refugee populations. However, results from smaller ethnic groups, such as American Indians and Asian Americans, are based on limited study samples from which to draw concrete conclusions (U.S. Department of Health and Human Services, 2001).

Given the distribution of depression among segments of women and its association with comorbid conditions such as substance use and HIV infection, a public health approach to etiologic research, treatment, and prevention is necessary (World Health Organization, 2000). Research on mood disorders such as depression must involve a multifaceted social model that takes into account the impact of additional physical, psychological, social, and economic factors at play. Furthermore, the consequences of depression are numerous (described below) and also require an interdisciplinary approach to research, treatment, and prevention.

VULNERABILITY FOR DEPRESSION

Numerous factors may contribute to depression in women, including developmental, reproductive, hormonal, genetic, and other biological differences (Gershon et al., 1982; Kendler et al., 2000; National Institute of Mental Health, 1995; Paykel, 1991). Cultural and social contexts appear to play a greater role in the development of depression compared to other mental disorders (U.S. Department of Health and Human Services, 2001). As presented earlier, not all segments of society are equally afflicted by depression. Although the increased risk for depression among women persists across cultures, races, and socioeconomic levels (Kessler, 2003; McGrath, Keita, Strickland, & Russo, 1990), many preventable risk factors occur among women who are at the greatest disadvantage in our society. Furthermore, women affected by depression in these communities often lack the resources that may minimize or modify depressive symptoms (Hobfoll, Johnson, Ennis, & Jackson, 2003).

Researchers have attributed the increased prevalence of depression among the economically disadvantaged to life stress and trauma, such as exposure to chronic negative events, less control over the environment and the occurrence of daily stressors, and increased vulnerability to the adverse impact of stressful events (Hobfoll et al., 1995; Ritchey et al., 1990; U.S. Department of Health and Human Services, 2001). Women living in impoverished, high-stress environments are highly vulnerable to a number of acute and chronic stressors. Their ability to cope with such stressors may impact their likelihood of developing depression and other mental health conditions.

Social Stressors

Women appear to experience and cope with stress differently than their male counterparts (Leibenluft, 1997; McGrath et al., 1990; Nolen-Hoeksema & Girgus, 1994; Stoppard, 2000). Women in inner-city communities are often faced with multiple social stressors, including poverty, unemployment, low education, child-care responsibilities, gender and racial discrimination, drug use, victimization and violence, sexual exploitation, and the strong presence of the HIV epidemic (Brown & Moran, 1997; Coyle, 1998; Dancy, 2000; Eisenberg, 1997; Hobfoll et al., 1995; Kessler, Mickelson, & Williams, 1999; Sterk, 1999; Wechsberg, Dennis, & Stevens, 1998). The interaction between social stressors such as drug use, victimization, and risk for HIV has emerged as an underlying threat to the health and well-being of many women in these communities. This ecological concentration of synergistic health and social problems has been referred to as "syndemic" (Singer, 1994). Such stressors can have a tremendous impact on the development and maintenance of mental health problems such as depression.

Discrimination and Employment. In a nationally representative sample, Kessler and colleagues (1999) found that major discrimination (as dramatic events such as being fired from a job) was associated with psychological distress and major depression in both African American and Caucasian respondents. Day-to-day discrimination was associated with distress and the diagnosis of anxiety and depression. However, the lifetime prevalence of major discrimination and day-to-day discrimination was reported in 50% and 25% of African American respondents as opposed to 30% and 3% of Caucasians, respectively. Among African Americans, perceived discrimination has been associated with psychological distress, lower well-being, and self-reported ill health (Ren, Amick, & Williams, 1999; Williams, 1997).

Discrimination toward women has also been shown to be associated with inequalities in employment opportunities and research has shown that unemployment may have a greater impact on women's mental health than men's (Bebbington, 1998; Popay, Bartley, & Owen, 1993). Pay inequities, inadequate provision of maternity leave and child-care services, and a lack of or minimal fringe benefits for part-time work can all contribute to added financial and emotional stress for women who are employed. For those who are unemployed, such factors may be deterrents to employment. Studies have documented a high prevalence of depression among welfare recipients (Jayakody, Danziger, & Pollack 2000; Olson & Pavetti, 1996).

Although employment can have a beneficial impact on health, the extra stresses and anxiety associated with employment may make some women more vulnerable to depression (Bebbington, 1998; Walters, 1993). This is particularly true for single inner-city women who are also mothers (Brown

& Moran, 1997). A majority of inner-city families have a female head of household, which adds an additional burden (Richman, Chapman, & Bowen, 1995). Brown and Moran (1997) found a twofold risk of onset of depression among single low-income, inner-city mothers compared to their married counterparts. Single mothers were also more likely to be in financial hardship, despite being twice as likely to be in full-time employment. Beyond the risk for postpartum depression, motherhood itself may be a time of heightened risk for depression because of the stress and demands it imposes (Bernazzani & Bifulco, 2003).

Victimization. One of the most devastating stressors experienced by inner-city women is that of abuse or victimization. Victimization and visible violence and crime are highly prevalent in many inner-city, often impoverished communities (Acierno, Resnick, & Kilpatrick, 1997; Bachman & Saltzman, 1995; Falck, Wang, Carlson, & Siegal, 2001; Miethe & Meier, 1994; Sampson & Lauritsen, 1990). The overwhelming majority of victims of interpersonal violence are female (Bachman & Saltzman, 1995; Koss, 1993). Both childhood and adult victimization have been linked to women's vulnerability to depression (Koss, 1993; Morrill, Kasten, Urato, & Larson, 2001; Weiss, Longhurst, & Mazure, 1999).

Researchers have demonstrated a link between childhood sexual abuse and a history of depression in adulthood (Beitchman et al., 1992; Cheasty, Clare, & Collins, 1998; Hall, Sachs, Rayens, & Lutenbacher, 1993; Whiffen & Clark, 1997). Increased symptoms of depression are reported among women who have been physically abused by a male partner and among those who have been sexually assaulted (Golding, 1999; McCauley et al., 1995; McGrath et al., 1990). Sexual rape can have a severe psychological impact, including major depression and long-term depressive symptoms, as well as posttraumatic stress disorder (Koss, 1993).

Abuse by a male partner has been reported as the greatest single cause of injury to women requiring emergency medical treatment (Stark & Flintcraft, 1985). Emotional abuse, particularly by a male partner, has also been associated with depression (Koss, 1993). The most damaging aspect of interpersonal abuse is its repetitive nature, which serves to exacerbate the severity of depressive symptoms (Stark & Flintcraft, 1985) and can weaken a woman's attempts to leave an abusive relationship.

Drug Use. An additional and common stressor in the lives of many inner-city women is their own drug use or the prevalence of drug use in their community. As mentioned previously, the co-occurrence of depression and other mental health conditions and alcohol and other drug use is highly prevalent among women (Grant & Pickering, 1996; Kessler et al., 1997; Regier et al., 1990). Female substance users are more likely than male users to experience serious depression (Brady & Randall, 1999).

For some women depression may be a consequence of a primary substance-use disorder, either as a side effect of the substance or as a condition exacerbated by substance use. Depression has been indicated as a side effect in certain substances of abuse. Cocaine and other stimulants, such as amphetamines and hallucinogens such as LSD (lysergic acid diethylamide) and PCP (phencyclidine) have been shown to increase the risk of depression (Abadinsky, 1997). Recent research has also shown that heavy ecstasy users are at risk of becoming clinically depressed, likely due to the effect of the drug on mood regulatory systems in the brain (Parrott, 2000). Substance use may also be a precipitating factor that causes mental health problems such as depression to manifest (Abraham & Fava, 1999).

Genetic makeup may also play a role in the development of both mental health and substance use disorders (Kendler et al., 2000; Merikangas et al., 1998; Prescott, Neale, Corey, & Kendler, 1997). Many women have often experienced a lifetime of substance abuse in their families, providing them with a genetic link to such disorders. Research has shown that drug use in the early teen and adolescent years may lead to major depression and other psychiatric disorders in the late 20s.

The consequences of substance-use disorders can greatly impact the development of depression (Marshall, 2000). Women who abuse or are dependent on certain substances often experience a lack of family and social support networks and more child-care responsibilities, which can cause greater conflicts and stressors for these women. Depression has been found to be associated with poor social support (Coiro, 2001; Collins, Dunkel-Schetter, Lobel, & Scrimshaw, 1993; Oxman, Berkman, Kasl, Freeman, & Barrett, 1992; Paykel, 1991; Ritter, Hobfoll, Lavin, Cameron, & Hulsizer, 2000). Women with substance-abuse problems also experience more stigma and social disapproval than men, resulting in lower feelings of self-worth, self-esteem and other traits that may make a woman vulnerable to depression.

Beyond the individual factors associated with depression among female substance users, the ever-present drug economy in many poor, inner-city communities can be an additional source of stress and strain in these women's lives.

HIV Infection and the HIV/AIDS Epidemic. The presence of the HIV epidemic is now a salient part of many inner-city women's lives. Despite considerable gains in treatment and prevention, the incidence of HIV continues to rise among ethnic minority women (Centers for Disease Control and Prevention, 2002). Over the past decade, studies have documented substantial mental health problems in persons with HIV (Low-Beer et al., 2000; Lyketsos, Hutton, Fishman, Schwartz, & Treisman, 1996; Rabkin, Ferrando, van Gorp et al., 2000; Starace, Bartoli et al., 2002). HIV infection increases a patient's risk for various psychiatric disorders, including depression, mania, psychosis, and substance abuse (Treisman et al., 2001). In a meta-analysis, Ciesla and Roberts (2001) reported a twofold increase in risk for depression among patients with HIV relative to those at risk for HIV infection. For those with HIV, having a life-threatening chronic condition, stigma, decreased physical functioning, financial difficulties, and decreased ability to meet role obligations add to experienced stress in their lives. Physical and functional decline presents an immediate challenge but numerous underlying predisposing personal and current psychosocial problems may account for depressive symptomatology and overall emotional distress. Late-stage HIV disease also has been associated with an increased risk for depression.

Among women with or without HIV infection, the epidemic itself is very present in many inner-city communities. Many women have lost family members, partners, and friends to HIV or AIDS. The loss of an attachment to a loved person or of some other significant attachment can lead to a prolonged period of distress and disability. The risk of HIV infection or of transmitting HIV is heightened for female drug users who support themselves or their drug habit with behaviors that place them at risk of HIV acquisition (Sterk, 2000). Factors associated with the HIV epidemic can cause additional stress in the lives of women already experiencing a multitude of stressors.

Individual and Community Strain and Resources

Much of the research linking psychosocial and biological conditions to depression has been based on the stress or strain model, or diathesis-stress model, in which characteristics of the social environment and the way an individual perceives this environment are regarded as threatening to their well-being and integrity of the self (Beck, 1991; Clark & Steer, 1994). That is, individuals prone to depression are characterized by personality traits that predispose them to becoming depressed. Studies of depressive disorder or depressive symptoms have used the diathesis-stress model approach, although not for depression conceptualized as depressive experiences (Clark & Steer, 1994; Coyne & Whiffen, 1995). In the model, diatheses such as dysfunctional beliefs, learned helplessness, and hopelessness have all been identified as being provoked by the stressor(s) (Beck, 1991; Ingram & Ritter, 2000).

Among women, specific cognitive styles have been considered as part of the diathesis-stress model in an attempt to explain gender differences in depression. Rumination or ruminative thinking (repetitive and passive mental focus on depressive or distressing symptoms) has been associated with an increased risk for depression and major depressive disorder, and with longer and more severe episodes of depression (Nolen-Hoeksema, Parker, & Larson, 1994). Other styles and personality characteristics associated with femininity, such as interpersonal dependency and an affiliative orientation, have been explored (Whitley, 1985).

Although the effects of stress on individual characteristics are important, actual exposure to stressful events and the context in which these events occur are extremely important when considering the potential causes of depression. Researchers have hypothesized that the experiences related to gender differences in social roles may play a strong role in women's greater risk of depression. Nazroo and colleagues (1998) found that when gender role differences in certain life crises are apparent, such as an event involving children, housing, or reproduction, the risk of depression among women was greater (Nazroo, Edwards, & Brown, 1998).

Among disadvantaged women, the number and types of stressors as well as the mechanisms for coping with stress may all contribute to the development of depression or depressive symptoms. Ennis and colleagues (2000) have shown that acute economic stress may be more detrimental than chronic stress on depressive mood (Ennis, Hobfoll, & Schroder, 2000). In addition to the type of stress or context in which it occurs, the available resources during times of stress are essential to understanding how individuals cope with these situations.

Individuals with greater resources available to them are less negatively impacted by stressful life circumstances (Cowen, Wyman, & Work, 1996; Coyne & Whiffen, 1995; Hobfoll & Jackson, 1991; Hobfoll et al., 2003; Norris & Kaniasty, 1996). The lack of personal and social resources during times of crisis or under stressful life circumstances has been associated with an increased risk of depression among women. Research has also demonstrated that the type and duration of resources available to women are important to understanding the link between available resources and the emotional response to stress (Hobfoll, 1989; Holahan, Moos, Holahan, & Cronkite, 1999). Protective resources or factors that may be protective against depression and other mental health conditions include a positive temperament, above-average intelligence, social competence, spirituality or religion, a smaller household structure, supportive relationships with parents, good sibling relationships, adequate rule setting and monitoring by parents, and in the community, commitment to schools, availability of health and social services, and social cohesion (U.S. Department of Health and Human Services, 2001).

Hobfoll and colleagues have proposed a conservation of resources theory to explain the role of resources and response to stress (Hobfoll, 1989). Key aspects of the theory are that a loss of resources can trigger a negative reaction to stress, that persons experiencing stress will activate remaining resources available to them, and that resource gain is of less impact than is resource loss. Loss of resources has been shown to increase the vulnerability to depression, whereas resource gain has been shown to decrease the risk (Holahan et al., 1999). In a cohort of patients with unipolar depression, Holahan and colleagues (2000) found a decrease in psychosocial resources as a result of negative life events, which in turn increased the patient's depressive symptoms (Holahan, Moos, Holahan, & Cronkite, 2000).

Many inner-city women experience tremendous resource loss due to characteristics in their personal lives and characteristics of the communities in which they reside (Wandersman & Nation, 1998). It is becoming imminent that, in addition to those at the individual level, resources at the community level are addressed.

Social capital can serve as a useful construct and it refers to community-level resources (Putnam, 2000). Studies have shown that the more social capital is present in a community, the better the health measures are (Kawachi & Berkman, 2000; Putnam, 2000). At the heart of social capital is collaboration among community members, based on trust, reciprocity, and shared interests. The relationship between social capital and depression among women residing in impoverished communities is worthwhile exploring. Whereas social capital builds a social infrastructure, including social cohesion, at the community level, improved social relationships among community members can lead to increased opportunities to prevent triggers for depression, especially among those whose biological vulnerability is high.

CONSEQUENCES OF DEPRESSION

Individuals experiencing depression or symptoms of depression may also experience declines in physical health and overall social and economic well-being. Depression has been shown to impair cellular immunity (Evans et al., 1992; Evans et al., 1989; Herbert & Cohen, 1993; Reichlin, 1993), resulting in less physical health and susceptibility to infection. Being afflicted by a mental disorder significantly reduces the wage rate, the probability of participating in the labor force, and the number of hours worked per week (Bartel & Taubman, 1979). With respect to personal relationships, there is evidence that depressed individuals are more vulnerable to interpersonal problems (Tweed, 1993).

All consequences of depression are too numerous to cover in this chapter. Therefore we present key consequences in the lives of inner-city women afflicted with depression, namely, risk behaviors for HIV and the effect of depression on HIV disease. Both of these are amicable to public health interventions.

Risk Behavior

The importance of identifying the role that depression and other psychosocial factors play in risk behavior has been emphasized (Kail, Watson, & Scott, 1995; Morrill et al., 2001; Schilling, el-Bassel, & Gilbert, 1993), particularly among women who use or abuse substances, who are at increased risk for depression, and who are a unique group (Rosenbaum, 1981; Sterk, 1999). Research has shown that depression and other mental health conditions are positively associated with overall HIV and other infectious disease risk behavior.

In a nationally representative sample of noninstitutionalized adults in the United States, Blumberg and Dickey (2003) found that adults with at least one of three psychiatric conditions (depression, generalized anxiety disorder, and panic attacks) in the previous 12 months were more likely to have engaged in HIV risk behaviors. Depression has been linked to drug- and injection-related risk such as needle and injection work sharing and frequency of injection drug use (Mandell, Kim, Latkin, & Suh, 1999; McKusker, Goldstein, Bigelow, & Zorn, 1995; Simpson, Knight, & Ray, 1993), as well as nonfatal drug overdose (Tobin & Latkin, 2003). High levels of neuroticism and anxiety have been associated with more severe drug addition (Ball & Schottenfeld, 1997). Depression has also been linked to women's sexual risk-taking practices such as multiple sex partners, sex exchanging, and condom use (El-Bassel, Simoni, Cooper, Gilbert, & Schilling, 2001; Klein, Elifson, & Sterk, 2003; Mandell et al., 1999; McKusker et al., 1995; Orr, Celentano, Santelli, & Burwell, 1994; Roberts, Wechsberg, Zule, & Burroughs, 2003; Simpson et al., 1993).

The relationship between depression and risk behavior in many inner-city women is often directed by or mediated by violence and substance use (Johnson, Cunningham-Williams, & Cottler, 2003). The "syndemics" of HIV, violence, and drug and other substance abuse are highly prevalent in the lives of many inner-city women and may play a large role in the development and consequences of depression and other mental health conditions. Higher than average rates of victimization have been reported among women with, or at risk for, HIV infection, and victimization or abuse is a risk factor for depression and substance use disorder (Boyd, 1993; El-Bassel et al., 1998; Grella, Anglin, & Annon, 1996; Johnson et al., 2003; Morrill et al., 2001). Understanding the complex relationship between victimization, substance use, mental health, and risk behavior is necessary for prevention and risk reduction efforts and for increasing personal health and safety and the quality of life for women.

There are many potential pathways leading from victimization experiences to HIV risk behavior and from involvement in risk behaviors to possible victimization. Women who have encountered abuse or who continue to encounter abuse in their daily lives may be more prone to depression as a natural reaction to the abusive experiences. They may also turn to drugs or alcohol or engage in behaviors that put them at risk for HIV as well as other blood-borne and sexually transmitted infections as a means of coping or due to feelings of hopelessness, low self-efficacy, impulsivity, or codependence (Morrill et al., 2001; Nyamathi, 1991; Schilling et al., 1993; Sterk, 2000). Conversely, a woman's substance using behavior may place her at increased risk for victimization experiences.

Violence in the domestic setting, drug-using setting, and community setting is not uncommon in the lives of female drug users. Many women who use or abuse drugs and alcohol often place themselves in certain situations in which abuse is more likely to occur. Drug use in shooting galleries, crack houses, or on the streets can all increase the risk for a violent encounter (Goldstein, 1998; Jensen & Brownfield, 1986; Rosenbaum, 1981; Sampson & Lauritsen, 1990; Schreck, 1999; Sterk-Elifson & Elifson, 1993). The exchange of sex for drugs or money, as well as many domestic sexual relationships, can place a woman at risk for victimization. A woman's position in intimate relationships often makes it difficult to initiate conversations about or to propose safer sex or needle use (Amaro, 1995; Deren et al., 1995; Kane, 1991; Rosenbaum, 1981; Sterk, 1999). For example, a female injection drug user who proposes each partner use a new syringe may risk verbal and physical abuse by her partner (Boyd, 1993). Furthermore, drug and alcohol use can impair one's judgment, leading to greater participation in certain risk behaviors that may place one at risk for victimization.

Due to the interconnected relationship between depression, violence, substance use, and HIV risk, several researchers have addressed the need to develop HIV prevention, intervention, and educational efforts that also focus on the social and economic contexts of women at risk for infection (Compton, Cottler, Ben-Abdallah, Cunningham-Williams, & Spitznagel, 2000; Ehrhardt & Exner, 2000; Ickovics & Yoshikawa, 1998; Sterk, Theall, & Elifson, 2003). In our own research among female drug users, we evaluated whether women participating in an HIV intervention ($N = 333$, aged 18–59 years) who inject drugs, smoke crack cocaine, or both and who had greater symptoms of depression and anxiety reduced HIV risk behaviors less than drug users with lower levels of symptoms (Sterk et al., 2003; Sterk, Theall, Elifson & Kidder, 2003). We also examined whether an enhanced intervention was more effective than a briefer standardized intervention for those with higher levels of symptoms. Women in the enhanced motivation intervention had significantly lower levels of depression and anxiety at 6-month followup compared to those in the standard intervention. Higher levels of depressive symptoms, anxiety, and low self-control were associated with less drug using and sexual behavioral improvement over time, even after controlling for important covariates.

HIV Progression, Treatment Failure, and Adherence

Additional consequences of depression exist for women infected with HIV. There is a growing body of literature on the relationship between depressive symptoms and HIV progression and treatment success among women, but much remains to be understood. In the HIV Epidemiologic Research Study, Ickovics and colleagues (2001) reported the mortality rate among women with chronic depressive symptoms to be twice that of women with limited or no depressive symptoms. In a cohort of HIV-infected injection drug users, Golub and colleagues (2003) found psychological distress to be associated independently with a shorter time to progressing to AIDS, especially among those with the lowest CD4 cell counts. In a longitudinal cohort of HIV-infected women, Evans et al. (2002) determined depressive and anxiety symptoms to be associated significantly with higher activated CD8 T-lymphocyte counts, higher viral loads, and lower natural killer cell activity. The authors concluded that depression may result in an increase in activated CD8 T lymphocytes and viral load.

Triple combination therapy has been associated with significant improvement in HIV-related depressive symptoms (Carpenter et al., 1997; Holzemer et al., 1999; Safren, Radomsky, Otto, & Salomon, 2002). Despite improvements, depressive symptomatology may be evident in individuals starting highly active antiretroviral therapy (HAART), may be a consequence of early side effects, and may adversely affect adherence and drug efficacy. Recently, studies have demonstrated declines in depression associated with HAART therapy (Brechtl, Breitbart, Galietta, Krivo, & Rosenfeld, 2001; Chan et al., 2003; Judd et al., 2000; Low-Beer et al., 2000; Rabkin, Ferrando, Lin, Sewell, & McElhiney, 2000). Successful treatment resulting in improved physical health is likely to have a positive affect on mental health. However, complicated treatment regimens, disease stage at the time of treatment, the need to make future plans, and changing social support systems may have a negative impact on mental health (Rabkin & Ferrando, 1997).

Fewer studies have focused on the association between depression and effectiveness of HAART. In the Italian NeuroICONA (Italian Cohort Naïve Antiretrovirals) cohort, Starace, Bartoli, and colleagues (2002) uncovered an elevated prevalence of depressive symptomatology among patients naïve to HAART (15.5% of patients) and found that such symptomatology varied according to regimen. Depressed mood was significantly associated with clinical stage and a higher viral load, but not with CD4 cell count. Substance use was a strong predictor of depression in both bivariate and multivariate analyses (Starace, Bartoli et al., 2002). In an HIV clinic Veterans Administration population, Paterson and colleagues (2000) reported a 40% increased relative risk of virologic failure in patients with active depression and a significant association between optimal adherence and lower psychiatric morbidity.

Depression and substance-use disorders may impact mortality for women with HIV directly via effects on immune suppression or indirectly via effects on health care utilization or adherence to treatment recommendations (Cohen et al., 2002). Adherence to antiretroviral therapy is strongly associated with treatment effectiveness (Le Moing et al., 2001; Paterson et al., 2000). Substance use (Carrieri et al., 2003; Casado et al., 1999; Daftary, Goolsby, Dutta, & Delapenha, 2002; Lucas, Gebo, Chaisson, & Moore, 2002; Witteveen & van Ameijden, 2002) and depression (Avants, Margolin, Warburton, Hawkins, & Shi, 2001; Catz, Kelly, Bogart, Benotsch, & McAuliffe, 2000; Daftary et al., 2002; Kalichman, Ramachandran, & Catz, 1999; Starace, Ammassari et al., 2002; Turner, Laine, Cosler, & Hauck, 2003) have been shown to be negatively associated with antiretroviral therapy adherence. Early recognition and treatment of depression and substance-use disorders may improve HIV treatment outcomes, quality of life and social functioning, and disease course for all individuals infected with HIV.

CONCLUSION

The primary aim of this chapter was to present the complex needs and the harmful effects of mental illness – namely depression – among disadvantaged women. Socioeconomic status has one of the strongest associations with the prevalence of mental disorders, as well as many physical conditions, but the causal pathways involved are complex, multidimensional, and incompletely understood. Nevertheless, it is fairly clear that social structures shape psychiatric morbidity through social stressors (often strongly influenced by socioeconomic status), culture, and gender.

Future research must focus on deciphering the complex interaction between social and environmental factors that interact with acquired and inherited vulnerabilities to have an effect on mental health outcomes (Hobfoll et al., 2003; Kessler, 2003; World Health Organization, 2000). Additional research is also needed on the biological and social consequences of psychosocial stress. Studying such interrelationships between macro-level or social contextual factors and individual factors will likely require multilevel analyses (Diez-Roux, 1998). Ongoing research will require collaborative investigations that employ a variety of research methods and the talents of many disciplines.

Additional barriers to understanding the cumulative burden and causes of depression among disadvantaged women and disadvantaged societies include proper assessment and detection of depression and comorbid conditions (World Health Organization, 2000). Ensuring adequate care and access to care will aid in removing these barriers.

The growing recognition of the public health burden of depression has led to the development and evaluation of models for detection and treatment in primary care settings (Katon et al., 1995; Wells et al., 2000). Unfortunately, for many women socioeconomic and cultural factors can contribute to misdiagnosis and inadequate care. In addition to detection and treatment, there has been a national focus on the prevention of depression and other mental disorders (Institute of Medicine, 1994; National Institute of Mental Health, 1993). However, much remains to be done in addressing prevention among disadvantaged women facing multiple adversities, particularly women of ethnic minorities. (Podorefsky, McDonald-Dowdell, & Beardslee, 2001). Cultural identity and cultural sensitivity are key issues that must be considered in the diagnosis, treatment, and prevention of depression among inner-city women.

Despite recognition of depression as a public health problem, there is ongoing concern that the condition is underrecognized and undertreated (Davidson & Meltzer-Brody, 1999; Kessler et al., 2003; Maj et al., 1994; Morris-Rush, Freda, & Bernstein, 2003). Depression in women is misdiagnosed approximately 30 to 50% of the time (McGrath et al., 1990; Morris-Rush et al., 2003; National Institute of Mental Health, 1999). Fewer than half of the women who experience clinical depression will ever seek care (Rupp, Gause, & Regier, 1998).

A recent report from the surgeon general points out that the mental health system in the United States is not well equipped to meet the needs of racial and ethnic minority populations (Satcher, 1999; U.S. Department of Health and Human Services, 2001). Minority women who suffer from depression and other mental health and comorbid conditions are less likely to seek help, and among those who do, many encounter inadequate treatment programs with little or no sensitivities to specific gender, racial, and cultural issues (Brown & Moran, 1997; Grella, Annon, & Anglin, 1995; U.S. Department of Health and Human Services, 2001; Wang, Demler, & Kessler, 2002). Research has demonstrated that racial and ethnic minority patients are less likely than Whites to receive the best available treatments for depression (Wang et al., 2000).

In addition to the barriers faced within the treatment system, many minority women may be less likely to seek care or to recognize depression due to cultural characteristics. This has been demonstrated among African American populations more than other racial and ethnic minorities. Research has shown that African Americans are less likely to seek treatment for mental health conditions, to drop out of services at a higher rate than Whites, and to use fewer treatment sessions for mental health services (Satcher, 1999; U.S. Department of Health and Human Services, 2001). African Americans may be less likely to perceive depression as a health problem and to recognize the symptoms of depression and more likely to think of depression as a personal weakness (National Mental Health

Association, 2000). Such beliefs can also act as barriers to adequate care and may promote misdiagnosis because symptoms of depression may be masked by substance-use disorder or other comorbid mental conditions (U.S. Department of Health and Human Services, 2001).

Nonetheless, the efficacy of treatment with medication or psychotherapy among low-income, minority women has been demonstrated (Miranda et al., 2003). Prevention efforts among similar populations have also been reported as successful (Podorefsky et al., 2001). Among women with HIV, receipt of mental health services has been shown to improve antiretroviral therapy adherence among women (Turner et al., 2003). Triple combination therapy has also been associated with significant improvement in HIV-related depressive symptoms and mental health functioning (Chan et al., 2003; Low-Beer et al., 2000; Rabkin, Ferrando, Lin et al., 2000). Despite the potential for successful treatment and intervention, many barriers still exist in accessing appropriate health care.

Systemic, economic, social, and cultural barriers can reduce the use of all health care services by inner-city women, particularly minority women. Women are more likely to encounter financial and nonfinancial barriers to care than do their male counterparts. These differences are accentuated among low income and minority women. Women living in impoverished, high-stress environments may have reduced access to medical care, which has serious implications for physical and mental health. Access to appropriate health services and prevention messages can contribute to a woman's vulnerability for depression and other comorbid conditions.

Although cultural stigma, stereotyping, lack of culturally sensitive programs, and mistrust of the health system may play a role in accessing care among inner-city women (particularly minorities), mental health may not be a top priority for women facing multiple adversities. Optimal mental and physical health are often secondary to day-to-day survival, child care, family, and other social needs of inner-city women. Women with HIV may be more integrated into the health and social service system due to their infection, but the potential for barriers to adequate care still exist (Kalichman, Catz, & Ramachandran, 1999; Levine, 2002; Mocroft, Gill, Davidson, & Phillips, 2000).

In both treatment and prevention efforts for depression and other mood disorders, researchers have found improved outcomes when therapy or intervention messages took into account the social and cultural contexts of participants. Miranda and colleagues (2003) found the greatest treatment success when treatments were paired with intensive outreach and encouragement to support the interventions. Podorefsky and colleagues (2001) refined a prevention intervention for mood disorders within families to include a three-tier engagement between the community, family, and caregiver. Greater emphasis was placed on daily needs and resilience and on awareness of cultural issues and experiences of violence. Incorporating the social context of these women's lives may be necessary for adequate detection, treatment, and prevention of depression among inner-city women. Additionally, training for health care providers is essential to improve culturally competent, psychological assessments of health issues for these women (Weissman, Campbell, Gokhale, & Blumenthal, 2001).

An integration of mental health treatment and services into the general health system, particularly into primary health care, is necessary if we are to combat the growing burden of mental health disorders in this population and among similar populations. Health care workers, working within the existing urban mental health and human service system, must attempt to address the multifaceted needs of these women. However, comprehensive sensitive, coordinated, mental health community support combined with case management is often nonexistent in many urban communities. Using a public health model, researchers and caregivers working in underserved areas must continue to advocate with policy makers for the development of accessible community support programs that provide coordinated services to those without the financial resources or families to assist them.

References

Abadinsky, H. (1997). *Drug abuse: An introduction* (3rd ed.). Chicago: Nelson-Hall.

Abraham, H. D., & Fava, M. (1999). Order of onset of substance abuse and depression in a sample of depressed outpatients. *Comprehensive Psychiatry, 40*, 44–50.

Acierno, R., Resnick, H. S., & Kilpatrick, D. G. (1997). Health impact of interpersonal violence. 1: Prevalence rates, case identification, and risk factors for sexual assault, physical assault, and domestic violence in men and women. *Behavioral Medicine, 23* (2), 53–64.

Akiskal, H. (2000). Mood disorders: Clinical features. In B. Sadock & V. Sadock (Eds.), *Comprehensive textbook of psychiatry* (Vol. 1, pp. 1338–1377). Philadelphia: Lippincott Williams & Wilkins.

Amaro, H. (1995). Love, sex, and power. Considering women's realities in HIV prevention. *American Psychologist, 50*, 437–447.

Angst, J., & Merikangas, K. (1997). The depressive spectrum: Diagnostic classification and course. *Journal of Affective Disorders, 45*, 31–39.

Avants, S. K., Margolin, A., Warburton, L. A., Hawkins, K. A., & Shi, J. (2001). Predictors of nonadherence to HIV-related medication regimens during methadone stabilization. *American Journal on Addictions, 10*, 69–78.

Bachman, R., & Saltzman, L. (1995). *Violence against women: Estimates from the redesigned survey Bureau of Justice Statistics Special Report* (Publication No. NCJ 154338). Washington, DC: Department of Justice.

Ball, S. A., & Schottenfeld, R. S. (1997). A five-factor model of personality and addiction, psychiatric, and AIDS risk severity in pregnant and postpartum cocaine misusers. *Substance Use & Misuse, 32*, 25–41.

Bartel, A., & Taubman, P. (1979). Health and labor market success: The role of various diseases. *Review of Economics and Statistics, 61*, 1–8.

Bassuk, E. L., Buckner, J. C., Perloff, J. N., & Bassuk, S. S. (1998). Prevalence of mental health and substance use disorders among homeless and low-income housed mothers. *American Journal of Psychiatry, 155*, 1561–1564.

Bebbington, P. E. (1998). Sex and depression. *Psychological Medicine, 28*, 1–8.

Beck, A., Ward, C., Mendelson, M., Mock, J., & Erbaugh, J. (1961). An inventory for measuring depression. *Archives of General Psychiatry, 4*, 561–571.

Beck, A. T. (1991). Cognitive therapy: A 30-year retrospective. *American Psychologist, 46*, 368–375.

Beitchman, J. H., Zucker, K. J., Hood, J. E., DaCosta, G. A., Akman, D., & Cassavia, E. (1992). A review of the long-term effects of child sexual abuse. *Child Abuse & Neglect, 16*, 101–118.

Belle, D. (1990). Poverty and women's mental health. *American Psychologist, 45*, 385–389.

Bernazzani, O., & Bifulco, A. (2003). Motherhood as a vulnerability factor in major depression: the role of negative pregnancy experiences. *Social Science & Medicine, 56*, 1249–1260.

Bing, E., Burnam, M., & Longshore, D. (2001). Psychiatric disorders and drug use among human immunodeficiency virus-infected adults in the United States. *Archives of General Psychiatry, 58*, 721–728.

Blazer, D., George, L. K., Landerman, R., Pennybacker, M., Melville, M. L., Woodbury, M., et al. (1985). Psychiatric disorders. A rural/urban comparison. *Archives of General Psychiatry, 42*, 651–656. [Erratum appears in *Archives General Psychiatry, 43*, 1142].

Blumberg, S. J., & Dickey, W. C. (2003). Prevalence of HIV risk behaviors, risk perceptions, and testing among US adults with mental disorders. *Journal of Acquired Immune Deficiency Syndromes, 32*, 77–79.

Bovasso, G. B., Eaton, W. W., & Armenian, H. K. (1999). The long-term outcomes of mental health treatment in a population-based study. *Journal of Consulting & Clinical Psychology, 67*, 529–538.

Boyd, C. J. (1993). The antecedents of women's crack cocaine abuse: Family substance abuse, sexual abuse, depression and illicit drug use. *Journal of Substance Abuse Treatment, 10*, 433–438.

Brady, K. T., & Randall, C. L. (1999). Gender differences in substance use disorders. *Psychiatric Clinics of North America, 22*, 241–252.

Brady, K. T., & Sonne, S. C. (1995). The relationship between substance abuse and bipolar disorder. *Journal of Clinical Psychiatry, 56*(Suppl 3), 19–24.

Brechtl, J. R., Breitbart, W., Galietta, M., Krivo, S., & Rosenfeld, B. (2001). The use of highly active antiretroviral therapy (HAART) in patients with advanced HIV infection: Impact on medical, palliative care, and quality of life outcomes. *Journal of Pain & Symptom Management, 21*, 41–51.

Brown, G. W., Harris, T. O., & Eales, M. J. (1996). Social factors and comorbidity of depressive and anxiety disorders. *British Journal of Psychiatry – Supplementum,* (30), 50–57.

Brown, G. W., & Moran, P. M. (1997). Single mothers, poverty and depression. *Psychological Medicine, 27*, 21–33.

Carpenter, C. C., Fischl, M. A., Hammer, S. M., Hirsch, M. S., Jacobsen, D. M., Katzenstein, D. A., et al. (1997). Antiretroviral therapy for HIV infection in

1997. Updated recommendations of the International AIDS Society-USA panel. [Comment]. *Journal of the American Medical Association, 277*, 1962–1969.

Carrieri, M. P., Chesney, M. A., Spire, B., Loundou, A., Sobel, A., Lepeu, G., et al. (2003). Failure to maintain adherence to HAART in a cohort of French HIV-positive injecting drug users. *International Journal of Behavioral Medicine, 10*, 1–14.

Casado, J. L., Sabido, R., Perez-Elias, M. J., Antela, A., Oliva, J., Dronda, F., et al. (1999). Percentage of adherence correlates with the risk of protease inhibitor (PI) treatment failure in HIV-infected patients. *Antiviral Therapy, 4*(3), 157–161.

Catz, S. L., Kelly, J. A., Bogart, L. M., Benotsch, E. G., & McAuliffe, T. L. (2000). Patterns, correlates, and barriers to medication adherence among persons prescribed new treatments for HIV disease. *Health Psychology, 19*(2), 124–133.

Centers for Disease Control and Prevention. (2002). *HIV/AIDS surveillance report. Year-end 2001 edition* (Vol. 13(2)). Atlanta, GA.

Chan, K. S., Orlando, M., Joyce, G., Gifford, A. L., Burnam, M. A., Tucker, J. S., et al. (2003). Combination antiretroviral therapy and improvements in mental health: Results from a nationally representative sample of persons undergoing care for HIV in the United States. *Journal of Acquired Immune Deficiency Syndromes, 33*, 104–111.

Cheasty, M., Clare, A. W., & Collins, C. (1998). Relation between sexual abuse in childhood and adult depression: case-control study. *British Medical Journal, 316*(7126), 198–201.

Ciesla, J. A., & Roberts, J. E. (2001). Meta-analysis of the relationship between HIV infection and risk for depressive disorders. *American Journal of Psychiatry, 158*, 725–730.

Clark, D. A., & Steer, R. A. (1994). Use of nonsomatic symptoms to differentiate clinically depressed and nondepressed hospitalized patients with chronic medical illnesses. *Psychological Reports, 75*(3 Pt. 1), 1089–1090.

Cohen, M. H., French, A. L., Benning, L., Kovacs, A., Anastos, K., Young, M., et al. (2002). Causes of death among women with human immunodeficiency virus infection in the era of combination antiretroviral therapy. *American Journal of Medicine, 113*, 91–98.

Coiro, M. J. (2001). Depressive symptoms among women receiving welfare. *Women & Health, 32*, 1–23.

Collins, N. L., Dunkel-Schetter, C., Lobel, M., & Scrimshaw, S. C. (1993). Social support in pregnancy: Psychosocial correlates of birth outcomes and postpartum depression. *Journal of Personality & Social Psychology, 65*, 1243–1258.

Compton, W. M., Cottler, L. B., Ben-Abdallah, A., Cunningham-Williams, R., & Spitznagel, E. L. (2000). The effects of psychiatric comorbidity on response to an HIV prevention intervention. *Drug & Alcohol Dependence, 58*, 247–257.

Cowen, E. L., Wyman, P. A., & Work, W. C. (1996). Resilience in highly stressed urban children: Concepts and findings. *Bulletin of the New York Academy of Medicine, 73*, 267–284.

Coyle, S. L. (1998). Women's drug use and HIV risk: Findings from NIDA's Cooperative Agreement for Community-Based Outreach/Intervention Research Program. *Women & Health, 27*, 1–18.

Coyne, J. C., & Whiffen, V. E. (1995). Issues in personality as diathesis for depression: the case of sociotropy-dependency and autonomy-self-criticism. *Psychological Bulletin, 118,* 358–378.

Daftary, M. N., Goolsby, T., Dutta, A., & Delapenha, R. A. (2002). Possible factors influencing nonadherence to antiretrovirals in an ambulatory HIV population. *ASHP Midyear Clinical Meeting, 37*(DEC).

Dalaker, J. (2001). *Poverty in the United States: 2000* (P60–214). Washington, DC: U.S. Census Bureau.

Dancy, B. (2000). HIV risk reduction strategies for low-income African American women. *Nurse Practitioner Forum, 11*(2), 109–115.

Davidson, J. R., & Meltzer-Brody, S. E. (1999). The underrecognition and under-treatment of depression: What is the breadth and depth of the problem? *Journal of Clinical Psychiatry, 60*(Suppl 7), 4–9; discussion 10–11.

Deren, S., Davis, W. R., Tortu, S., Beardsley, M., Ahluwalia, I., & National, A. R. C. (1995). Women at high risk for HIV: Pregnancy and risk behaviors. *Journal of Drug Issues, 25,* 57–71.

Diez-Roux, A. V. (1998). On genes, individuals, society, and epidemiology. *American Journal of Epidemiology, 148,* 1027–1032.

Eaton, W. W., Anthony, J. C., Gallo, J., Cai, G., Tien, A., Romanoski, A., et al. (1997). Natural history of Diagnostic Interview Schedule/DSM-IV major depression: The Baltimore epidemiologic catchment area follow-up. *Archives of General Psychiatry, 54,* 993–999.

Ehrhardt, A. A., & Exner, T. M. (2000). Prevention of sexual risk behavior for HIV infection with women. *AIDS, 14*(Suppl 2), S53–58.

Eisenberg, L. (1997). Psychiatry and health in low-income populations. *Comprehensive Psychiatry, 38,* 69–73.

El-Bassel, N., Gilbert, L., Krishnan, S., Schilling, R., Gaeta, T., Purpura, S., et al. (1998). Partner violence and sexual HIV-risk behaviors among women in an inner-city emergency department. *Violence & Victims, 13,* 377–393.

El-Bassel, N., Simoni, J. M., Cooper, D. K., Gilbert, L., & Schilling, R. F. (2001). Sex trading and psychological distress among women on methadone. *Psychology of Addictive Behaviors, 15*(3), 177–184.

Ellason, J. W., Ross, C. A., & Fuchs, D. L. (1995). Assessment of dissociative identity disorder with the Million Clinical Multiaxial Inventory-II. *Psychological Reports, 76*(3 Pt 1), 895–905.

Ennis, N. E., Hobfoll, S. E., & Schroder, K. E. (2000). Money doesn't talk, it swears: How economic stress and resistance resources impact inner-city women's depressive mood. *American Journal of Community Psychology, 28,* 149–173.

Evans, D. L., Folds, J. D., Petitto, J. M., Golden, R. N., Pedersen, C. A., Corrigan, M., Gilmore, J. H., Silva, S. G., Quade, D., & Ozer, H. (1992). Circulating natural killer cell phenotypes in men and women with major depression. Relation to cytotoxic activity and severity of depression. *Archives of General Psychiatry, 49,* 388–395.

Evans, D. L., Leserman, J., Pedersen, C. A., Golden, R. N., Lewis, M. H., Folds, J. A., et al. (1989). Immune correlates of stress and depression. *Psychopharmacology Bulletin, 25,* 319–324.

Evans, D. L., Ten Have, T. R., Douglas, S. D., Gettes, D. R., Morrison, M., Chiappini, M. S., et al. (2002). Association of depression with viral load, CD8 T lymphocytes, and natural killer cells in women with HIV infection. *American Journal of Psychiatry, 159,* 1752–1759.

Falck, R. S., Wang, J., Carlson, R. G., & Siegal, H. A. (2001). The epidemiology of physical attack and rape among crack-using women. *Violence & Victims, 16,* 79–89.

Gebo, K. A., Keruly, J., & Moore, R. D. (2003). Association of social stress, illicit drug use, and health beliefs with nonadherence to antiretroviral therapy. *Journal of General Internal Medicine, 18*(2), 104–111.

Gershon, E. S., Hamovit, J., Guroff, J. J., Dibble, E., Leckman, J. F., Sceery, W., et al. (1982). A family study of schizoaffective, bipolar I, bipolar II, unipolar, and normal control probands. *Archives of General Psychiatry, 39,* 1157–1167.

Gilligan, C. (1982). In a different voice: Psychological theory and women's development. Cambridge, MA: Harvard University Press.

Golding, J. M. (1999). Sexual assault history and headache: Five general population studies. *Journal of Nervous & Mental Disease, 187*(10), 624–629.

Goldstein, P. J. (1998). Drugs, violence, and federal funding: A research odyssey. *Substance Use & Misuse, 33,* 1915–1936.

Golub, E., Astemborski, J., Hoover, D., Anthony, J., Vlahov, D., & Strathdee, S. (2003). Psychological distress and progression to AIDS in a cohort of injection drug users. *Journal of AIDS, 32*(4), 429–434.

Goodwin, F. K. (1990). From the Alcohol, Drug Abuse, and Mental Health Administration. *Journal of the American Medical Association, 264,* 1389.

Gotlib, I., Whiffen, V., Mount, J., Milne, K., & Cordy, N. (1989). Prevalence rates and demographic characteristics associated with depression in pregnancy and the postpartum. *Journal of Consulting and Clinical Psychology, 57,* 269–274.

Grant, B., & Pickering, M. (1996). Results from the National Longitudinal Alcohol Epidemiologic Study. *Alcohol Health and World Research, 20,* 67–72.

Grella, C. E., Anglin, D., & Annon, J. J. (1996). HIV risk behaviors among women in methadone maintenance treatment. *Substance Use & Misuse, 31,* 277–301.

Grella, C. E., Annon, J. J., & Anglin, M. D. (1995). Ethnic differences in HIV risk behaviors, self-perceptions, and treatment outcomes among women in methadone maintenance treatment. *Journal of Psychoactive Drugs, 27,* 421–433.

Hall, L. A., Sachs, B., Rayens, M. K., & Lutenbacher, M. (1993). Childhood physical and sexual abuse: Their relationship with depressive symptoms in adulthood. *Image – the Journal of Nursing Scholarship, 25,* 317–323.

Harding, J. (1989). Postpartum psychiatric disorders: A review. *Comprehensive Psychiatry, 30,* 109–112.

Hasin, D. S., Tsai, W. Y., Endicott, J., Mueller, T. I., Coryell, W., & Keller, M. (1996). Five-year course of major depression: Effects of comorbid alcoholism. *Journal of Affective Disorders, 41,* 63–70.

Herbert, T. B., & Cohen, S. (1993). Depression and immunity: A meta-analytic review. *Psychological Bulletin, 113,* 472–486.

Hobfoll, S. E. (1989). Conservation of resources. A new attempt at conceptualizing stress. *American Psychologist, 44,* 513–524.

Hobfoll, S. E., & Jackson, A. P. (1991). Conservation of resources in community intervention. *American Journal of Community Psychology, 19,* 111–121.

Hobfoll, S. E., Johnson, R. J., Ennis, N., & Jackson, A. P. (2003). Resource loss, resource gain, and emotional outcomes among inner city women. *Journal of Personality & Social Psychology, 84*, 632–643.

Hobfoll, S. E., Ritter, C., Lavin, J., Hulsizer, M. R., & Cameron, R. P. (1995). Depression prevalence and incidence among inner-city pregnant and postpartum women. *Journal of Consulting & Clinical Psychology, 63*, 445–453.

Holahan, C. J., Moos, R. H., Holahan, C. K., & Cronkite, R. C. (1999). Resource loss, resource gain, and depressive symptoms: A 10-year model. *Journal of Personality & Social Psychology, 77*, 620–629.

Holahan, C. J., Moos, R. H., Holahan, C. K., & Cronkite, R. C. (2000). Long-term posttreatment functioning among patients with unipolar depression: An integrative model. *Journal of Consulting & Clinical Psychology, 68*, 226–232.

Holzemer, W. L., Corless, I. B., Nokes, K. M., Turner, J. G., Brown, M. A., Powell-Cope, G. M., et al. (1999). Predictors of self-reported adherence in persons living with HIV disease. *AIDS Patient Care & STDs, 13*, 185–197.

Ickovics, J. R., Hamburger, M. E., Vlahov, D., Schoenbaum, E. E., Schuman, P., Boland, R. J., et al. (2001). Mortality, CD4 cell count decline, and depressive symptoms among HIV-seropositive women: Longitudinal analysis from the HIV Epidemiology Research Study. *Journal of the American Medical Association, 285*, 1466–1474.

Ickovics, J. R., & Yoshikawa, H. (1998). Preventive interventions to reduce heterosexual HIV risk for women: Current perspectives, future directions. *AIDS, 12*(Suppl A), S197–208.

Ingram, R. E., & Ritter, J. (2000). Vulnerability to depression: Cognitive reactivity and parental bonding in high-risk individuals. *Journal of Abnormal Psychology, 109*, 588–596.

Institute of Medicine. (1994). *Reducing risks for mental disorders: Frontiers for preventive intervention research.* Washington, DC: National Academy Press.

Jayakody, R., Danziger, S., & Pollack, H. (2000). Welfare reform, substance use, and mental health. *Journal of Health Politics, Policy and Law, 25*, 623–651.

Jensen, G. F., & Brownfield, D. (1986). Gender, lifestyles, and victimization: Beyond routine activity. *Violence & Victims, 1*(2), 85–99.

Johnson, S. D., Cunningham-Williams, R. M., & Cottler, L. B. (2003). A tripartite of HIV-risk for African American women: The intersection of drug use, violence, and depression. *Drug & Alcohol Dependence, 70*, 169–175.

Judd, F. K., Cockram, A. M., Komiti, A., Mijch, A. M., Hoy, J., & Bell, R. (2000). Depressive symptoms reduced in individuals with HIV/AIDS treated with highly active antiretroviral therapy: A longitudinal study. *Australian & New Zealand Journal of Psychiatry, 34*, 1015–1021.

Kahn, R. S., Wise, P. H., Kennedy, B. P., & Kawachi, I. (2000). State income inequality, household income, and maternal mental and physical health: Cross sectional national survey. *British Medical Journal 321*(7272), 1311–1315.

Kail, B., Watson, D., & Scott, R. (1995). Needle-using practices within the sex industry. *American Journal of Drug & Alcohol Abuse, 21*, 241–255.

Kalichman, S. C., Catz, S., & Ramachandran, B. (1999). Barriers to HIV/AIDS treatment and treatment adherence among African-American adults with disadvantaged education. *Journal of the National Medical Association, 91*(8), 439–446.

Kalichman, S. C., Ramachandran, B., & Catz, S. (1999). Adherence to combination antiretroviral therapies in HIV patients of low health literacy. [Comment.] *Journal of General Internal Medicine, 14*(5), 267–273.

Kane, S. (1991). HIV, heroin and heterosexual relations. *Social Science & Medicine, 32*, 1037–1050.

Katon, W., Von Korff, M., Lin, E., Walker, E., Simon, G. E., Bush, T., et al. (1995). Collaborative management to achieve treatment guidelines. Impact on depression in primary care. *Journal of the American Medical Association, 273*, 1026–1031.

Kawachi, I., & Berkman, L. (2000). *Social epidemiology.* New York: Oxford University Press.

Kendler, K. S., Bulik, C. M., Silberg, J., Hettema, J. M., Myers, J., & Prescott, C. A. (2000). Childhood sexual abuse and adult psychiatric and substance use disorders in women: An epidemiological and cotwin control analysis. *Archives of General Psychiatry, 57*, 953–959.

Kessler, R., Crum, R., & Warner, L. (1997). Lifetime co-occurrence of DSM-III-R alcohol abuse and dependence with other psychiatric disorders in the national comorbidity survey. *Archives of General Psychiatry, 54*, 313–321.

Kessler, R. C. (2003). Epidemiology of women and depression. *Journal of Affective Disorders, 74*, 5–13.

Kessler, R. C., Barker, P. R., Colpe, L. J., Epstein, J. F., Gfroerer, J. C., Hiripi, E., et al. (2003). Screening for serious mental illness in the general population. *Archives of General Psychiatry, 60*, 184–189.

Kessler, R. C., McGonagle, K. A., Nelson, C. B., Hughes, M., Swartz, M., & Blazer, D. G. (1994). Sex and depression in the National Comorbidity Survey. II: Cohort effects. *Journal of Affective Disorders, 30*, 15–26.

Kessler, R. C., McGonagle, K. A., Zhao, S., Nelson, C. B., Hughes, M., Eshleman, S., Wittchen, H. U., et al. (1994). Lifetime and 12-month prevalence of DSM-III-R psychiatric disorders in the United States. Results from the National Comorbidity Survey. *Archives of General Psychiatry, 51*, 8–19.

Kessler, R. C., Mickelson, K. D., & Williams, D. R. (1999). The prevalence, distribution, and mental health correlates of perceived discrimination in the United States. *Journal of Health & Social Behavior, 40*, 208–230.

Kessler, R. C., & Zhao, S. (1999). Overview of descriptive epidemiology of mental disorders. In C. S. Aneshensel & J. C. Phelan (Eds.), *Handbook of sociology of mental health* (pp. 127–150). New York: Plenum.

Klein, H., Elifson, K., & Sterk, C. (2003, August 14–17). *Depression and HIV risk behavior practices among "at risk" women.* Paper presented at the annual meeting of the American Sociological Association, Atlanta, GA.

Komiti, A., Judd, F., Grech, P., Mijch, A., Hoy, J., Lloyd, J. H., et al. (2001). Suicidal behaviour in people with HIV/AIDS: A review. *Australian & New Zealand Journal of Psychiatry, 35*, 747–757.

Koss, M. P. (1993). Rape. Scope, impact, interventions, and public policy responses. *American Psychologist, 48*, 1062–1069.

Le Moing, V., Chene, G., Carrieri, M. P., Besnier, J. M., Masquelier, B., Salamon, R., et al. (2001). Clinical, biologic, and behavioral predictors of early immunologic and virologic response in HIV-infected patients initiating protease inhibitors. *Journal of Acquired Immune Deficiency Syndromes, 27*, 372–376.

Leibenluft, E. (1997). Issues in the treatment of women with bipolar illness. *Journal of Clinical Psychiatry, 58*(Suppl 15), 5–11.

Leopold, K., & Zoschnick, L. (1997). Postpartum depression. *The Female Patient, 22*, 40–49.

Levine, A. M. (2002). Evaluation and management of HIV-infected women. *Annals of Internal Medicine, 136*, 228–242.

Liewellyn, A., Stowe, Z., & Nemeroff, C. (1997). Depression during pregnancy and the puerperium. *Journal of Clinical Psychiatry, 58*, 26–32.

Low-Beer, S., Chan, K., Yip, B., Wood, E., Montaner, J. S., O'Shaughnessy, M. V., et al. (2000). Depressive symptoms decline among persons on HIV protease inhibitors. *Journal of Acquired Immune Deficiency Syndromes, 23*, 295–301.

Lucas, G. M., Gebo, K. A., Chaisson, R. E., & Moore, R. D. (2002). Longitudinal assessment of the effects of drug and alcohol abuse on HIV-1 treatment outcomes in an urban clinic. *AIDS, 16*, 767–774.

Lyketsos, C. G., Hoover, D. R., Guccione, M., Dew, M. A., Wesch, J., Bing, E. G., et al. (1996). Depressive symptoms over the course of HIV infection before AIDS. *Social Psychiatry & Psychiatric Epidemiology, 31*(3–4), 212–219.

Lyketsos, C. G., Hutton, H., Fishman, M., Schwartz, J., & Treisman, G. J. (1996). Psychiatric morbidity on entry to an HIV primary care clinic. *AIDS, 10*, 1033–1039.

Maj, M., Satz, P., Janssen, R., Zaudig, M., Starace, F., D'Elia, L., et al. (1994). WHO Neuropsychiatric AIDS study, cross-sectional phase II. Neuropsychological and neurological findings. *Archives of General Psychiatry, 51*, 51–61.

Mandell, W., Kim, J., Latkin, C., & Suh, T. (1999). Depressive symptoms, drug network, and their synergistic effect on needle-sharing behavior among street injection drug users. *American Journal of Drug & Alcohol Abuse, 25*, 117–127.

Marshall, J. (2000). Alcohol and drug misuse in women. In D. Kohen (Ed.), *Women and mental health*. Philadelphia: Routledge.

McCauley, J., Kern, D. E., Kolodner, K., Dill, L., Schroeder, A. F., DeChant, H. K., et al. (1995). The "battering syndrome": Prevalence and clinical characteristics of domestic violence in primary care internal medicine practices. [Comment.] *Annals of Internal Medicine, 123*, 737–746.

McGrath, E., Keita, G., Strickland, B., & Russo, N. (1990). *Women and depression: Risk factors and treatment issues.* Washington, DC: American Psychological Association.

McKusker, J., Goldstein, R., Bigelow, C., & Zorn, M. (1995). Psychiatric status and HIV risk reduction among residential drug abuse treatment clients. *Addiction, 90*, 1377–1387.

Merikangas, K. R., Mehta, R. L., Molnar, B. E., Walters, E. E., Swendsen, J. D., Aguilar-Gaziola, S., et al. (1998). Comorbidity of substance use disorders with mood and anxiety disorders: Results of the International Consortium in Psychiatric Epidemiology. *Addictive Behaviors, 23*, 893–907.

Miethe, T. D., & Meier, R. F. (1994). *Crime and its social context: Toward an integrated theory of offenders, victims, and situations.* Albany, NY: State University of New York Press.

Miranda, J., Duan, N., Sherbourne, C., Schoenbaum, M., Lagomasino, I., Jackson-Triche, M., et al. (2003). Improving care for minorities: Can quality improvement interventions improve care and outcomes for depressed minorities? Results of a randomized, controlled trial. *Health Services Research, 38*, 613–630.

Mocroft, A., Gill, M. J., Davidson, W., & Phillips, A. N. (2000). Are there gender differences in starting protease inhibitors, HAART, and disease progression despite equal access to care? *Journal of Acquired Immune Deficiency Syndromes, 24*, 475–482.

Morrill, A. C., Kasten, L., Urato, M., & Larson, M. J. (2001). Abuse, addiction, and depression as pathways to sexual risk in women and men with a history of substance abuse. *Journal of Substance Abuse, 13*, 169–184.

Morrison, M. F., Petitto, J. M., Ten Have, T., Gettes, D. R., Chiappini, M. S., Weber, A. L., et al. (2002). Depressive and anxiety disorders in women with HIV infection. *American Journal of Psychiatry, 159*, 789–796.

Morris-Rush, J. K., Freda, M. C., & Bernstein, P. S. (2003). Screening for postpartum depression in an inner-city population. *American Journal of Obstetrics & Gynecology, 188*, 1217–1219.

Muntaner, C., Eaton, W. W., Diala, C., Kessler, R. C., & Sorlie, P. D. (1998). Social class, assets, organizational control and the prevalence of common groups of psychiatric disorders. *Social Science & Medicine, 47*, 2043–2053.

Murray, C., & Lopez, A. D. (1996). Evidence–based health policy–lessons from the global burden of disease study. *Science, 274*, 740–743.

National Institute of Mental Health. (1993). *The prevention of mental disorders: A national research agenda*. Bethesda, MD.

National Institute of Mental Health. (1995). *D/ART Campaign: Depression: What every woman should know* (95–3871). Bethesda, MD.

National Institute of Mental Health. (1999). *Depression: Treat it, defeat it*. Bethesda, MD.

National Institute on Drug Abuse. (1998). *Treatment of drug-dependent individuals with comorbid mental disorders*. Rockville, MD.

National Mental Health Association. (2000). *Depression and African Americans (factsheet)*. Alexandria, VA.

Nazroo, J. Y., Edwards, A. C., & Brown, G. W. (1998). Gender differences in the prevalence of depression: Artefact, alternative disorders, biology or roles? *Sociology of Health & Illness, 20*, 312–330.

Nolen-Hoeksema, S., & Girgus, J. S. (1994). The emergence of gender differences in depression during adolescence. *Psychological Bulletin, 115*, 424–443.

Nolen-Hoeksema, S., Parker, L. E., & Larson, J. (1994). Ruminative coping with depressed mood following loss. *Journal of Personality & Social Psychology, 67*, 92–104.

Norquist, G. S., Regier, D. A., Burke, J. D., Jr., & Regier, D. A. (1996). The epidemiology of psychiatric disorders and the de facto mental health care system. *Annual Review of Medicine, 47*, 473–479.

Norris, F. H., & Kaniasty, K. (1996). Received and perceived social support in times of stress: A test of the social support deterioration deterrence model. *Journal of Personality & Social Psychology, 71*, 498–511.

Nyamathi, A. M. (1991). Relationship of resources to emotional distress, somatic complaints, and high-risk behaviors in drug recovery and homeless minority women. *Research in Nursing & Health, 14*(4), 269–277.

O'Hara, M., & Swain, A. (1996). Rates and risk of postpartum depression-A meta-analysis. *International Review of Psychiatry, 8*, 37–54.

Olson, K., & Pavetti, L. (1996). *Personal and family challenges to the successful transition from welfare to work*. Washington, DC: Urban Institute.

Orr, S. T., Celentano, D. D., Santelli, J., & Burwell, L. (1994). Depressive symptoms and risk factors for HIV acquisition among black women attending urban health centers in Baltimore. *AIDS Education & Prevention, 6,* 230–236.

Oxman, T. E., Berkman, L. F., Kasl, S., Freeman, D. H., Jr., & Barrett, J. (1992). Social support and depressive symptoms in the elderly. *American Journal of Epidemiology, 135,* 356–368.

Parrott, A. C. (2000). Human research on MDMA (3,4-Methylenedioxymeth-amphetamine) neurotoxicity: Cognitive and behavioural indices of change. *Neuropsychobiology, 42,* 17–24.

Paterson, D. L., Swindells, S., Mohr, J., Brester, M., Vergis, E. N., Squier, C., et al. (2000). Adherence to protease inhibitor therapy and outcomes in patients with HIV infection. *Annals of Internal Medicine, 133,* 21–30.

Paykel, E. S. (1991). Depression in women. *British Journal of Psychiatry – Supplementum,* (10) 22–29.

Podorefsky, D. L., McDonald-Dowdell, M., & Beardslee, W. R. (2001). Adaptation of preventive interventions for a low-income, culturally diverse community. *Journal of the American Academy of Child & Adolescent Psychiatry, 40,* 879–886.

Popay, J., Bartley, M., & Owen, C. (1993). Gender inequalities in health: Social position, affective disorders and minor physical morbidity. *Social Science & Medicine, 36,* 21–32.

Prescott, C. A., Neale, M. C., Corey, L. A., & Kendler, K. S. (1997). Predictors of problem drinking and alcohol dependence in a population-based sample of female twins. *Journal of Studies on Alcohol, 58,* 167–181.

Rabkin, J. G., & Ferrando, S. (1997). A 'second life' agenda. Psychiatric research issues raised by protease inhibitor treatments for people with the human immunodeficiency virus or the acquired immunodeficiency syndrome. *Archives of General Psychiatry, 54,* 1049–1053.

Rabkin, J. G., Ferrando, S. J., Lin, S. H., Sewell, M., & McElhiney, M. (2000). Psychological effects of HAART: A 2-year study. *Psychosomatic Medicine, 62,* 413–422.

Rabkin, J. G., Ferrando, S. J., van Gorp, W., Rieppi, R., McElhiney, M., & Sewell, M. (2000). Relationships among apathy, depression, and cognitive impairment in HIV/AIDS. *Journal of Neuropsychiatry & Clinical Neurosciences, 12,* 451–457.

Radloff, L. (1977). The CES-D Scale: A self-report depression scale for research in the general population. *Applied Psychological Measurement, 1,* 385–401.

Regier, D., Farmer, M., & Rae, D. (1990). Comorbidity of mental disorders with alcohol and other drug abuse. Results from the Epidemiologic Catchment Area (ECA) Study. *Journal of the American Medical Association, 264,* 2511–2518.

Reichlin, S. (1993). Neuroendocrine-immune interactions. *New England Journal of Medicine, 329,* 1246–1253.

Ren, X. S., Amick, B. C., & Williams, D. R. (1999). Racial/ethnic disparities in health: The interplay between discrimination and socioeconomic status. *Ethnicity & Disease, 9,* 151–165.

Richman, J. M., Chapman, M. V., & Bowen, G. L. (1995). Recognizing the impact of marital discord and parental depression on children. A family-centered approach. *Pediatric Clinics of North America, 42,* 167–180.

Ritchey, F., La Gory, M., Fitzpatrick, K., & Mullis, J. (1990). A comparison of home-less, community-wide, and selected distressed samples on the CES-depression scale. *American Journal of Public Health, 80*, 1384–1386.

Ritter, C., Hobfoll, S. E., Lavin, J., Cameron, R. P., & Hulsizer, M. R. (2000). Stress, psychosocial resources, and depressive symptomatology during pregnancy in low-income, inner-city women. *Health Psychology, 19*, 576–585.

Roberts, A. C., Wechsberg, W. M., Zule, W., & Burroughs, A. R. (2003). Contextual factors and other correlates of sexual risk of HIV among African-American crack-abusing women. *Addictive Behaviors, 28*, 523–536.

Rosenbaum, M. (1981). *Women on heroin*. New Brunswick, NJ: Rutgers University Press.

Ross, C. E., & Huber, J. (1985). Hardship and depression. *Journal of Health & Social Behavior, 26*, 312–327.

Rupp, A., Gause, E., & Regier, D. (1998). Research policy implications of cost-of-illness studies for mental disorders. *British Journal of Psychiatry – Supplementum, 36*, 19–25.

Safren, S. A., Radomsky, A. S., Otto, M. W., & Salomon, E. (2002). Predictors of psychological well-being in a diverse sample of HIV-positive patients receiving highly active antiretroviral therapy. *Psychosomatics, 43*, 478–485.

Sampson, R. J., & Lauritsen, J. L. (1990). Deviant lifestyles, proximity to crime, and the offender-victim link in personal violence. *Journal of Research in Crime & Delinquency, 27*, 110–139.

Satcher, D. (1999). *Mental health: A report of the surgeon general*. Paper presented at the 92nd Annual NAACP Convention, New Orleans, LA.

Schilling, R. F., el-Bassel, N., & Gilbert, L. (1993). Predictors of changes in sexual behavior among women on methadone. *American Journal of Drug & Alcohol Abuse, 19*, 409–422.

Schreck, C. (1999). Criminal victimization and low self-control: An extension and test of a general theory of crime. *Justice Quarterly, 16*, 633–654.

Simpson, D. D., Knight, K., & Ray, S. (1993). Psychosocial correlates of AIDS-risk drug use and sexual behaviors. *AIDS Education & Prevention, 5*, 121–130.

Singer, M. (1994). AIDS and the health crisis of the U. S. urban poor: The perspective of critical medical anthropology. *Social Science & Medicine, 39*, 931–948.

Spraggins, R. (2003). *Women and men in the U.S.: March 2002. Population characteristics*. Washington, DC: U.S. Census Bureau.

Starace, F., Ammassari, A., Trotta, M. P., Murri, R., De Longis, P., Izzo, C., et al. (2002). Depression is a risk factor for suboptimal adherence to highly active antiretroviral therapy. *Journal of Acquired Immune Deficiency Syndromes, 31*, (Suppl 3), S136–139.

Starace, F., Bartoli, L., Aloisi, M. S., Antinori, A., Narciso, P., Ippolito, G., et al. (2002). Cognitive and affective disorders associated to HIV infection in the HAART era: Findings from the NeuroICONA study. Cognitive impairment and depression in HIV/AIDS. The NeuroICONA study. *Acta Psychiatrica Scandinavica, 106*, 20–26.

Stark, E., & Flintcraft, A. (1985). Woman-battering, child abuse and social heredity: What is the relationship? *Sociological Review Monograph, 31*, 147–171.

Sterk, C. (1999). *Fast lives: Women who use crack cocaine*. Philadelphia: Temple University Press.

Sterk, C. (2000). *Tricking and tripping: Prostitution in the era of AIDS.* Putnam Valley, NY: Crime Prevention Bureau, Detroit Police Dept. 178.

Sterk, C., Theall, K., & Elifson, K. (2003). Effectiveness of an HIV risk reduction intervention for African American women who use crack cocaine. *AIDS Education & Prevention, 15,* 15–32.

Sterk-Elifson, C., & Elifson, K. W. (1993). The social organization of crack cocaine use: The cycle in one type of base house. *Journal of Drug Issues, 23,* 429–441.

Stoppard, J. (2000). *Understanding depression: Feminist social constructionist approaches.* New York: Routledge.

Stoskopf, C. H., Kim, Y. K., & Glover, S. H. (2001). Dual diagnosis: HIV and mental illness, a population-based study. *Community Mental Health Journal, 37,* 469–479.

Thomas, J. L., Jones, G. N., Scarinci, I. C., Mehan, D. J., & Brantley, P. J. (2001). The utility of the CES-D as a depression screening measure among low-income women attending primary care clinics. The Center for Epidemiologic Studies-Depression. *International Journal of Psychiatry in Medicine, 31,* 25–40.

Tobin, K. E., & Latkin, C. A. (2003). The relationship between depressive symptoms and nonfatal overdose among a sample of drug users in Baltimore, Maryland. *Journal of Urban Health, 80,* 220–229.

Treisman, G. J., Angelino, A. F., & Hutton, H. E. (2001). Psychiatric issues in the management of patients with HIV infection. *Journal of the American Medical Association, 286,* 2857–2864.

Turner, B. J., Laine, C., Cosler, L., & Hauck, W. W. (2003). Relationship of gender, depression, and health care delivery with antiretroviral adherence in HIV-infected drug users. *Journal of General Internal Medicine, 18*(4), 248–257.

Turnock, B. (2004). *Public health: What it is and how it works* (3rd ed.). Boston: Jones and Bartlett.

Tweed, D. (1993). Depression-related impairment: estimating concurrent and lingering effects. *Psychological Medicine, 23,* 373–386.

U.S. Department of Health and Human Services. (2001). *Mental health: Culture, race, ethnicity – A supplement to mental health: A report of the Surgeon General.* Rockville, MD: U.S. Department of Health and Human Services, Substance Abuse and Mental Health Services Administration, Center for Mental Health Services.

Wakschlag, L. S., Pickett, K. E., Cook, E., Jr., Benowitz, N. L., & Leventhal, B. L. (2002). Maternal smoking during pregnancy and severe antisocial behavior in offspring: a review. *American Journal of Public Health, 92,* 966–974.

Walters, V. (1993). Stress, anxiety and depression: Women's accounts of their health problems. *Social Science & Medicine, 36,* 393–402.

Wandersman, A., & Nation, M. (1998). Urban neighborhoods and mental health. Psychological contributions to understanding toxicity, resilience, and interventions. *American Psychologist, 53,* 647–656.

Wang, P. S., Demler, O., & Kessler, R. C. (2002). Adequacy of treatment for serious mental illness in the United States. *American Journal of Public Health, 92,* 92–98.

Wang, P. S., Gilman, S. E., Guardino, M., Christiana, J. M., Morselli, P. L., Mickelson, K., & Kessler, R. C. (2000). Initiation of and adherence to treatment for mental disorders: Examination of patient advocate group members in 11 countries. *Medical Care, 38,* 926–936.

Wechsberg, W. M., Dennis, M. L., & Stevens, S. J. (1998). Cluster analysis of HIV intervention outcomes among substance-abusing women. *American Journal of Drug & Alcohol Abuse, 24*, 239–257.

Weiss, E. L., Longhurst, J. G., & Mazure, C. M. (1999). Childhood sexual abuse as a risk factor for depression in women: Psychosocial and neurobiological correlates. *American Journal of Psychiatry, 156*, 816–828.

Weissman, J. S., Campbell, E. G., Gokhale, M., & Blumenthal, D. (2001). Residents' preferences and preparation for caring for underserved populations. *Journal of Urban Health, 78*, 535–549.

Weissman, M. M., Bland, R. C., Canino, G. J., Faravelli, C., Greenwald, S., Hwu, H. G., et al. (1996). Cross-national epidemiology of major depression and bipolar disorder. *Journal of the American Medical Association, 276*(4), 293–299.

Wells, K. B., Sherbourne, C., Schoenbaum, M., Duan, N., Meredith, L., Unutzer, J., et al. (2000). Impact of disseminating quality improvement programs for depression in managed primary care: A randomized controlled trial. *Journal of the American Medical Association, 283*, 212–220. [Erratum appears in *JAMA, 283*, 3204].

Whiffen, V. E., & Clark, S. E. (1997). Does victimization account for sex differences in depressive symptoms? *British Journal of Clinical Psychology, 36*(Pt. 2), 185–193.

Whitley, B. E. (1985). Sex-role orientation and psychological well-being: Two meta-analyses. *Sex Roles, 12*, 207–225.

Williams, D. R. (1997). Race and health: Basic questions, emerging directions. [Comment.] *Annals of Epidemiology, 7*(5), 322–333.

Witteveen, E., & van Ameijden, E. J. C. (2002). Drug users and HIV-combination therapy (HAART): Factors which impede or facilitate adherence. *Substance Use & Misuse, 37*, 1905–1925.

World Health Organization. (2000). *Women's mental health: An evidence based review.* Geneva, Switzerland: Department of Mental Health and Substance Dependence, WHO.

PART V

SYSTEMS AND PROCESSES OF TREATMENT, PREVENTION, AND POLICY

17

Services and Treatment for Depression

International Perspectives and Implications for a Gender-Sensitive Approach

Shekhar Saxena and Pratap Sharan

INTRODUCTION

The Global Burden of Disease data for the year 2002 shows that depression is the fourth leading cause of disability in women worldwide and accounts for 5.66% of disability-adjusted life years (DALYs) in women, whereas it is the seventh leading cause of disability in men and accounts for 3.45% of DALYs (World Health Organization [WHO], 2003a). It is projected that by 2020 depression will become the second leading cause of disability in women (Murray & Lopez, 1996). Traditionally, mental health programs have paid little attention to the special or unique needs, of women. But, the heavy burden placed by depression on the mental health of women; the association of depression with different phases of women's life cycle and other gender-specific determinants; and gender differences in clinical presentation, course, response to treatment, service needs, and response of care givers makes a gender-sensitive approach in service development and management of depression an imperative. This chapter includes information from WHO's Project Atlas regarding the availability of treatment services in the world and provides suggestions regarding methods by which the gap between need for and receipt of effective services for depressed women can be reduced. It also reviews the types and efficacy of physical and psychosocial treatment of depression in women and whether such treatments vary by gender. Suggestions from a life cycle perspective on the clinical management of depression in women and issues regarding prevention, disability, and rehabilitation are highlighted. We conclude with recommendations regarding future research, policy, and planning toward a gender-sensitive approach to the management of depression in women around the world. The focus of the chapter is at the macro level and issues in relation to prevention of depression and rehabilitation at the individual level are not discussed.

PUBLIC HEALTH IMPORTANCE OF DEPRESSION IN WOMEN

Depression is a major public health problem in rich (i.e., developed) and poor (i.e., developing) countries and is especially common in women, in particular in women who are socially disadvantaged (Araya, Rojas, Fritsch, Acuna, & Lewis, 2001; Murray & Lopez, 1996; WHO, 2003a). In an international study on common mental disorders in the primary health care settings, though there was a more than 12-fold difference in current prevalence rates for depression for women (Nagasaki, 2.8%; Santiago, 36.8%) between centers, in most centers women had higher rates of current episodes of depression than men (Ustün & Sartorius, 1995). Similar findings were reported in other multicountry studies on depression (Bland, 1997; Brown, 1998).

The higher rates of depression among women are detected at midpuberty through adult life and decrease in older cohorts (Piccinelli & Wilkinson, 2000). Increased risk of women varies by diagnostic subtypes and is substantial for major depression, dysthymia, atypical depression, and seasonal winter depression, although it does not occur in bipolar disorder (but women predominate in rapid cycling and mixed state episodes). Depressed women report higher rates of comorbid psychiatric disorder such as anxiety, panic disorder, somatization, and eating disorder, which can complicate evaluation and predict a worse outcome (Ustün & Sartorius, 1995; Young, Scheftner, Fawcett, & Klerman, 1990). Although artifactual determinants such as threshold for caseness (women report more symptoms), measurement procedure (women report more somatic symptoms), effect of recall, course of illness (women may have higher rates of first onset depression and chronic depression), help seeking and illness behavior (women are more likely to report symptoms and seek medical help), and developmental pathway (females suffer from preexisting anxiety disorders and males experience more externalizing disorders) may enhance a female preponderance to some extent, gender differences in depressive disorders appear to be genuine (Piccinelli & Wilkinson, 2000).

At present, adverse experiences in childhood (Harris, 2001), depression and anxiety disorders in childhood and adolescence (Veijola et al., 1998), sociocultural roles (Wilhelm, Parker, & Hadzi-Pavlovic, 1997) with related adverse experiences to life events (related directly to defeat, humiliation, and entrapment [Craig, 1996]) and coping skills (e.g., rumination [Hänninen & Aro, 1996]) are believed to be involved in gender differences in development of depression. Genetic (McGuffin, Katz, Watkins & Rutherford, 1996) and biological factors and poor social support (cost of caring [Bebbington, 1996]) seem to have fewer effects in the emergence of gender differences (Piccinelli & Wilkinson, 2000). The effect of life events and difficulties may be even more salient in developing countries. In a Zimbabwean population of depressed women, life events were reported to be more severe than those described in British research. In Zimbabwe the population attributable risk of life events and difficulties in the causation of depression in women was 0.94 (Abas & Broadhead, 1997; Broadhead & Abas, 1998).

Rates and determinants of completed suicide among women have proved difficult to determine because of underestimation and unavailability of systematic data. In industrialized countries there is an excess of male-to-female deaths by suicide (Brockington, 2001). However, in the countries of South and East Asia, the ratio is reversed, especially among younger women. It is suggested that in these settings it is linked to the lower social position occupied by women and to restrictions on their educational opportunities and individual autonomy (Brockington, 2001; Ji, Kleinman, & Becker, 2001). This trend was also observed in Asian ethnic minorities in United Kingdom and was thought to be related to social and cultural stress (Bhugra, Baldwin, Desai, & Baldwin, 1999; Bhugra, Desai, & Jacob 1999). Reviews have suggested that although suicide in pregnancy is not common, it exists and is primarily associated with the intolerable predicaments of unwanted pregnancy, entrapment in situations of sexual or physical abuse, or poverty (Brockington, 2001; Frautschi, Cerulli, & Maine, 1994). Parasuicide (thoughts of suicide and attempts to self-harm) is up to 20 times more common than suicide (Brockington, 2001) and is more prevalent in women than men in most countries. It is associated with low education and socioeconomic status, but predominantly with childhood sexual and physical abuse and sexual and domestic violence (Brockington, 2001).

Gender differences in disability and burden due to depression and associated comorbidity is considerable (WHO, 2002a) and the differences in cost of providing services are being evaluated (Huskamp, Azzone, & Frank, 1998; Oss, Yennie, & Birch, 1988). The impact of depression on outcome and management of other disorders is also gender linked. For example, studies have shown that persistent depression in HIV-infected women (who are particularly vulnerable to experiencing depressive symptoms [Hader, Smith, Moore, & Holmberg, 2001]) is associated with significantly poorer survival in comparison to men. The worse outcome may be related to gender differences in adherence to antiretroviral therapy (Ickovics et al., 2001). Contact with mental health services for treatment of depression had a favorable association with adherence that was stronger in women than in men (Turner, Laine, Cosler, & Hauck, 2003).

GENDER-BASED APPROACH

Gender refers to the socially constructed roles, behaviors, activities, and attributes that a given society considers appropriate for men and women (WHO, 2003b). The distinct roles and behaviors of men and women in a given culture, dictated by that culture's gender norms and values, give rise to *gender differences.* Some gender differences do not favor either group. Gender norms and values, however, also give rise to power-based *gen-*

der inequalities – that is, differences between men and women that systematically empower one group, often to the detriment of the other, for example, women around the world are disadvantaged in socioeconomic terms. Dijkstra and Hanmer (1997) developed a measure of socioeconomic gender inequality within countries (Gender-Related Development Index – GDI). The GDI is a composite indicator that uses the same variables as the Human Development Index (HDI), disaggregated by gender. The HDI is a summary composite index that measures a country's average achievements in three basic aspects of human development: longevity, knowledge, and a decent standard of living. Longevity is measured by life expectancy at birth; knowledge is measured by a combination of the adult literacy rate and the combined primary, secondary, and tertiary gross enrollment ratio; and standard of living by gross domestic product per capita (purchasing power parity US$). In the 146 countries for which the GDI was calculated (United Nations Development Program [UNDP], 1997), none had a GDI value higher than the HDI value. Some 41 countries had a GDI value of more than 0.8 but almost as many other countries (39) had a GDI value of less than 0.5. Some developing countries outperformed richer industrialized ones in achieving gender empowerment in political, economic, and professional activities. Women had lower incomes relative to men and were overrepresented among those living in absolute poverty, accounting for around 70% of the world's poor (UNDP, 1997).

Both gender differences and gender inequalities can give rise to inequities between men and women in health status and access to health care. Across continents and cultures, established gender norms and values mean that women typically control less power and fewer resources than men. Not surprisingly, this often gives men an advantage – in the economic, political, and educational arenas, but also with regard to health and health care. Gender roles and unequal gender relations interact with other social and economic variables, resulting in different and sometimes inequitable patterns of exposure to health risk, and in differential access to and utilization of health information, care, and services. For example, a married woman may contract HIV because societal standards encourage her husband's promiscuity while simultaneously preventing her from insisting on condom use, or a woman may not be able to access health services because norms in her community prevent her from traveling alone to a clinic (WHO, 2003b).

HEALTH SERVICES

Mental health services are one of the means by which effective interventions for mental health are delivered. The way these services are organized has an important bearing on their effectiveness and ultimately on whether they meet the aims and objectives of a mental health policy.

Gender Mainstreaming

If health care systems are to respond adequately to problems caused by gender inequality, it is not enough to add in a gender component late in a given project's development. Research, interventions, health system reforms, health education, health outreach, and health policies and programs must consider gender from their inception. Gender is not something that can be consigned to a single office or clinic, because no single office or clinic could possibly involve itself in all phases of an organization's activities. *All* health professionals must have knowledge and awareness of the ways gender affects health, so that they may address gender issues wherever appropriate and thus make their work more effective. The process of creating this knowledge and awareness of – and responsibility for – gender among all health professionals is called *gender mainstreaming*. Gender mainstreaming can contribute to increasing the coverage, effectiveness, efficiency, and ultimately the impact of health interventions for both women and men, while at the same time contributing to achievement of social justice (WHO, 2003b).

A quick look at the situation of mental health services in the world would help in understanding the enormity of the task of delivering gender appropriate mental health services to women.

Project Atlas: Mapping Mental Health Resources around the World

Project Atlas (WHO, 2001a, 2001b, 2002b) sought to map mental health resources around the world with the aim of providing policy makers and the WHO with data that could help in designing programs and prioritizing needs. Initiated in 2000, Project Atlas collected basic information on mental health resources from WHO's 191 member states. The initial data collection was conducted from October 2000 to March 2001. Member states were requested to provide information on mental health policy, legislation, financing and budget, mental health services, human services, information and data collection system, care for special populations, and drugs. The Atlas results highlighted the fact that resources and services for mental and behavioral disorders are disproportionately low compared to the burden caused by these disorders in both developing and developed countries.

National policies, programs, and legislation on mental health are basic requirements for mental health care in any country. Of the countries on which information is available, only 59.5% have a mental health policy and 69.4% have a substance abuse policy. About half of the African and Western Pacific region countries do not have a mental health or substance abuse policy. Sixty-nine percent of the countries have a national mental health program, with the majority (more than four fifths) having initiated it in the past decade. Only 74.7% of countries out of the 170 countries that provided information have mental health legislation. In about 15% of these countries, the legislation dates back to a period before 1960, when most of the currently used treatment methods were unavailable. A therapeutic drug policy or essential list of medicines is present in 88.4% of the countries, with about half having been formulated in the past 10 years, making it unlikely that the benefits of the policy have percolated down to the consumers.

Data from Atlas confirm that mental health budgets are a very small part of the overall budget for health. Only 72% of countries reported having a mental health budget specification in their total health budget. Details about how much they devoted to mental health was received from 91 countries. Of these, 37.4% of the countries allocate less than 1% of their health budgets for mental health. Although half of the countries in the European region allocated more than 5% of their total health budgets for mental health, more than four fifths of the countries in the African region spent less than 1% of their health budgets on mental health. A total of 171 countries provided information on the most important sources of mental health financing; tax-based financing is the primary financing source in 60%, followed by insurance (social, 19%, and private, 2%) in 21% of the countries. However, out-of-pocket financing is the primary method in 16% and grants from outside sources in 3% of countries. Of the 179 countries from which information on disability benefits is available, about a quarter do not provide any state or public disability benefit for mental illness. Many others provide very limited assistance (e.g., a small monetary allowance or pension benefits for government employees).

Although it is agreed that most mental disorders are best managed at the primary care level, this has been difficult to achieve in practice. Eighty-six percent of countries state that mental health services are available at the primary health care level, but 42% of countries have no regular program to train primary care personnel in mental health care. About 20% of countries do not have the three most common drugs – phenytoin (antiepileptic), amitriptylline (antidepressant), and chlorpromazine (antipsychotic) – available in their primary care services. Research and experience shows that community-based care achieves better treatment results than institutional care for chronic mental disorders such as schizophrenia and affective disorders. However, these facilities are not available in 37% of the countries. Even in countries that have community care, coverage is incomplete. Large mental institutions are no longer considered appropriate for the treatment and rehabilitation of people with mental disorders. Yet in most countries, mental health care continues to be confined to such institutions. More than 65% of all beds for patients with mental disorders are still in large mental hospitals. On the issue of equity, programs for the elderly and children are present in only 48% and 60% countries, respectively. Programs for refugees, minorities, and indigenous populations are not present in most of the countries.

Globally, about 70% of the world has access to less than one psychiatrist per 100,000 people, and about 44% to less than one psychiatric nurse per 100,000 people. Again, there are sharp regional variations. Although the countries in the African region of WHO have only about 1,200 psychiatrists and 12,000 psychiatric nurses for a population of around 620 million, the European region has more than 77,000 psychiatrists and 285,000 nurses for a population of around 841 million. The median number of psychiatrists for the group of lower income countries is 0.06 per 100,000, whereas the group of high income countries has 9 per 100,000, revealing a wide disparity between countries. Though mental hospitals with large number of beds are not recommended for mental health care due to the limitations of institutional care, a certain number of beds in general hospitals for emergency care is considered essential. However, there is a wide variation in general hospital beds available for mental health care. The median number of hospital beds for the world population is 1.6 per 10,000 people. Regional medians vary from 0.33 beds per 10,000 people in the Southeast Asia region to 8.7 beds per 10,000 people in the European region. The median number of beds per 10,000 population is 0.24 in the low-income-group countries and 8.7 in the high-income-group countries. More than 3.8 billion people have access to less than 1 bed per 10,000 people.

More than 27% of the countries have no system of reporting mental health data in their annual health report. Countries that have a system often lack the details that are important for planning future interventions. For example, 44% of the countries have no facilities for collection of epidemiological or service data at the national level. Poor information systems also mean that gender disaggregated data are often not available.

Very few countries have an optimal mix of services. Even within countries there are usually significant disparities between different regions. In developed countries, the process of deinstitutionalization during the past three decades has not been accompanied by sufficient provision of community-based and primary care services. In many developing countries there is gross underprovision of resources, personnel, and services. However, other countries such as Chile and Iran have made mental health services more widely available by integrating them into primary care services or by making mental health services available at general hospitals. In some countries there are good examples of intersectoral collaboration between informal care services (see Table 17.1), nongovernmental organizations (see Table 17.1), academic institutions, public sector health services, and users that have led to the development of community-based services (WHO, 2003c). Even within the resource constraints of health services in most countries, significant improvements in delivery are possible by redirecting resources toward services that are less expensive, have reasonably good outcomes, and benefit increased proportions of populations. For example, in developing countries the integration of mental health

TABLE 17.1. *Examples of Intersectoral Collaboration in Mental Health Service Delivery in Low- and Middle-Income Countries*

Sector	Country	Description
Community	Zimbabwe	The Harare City Health Department and the University of Zimbabwe Medical School are collaborating in a research project, in which members of the community and primary care nurses are looking for ways of treating depressed women. Local terminology for depression and ideas on treatment were established by interviewing traditional healers and key community figures including schoolteachers, police officers, church officials, organizers of women's cooperatives and policy makers. The group recommended that (1) there should be a private room in primary care clinics for counselling on emotional problems; (2) a directory should be created to improve communication between helping agencies; (3) traditional healers, church leaders, teachers, and the media should be used to provide education for living; and (4) the detection and treatment of depression in primary care clinics should be improved (Abas, 1995). The program of identification and treatment at the primary care level was integrated with a preexisting initiative, viz. a maternal and child health program that involved cooperation between the mental health sector and the general health sector. A more general mental health package was also integrated into the primary care system in Harare on the basis of a similar cooperative approach.
Nongovernmental organizations	India	Nongovernmental organizations have gained considerable experience in health care in India, notably in reproductive and child health services and research. In recent years a growing number of nongovernmental organizations have begun to develop innovative programs for mental health care. The Sangath Society is located in the state of Goa on the west coast of India. It was founded in 1996 by a team of health professionals working in the field of child and adolescent development and women's mental health. The society integrates research, training, and service delivery and collaborates actively with other sectors of the health and social welfare systems (Patel & Thara, 2003).

services into established physical health and social programs provides a feasible and affordable way of implementing mental health programs. Thus depression in women can be tackled within programs concerned with reproductive health, domestic violence, and HIV/AIDS (WHO, 2003c).

Gender Issues Related to Services

Studies reveal gender differences in perceptions of distress and in patterns of health-care seeking among those suffering from mental health problems. Many studies from industrialized countries show that women report higher levels of distress than men, women were more likely to perceive that they have an emotional problem than men who had a similar level of symptoms, and women are consistently more likely to use primary or general health care and outpatient mental health services than are men (Burns, Cain, & Husaini, 2001; Moller-Leimkuhler, 2002; Scherbourne, Dwight-Johnson, & Klap, 2001). Men may seek care at a later stage after the onset of symptoms or delay until symptoms become severe. However, once men recognize that they had a problem, they were as likely as women to use mental health services (Katz, Kessler, Frank, Leaf, & Lin, 1997). It is hypothesized that the low detection and referral rates for mental disorders in primary care and segregation of mental health care from general health care may affect women more than they affect men, because more women present to primary and general health rather than specific mental health facilities when they have a mental health problem (Huskamp et al., 1998). Gender-related experiences and stereotypes on the part of the physician may also influence service delivery, for example, higher rates of diagnosis of depression in and prescription of psychotropic drugs to women (WHO, 2002a). However, gender bias in assessment and treatment may not be universal (Olfson, Zarin, Mittman, & McIntyre, 2001).

Improving Services for Treatment of Depression in Women

Approximately 60 to 70% of patients enrolled in research programs that attempt to improve depression outcome in primary care are women, so these studies can provide information that is relevant to management of depression in women in the setting. Screening for depression alone increased accuracy of diagnosis but had little effect on outcomes (Katon & Gonzalez, 1994). Training physicians regarding guidelines of care also did not lead to improvement in outcomes (Katon & Ludman, 2003). Integration of mental health professionals into primary care improved symptomatic and functional outcomes through the provision of pharmacotherapy and various types of psychotherapies: interpersonal therapy (IPT; Schulberg et al., 1996), problem-solving therapy (Mynors-Wallis, Gath, Day, & Baker, 2000), cognitive behavioral therapy, and nondirective counseling based on Rogerian methods (Ward et al., 2000). Recent studies using cognitive behavior and motivational interviewing principles have proven successful in relapse prevention (Katon, Von Korff, Lin, & Simon, 2001) in a primary care setting.

A briefer model that has been highly successful in improving outcomes and cost-effectiveness of care for major depression is the collaborative care model with mental health specialists (Katon et al., 1996). Because mental health professionals are not available in many primary care clinics, this research has focused on utilizing alternative health professionals such as nurses to enhance self-management, including adherence to medication. Most of these studies showed improvement in depression outcomes in collaborative care patients in comparison to control patients who received usual care (Katon & Ludman, 2003). Such nurse (and case) manager interventions may be very useful in improving outcomes in underserved population that are less able to afford speciality mental health services. A study using this model showed that improvement in outcome was greater among minority women (Miranda et al., 2003).

The most recent innovation in the field is the stepped care model. Katon et al. (1999) randomized primary care patients with depression who did not improve after 8 weeks of acute phase treatment by primary care physicians to collaborative care with a psychiatrist (stepped care program) or to continued usual care. The stepped care program improved clinical and functional outcomes. A similar model was employed very successfully for management of depression of disadvantaged women in a primary care setting in Chile. Primary care clinics are the main source of care for the poor in Chile. These clinics are underfunded and understaffed, particularly with respect to physicians. Physicians spend less than 2 years in primary care posts, whereas other primary care workers stay much longer. Araya et al. (2003) designed a multicomponent stepped care program to improve treatment practices for depression and efficiency with which resources are used. The program was led by a trained nonmedical health worker (12 hours of training and 8 hours of supervision) and it included psychoeducational group intervention (9 sessions involving education, behavioral activation, problem-solving techniques, relapse prevention techniques), structured and systematic follow-up, and medication. A doctor was involved only if medication was needed for patients with severe depression. Despite few resources and marked deprivation, poor women with major depression responded well to a structured stepped care program. The success of the program led to its introduction across Chile. This study provides a model of how developed countries can learn much from developing countries in rationalization of resources when working with deprived populations under tight budgetary conditions. Use of the stepped care principle would allow primary care physicians to make the initial diagnosis. Patients with mild, subclinical disorders may be followed by primary care personnel with supportive counseling and monitoring (Step 1). Patients with major depression or those whose depression persists may be managed by a primary care physician with medication or psychotherapy with the help of mental health professionals (Step 2). Step 3 would involve referral to the specialist mental health setting.

Improvement of services for depression in women would also require evidence-based treatment strategies for individual patients and their efficient utilization in practice. The next section deals with the theoretical and empirical evidence in support of a gendered approach to treatment.

GENDER DIFFERENCES IN PHARMACOTHERAPY

Pharmacodynamics

Gender differences have been examined in brain structure and function, including neuroendocrine and neurotransmitter systems, genetic transmission, and reproductive function (Kornstein, 1997). Evidence from genetic studies suggests that major depression is equally heritable in men and women. However, genetic factors might indirectly increase vulnerability to depression in one gender through temperamental features (Kendler, Gardner, & Prescott, 1998; McGuffin et al., 1996). Both estrogen and progesterone appear to influence mood and behavior by affecting neurotransmitter synthesis, release, reuptake, and enzymatic inactivation (McEwen, 1991). Estrogen has been shown to modulate serotonin and norepinephrine concentrations and decrease monoamine oxidase concentrations, which in turn decreases the degradation of norepinephrine (Pearlstein, 1995; Sherwin & Suranyi-Cadotte, 1990). Progesterone has been shown to increase monoamine oxidase concentrations, which would promote the enzymatic degradation of neurotransmitters in the synaptic cleft (Luine & Rhodes, 1983).

The hypothalamic-pituitary-adrenal axis seems to be more reactive to stress in women than in men possibly due to the modulating role of gonadal hormones (Osuch et al., 2001). Age may have a differential effect on plasma cortisol level in men and women, with correlation in age seen in women of reproductive age but not in men or postmenopausal women in one study (Halbreich & Lumley, 1993), but not in another (Young, 1995). Cerebrospinal fluid TRH (thyrotropin releasing hormone) was lower in women compared to men in a study. This finding is important because TRH has a potential role as an endogenous antidepressant (Frye et al., 1999). One explanation for gender differences in seasonal affective disorder proposes that decreased concentrations of estradiol in the ventromedial hypothalamus shorten the circadian period, lengthen the sleep phase, advance sleep onset, and consolidate sleep, resulting in atypical depressive symptoms such as hypersomnia and increased daytime sleepiness (Partonen, 1995).

Pharmacokinetics

Few studies have examined gender differences in antidepressant metabolism and treatment response in humans because women were excluded from early clinical trials, such as Phase 1 pharmacokinetic studies (Dawkins & Potter, 1991). Rates of absorption, distribution, metabolism, and elimination of a drug in the body can be affected by gender-related factors such as lean body mass, hormone concentrations, gastric absorption and emptying, metabolic enzymes, protein binding, and cerebral blood flow (Bies, Bigos, & Pollock, 2003; Hamilton & Jensvold, 1995). Women have a slower gastrointestinal transit time than usual when their progesterone concentrations are increased during the luteal phase and during pregnancy (Hudson, Roehrkasse, & Wald, 1989). The predominant effect of these changes is to decrease drug levels premenstrually as was demonstrated with antidepressants such as desimipramine, trazadone, and nortryptiline (Kimmel, Gonsalves, Young, & Gidwani, 1992). Gender differences in pharmacokinetics of psychotropic medications can also be due to other factors, for example, smoking and alcohol intake, that are more common among men (Sutfin, Perini, & Molan, 1988).

Pregnancy and oral contraceptive use might affect the disposition and antidepressant dose requirements of women. Wisner, Perel, and Wheeler (1993) found that the doses of nortriptyline, clomipramine, and imipramine required to achieve remission of symptoms and therapeutic drug serum concentrations increased during the second half of pregnancy to 1.6 times the mean dose required when the patients were not pregnant. The progressive physiologic changes occurring in women during pregnancy that can alter the pharmacokinetics of various medications include a decrease in protein binding capacity, enhanced hepatic metabolism, hemodynamic changes such as an increase in cardiac output and renal plasma flow, and progesterone-induced decrease in gastrointestinal motility (Everson, 1992; Guay, Grenier, & Vann, 1998). Abernethy, Greenblatt, and Shader (1984) examined women taking long-term, low-dose oral contraceptives and reported that oral contraceptives decrease the hepatic metabolism of imipramine because of changes in hepatic blood flow. These data suggest that a dosage reduction to about two thirds should be done to achieve similar imipramine steady-state concentrations during oral dosing.

Several studies have examined the pharmacokinetic properties of antidepressants in men and women. Preskorn and Mac (1985) reported that women older than 50 years have higher plasma concentrations of amitriptyline than age-matched men and Gex-Fabry, Balant-Gorgia, Balant, and Garrone (1990) found that women have significantly lower hydroxylation clearance of clomipramine. Dahl, Bertilsson, and Nordin (1996) reported that female extensive metabolizers have significantly higher nortriptyline plasma concentrations than do male extensive metabolizers; however, gender did not influence the steady-state plasma concentrations of its main active metabolite 10-hydroxynortriptyline (10-OH-NT). Greenblatt et al. (1987) reported that the volume of distribution of Trazodone was larger in women than in men, probably due to higher relative amounts of adipose tissue in women and Barbhaiya, Buch, and

Greene (1996) observed that the concentrations of nefazodone and its metabolite are approximately 50% greater in elderly women than in elderly men, young women, or young men. Ronfeld, Tremaine, and Wilner (1997) found that the terminal elimination half-life of sertraline was comparable in young women and elderly men and women, but shorter in young men. However, gender did not significantly alter the disposition or tolerance of venlafaxine and its metabolite following single and multiple dosing (Klamerus, Parker, Rudolph, Derivan, & Chiang, 1996).

Most of the studies discussed above had small sample sizes. However, clinicians should still be aware that women, especially the elderly, may require a reduction in antidepressant dose.

Efficacy

A meta-analysis of 35 imipramine studies showed a modest but statistically significant difference favoring tricyclic response in men compared to women (Quitkin et al., 2002). In an earlier study (Raskin, Schulterbrandt, Reatig, Crook, & Odle, 1974) of gender differences, younger women responded preferentially to treatment with monoamine oxidase inhibitors. A randomized, double-blind study (Kornstein et al., 2000) showed that women have a more favorable response to sertraline than imipramine, whereas the opposite was true for men. Premenopausal women responded similarly to sertraline and imipramine. Martenyi, Dossenbach, Mraz, and Metcalfe (2000) also found a gender–age treatment type interaction in a 6-week, double-blind study that compared response to fluoxetine and maprotiline. Similar findings were reported by Gijsbers (2002) and Gijsbers, Dekker, Peen, & De Jonghe (2002). These findings may reflect differential efficacy of selective serotonin reuptake inhibitors (SSRIs) in treating different subtypes of depression (atypical versus melancholic features) presented by women and men (Pande, Birkett, Fechner-Bates, Haskett, & Greden, 1996) or differential effects of gonadal hormones on the distribution and response to these antidepressants (Halbreich et al., 1995).

Thase, Entsuah, and Rudolph (2001) pooled original patient data from eight randomized, double-blind studies comparing venlafaxine and SSRIs (fluoxetine, paroxetine, or fluvoxamine) in the treatment of major depression. In the overall patient population, significantly more venlafaxine-treated patients achieved remission. In depression, full remission is defined as the virtual elimination of symptoms. Recovery represents the combination of sustained symptom remission and return to full functioning. Younger (less than 50 years old) women responded somewhat better to SSRIs compared with older women (more than 50 years old) (Sloan & Kornstein, 2003). The advantage for venlafaxine compared with SSRIs was 2.5 times greater among the older women than the younger women. These findings suggests that the noradrenergic effects of venlafaxine might convey added benefit for depressed older women.

There is no consensus about gender differences in antidepressant efficacy. Results from a large pooled analysis by Quitkin et al. (2002) showed
no overall gender differences between men and women in response to tricyclics and fluoxetine. However, when stratified by age, results showed
that older women responded significantly better than younger women to
tricyclics, a finding consistent with previous research. Lewis-Hall, Wilson,
Tepner, and Koke (1997) had noted earlier that fluoxetine was better toler-
ated but no more effective than tricyclics in depressed women.

Recent studies suggest that estrogen alone may be efficacious in the
treatment of perimenopausal depression (Halbreich & Kahn, 2001; Soares,
Almeida, Joffe, & Cohen, 2001). It may also be useful as an adjunct to
SSRIs in treatment-resistant depression (Halbreich & Kahn, 2001; Sharan
& Saxena, 1998). Elderly depressed women patients had a significantly
greater improvement on combination of estrogen replacement therapy
(ERT) and fluoxetine than with fluoxetine alone or ERT with placebo
(Schneider et al., 1997). Similar findings were reported for sertraline
(Schneider, Small, & Clary, 2001). In contrast to these results, a study
examining estrogen as an adjunct therapy to fluoxetine showed no additional benefit (Amsterdam et al., 1999). Sloan and Kornstein (2003) also
report that hormone replacement therapy augmentation was not needed
for venlafaxine. Given the recent Women's Health Initiative study (2002)
reporting increased risks of breast cancer, myocardial infection, and stroke
in women treated with hormone replacement therapy, antidepressants
should remain the treatment of first choice in depressed peri- and postmenopausal women.

At this point it is important to remember that efficacy (the potential of a
treatment) should be distinguished from effectiveness (the results obtained
under clinical conditions). Studies of effectiveness or naturalistic studies
show poorer results than efficacy studies in all areas of health care. This is
true both for pharmacological and nonpharmacological therapies.

Adverse Events

There are few studies examining the incidence and prevalence of adverse
events of antidepressants in males and females. Kornstein et al. (2000)
found that women reported more nausea and men more sexual dysfunction
and urinary frequency with both imipramine and sertraline. Piazza et al.
(1997) reported that with SSRI treatment, sexual functioning in women
improved significantly, whereas in men it worsened significantly. They
suggested that sexual adverse events in women taking SSRIs may be less
common due to a slower onset in their appearance, failure by women
to spontaneously report them, or the fact that the sexual adverse events
caused by SSRIs are overshadowed by the positive effects on sexual functioning that are associated with treating depression. Shen and Hsu (1995)

retrospectively reviewed clinic records of SSRI-treated patients and found that SSRI-associated female sexual dysfunction occurs at a higher rate than was previously thought, and they suggested that buproprion might be an appropriate antidepressant in female patients who develop sexual adverse events secondary to SSRI treatment (Walker et al., 1993). Amsterdam, Garcia-Espana, Goodman, Hooper, and Hornig-Rohan (1997) found that 39% of women patients treated with an SSRI or venlafaxine reported some degree of breast engorgement, which was significantly more common in patients treated with SSRIs. Unfortunately, gender differences with regard to other adverse events such as weight gain and sedation have not been investigated.

PSYCHOSOCIAL THERAPIES FOR MANAGEMENT
OF DEPRESSION IN WOMEN

There is little doubt that psychotherapies such as interpersonal psychotherapy, problem-solving, and cognitive-behavioral therapy (CBT) are effective in the treatment of depression. Improvement of depression was also noted in the single randomized controlled trials that have tested other psychotherapies – the behavioral activation component of CBT without cognitive restructuring and couple therapy for people with depression who live with a critical partner (Marks, 2002). Psychosocial therapies have also been tried specifically in samples of women with depression and found to be efficacious. Zust (2000) tried out a 20-week cognitive therapy intervention in rural battered women and found it a promising candidate for randomized trials. Frank et al. (2000) found the sequential strategy of offering IPT to women with recurrent unipolar depression, and in the absence of remission, adding antidepressants to be significantly more effective in bringing about remission in comparison to the strategy of initiating IPT and pharmacotherapy at the outset of treatment. The high remission rates achieved through the sequential strategy made it particularly attractive for women in childbearing years. Childhood maltreatment and adult personality pathology did not predict treatment efficacy of IPT or combination therapy (IPT with SSRI), suggesting that the presence of these were not a hindrance for IPT.

Koerner, Prince, and Jacobson (1994) have suggested that integrated behavior–couple therapy may enhance the treatment and prevention of depression in women. Volunteer befriending had a significant effect on remission of symptoms after 1 year in women with chronic depression (Harris, Brown, & Robinson, 1999). Murray, Cooper, Wilson, and Romaniuk (2003) reported that nondirective counseling and therapies with cognitive behavioral and dynamic psychotherapy components were equally efficacious at relieving depression in women but had lesser influence on mother–child interaction. Similar findings had emerged in a study by

Seeley, Murray, and Cooper (1996), but not in an earlier study by a Canadian group (Fleming, Klein, & Corter, 1992).

Gender differences in response to combined treatment with medication and psychotherapy have been noted. Frank, Carpenter, and Kupfer (1988) found that women were slower than men to respond to treatment with imipramine and IPT. Thase et al. (1997) pointed out that this was particularly true for younger women. On the other hand, addition of short-term psychodynamic psychotherapy was needed to improve response to SSRIs in depressed men (Gijsbers et al., 2002). These findings probably reflect the pattern of gender differences in response to Tricyclic antidepressants (TCAs) and SSRIs. Sherpa (2001) did not find any impact of gender on treatment by Davanloo's partial trial therapy in depressed subjects.

TREATMENT GUIDELINES

A variety of models for the treatment of depression exist, for example, those developed by Agency for Health Care Policy and Research (AHCPR) for treatment of depression in primary health care (AHCPR, 1993) and American Psychiatric Association (APA, 2000a); however, they do not provide detailed gender-specific guidance. An expert panel (Altshuler et al., 2001) recommended that for women with severe depression, the first line of treatment should be antidepressants combined with other modalities (generally psychotherapy). For initial treatment of milder symptoms, the expert group either gave equal endorsement to other treatment modalities (e.g., hormone replacement in premenopause) or preferred psychotherapy over medication (during conception, pregnancy, or lactation). In milder cases, however, antidepressants were recommended as at least second-line options. Among antidepressants, SSRIs as a class were recommended as a first-line treatment in all situations and TCAs were highly rated alternatives to SSRIs in pregnancy and lactation.

CLINICAL MANAGEMENT OF DEPRESSION IN WOMEN

The diagnostic criteria for major depression, as established in the *Diagnostic and Statistical Manual of Mental Disorders* (*DSM-IV*), are the same for women and men (APA, 2000b). However, the presentation and course of depression are sometimes different in women (Kornstein, 1997; Seeman, 1997). Compared with men, women experience seasonal depression (Leibenluft, Harding, & Rosenthal, 1995) and symptoms of atypical depression (i.e., hypersomnia, hyperphagia, carbohydrate craving, weight gain, a heavy feeling in the arms and legs, evening mood exacerbations, and initial insomnia) (APA, 2000b) more often. In addition, women more frequently have symptoms of anxiety, panic, phobia, eating disorders, and dependent personality. Women also have a higher incidence of hypothyroidism, a condition that can cause depression (Pajer, 1995). Thus, it is important to screen depressed female patients for hypothyroidism.

Women, especially those younger than 30 years of age, more often attempt suicide, whereas men more often complete the act of self-destruction possibly because women frequently choose less lethal methods such as self-poisoning (AHCPR, 1993). It is thus prudent to prescribe antidepressants (especially those with potential lethality in overdose, such as a tricyclic agents) for a short duration and to enlist the support of family members or friends to monitor intake of the prescribed antidepressant.

The diagnostic assessment should include a detailed inquiry regarding reproductive life history, menstrual cycle, menopause, birth control, and abortions. History of experiences of sexual and physical abuse, post-traumatic stress disorder, and treatment, if any, should be obtained (APA, 2000a; Pajer, 1995). Dysthymia and major depression commonly display premenstrual exacerbation (Pearlstein, 1995), therefore, clinical interviews are recommended during the follicular as well as luteal phase to document functional impairment and severity of symptoms at different times of the menstrual cycle and to clarify the diagnosis (Pearlstein, 1995).

Psychosocial and pharmacologic treatments may be considered. Psychosocial therapies should address issues that particularly affect women, such as competing roles and conflicts (McGrath, 1990). Commonly used treatments include psychotherapy to correct interpersonal conflicts and to help women develop interpersonal skills, cognitive-behavioral therapy to correct negative thinking and associated behavior, and couples therapy to reduce marital conflicts. In patients with mild to moderate depression, psychosocial therapies may be used alone for a limited period, or they may be used in conjunction with antidepressant medication (Bhatia & Bhatia, 1999).

Because of many biologic differences, antidepressant plasma concentrations may be higher in women. Thus, female patients with depression may require lower dosages of antidepressants than their male counterparts. Women may also experience drug side effects more frequently. Although women frequently experience sexual side effects, they generally do not report these effects unless specifically asked (Olfson et al., 2001). In monitoring adverse events, clinicians should be aware that gender-related roles may differentially affect response to certain drug-adverse events. For example, because the societal emphasis on thinness is greater for women, they may not be willing to tolerate antidepressant weight gain (Hamilton & Jensvold, 1995).

Patient preferences should be taken into account while deciding on treatment strategy. Women were less likely to have discussed their treatment preferences with their psychiatrist (Olfson et al., 2001). Hood, Egan, Gridley, and Brew (1999) found that women attending primary care clinics in Australia expressed a personal responsibility for both the occurrence and management of depression and felt that psychosocial interventions were critical for getting well. Medications were viewed positively but as a means of relieving the extremes of symptoms of the illness and thereby providing space for psychosocial interventions. On the other hand, Patel et al. (2003) described a strong expectation of receiving physical treatment in a sample comprised mainly of women, in their study on common mental disorders in a general medical setting in Goa, India.

Premenstrual Dysphoric Disorder

Between 3 and 5% of women meet the diagnostic criteria for premenstrual dysphoric disorder (PMDD), which presents with depression, anxiety, and cognitive and physical symptoms that do not represent the exacerbation of preexisting anxiety, depression, or personality disorder. Many investigators have concluded that women with this disorder have an abnormal response to normal gonadal steroids (Schmidt, Nieman, Danaceau, Adams, & Rubinow, 1998). PMDD is a severely distressing and debilitating condition that requires treatment. Nonpharmacologic treatments such as aerobic exercise, caffeine restriction, complex carbohydrate consumption, and moderation of alcohol intake have not been consistently beneficial in alleviating the symptoms of the disorder (Pearlstein, 1996). Pharmacologic treatments for PMDD have included progesterone, antidepressants, and antianxiety drugs. Serotonergic and tricyclic antidepressants have been beneficial, but side effects limit the use of tricyclic drugs (Ensom, 2000; Gelenberg, 1998). The results achieved with benzodiazepines, especially alprazolam, have been mixed (Harrison, Endicott, & Nee, 1990; Schmidt, Grover, & Rubinow, 1993). Progesterone therapy is relatively ineffective (Freeman, Rickels, Sondheimer, & Polansky, 1990). Gonadotropin-releasing hormone agonists are effective (Mortola, Girton, & Fisher, 1991) but they should be used only in patients who are resistant to other forms of therapy because they can cause menopausal symptoms. The use of calcium supplementation to reduce symptoms has attracted interest (Bhatia & Bhatia, 1999). If simpler, nonpharmacologic options are ineffective, selective SSRIs may be used to treat patients with PMDD (Gelenberg, 1998).

Depression during Pregnancy

Factors such as a history of depression or PMDD, younger age, limited social support, living alone, greater number of children, marital conflict, and ambivalence about pregnancy increase the risk of depression during pregnancy and the postpartum period (Altshuler, Hendrick, & Cohen, 1998). The risks of treatment should be compared with the risks of not treating depression, which may include suicide, poor maternal and fetal nutrition, an adverse neonatal obstetric outcome, continuation of depression into the postpartum period, impaired mother–child bonding, marital disharmony and relationship with the spouse, and treatment resistance (APA, 2000a; Murray et al., 2003).

Interpersonal and cognitive-behavioral therapies may have a special advantage in pregnant patients with less severe depression. These techniques can be helpful in resolving interpersonal and psychosocial conflicts, resulting in a positive outcome without exposing the mother or fetus to drugs. The considerations for the use of psychotherapy during pregnancy are identical to those relevant to nonpregnant patients, with the caveat that the risks of a delay in effectiveness may need to be considered in the context of the mother's safety as well as the safety of her fetus (APA, 2000a).

Pharmacotherapy for depression during pregnancy requires an assessment of the risks and benefits of treatment for both mother and fetus. Women of childbearing potential in psychiatric treatment should be carefully counselled as to the risks of becoming pregnant while taking psychotropic medications. Whenever possible, a pregnancy should be planned in consultation with the psychiatrist so that medication may be discontinued before conception, if feasible. Antidepressant medication treatment should be considered for pregnant women who have major depressive disorder, as well as for those women who are in remission from major depressive disorder, receiving maintenance medication, and deemed to be at high risk for a recurrence if the medication is discontinued (APA, 2000a). McElhatton et al., (1996) and Cohen and Rosenbaum (1998) found no causal relationship between in utero exposure to therapeutic dosages of tricyclic and noncyclic antidepressants during the first trimester of pregnancy and adverse pregnancy outcomes. No increase in teratogenic risk was noted to be due to in utero exposure to SSRIs but perinatal neurobehavioral effects were seen (Kulin et al., 1998). Nulman et al. (1997) observed no effect on global intelligence quotient, language, or behavioral development in preschool children exposed in utero to either tricyclic antidepressants or fluoxetine. However, replication studies, as well as data regarding other newer antidepressants, are needed. Neonatal withdrawal syndromes have been reported in babies exposed, in utero, to tricyclic antidepressants, fluoxetine, and sertraline. Given these data, it is recommended that consideration be given to using either a tricyclic antidepressant or an SSRI that has been studied in pregnant women.

Dose requirements change during pregnancy because of changes in volume of distribution, hepatic metabolism, protein binding, and gastrointestinal absorption. Tapering the medication 10–14 days before the expected date of delivery should be considered. The medication can be restarted following delivery if the mother is considered to be at risk for a relapse, at the dose that was required before pregnancy (APA, 2000a). In selected cases not responding to or unsuitable for medication, and those with psychotic features or at high risk for suicide, electro-convulsive treatment (ECT) may be used as an alternative treatment. The current literature supports its safety for mother and fetus, as well as its efficacy during pregnancy (APA, 2000a; Miller, 1994).

Depression in the Postpartum Period

Between 30 and 75% of women experience mild postpartum blues last-ing 4 to 10 days. This postpartum depression is characterized by labile mood, tearfulness, irritability, anxiety, and sleep and appetite disturbances (Nonacs & Cohen, 1998). Patient education and reassurance are generally adequate treatment measures. If symptoms persist for 2 weeks, patients should be evaluated for postpartum major depression, particularly if they have a history of depression.

Postpartum major depression is relatively common, with a prevalence rate approximately the same as that for major depression in nonpregnant women. The risk of postpartum depression is increased in women with pregravid depression, a history of PMDD during pregnancy, primiparous status, negative life events during pregnancy, an inadequate social sup-port system, and marital problems (Bhatia & Bhatia, 1999). Nonpuerperal and puerperal depression are treated similarly unless the mother is breast-feeding. The effect of antidepressants on nursing infants, however, is not well known. Electroconvulsive therapy may be of value in patients who have severe depression with psychosis and a high risk of suicide.

Perimenopausal Depression

Depressive disorders do not seem to cluster around this period of the repro-ductive cycle (Kornstein, 1997). Symptoms of menopausal depression often occur in conjunction with vasomotor instability as a result of declining ovarian function. Estrogen replacement alone can provide relief of vaso-motor symptoms and minor cognitive and mood symptoms. This therapy is also useful in preventing osteoporosis. However, hormone replacement therapy has limited benefit in the treatment of major depression unless patients receive concomitant antidepressant drug therapy or psychother-apy (Bhatia & Bhatia, 1999).

MAINTENANCE THERAPY

Over 70% of patients experience a recurrence of depression during their lifetime; hence, maintenance phase of treatment is important for the treat-ment and prevention of future depressive episodes (Mueller et al., 1999; Remick, 2002). Current guidelines suggest that, after recovery occurs, antidepressant therapy should be continued at the therapeutic dosage for at least 6 months to significantly lessen the chance of relapse (APA, 2000a; Kennedy, Lam, Cohen, Ravindran, & the CANMAT Depression Work Group, 2001). Indefinite antidepressant maintenance therapy should be discussed with patients with additional risk factors (two or more episodes of depression in 5 years, episodes after the age of 50, and difficult-to-treat episodes) (Remick, 2002). The clinician should review with the patient the benefits (prevention of recurrence) and risks (e.g., cost, side effects, inconvenience of taking medication) of treatment. If antidepressant therapy is stopped, the dosage should be tapered gradually to avoid discontinuation symptoms (Haddad, 1988). Specific psychotherapeutic interventions are as effective as antidepressant therapy in mild to moderate major depression and dysthymic disorders. Maccarelli (2002) found that maintenance IPT led to low rates of recurrence in women (22%). The choice of psychotherapy or pharmacotherapy should be based on patient preference, clinician judgment, and cost as well as the practical issue of availability of psychotherapy. Furthermore, patients who have not responded to their preferred treatment modality should be encouraged to try other interventions. Psychological treatment combined with antidepressant therapy was shown to be associated with a higher improvement rate than drug treatment alone in a meta-analysis involving patients of both genders. In longer therapies, the addition of psychotherapy helps to keep patients in treatment. Further studies are needed to investigate whether the improvement in response attributable to the combination of drug treatment and psychotherapy can be achieved by a combination of pharmacotherapy and a compliance-enhancing intervention (Pampallona, Bollini, Tibaldi, Kupelnick, & Munizza, 2004).

DEPRESSION, DISABILITY, AND REHABILITATION

Disability is defined as (a) the inability to perform certain behaviors (physical disability), (b) the inability to successfully perform major life roles (social disability), (c) the disruption of one's daily schedule due to illness (economic disability), or (d) the receipt of disability income support (income support disability). Rehabilitation in this context is defined as services and supports designed to restore an individual's behaviors, role functioning, pursuit of daily activities, or income to optimal levels (Cook, 2003).

Kornstein et al. (1995) found that women with clinical depression self-reported more social impairment than men in marital, family, and social roles, but clinicians did not echo this gender difference in their own single-item Global Assessment of Functioning (GAF) score. There are problems with self-rating in depressed individuals, and also with assessment by GAF, which conflates clinical and functional aspects besides being unidimensional (Cook, 2003). Holstein and Harding (1992) noted that specially for women, single-dimension scales fail to capture differential performance in multiple social and work roles, as paid work roles are used to rate work function ahead of housework and child care responsibilities of women. Often, assessment of disability also fails to measure the effects of a series of prior events and difficulties that affect depression in women: abuse, partner violence, discrimination, harassment, poverty, and criminal victimization. Further, women's pattern of labor force participation (lower rates of employment and higher likelihood of entry and exit) make the use of job-defined disability days a poor indicator for them (Cook, 2003). A consideration of these issues suggests that the magnitude of disability due to mental disorders in women is underestimated in the literature.

Kornstein et al. (1995) found higher levels of depression-related disability among women compared to men. Often this greater disability occurs in more gender-stereotypical areas of social, marital, and familial functioning (Kornstein et al., 1995; Gniwesch, 1999), but greater disability has also been documented in the educational and employment realm (Marcotte, Wilcox-Gok, & Redmon, 2000; Berndt et al., 2000), as well as in activities of daily living (Lyness, Caine, & Yeates, 1993). Single parenthood combined with economic hardship can compound this association (Baker & North, 1999). Moreover, major depression among women seriously impairs their children's functioning (Weissman et al., 1987). Comorbid anxiety disorder leads to additional disability that affects social and vocational roles necessary for women to become productive members of society (Scherbourne et al., 2001).

Traditional rehabilitation systems are not oriented toward the needs of women. Vocational options offered to women seldom include skilled placements or skills training (Razzano & Cook, 1994). More recent psychosocial rehabilitation (PSR) models that emphasize the learning of adult role skills in natural environments and provision of support from professionals, families, and peers also have failed to address the functional impairments that are more characteristic of women. For example, most PSR programs fail to recognize the necessity of offering parenting skills training and parenting support groups for clients who are mothers, especially those who are single parents (Mowbray, Oyserman, & Ross, 1995). Similarly, PSR programs seldom offer training in community safety skills to help women avoid victimization in the community (Mowbray & Chamberlain, 1986), sexuality and contraception (Jonikas, Bamberger, & Laris, 1998), gender-sensitive career education, and job placement (Razzano & Cook, 1994). Given the intersection of poverty, unemployment, labor force discrimination, single parenthood, and role overload among women with severe depression, failure to develop and evaluate rehabilitation services for this group has the potential for serious societal consequences.

PREVENTION OF DEPRESSION IN WOMEN

Preventive efforts may be particularly important in groups who underutilize mental health services, for example, women from ethnic minorities (Cornelius & Collins, 2000; Jacob, Bhugra, Lloyd, & Mann, 1998; Wells, Klap, Koike, & Sherbourne, 2001), because they may be more likely to seek care when the clinical syndrome is more severe, after having experienced recurrent episodes (Brown, Abe-Kim, & Barrio, 2003). Also, in societies in which mental health services are poorly developed, preventive strategies aimed at strengthening protective factors in local communities may be a sensible investment of scarce resources rather than trying to duplicate extensive mental health care systems of richer societies. To ensure that multiple determinants of mental health are taken care of, it may be essential to work through intersectoral collaboration across government departments and nongovernmental organizations with gender-sensitive policy making in education, transport, housing, and employment. An example of such an approach would be to integrate mental health prevention issues with preexisting maternal and child health and HIV programs.

DISCUSSION

Depression in women is a significant public health concern. Rates of depression in women are approximately twice those seen in men. Observations of gender-based differences in prevalence, presentation, and treatment response (e.g., SSRIs in women in the reproductive age group and venlafaxine in women over the age of 50 years) in depression and vulnerability to reproductive-associated mood disturbance, including premenstrual dysphoric disorder, postpartum depression, and perimenopausal mood disturbance, as well as differences in social roles and power have implications for clinical management and service provision. The management of women with depression can be done safely and effectively using antidepressants and psychosocial interventions throughout the life cycle. Clinicians should consider gender differences in the phenomenology of depression and response to antidepressant treatment when screening for depressive illness, selecting appropriate treatment, and assessing therapeutic response. Appropriate service development would involve gender sensitivity being built into mainstream services.

Mental health policies and programs should incorporate an understanding of gender issues related to depression and be developed in consultation with women from communities and families and from among service users. Gender-based barriers to accessing mental health care for depression need to be addressed in program planning. A public health approach to address risk factors of depression, many of which are gender specific, is needed. The gender discrimination, gender-based violence, and gender-role stereotyping that underlie at least some part of depression need to be addressed through legislation and specific policies, programs, and interventions. Training for building health providers' capacity to identify and to treat depression in primary health care services needs to integrate a gender analysis. The training should also raise awareness about specific risk factors such as gender-based violence. Primary care and maternal health services that are responsive to psychosocial issues and are sensitive to gender differences are well placed to provide cost-effective mental health services for depression. In this context, it may be important to promote the concept of meaningful assistance for mental health care needs, including psychosocial counseling and support to cope better with difficult life situations and not just prescription of drugs. Provision of community-based care for chronic depression should be organized to ensure that facilities meet the specific needs of women (e.g., rehabilitative services that focus on social as well as economic issues) and that the burden of caring does not fall disproportionately on women (WHO, 2002a).

In research, it is important to go beyond documenting sex differences in rates of depression. There is a need to examine how gender differences influence women's risk and vulnerability, their access to health services, and the social and economic consequences of depression, in different settings and social groups and at different points in the life cycle. More research is needed on factors that can facilitate management of depression at the community and primary care level and on how women's reproductive biology influences depression and modifies the effects of different pharmacologic and psychosocial treatments. More systematic evidence is also needed on how the mental health consequences of intimate partner violence and sexual abuse in women can be addressed, especially in setting in which resources are scarce and social norms condone violence.

References

Abas, M., Broadhead, J. (1997). Depression and anxiety among women in an urban setting in Zimbabwe. *Psychological Medicine, 27*, 59–71.

Abas, M., Broadhead, J., & Blue I. (1995). Health service and community-based responses to mental ill-health in urban areas. In T. Harpham & I. Blue (Eds.), *Urbanization and mental health in developing countries* (pp. 227–248). Brookfield, VT: Ashgate.

Abernethy, D. R., Greenblatt, D. J., & Shader, R. I. (1984). Imipramine disposition in users of oral contraceptive steroids. *Clinical Pharmacology & Therapy, 35*, 792–797.

Agency for Health Care Policy and Research. (1993). *Depression in primary care: Detection, diagnosis, and treatment* (AHCPR Publication No. 93-0552). Rockville, MD: U.S. Department of Health and Human Services, Public Health Service, Agency for Health Care Policy and Research.

Altshuler, L. L., Cohen, L. S., Moline, M. L., Kahn, D. A., Carpenter, D., Docherty, J. P., et al. (2001). *The Expert Consensus Guidelines Series: Treatment of depression in women.* New York: McGraw-Hill.

Altshuler, L. L., Hendrick, V., & Cohen, L. S. (1998). Course of mood and anxiety disorders during pregnancy and the postpartum period. *Journal of Clinical Psychiatry, 59* (Supplement 2), 29–33.

American Psychiatric Association. (2000a). Practice guideline for the treatment of patients with major depressive disorder (revision). *American Journal of Psychiatry, 157* (4 Supplement), 1–45.

American Psychiatric Association. (2000b). *Diagnostic and statistical manual of mental disorders (4th ed.).* Washington, DC: American Psychiatric Association.

Amsterdam, J. D., Garcia-Espana, F., Fawcett, J., Quitkin, F., Reimherr, F., Rosenbaum, J., et al. (1999). Fluoxetine efficacy in menopausal women with and without estrogen replacement. *Journal of Affective Disorder, 55*, 11–17.

Amsterdam, J. D., Garcia-Espana, F., Goodman, D., Hooper, M., & Hornig-Rohan, M. (1997). Breast enlargement during chronic antidepressant therapy. *Journal of Affective Disorder, 46,* 151–156.

Araya, R., Rojas, G., Fritsch, R., Acuna, J., & Lewis, G. (2001). Common mental disorders in Santiago, Chile: Prevalence and socio-demographic correlates. *British Journal of Psychiatry, 178,* 228–233.

Araya, R., Rojas, G., Fritsch, R., Gaete, J., Rojas, M., Simon, G., et al. (2003). Treating depression in primary care in low-income women in Santiago, Chile: A randomised controlled trial. *Lancet, 361,* 995–1000.

Baker, D., & North, K. (1999). Does employment improve the health of lone mothers? *Social Science & Medicine, 49,* 121–131.

Barbhaiya, R. H., Buch, A. B., & Greene, D. S. (1996). A study of the effect of age and gender on the pharmacokinetics of nefazodone after single and multiple doses. *Journal of Clinical Psychopharmacology, 16,* 19–25.

Bebbington, P. E. (1996). The origins of sex differences in depressive disorder: Bridging the gap. *International Review of Psychiatry, 8,* 295–332.

Berndt, E., Koran, L., Finkelstein, S., Gelenberg, A. J., Kornstein, S. G., Miller, I. M., et al. (2000). Lost human capital from early-onset chronic depression. *American Journal of Psychiatry, 157,* 940–947.

Bhatia, S. C., & Bhatia, S. K. (1999). Depression in women: diagnostic and treatment considerations. *American Family Physician, 60,* 225–240.

Bhugra, D., Baldwin, D., Desai, M., & Jacob, K. S. (1999). Attempted suicide in West London, II. Inter-group comparisons. *Psychological Medicine, 29,* 1131–1139.

Bhugra, D., Desai, M., & Baldwin, D. (1999). Attempted suicide in West London, I. Rates across ethnic communities. *Psychological Medicine, 29,* 1125–1130.

Bies, R. R., Bigos, K. L., & Pollock, B. G. (2003). Gender differences in the pharmacokinetics and pharmacodynamics of antidepressants. *Journal of Gender Specific Medicine, 6,* 12–20.

Bland, R. C. (1997). Epidemiology of affective disorders: A review. *Canadian Journal of Psychiatry, 42,* 367–377.

Broadhead, J. C., & Abas, M. A. (1998). Life events, difficulties and depression among women in an urban setting in Zimbabwe. *Psychological Medicine, 28,* 29–38.

Brockington, L. (2001). Suicide in women. *International Clinical Psychopharmacology, 16* (Supplement 2), S7–S19.

Brown, C., Abe-Kim, J. S., & Barrio, C. (2003). Depression in ethnically diverse women: Implications for treatment in primary care. *Professional Psychology – Research & Practice, 34,* 10–19.

Brown, G. W. (1998). Genetic and population perspectives on life events and depression. *Social Psychiatry & Psychiatric Epidemiology, 33,* 363–372.

Burns, M. J., Cain, V. A., & Husaini, B. A. (2001). Depression, service utilization, and treatment costs among Medicare elderly: Gender differences. *Home Health Care Services Quarterly, 19,* 35–44.

Cohen, L. S., & Rosenbaum, J. F. (1998). Psychotropic drug use during pregnancy: Weighing the risks. *Journal of Clinical Psychiatry, 59* (Supplement 2), 18–28.

Cook, J. A. (2003). Depression, disability and rehabilitative services for women. *Psychology of Women Quarterly, 27,* 121–129.

Cornelius, L. J., & Collins, K. S. (2000). Financial barriers for working-age minority populations: Poverty and beyond. In C. J. R. Hogue, M. A. Hargaves, & K. S. Collins (Eds.), *Minority health in America: Findings and policy implications from the Commonwealth Fund Minority Health Survey* (pp. 124–141). Baltimore: Johns Hopkins University Press.

Craig, T. K. J. (1996). Adversity and depression. *International Review of Psychiatry, 8*, 341–353.

Dahl, M., Bertilsson, L., & Nordin, C. (1996). Steady-state plasma levels of nortriptyline and its 10-hydroxy metabolite: Relationship to the CYP2D6 geno-type. *Psychopharmacology (Berlin), 123*, 315–319.

Dawkins, K., & Potter, W. Z. (1991). Gender differences in pharmacokinetics and pharmacodynamics of psychotropics: Focus on women. *Psychopharmacology Bulletin, 27*, 417–426.

Dijkstra, A. G., & Hanmer, L. C. (1997). *Measuring socioeconomic gender inequality: Towards an alternative to the UNDP Gender-Related Development Index* (Working paper). The Hague: Institute of Social Studies.

Ensom, M. H. (2000). Gender-based differences and menstrual cycle-related change in specific diseases: Implications for pharmacotherapy. *Pharmacotherapy, 20*, 523–539.

Everson, G. T. (1992). Gastrointestinal motility in pregnancy. *Gastroenterology Clinics of North America, 21*, 751–776.

Fleming, A. S., Klein, E., & Corter, C. (1992). The effects of a social support group on depression, maternal attitudes and behaviour in new mothers. *Journal of Child Psychology & Psychiatry, 33*, 685–698.

Frank, E., Carpenter, L. L., & Kupfer, D. J. (1988). Sex differences in recurrent depression: Are there any that are significant? *American Journal of Psychiatry, 145*, 41–45.

Frank, E., Grochocinski, V. J., Spanier, C. A., Buysse, D. J., Cherry, C. R., Houck, P. R., et al. (2000). Interpersonal psychotherapy and antidepressant medication: Evaluation of a sequential treatment strategy in women with recurrent major depression. *Journal of Clinical Psychiatry, 61*, 51–57.

Frautschi, S., Cerulli, A., & Maine, D. (1994). Suicide during pregnancy and its neglect as a component of maternal mortality. *International Journal of Gynaecology and Obstetrics, 47*, 275–284.

Freeman, E., Rickels, K., Sondheimer, S. J., & Polansky, M. (1990). Ineffectiveness of progesterone suppository treatment for premenstrual syndrome. *Journal of American Medical Association, 264*, 349–353.

Frye, M. A., Gary, K. A., Marangell, L. B., George, M. S., Callahan, A. M., Little, J. T., et al. (1999). CSF thyrotropin-releasing hormone gender difference: Implications for neurobiology and treatment of depression. *Journal of Neuropsychiatry & Clinical Neurosciences, 11*, 349–353.

Gelenberg, A. J. (1998). SSRIs for PMDD. *Biological Therapy Psychiatry, 21*, 13–14.

Gex-Fabry, M., Balant-Gorgia, A. E., Balant, L. P., & Garrone, G. (1990). Clomipramine metabolism: Model-based analysis of variability factors from drug monitoring data. *Clinical Pharmacokinetics, 19*, 241–255.

Gijsbers, V. W. C. M. T. (2002). Depression: Gender differences in prevalence, clinical features and treatment response. *Tijschrift voor Pcyhiatrie, 44*, 377–382.

Gijsbers V. W. C. M. T., Dekker, J., Peen, J., & De Jonghe, F. (2002). Depression in men and women: Sex differences in treatment outcomes of farmacotherapy and combined therapy. *Tijdschrift voor Psychistrie, 44,* 301–311.

Gniwesch, L. (1999). Social and vocational functioning in chronic depression. *Dissertation Abstracts International. 60* (2-B), 0829.

Greenblatt, D. J., Friedman, H., Burstein, E. S., Scavone, J. M., Blyden, G. T., Ochs, H. R., et al. (1987). Trazodone kinetics: Effects of age, gender, and obesity. *Clinical Pharmacology & Therapy, 42,* 193–200.

Guay, J., Grenier, Y., & Vann, F. (1998). Clinical pharmacokinetics of neuromuscular relaxants in pregnancy. *Clinical Pharmacokinetics, 34,* 483–496.

Haddad, P. (1988). The SSRI discontinuation syndrome. *Journal of Psychopharmacology, 12,* 305–313.

Hader, S. L., Smith, D. K., Moore, J. S., & Holmberg, S. D. (2001). HIV infection in women in the United States: Status at the millennium. *Journal of American Medical Association, 285,* 1186–1192.

Halbreich, U., & Kahn, L. S. (2001). Role of estrogen in the aetiology and treatment of mood disorders. *CNS Drugs, 15,* 797–817.

Halbreich, U., & Lumley, L. A. (1993). The multiple interactional biological processes that might lead to depression and gender differences in its appearance. *Journal of Affective Disorders, 29,* 159–173.

Halbreich, U., Rojansky, N., Palter, S., Tworek, H., Hissin, P., & Wang, K. (1995). Estrogen augments serotonergic activity in postmenopausal women. *Biological Psychiatry, 37,* 434–441.

Hamilton, J. A., & Jensvold, M. F. (1995). Sex and gender as critical variables in feminist psychopharmacology research and pharmacotherapy. *Women's Therapy, 16,* 9–30.

Hänninen, V., & Aro, H. (1996). Sex differences in coping and depression among young adults. *Social Science & Medicine, 43,* 1453–1460.

Harris, T. (2001). Recent developments in understanding the psychosocial aspects of depression. *British Medical Bulletin, 57,* 17–32.

Harris, T., Brown, G. W., & Robinson, R. (1999). Befriending as an intervention for chronic-depression among women in an inner city. I. Randomized controlled trial. *British Journal of Psychiatry, 174,* 219–224.

Harrison, W. M., Endicott, J., & Nee, J. (1990). Treatment of premenstrual dysphoria with alprazolam: A controlled study. *Archives of General Psychiatry, 47,* 270–275.

Holstein, A. R., & Harding, C-M. (1992). Omissions in assessment of work roles: Implications for evaluating social functioning and mental illness. *American Journal of Orthopsychiatry, 62,* 469–474.

Hood, B., Egan, R., Gridley, H., & Brew, C. (1999). Treatment options for depression: Women and primary service providers. *Australian Journal of Primary Health – Interchange, 5,* 38–52.

Hudson, W. R., Roehrkasse, R. L., & Wald, A. (1989). Influence of gender and menopause on gastric emptying and motility. *Gastroenterology, 96,* 11–17.

Huskamp, H. A., Azzone, V., & Frank, R. G. (1998). Carve-outs, women, and the treatment of depression. *Women's Health Issues, 8,* 267–282.

Ickovics, J. R., Hamburger, M. E., Vlahov, D., Schoenbaum, E. E., Schuman, P., Boland, R. J., et al. (2001). Mortality. CD4 cell count decline, and depressive symptoms among HIV-seropositive women: Longitudinal analysis from the HIV

Epidemiology Research Study. *Journal of American Medical Association, 285*, 1466–1474.

Jacob, K. S., Bhugra, D., Lloyd, K. R., & Mann, A. H. (1998). Common mental health disorders, explanatory models and consultation behaviour among Indian women living in the UK. *Journal of the Royal Society of Medicine, 9*, 66–71.

Ji, J., Kleinman, A., & Becker, A. E. (2001). Suicide in contemporary China: A review of China's distinctive suicide demographics in their sociocultural context. *Harvard Review of Psychiatry, 9*, 1–12.

Jonikas, J., Bamberger, E., & Laris, A. (1998). *Having our say: Women mental health consumers/ survivors identify their needs and strengths.* Chicago: University of Illinois at Chicago, National Research and Training Center on Psychiatric Disability.

Katon, W., & Gonzales, J. (1994). A review of randomized trials of psychiatric consultation-liaison studies in primary care. *Psychosomatics, 35*, 268–278.

Katon, W. J., & Ludman, E. J. (2003). Improving services for women with depression in primary care setting. *Psychology of Women Quarterly, 27*, 114–120.

Katon, W., Robinson, P., Von Korff, M., Lin, E., Bush, T., Ludman, E., et al. (1996). A multi-faceted intervention to improve treatment of depression in primary care. *Archives of General Psychiatry, 53*, 924–932.

Katon, W., Von Korff, M., Lin, E., Simon, G., Walker, E., Unutzer, J., et al. (1999). Stepped collaborative care for primary care patients with persistent symptoms of depression: A randomized trial. *Archives of General Psychiatry, 56*, 1109–1115.

Katon, W., Von Korff, M., Lin, E., & Simon, G. (2001). Rethinking practitioner roles in chronic illness: The specialist, primary care physician, and the practice nurse. *General Hospital Psychiatry, 23*, 138–144.

Katz, S. J., Kessler, R. C., Frank, R. G., Leaf, P., & Lin, E. (1997). Mental health care use, morbidity, and socioeconomic status in the United States and Ontario. *Inquiry, 34*, 38–49.

Kendler, K. S., Gardner, C. O., & Prescott, C. A. (1998). A population-based twin study of self-esteem and gender. *Psychological Medicine, 28*, 1403–1409.

Kennedy, S. H., Lam, R. W., Cohen, N. L., Ravindran, A. V., & the CANMAT Depression Work Group. (2001). Clinical guidelines for the treatment of depressive disorders. IV. Medications and other biological treatments. *Canadian Journal of Psychiatry, 46* (Supplement 1), 38S–58S.

Kimmel, S., Gonsalves, L., Young, D., & Gidwani, G. (1992). Fluctuating levels of antidepressants premenstually. *Journal of Psychosomatic Obstetrics & Gynecology, 13*, 277–280.

Klamerus, K. J., Parker, V. D., Rudolph, R. L., Derivan, A. T., & Chiang, S. T. (1996). Effects of age and gender on venlafaxine and O-desmethylvenlafaxine pharmacokinetics. *Pharmacotherapy, 16*, 915–923.

Koerner, K., Prince, S., & Jacobson, N. S. (1994). Enhancing the treatment and prevention of depression in women: The role of integrative behavioural couple therapy. *Behaviour Therapy, 25*, 373–390.

Kornstein, S. G. (1997). Gender differences in depression: Implications for treatment. *Journal of Clinical Psychiatry, 58* (Supplement 15), 12–18.

Kornstein, S. G., Schatzberg, A. F., Thase, M. E., Yonkers, K. A., McCullough, J. P., Keitner, G. I., et al. (2000). Gender differences in treatment response to sertraline versus imipramine in chronic depression. *American Journal of Psychiatry, 157*, 1445–1452.

Kornstein, S. G., Schatzberg, A. F., Yonkers, K. A., Thase, M. E., Keitner, G. I., Ryan, C. E., et al. (1995). Gender differences in presentation of chronic major depression. *Psychopharmacology Bulletin, 31*, 711–718.

Kulin, N. A., Pastuszak, A., Sage, S. R., Schick-Boschetto, B., Spivey, G., Feldkamp, M., et al. (1998). Pregnancy outcome following maternal use of new selective serotonin reuptake inhibitors: A prospective controlled multi-center study. *Journal of American Medical Association, 279*, 609–610.

Leibenluft, E., Harding, T. A., & Rosenthal, N. E. (1995). Gender differences in seasonal affective disorder. *Depression, 3*, 13–19.

Lewis-Hall, F. C., Wilson, M. G., Tepner, R. G., & Koke, S. C. (1997). Fluoxetine vs tricyclic antidepressants in women with major depressive disorder. *Journal of Women's Health, 6*, 337–343.

Luine, V. N., & Rhodes, J. C. (1983). Gonadal hormone regulation of MAO and other enzymes in hypothalamic areas. *Neuroendocrinology, 36*, 235–241.

Lyness, J. M., Caine, E. D., & Yeates, C. (1993). Depressive symptoms, medical illness, and functional status in depressed psychiatric inpatients. *American Journal of Psychiatry, 150*, 910–915.

Maccarelli, L. M. (2002). Maintenance interpersonal psychotherapy (IPT-M) treatment specificity: The impact on length of remission in women with recurrent depression. *Dissertation Abstracts International, 63*(1-B), 536.

Marcotte, D. E., Wilcox-Gok, V., & Redmon, D. P. (2000). The labor market effects of mental illness: The case of affective disorders. In D. Salever & A. Sorokinn (Eds.), *Economics of disability* (pp. 181–210). Greenwich, CT: JAI Press.

Marks, I. M. (2002). The maturing of therapy: Some brief psychotherapies help anxiety/depressive disorders but mechanisms of action are unclear. *British Journal of Psychiatry, 180*, 200–204.

Martenyi, F., Dossenbach, M., Mraz, K., & Metcalfe, S. (2000). Gender differences in the efficacy of fluoxetine and maprotiline in depressed patients: A double-blind trial of antidepressants with serotonergic or norepinephrinergic reuptake inhibition profile. *European Neuropsychopharmacology, 11*, 227–232.

McElhatton, P. R., Garbis, H. M., Elefant, E., Vial, T., Bellemin, B., Mastroiacovo, P., et al. (1996). The outcome of pregnancy in 689 women exposed to therapeutic doses of antidepressants: A collaborative study of the European Network of Teratology Information Service (ENTIS). *Reproductive Toxicology, 10*, 285–294.

McEwen, B. S. (1991). Nongenomic and genomic effects of steroids on neural activity. *Trends in Pharmacological Science, 12*, 141–147.

McGrath, E. (Ed.). (1990). *Women and depression: Risk factors and treatment issues.* Washington, DC: American Psychological Association.

McGuffin, P., Katz, R., Watkins, S., & Rutherford, J. (1996). A hospital-based twin register of the heritability of DSM-IV unipolar depression. *Archives of General Psychiatry, 53*, 129–136.

Miller, L. J. (1994). Use of electroconvulsive therapy during pregnancy. *Hospital Community Psychiatry, 45*, 444–450.

Miranda, J., Duan, N., Sherbourne, C., Schoenbaum, M., Lagomasino, I., Jackson-Triche, M., et al. (2003). Improving care for minorities: Can quality improvement interventions improve care and outcomes for depressed minorities? Results of a randomized, controlled trial. *Health Services Research, 38*, 613–630.

Moller-Leimkuhler, A. M. (2002). Barriers to help-seeking by men: A review of sociocultural a clinical literature with particular reference to depression. *Journal of Affective Disorder, 71,* 1–9.

Mortola, J. F., Girton, L., & Fisher, U. (1991). Successful treatment for severe premenstrual syndrome by combined use of gonadotropin-releasing hormone agonist and estrogen/progestin. *Journal of Clinical Endocrinology, 72,* 252A-F.

Mowbray, C. T., & Chamberlain, P. (1986). Sex differences among the long-term mentally disabled. *Psychology of Women Quarterly, 10,* 383–392.

Mowbray, C. T., Oyserman, D., & Ross, S. (1995). Parenting and the significance of children for women with a serious mental illness. *Journal of Mental Health Administration, 22,* 189–200.

Mueller, T. I., Leon, A. C., Keller, M. B., Solomon, D. A., Endicott, J., Coryell, W., et al. (1999). Recurrence after recovery from major depressive disorder during 15 years of observational follow-up. *American Journal of Psychiatry, 156,* 1000–1006.

Murray, C. J. L., & Lopez, A. D. (1996). *The global burden of diseases: A comprehensive assessment of mortality and disability from diseases, injuries and risk factors in 1990 and projected to 2020.* Boston: Harvard School of Public Health, WHO, and World Bank.

Murray, L., Cooper, P. J., Wilson, A., & Romaniuk, H. (2003). Controlled trial of the short and long term effect of psychological treatment of post partum depression. *British Journal of Psychiatry, 182,* 420–427.

Mynors-Wallis, L. M., Gath, D. H., Day, A., & Baker, F. (2000). Randomized controlled trial of problem solving treatment, antidepressant medication, and combined treatment for major depression in primary care. *British Medical Journal, 320,* 26–30.

Nonacs, R., & Cohen, L. S. (1998). Postpartum mood disorders: diagnosis and treatment guidelines. *Journal of Clinical Psychiatry, 59* (Supplement 2), 34–40.

Nulman, I., Rovet, J., Stewart, D. E., Wolpin, J., Gardner, H. A., Theis, J. G., et al. (1997). Neurodevelopment of children exposed in utero to antidepressant drugs. *New England Journal of Medicine, 336,* 258–262.

Olfson, M., Zarin, D. A., Mittman, B. S., & McIntyre, J. S. (2001). Is gender a factor in psychiatrists' evaluation and treatment of patients with major depression? *Journal of Affective Disorders, 63,* 149–157.

Oss, M. E., Yennie, H., & Birch, S. (1998). Managed care approaches and models for the treatment and management of depression: Specific issues for women. *Women's Health Issues, 8,* 283–292.

Osuch, E. A., Cora-Locatelli, G., Frye, M. A., Huggins, T., Kimbrell, T. A., Ketter, T. A., et al. (2001). Post-dexamethasone cortisol correlates with severity of depression before and during carbamazepine treatment in women but not men. *Acta Psychiatrica Scandinavica, 104,* 397–401.

Pajer, K. (1995). New strategies in the treatment of depression in women. *Journal of Clinical Psychiatry, 56* (Supplement 2), 30–37.

Pampallona, S., Bollini, P., Tibaldi, G., Kupelnick, B., & Munizza, C. (2004). Combined pharmacotherapy and psychological treatment for depression: A systematic review. *Archives of General Psychiatry, 61,* 714–719.

Pande, A., Birkett, M., Fechner-Bates, S., Haskett, R. F., & Greden, J. F. (1996). Fluoxetine versus phenelzine in atypical depression. *Biological Psychiatry, 40,* 1017–1020.

Partonen, T. (1995). Estrogen could control photoperiodic adjustment in seasonal affective disorder. *Medical Hypotheses, 45,* 35–36.

Patel, V., & Thara, R. (Eds.). (2003). *Meeting mental health needs in developing countries: NGO innovations in India.* New Delhi: Sage India.

Patel, V., Chisholm, D., Rabe-Hesketh, S., Dias-Saxena, F., Andrew, G., & Mann, A. (2003). Efficacy and cost-effectiveness of drug and psychological treatments for common mental disorders in general health care in Goa, India: A randomised, controlled trial. *Lancet, 361,* 33–39.

Pearlstein, T. (1996). Nonpharmacologic treatment of premenstrual syndrome. *Psychiatric Annals, 26,* 590–594.

Pearlstein, T. B. (1995). Hormones and depression: What are the facts about premenstrual syndrome, menopause, and hormone replacement therapy? *American Journal of Obstetrics & Gynecology, 173,* 646–653.

Piazza, L. A., Markowitz, J. C., Kocsis, J. H., Leon, A. C., Portera, L., Miller, N. L., et al. (1997). Sexual functioning in chronically depressed patients treated with SSRI antidepressants: A pilot study. *American Journal of Psychiatry, 154,* 1757–1765.

Piccinelli, M., & Wilkinson, G. (2000). Gender differences in depression. Critical review. *British Journal of Psychiatry, 177,* 486–492.

Preskorn, S. H., & Mac, D. S. (1985). Plasma levels of amitriptyline: Effect of age and sex. *Journal of Clinical Psychiatry, 46,* 276–277.

Quitkin, F. M., Stewart, J. W., McGrath, P. J., Taylor, B. P., Tisminetzky, M. S., Petkova, E., et al. (2002). Are there differences between women's and men's antidepressant responses? *American Journal of Psychiatry, 159,* 1848–1854.

Raskin, A., Schulterbrandt, J. G, Reatig, N., Crook, T. H., & Odle, D. (1974). Depression subtypes and response to phenelzine, diazepam, and a placebo: Results of a nine hospital collaborative study. *Archives of General Psychiatry, 30,* 66–75.

Razzano, L., Cook, J. A. (1994). Gender and vocational assessment: What works for men may not work for women. *Journal of Applied Rehabilitation Counselling, 25,* 22–31.

Remick, R. A. (2002). Diagnosis and management of depression in primary care: A clinical update and review. *Canadian Medical Association Journal, 167,* 1253–1260.

Ronfeld, R. A., Tremaine, L. M., & Wilner, K. D. (1997). Pharmacokinetics of sertraline and its N-desmethyl metabolite in elderly and young male and female volunteers. *Clinical Pharmacokinetics, 32* (Supplement 1), 22–30.

Scherbourne, C., Dwight-Johnson, M., & Klap, R. (2001). Psychological distress, unmet need and barriers to mental health care: Results from the Commonwealth Fund's 1998 Survey of Women's Health. *Women's Health Issues, 11,* 231–243.

Schmidt, P. J., Grover, G. N., & Rubinow, D. R. (1993). Alprazolam in the treatment of premenstrual syndrome: A double-blind, placebo-controlled trial. *Archives of General Psychiatry, 50,* 467–473.

Schmidt, P. J., Nieman, L. K., Danaceau, M. A., Adams, L. F., & Rubinow, D. R. (1998). Differential behavioural effects of gonadal steroids in women with and in those without premenstrual syndrome. *New England Journal of Medicine, 338,* 209–216.

Schneider, L. S., Small, G. W., & Clary, C. M. (2001). Estrogen replacement therapy and antidepressant response to sertraline in older depressed women. *American Journal of Geriatric Psychiatry, 9,* 393–399.

Schneider, L. S., Small, G. W., Hamilton, S. H., Bystritsky, A., Nemeroff, C. B., & Meyers, B. S. (1997). Estrogen replacement and response to fluoxetine in a multicenter geriatric depression trial: Fluoxetine Collaborative Study Group. *American Journal of Geriatric Psychiatry, 5*, 97–106.

Schulberg, H., Block, M., Madonia, M., Scott, C. P., Rodriguez, E., Imber, S. D., et al. (1996). Treating major depressions in primary care practice: 8-month clinical outcomes. *Archives of General Psychiatry, 53*, 913–919.

Seeley, S., Murray, L., & Cooper, P. J. (1996). The outcome for mothers and babies of health visitor intervention. *Health Visitor, 69*, 135–138.

Seeman, M. V. (1997). Psychopathology in women and men: Focus on female hormones. *American Journal of Psychiatry, 154*, 1641–1647.

Sharan, P., & Saxena, S. (1998). Treatment-resistant depression: clinical significance, concept and management. *National Medical Journal of India, 11*, 69–79.

Shen, W. W., & Hsu, J. H. (1995). Female sexual side effects associated with selective serotonin reuptake inhibitors: A descriptive clinical study of 33 patients. *International Journal of Psychiatry & Medicine, 25*, 239–248.

Sherpa, L. R. (2001). The impact of acculturation and gender on how Japanese-American subjects respond to Davanloo's Partial Trial Therapy as measured by levels of anxiety, depression, and self-esteem. *Dissertation Abstracts International, 62*(1-B), 564.

Sherwin, B. B., & Suranyi-Cadotte, B. E. (1990). Up-regulatory effect of estrogen on platelet 3H-imipramine binding sites in surgically menopausal women. *Biological Psychiatry, 28*, 339–348.

Sloan, D. M. E., & Kornstein, S. G. (2003). Gender differences in depression and response to antidepressant treatment. *Psychiatric Clinics of North America, 26*, 581–594.

Soares, C. N., Almeida, O. P., Joffe, H., & Cohen, L. S. (2001). Efficacy of estradiol for the treatment of depressive disorders in perimenopausal women: A double blind, randomized, placebo-controlled trial. *Archives of General Psychiatry, 58*, 529–534.

Sutfin, T. A., Perini, G. I., & Molan, G. (1988). Multiple-dose pharmacokinetics of imipramine and its major active and conjugated metabolites in depressed patients. *Journal of Clinical Psychopharmacology, 8*, 48–53.

Thase, M. E., Greenhouse, J. B., Frank, E., Reynolds, C. F., 3rd, Pilkonis, P. A., Hurley, K., et al. (1997). Treatment of major depression with psychotherapy or pharmacotherapy-psychotherapy combination. *Archives of General Psychiatry, 54*, 1009–1015.

Thase, M. E., Entsuah, A. R., & Rudolph, R. L. (2001). Remission rates during treatment with venlafaxine or selective serotonin reuptake inhibitors. *British Journal of Psychiatry, 178*, 234–241.

Turner, B. J., Laine, C., Cosler, L., & Hauck, W. W. (2003). Relationship of gender, depression, and health care delivery with antiretroviral adherence in HIV-infected drug users. *Journal of General Internal Medicine, 18*, 248–257.

United Nations Development Program. (1997). *Human development report.* New York: Oxford University Press.

Ustün, T. B., & Sartorius, N. (Eds.). (1995). *Mental illness in general health care: An international study.* Chichester: Wiley.

Veijola, J., Puukka, P., Lehtinen, V., Moring, J., Lindholm, T., & Vaisanen, E. (1998). Sex differences in the association between childhood experiences and adult depression. *Psychological Medicine, 28*, 21–27.

Walker, P. W., Cole, J. O., Gardner, E. A., Hughes, A. R., Johnston, J. A., Batey, S. R., et al. (1993). Improvement in fluoxetine-associated sexual dysfunction in patients switched to buproprion. *Journal of Clinical Psychiatry, 54*, 459–465.

Ward, E., King, M., Lloyd, M., Bower, P., Sibbald, B., Farrelly, S., et al. (2000). Randomized controlled trial of non-directive counseling, cognitive-behavioral therapy and usual general practitioner care for patients with depression: I. Clinical effectiveness. *British Medical Journal, 321*, 1383–1388.

Weissman, M. M., Gammon, G. D., John, K., Merikangas, K. R., Warner, W. V., Prusoff, B. A., et al. (1987). Children of depressed parents: Increased psychopathology and early onset of major depression. *Archives of General Psychiatry, 44*, 847–853.

Wells, K., Klap, R., Koike, A., & Sherbourne, C. (2001). Ethnic disparities in unmet needs for alcoholism, drug abuse, and mental health care. *American Journal of Psychiatry, 158*, 2027–2032.

Wilhelm, K., Parker, G., & Hadzi-Pavlovic, D. (1997). Fifteen years on: Evolving ideas in researching sex differences in depression. *Psychological Medicine, 27*, 875–883.

Women Health Initiative Study. (2002). Risks and benefits of estrogen plus progestin in healthy postmenopausal women. *Journal of American Medical Association, 288*, 321–333.

Wisner, K. L., Perel, J. M., Wheeler, S. B. (1993). Tricyclic dose requirements across pregnancy. *American Journal of Psychiatry, 150*, 1541–1542.

World Health Organization. (2001a). *Atlas: Mental health resources in the world 2001* (WHO/NMH/MSD/MDP/01.1). Geneva.

World Health Organization. (2001b). *Country profiles on mental health resources 2001* (WHO/NMH/MSD/MDP/01.3). Geneva.

World Health Organization. (2002a). *Gender and mental health.* Geneva: Author. Retrieved September 19, 2004, from www.who.int/gender/other_health/en/genderMH.pdf.

World Health Organization. (2002b). *Atlas: Mapping mental health resources in the world.* Geneva: Author. Retrieved September 19, 2004, from www.cvdinfobase.ca/mh-atlas/.

World Health Organization. (2003a). *World health report 2003: Shaping the future.* Geneva: Author.

World Health Organization. (2003b). *What is "gender mainstreaming"?* Geneva. Retrieved September 19, 2004, from www.who.int/gender/mainstreaming/en/.

World Health Organization. (2003c). *Organization of services for mental health (Mental health policy and service guidance package)* (WHO/NMH/MSD/MDP/01.1). Geneva.

Young, E. A. (1995). Glucocorticoid cascade hypothesis revisited: Role of gonadal steroids. *Depression, 3*, 20–27.

Young, M. A., Scheftner, W. A., Fawcett, J., & Klerman, G. L. (1990). Gender differences in the clinical features of unipolar major depressive disorder. *Journal of Nervous & Mental Disease, 178*, 200–203.

Zust, B. L. (2000). Effect of cognitive therapy on depression in rural, battered women. *Archives of Psychiatric Nursing, 14*, 51–63.

18

Prevention of Depression in Women

Tamar Mendelson and Ricardo F. Muñoz

THE CASE FOR DEPRESSION PREVENTION IN WOMEN

Depression is currently a major public health problem for women worldwide. Epidemiological data indicate that lifetime prevalence rates of major depressive disorder (MDD) in the United States are 1.7 to 2.7 times greater for women than for men (Burt & Stein, 2002), and rates of MDD have been found to be approximately twice as high among women as men across a range of cultures and countries (Bebbington et al., 1998; Weissman et al., 1996). The negative impact of depression on women is far-reaching. The 2000 census figures estimate the U.S. population at 281,421,906, of whom 143,368,343 (50.9%) are female (U.S. Census Bureau, 2004). Nationally representative data indicate that women have a 21.3% lifetime prevalence of major depressive episodes (MDE; Burt & Stein, 2002), suggesting that over 30 million females will have at least one episode during their lives in the U.S. alone.

Depression produces significant impairments, causing dysfunction that equals or exceeds chronic physical illness (Hays, Wells, Sherbourne, Rogers, & Spritzer, 1995). In addition, researchers underestimate the negative impact of depression when they look only at its direct effects without attending to the concept of *attributable risk* (Muñoz, 2001), that is, the empirically documented contribution of depression to major causes of preventable death, such as smoking, poor diet and exercise, substance use, firearms, risky sexual behavior, and car accidents (see, e.g., McGinnis & Foege, 1993). Similarly, depression has been linked with higher risk for serious medical conditions. For instance, risk for cardiovascular disease (CVD) was found to be 1.7 times higher among depressed as compared with nondepressed individuals, and the association between depression and CVD was strongest for women 45 years or older (Keyes, 2004).

Furthermore, despite encouraging advances in our ability to diagnose and treat major depression, large numbers of depressed individuals do not currently receive treatment, and those who are treated often receive inadequate or inappropriate care. For instance, according to the National Comorbidity Study, only 21.6% of individuals in the United States who reported a major depressive episode over the past 12 months had received treatment judged to be at least minimally adequate during that time period (Kessler et al., 2003). At the same time, rates of major depression appear to be increasing, both in the United States and worldwide (Kessler et al., 2003). Treatment services are not currently adequate – nor, most likely, will they ever be – to address the alarming prevalence and growth of depression among women. Researchers need to augment treatment services with aggressive prevention programs (Muñoz & Ying, 1993).

Depression prevention for both sexes merits national health priority status, but prevention research and policy initiatives among women are particularly critical for two reasons. First, women are at twice the risk of developing depression compared with men across diverse cultural and social environments. Second, depression in women has a potentially major impact on intergenerational transmission of risk (see Goodman & Tully, this volume). Despite dramatic social role changes for women in the past century, women continue to serve as the primary caretakers for infants and children, even in Western countries. Research indicates that the children of depressed women are themselves at risk for developing depression and other psychological disorders. Thus, preventing depression – and promoting well-being – in women may contribute significantly to disease prevention and health promotion for subsequent generations of women and men.

In the remainder of this chapter, we summarize theoretical and empirical work relevant to depression prevention in women and discuss potentially beneficial directions for the field. We first define the concept of prevention and then review research on depressive risk factors among women. Next, we outline prior research on depression prevention with women and conclude with recommendations for how future prevention research might best be developed and implemented.

PREVENTION: WHAT AND HOW?

The Institute of Medicine (IOM) Committee on Prevention of Mental Disorders (Mrazek & Haggerty, 1994) proposed that the term *prevention* be reserved for interventions administered before the onset of a clinically diagnosable disorder. By contrast, *treatment* refers to interventions aimed at curing or improving an existing disorder. The IOM report specified three levels of preventive interventions. *Universal preventive interventions* target an entire population group (e.g., schoolwide coping skills training to prevent depression). *Selective preventive interventions* target high-risk groups within a community, using biological, psychological, or social factors empirically associated with the onset of a disorder to determine risk status (e.g., coping skills training for children of depressed parents). *Indicated preventive interventions* target individuals with early signs or symptoms of a disorder who do not yet meet full diagnostic criteria (e.g., a mood management program for high school students with elevated depression symptoms).

Distinct advantages and difficulties are associated with each level of prevention. For instance, expensive and potentially burdensome or risky interventions are more appropriately reserved for indicated prevention efforts, whereas inexpensive, safe, and easily implemented interventions can be more readily disseminated on a universal scale. To choose the appropriate level of intervention, the researcher must carefully define the scope and probable cause of the problem, as well as the most effective means of combating it.

Muñoz and Ying (1993) propose five key steps in mounting an effective prevention program. They suggest the following: (1) identify the target of prevention (e.g., major depressive episodes), (2) choose a theory to guide the intervention (e.g., diathesis-stress model specifying the interaction of life stress and negative thinking), (3) identify a high-risk group for whom the intervention is appropriate (e.g., individuals who display negative thought patterns on screening measures), (4) design the intervention (e.g., develop a skills-training course to modify negative cognitions), and (5) design the study (e.g., select appropriate outcome measures and method for delivering and testing the intervention). To carry out these steps effectively, it is important to obtain data on relevant risk factors and their causal mechanisms.

A great deal of research has been done on vulnerability to depression and, increasingly, researchers are investigating potential gender differences in depressive vulnerability. However, there is still a relative lack of well-elaborated theoretical models that map the way in which multi-level factors interact across diverse developmental time-points and cultural settings to create gender differences in vulnerability. More often, research has targeted individual-level risk factors without specifying the relation of such factors to the broad spectrum of influences that impinge on the individual. Biological, psychological, cultural, and social factors have often been investigated separately, due to the frequent lack of communication across disciplines and the difficulty inherent in conducting integrative research. The following section offers an overview of research on biological and psychosocial risk factors for depression in women. Recent attempts to create more integrative models of risk are also reviewed.

SOURCES OF FEMALE VULNERABILITY TO DEPRESSION

Genetic and Hormonal Factors

Although certain data suggest gender differences in genetic vulnerability to depression, findings are not entirely consistent, and their implications are not yet clear (see Hankin & Abramson, 2001). Indeed, some researchers argue that studies have failed to identify a direct link between genetic factors and increased rates of depression among women (Piccinelli & Wilkinson, 2000). For example, recent genetics research identified a functional polymorphism in the serotonin transporter (5-HTT) that interacted with stressful life events to predict increased depression symptoms and incidence rates (Caspi et al., 2003), but sex differences in genetic vulnerability were not reported. The gene-by-environment interaction may well be found in future research to differ by sex, as there are data to suggest that women may experience greater numbers of stressful life events, and women may respond differently to such events than men.

Because gender differences in depression emerge at puberty and, by most accounts, diminish after menopause (Steiner, Dunn, & Born, 2003), some researchers argue that the preponderance of depression among women derives from increased gonadal hormone activity during female childbearing years. Gonadal hormones such as estrogen regulate the transcription of serotonin, a neurotransmitter that is strongly associated with depression (Steiner et al., 2003). However, given that not all adolescent females become depressed, hormonal changes per se are not likely to cause depression (Steiner et al., 2003), and empirical evidence does not support a direct link between hormones and depression (Hankin & Abramson, 2001). Hormonal activity more likely interacts with preexisting genetic vulnerabilities and psychosocial factors to create vulnerability.

Cognitive Factors

A number of researchers have hypothesized that cognitive factors, such as attributional style and coping strategies, contribute to increased depressive vulnerability among females. According to attribution theories, individuals who ascribe negative life events to internal, stable, and global causes are more likely to become depressed (Abramson, Alloy, & Metalsky, 1988). Although research has not consistently indicated gender differences in attributional style, adolescent girls exhibited a more depressogenic attributional style than adolescent boys in a study that employed a psychometrically improved measure of attributional style (Hankin & Abramson, 2002). With respect to coping strategies, Nolen-Hoeksema's research indicates that ruminative coping (i.e., a passive focus on distress-ing thoughts) is linked to the onset and maintenance of depression and that females are more likely to engage in ruminative coping than men (Nolen-Hoeksema, 1990; Nolen-Hoeksema, Morrow, & Fredrickson, 1993). Similarly, self-focused attention (i.e., attention directed internally toward thoughts and feelings) has been found to predict depression (Ingram, Lumry, Cruet, & Sieber, 1987). Data suggest that women are more inclined to self-focus than men, and women who embody feminine sex roles have been found to display the greatest propensity for self-focusing (Ingram, Cruet, Johnson, & Wisnicki, 1988).

Stressful Life Events

Extensive research has documented the link between stressful life events and onset or recurrence of depression for both women and men (Mazure, Keita, & Blehar, 2002). A number of studies have found that women experi-ence more adverse life events than men, and research also tends to suggest that women display greater sensitivity, either to all classes of stressful life events or to specific classes of events, such as those in an interpersonal domain (see Kendler, Thornton, & Prescott, 2001, for a review). For instance, Maciejewski, Prigerson, and Mazure (2001) reported that women in their sample were three times more likely than men to experience depression as a result of stressful events.

Childhood Sexual Abuse, Adult Sexual Assault, and Intimate Partner Violence

Trauma – an extreme variant of life stress – has extremely deleterious effects on an individual's emotional functioning and risk for psychopathology. Women experience childhood sexual abuse, adult sexual assault, and inti-mate partner violence at far higher rates than men (Fergusson, Swain-Campbell, & Horwood, 2002; Weiss, Longhurst, & Mazure, 1999). In their review of 21 studies on child sexual abuse, Weiss et al. (1999) found only one study that did *not* report a higher incidence of adult depression among survivors of childhood sexual abuse. In addition, extensive evidence sug-gests that adult sexual assault and partner violence predict depression among women (Mazure et al., 2002).

Gender Roles and Role Strain

A large literature has addressed the mental health implications of women's traditional roles (homemaker, wife, mother) as well as the multiple roles filled by employed women, who are often disproportionately burdened with child care and household chores at home while facing financial and social inequities on the job (see Mirowsky and Ross, 1989). Research tends to indicate that multiple roles (e.g., partner, parent, wage earner) are psychologically beneficial for both women and men (Barnett & Hyde, 2001). A number of studies have reported lower depression levels among employed than unemployed women, whereas no studies have reported the reverse (Barnett & Hyde, 2001). These findings indicate that married women who are not employed may, in some cases, be at higher risk for depression, although moderating factors (e.g., quality of spousal support) likely impact the nature and extent of this association.

Poverty and Income Inequality

A recent meta-analysis indicates that low socioeconomic status (SES) predicts slight increases in risk of episode onset and moderate increases in the persistence of depression (Lorant et al., 2003). Not only are women more likely than men to have incomes below the poverty line (Belle & Doucet, 2003), but some data suggest that low socioeconomic status in adulthood and childhood may be more strongly associated with adult depression among females than males (Gilman, Kawachi, Fitzmaurice, & Buka, 2002; Gore, Aseltine, & Colton, 1992; Lorant et al., 2003). Furthermore, state income inequality (i.e., being a low-income resident of a high-income state) was found to be associated with a 60% greater risk of depressive symptoms among women, above and beyond the effects of household income (Kahn, Wise, Kennedy, & Kawachi, 2000).

Sexism and Discrimination

A number of researchers have investigated the hypothesis that women's higher rates of depression result from their greater exposure to discrimination and devaluation on the basis of their gender. For instance, undergraduate women were found more likely than men to experience sexist hassles, which were associated with depression, anger, and lower self-esteem (Swim, Hyers, Cohen, & Ferguson, 2001). In addition, female students who experienced frequent sexism reported more depression symptoms than male students, whereas female students exposed to low levels of sexism did not report more depression than men (Klonoff, Landrine, & Campbell, 2000). Katz, Joiner, and Kwon (2002) reported that the association between female gender and depression was fully mediated by women's perception that others evaluated their gender negatively. In addition, status differentials between men and women may increase women's risk for depression both directly and indirectly, via increases in negative health outcomes and intimate partner violence against women (Belle & Doucet, 2003).

Racial, Ethnic, and Cultural Factors

Racial discrimination is associated with increases in depressive symptoms, and women of color face the dual challenge of discrimination on the basis of both race and gender (Belle & Doucet, 2003). Acculturation can also be a risk factor for ethnic minority women. For instance, research indicates that Mexican Americans who were born in the United States have higher levels of depressive symptoms than those born in Mexico (e.g., Heilemann, Lee, & Kury, 2002; Vega et al., 1998). Moreover, women of color experience a higher burden from depression than Caucasian women because they are less likely to receive accurate diagnoses and appropriate care (e.g., Borowsky et al., 2000).

Interdisciplinary Models

Theorists are becoming increasingly aware of the need for models that address the interaction of biological, psychological, and sociocultural factors, but surprisingly few such models have been articulated or tested. Cyranowski, Frank, Young, and Shear (2000) propose a model in which female gender socialization processes interact with female pubertal increases in oxytocin (a neurohormone that regulates affiliative behaviors) to increase the importance of interpersonal relationships for adolescent girls. Insecure attachments, anxious and inhibited temperament, and low instrumental coping skills act as risk factors, which increase an adolescent girl's likelihood of experiencing interpersonal stress and amplify its depressogenic effects.

Hankin and Abramson (2001) propose that, for both males and females, depression results from reciprocal interactions among five causal factors: preexisting vulnerabilities (genetic, personality, and environmental), learned cognitive vulnerabilities, negative life events, initial negative affect, and increases in depression over time. They further suggest that, starting in adolescence, females are more vulnerable to depression than boys because they are at greater risk of elevations in all five causal factors. For instance, adolescent girls are more likely than boys to have preexisting vulnerabilities (e.g., sexual abuse, neuroticism); they are more likely to encounter negative life events, leading to greater elevations of initial negative affect; and they are more likely to be cognitively vulnerable (e.g., frequent rumination), leading to greater increases in depression over time.

In addition, Kendler, Gardner, and Prescott (2002) recently constructed a developmental model for predicting major depression in women using data from 1,942 adult female twins who had been assessed longitudinally at multiple time points. Their model incorporated 18 risk factors, including genetic risk and family environment. Their findings suggest that major depressive disorder results from complex interactions among multiple factors with three major pathways emerging as primary: internalizing symptoms (e.g., neuroticism, early-onset anxiety disorders), externalizing symptoms (e.g., conduct disorder, substance misuse), and psychosocial adversity (e.g., childhood and adult trauma and interpersonal difficulties).

Although these models represent a substantial advance in the study of gender differences in vulnerability to depression, they still require further empirical validation. In addition, models with a more explicit focus on social, cultural, economic, and political factors are sorely needed. Integrative, multilevel models are critical not only for understanding the etiology of gender differences in depression but also for the creation of prevention programs capable of positively impacting women's mental health at the population level.

CURRENT STATUS OF DEPRESSION PREVENTION EFFORTS WITH WOMEN

The Institute of Medicine's Committee for Prevention of Mental Disorders (Muñoz, Mrazek, & Haggerty, 1996) issued an urgent call for depression prevention research, and participants at the Summit on Women and Depression argued that depression prevention in women should be recognized as a national priority (Le, Muñoz, Ghosh Ippen, & Stoddard, 2003; Mazure et al., 2002). However, depression prevention research still lags far behind treatment research, and empirically based interventions that specifically target depression prevention in women are indeed scarce. In this section, we summarize some of the depression prevention research that has been undertaken to date (excluding studies focused on prevention of relapse or recurrence following treatment). We highlight randomized controlled trials of psychosocial interventions specifically aimed at reducing incidence rates of major depression (see Jané-Llopis, Hosman, Jenkins, & Anderson (2003)'s meta-analysis for a more extensive review).

Randomized Controlled Trials Aimed at Preventing Incidence of Major Depression

Thus far, and to our knowledge, only four randomized controlled trials have been conducted on psychosocial interventions to prevent (non-postpartum) major depressive episodes among adolescents and adults (see Table 18.1). These trials tested the efficacy of mood management groups, and they included both men and women. Whereas the two studies using adult samples (Muñoz et al., 1995; Seligman, Schulman, DeRubeis, & Hollon, 1999) did not report statistically significant differences in incidence rates of major depression, the two trials conducted with adolescents did achieve this goal (Clarke et al., 1995, 2001).

TABLE 18.1. *Randomized Controlled Trials Aimed at Prevention of Major Depressive Episodes (MDEs) in Adults and Adolescents*

Study	Participants	Intervention	Depression measure(s)	Findings
Adult trials:				
Muñoz et al. (1995)	150 primary care adult patients	8-session cognitive behavioral (CBT) course vs. no intervention vs. brief CBT videotape	Diagnostic Interview Schedule (DIS)	No statistically significant group differences in cumulative MDE incidence at 1 year
Seligman, Schulman, & DeRubeis (1999)	231 undergraduates at risk for depression	8-week CBT group vs. assessment-only control group	Structured Clinical Interview for the DSM-III-R (SCID)	No statistically significant group differences in cumulative MDE incidence at 3 years ($p < .08$)
Adolescent trials:				
Clarke et al. (1995)	150 adolescents with subsyndromal depressive symptoms	15-session cognitive therapy group vs. usual care control	SADS for School-Age Children, Epidemiological Version (K-SADS-E) and Longitudinal Interval Follow-up Evaluation (LIFE)	14.5% cumulative MDE incidence for experimental group vs. 25.7% for control group at 1 year ($p < .005$)
Clarke et al. (2001)	94 adolescents with subsyndromal depressive symptoms and depressed parents	15-session cognitive therapy group vs. usual HMO care	K-SADS-E	9.3% cumulative MDE incidence for experimental group vs. 28.8% for usual-care control group at 1 year ($p = .003$)

The trials conducted by Clarke and his colleagues indicate that it is possible to prevent onset of major depression among high-risk adolescents. In addition, Seligman et al. (1999) reported a nonsignificant trend toward prevention of depressive episodes ($p < .08$), which achieved significance when episodes of moderate severity were analyzed separately ($p < .03$). Furthermore, significant reduction in depressive symptomatology was achieved in all trials except Clarke et al. (1995), highlighting that prevention programs had positive effects on mood even in cases for which reductions in the incidence of depression did not attain statistical significance. Seligman and colleagues, the only investigators to report gender differences in outcome, indicated that fewer women than men developed depressive episodes following the intervention (46% vs. 60%), although gender differences in depressive symptoms were not obtained. In addition, it is noteworthy that the majority of participants in both adolescent trials (Clarke et al., 1995, 2001) were female (70% and 59% respectively), indicating that it is feasible to recruit, retain, and positively impact both symptoms and MDE incidence among adolescent girls.

Randomized Controlled Trials Aimed at Antenatal Prevention of Postpartum Depressive Disorder

Major depression with postpartum onset has its own implications for prognosis and treatment, and there is growing interest in preventing postpartum depression using antenatal (i.e., delivered during pregnancy) interventions. A major reason for preventing maternal depression during early childhood is the potential for reducing lifetime risk of depression in the offspring (see section on identifying critical periods, below). We are aware of five published randomized controlled studies of antenatal psychosocial interventions aimed at reducing postpartum depression (PPD) incidence rates among pregnant women at risk for PPD (see Table 18.2), Four tested the efficacy of psychosocial group interventions delivered during pregnancy, while the fifth (Marks, Siddle, & Warwick, 2003) assessed the impact of individual visits with a midwife designed to provide continuity of care and social support. Two of those studies (Elliott et al., 2000; Zlotnick, Johnson, Miller, Pearlstein, & Howard, 2001) reported statistically significant effects in preventing postpartum major depressive episodes. In addition, our research group recently conducted a pilot randomized controlled trial of a preventive intervention for low-income, primarily Latina pregnant women (The Mamás y Bebés/Mothers and Babies Course; Muñoz et al., 2001a, 2001b), which indicated that it is feasible to recruit and retain this population in research trials (Mendelson, Le, Soto, Lieberman, & Muñoz, 2003). Significant prevention effects were not obtained, but it is difficult to interpret this finding given that the study lacked sufficient power to detect effects due to small sample size.

TABLE 18.2. *Randomized Controlled Trials Aimed at Antenatal Prevention of Postpartum Depression (PPD)*

Study	Participants	Intervention	Depression measure(s)	Findings
Stamp, Williams, & Crowther (1995)	139 women in their first pregnancy at risk for PPD	2 antenatal support groups + 1 postnatal support group vs. routine care only	Edinburgh Postnatal Depression Scale (EPDS) score > 12	No significant group differences at 6 weeks, 12 weeks, or 6 months postpartum
Brugha et al. (2000)	190 women in their first pregnancy at risk for PPD	6 weekly antenatal classes using cognitive and problem-solving approaches vs. routine antenatal care	General Health Questionnaire Depression Scale (GHQ-D) score ≥ 2	No significant group differences at 3 months postpartum
Elliott et al. (2000)[a]	99 women in their first or second pregnancy at risk for PPD	5 antenatal sessions + 6 postnatal sessions of psychoeducational group vs. control group	Present State Examination (PSE)	Group differences found for first-time mothers at 3 months postpartum
Zlotnick, Johnson, Miller, Pearlstein, & Howard (2001)	35 low-income pregnant women at risk for PPD	4-session interpersonal-therapy-oriented antenatal group vs. routine antenatal care	Structured Clinical Interview for DSM-IV Diagnoses (SCID) depression module	0% PPD incidence in intervention group vs. 33% in control group at 3 months postpartum ($p = .02$)
Marks, Miller, & Warwick (2003)	98 pregnant women at risk for PPD	8–12 antenatal visits with midwife + postpartum visits as required vs. usual care	Structured Clinical Interview for DSM-III-R (SCID) & EPDS	No significant group differences at 4 weeks or 3 months postpartum

[a] First-time and second-time mothers were allocated to different groups. Group allocation does not appear to have been fully randomized.

As Austin (2003) points out, it is difficult to draw firm conclusions regarding the efficacy of postpartum depression prevention given methodological limitations in most of the published studies in this area, including lack of a validated assessment of risk status, lack of power analyses, high attrition rates, and small sample sizes. Moreover, only three published studies (Elliott et al., 2000; Marks et al., 2003; Zlotnick et al., 2001) employed structured diagnostic interviews to assess depression incidence. However, Zlotnick et al.'s findings suggest that interpersonal therapy may be a promising intervention for preventing postpartum depression and merits further study with larger samples. Elliott et al.'s findings point to the potential importance of tailoring interventions to first-time versus second-time mothers, as these two groups appear to differ in their response.

Randomized Controlled Prevention Trials Aimed at Depression Symptom Reduction

A number of prevention programs have aimed to reduce the severity of depressive symptoms, rather than the number of major depressive episodes. Controlled trials evaluating those programs with adult, adolescent, and child samples have generally yielded favorable results (see Jané-Llopis et al., 2003). Prevention of depression symptoms has been attempted with some success with a variety of populations, such as women recently diagnosed with gynecological cancer (e.g., Petersen & Quinlivan, 2002). Encouragingly, some studies have begun to report on the efficacy of such programs with culturally diverse samples, such as Chinese children (Yu & Seligman, 2002) and low-income Latino and African American children (Cardemil, Reivich, & Seligman, 2002). Programs targeting reduction of depressive symptoms have included indicated, selective, and universal approaches. Characteristics of such programs that have been found to be associated with larger effect sizes include a multicomponent design, competence techniques (e.g., skills training), more than eight sessions, sessions 60–90 minutes long, and a high quality of research design (Jané-Llopis et al., 2003).

In sum, the data thus far show that we can prevent major depressive episodes in high-risk adolescents, and two antenatal interventions also show promise in preventing postpartum depressive episodes. With respect to depressive symptom outcomes, data indicate that indicated, selective, and universal approaches can each produce reductions in depressive symptoms. Thus, depression prevention research has yielded a number of positive outcomes. However, this research is still in the early stages of its development. A greater number of rigorous randomized controlled prevention trials aimed at reducing rates of major depressive disorder and severity of depressive symptoms would be desirable. It is important that researchers routinely test for possible moderating effects of gender to facilitate development of gender-sensitive interventions. Moreover, we need studies with large enough sample sizes to allow for assessment of a range of possible moderators and mediators. Fine-grained analysis of change mechanisms will aid the process of intervention development for various female populations. In addition, to date most prevention trials have employed selective or indicated designs, rather than a universal approach. We need to find ways to expand upon the benefits of research with targeted samples to reach a larger number of individuals. Ultimately, we need to extend existing research to create and test effective programs capable of preventing depression in women across a range of ages, ethnicities, and life circumstances.

FUTURE DIRECTIONS FOR DEPRESSION PREVENTION IN WOMEN

In the remainder of the chapter we outline our recommendations regarding depression prevention for women. We have identified seven areas we believe are important to the success of this endeavor: (1) identification of critical periods to intervene, (2) identification of women in urgent need of preventive services, (3) inclusion of a broad range of outcome measures, (4) use of multiple prevention levels, (5) adoption of interdisciplinary prevention frameworks, (6) increased attention to race, ethnicity, culture, and community, and (7) nontraditional delivery of prevention services.

Identifying Critical Periods for Intervention: Adolescence and Pregnancy

Based on what we know about lifespan development, depression risk factors, and depression course, it appears likely that certain points in the lifespan represent windows of opportunity for preventive intervention. Theory and research suggest that individuals may be more responsive to prevention strategies if they are implemented during sensitive or critical periods when cognitive, social, or biological patterns are undergoing transition and development than they are once such patterns have become entrenched (Howe, 2003). Adolescence represents such a period because during the teen years social and biological aspects of identity are actively developing and have not yet stabilized. The success of Clarke and his colleagues at preventing major depressive episodes among high-risk adolescents (Clarke et al., 1995, 2001) suggests that adolescence may be a promising developmental period for implementing preventive interventions. In addition, we know that adolescence is a particularly hazardous developmental transition for girls. Whereas girls and boys manifest similar rates of depression prior to adolescence, the roughly 2:1 female-to-male predominance of depression begins to emerge between the ages of 11 and 15 (Kessler et al., 1994).

Preventing first episodes of depression among adolescent girls may be the most effective way to reduce the gender gap in depression because, although most evidence suggests females are not at greater risk for *recurrence* of depression than males, they are far more likely to experience a first episode (Kessler, 2003). Research suggests that experience of a single major depressive episode confers a 50% risk of recurrence, two episodes confer a 70% recurrence risk, and three episodes a 90% recurrence risk (Depression Guideline Panel, 1993). Given this high likelihood of recurrence, the increased risk for first episodes among women – as well as women's longer average life expectancy compared with men – suggests that women will experience more episodes than men throughout their lifespan. If we intervene before an adolescent girl first becomes depressed, we may be better able to reduce nascent cognitive, behavioral, and interpersonal risk factors, as well as to enhance protective and resiliency factors. In turn, such interventions may serve both to reduce a girl's chances of experiencing life stress and also to increase her capacity for dealing effectively with stress.

We believe that pregnancy is another critical period to intervene in preventing depression in women. The birth of a child, especially a first child, represents a significant emotional, biological, and physical stressor for the mother. Whereas rates of depression during the antenatal and postnatal periods do not appear to differ substantially from depression rates among women of childbearing age who do not become pregnant (see Le et al., 2003), the impact of postnatal depression can have far-reaching effects on the mother, on the formation of a secure mother–infant attachment bond, and on the infant's development.

Depressed mothers often feel less efficacious and competent in their parenting than nondepressed mothers and have greater difficulty with the demands of parenthood (O'Hara, 1994; Teti & Gelfand, 1991). In addition, research indicates that postpartum depression can have profound negative effects on infants and children (see Le et al., 2003, for a review). Infants of depressed mothers are at higher risk than infants of nondepressed mothers for slowed cognitive development, insecure attachment, negative mother–infant interactions, and behavioral difficulties. Furthermore, children of depressed mothers are at higher risk for developing major depression or other psychiatric disorders in their lifetime. Indeed, recent evidence indicates that the *grandchildren* of depressed women are also at significantly increased risk for depression (Weissman & Jensen, 2002). Thus, successful preventive intervention with pregnant women has the potential not only to reduce such women's suffering but also to reduce the intergenerational transmission of risk for depression.

Identifying Women in Urgent Need of Preventive Services

As noted earlier, certain women are at particularly high risk for depression based on their psychological and sociocultural circumstances. We believe two subgroups of women who would greatly benefit from prevention efforts are women living in poverty and female survivors of trauma and abuse. These subpopulations generally experience significant suffering and are at risk for developing multiple mental and physical disorders. In addition, they are often overlooked by prevention and treatment researchers due to difficulties of recruitment and retention.

As outlined in our discussion of vulnerability factors, growing evidence indicates that socioeconomic status (SES) is inversely associated with depression across all SES levels and that women living in poverty may be at particular risk (Lorant et al., 2003). Women of low SES are likely to be exposed to more frequent and more adverse life stressors, including neighborhood violence and concerns about basic necessities, such as food and shelter. These women may also be more likely to face sources of interpersonal stress such as teen pregnancy, single parenthood, partner violence, and discrimination (Belle & Doucet, 2003).

Low-income women are often more difficult to recruit and retain in research studies due to difficulties with child care and transportation, as well as pressing psychosocial stressors (e.g., homelessness) that may interfere with their ability to attend scheduled meetings. However, recent pilot studies indicate that it is feasible to include low-income women in randomized controlled trials on depression prevention (Mendelson et al., 2003; Zlotnick et al., 2002). Indeed, focus groups conducted with low-income, Latina women who participated in the group preventive intervention conducted by our research group at San Francisco General Hospital indicated that they found the experience to be of great personal benefit. In addition, we believe prevention research with low-income women can be expanded to include community-based interventions conducted in partnership with churches or clinics in low-income neighborhoods.

Women who have survived prolonged or acute traumatic experiences are also at high risk for developing depression, often in combination with posttraumatic stress symptomatology. Low-income women are at higher risk for experiencing multiple incidents of physical and sexual abuse, and women saddled with these dual risk factors are likely to be in greatest need of services but may also have the greatest barriers to accessing such services. However, a growing number of programs are attempting to improve outreach efforts with trauma survivors, including low-income women. For instance, the University of California San Francisco's Trauma Recovery Center (TRC), which was established by the Department of Psychiatry at San Francisco General Hospital, conducts extensive outreach efforts, including visits to homeless women in the community who are not reachable by phone or mail (V. Kelly, personal communication, October 20, 2003). Efforts to involve trauma survivors in prevention efforts may be greatly enhanced by having a therapist available to speak with women during their medical assessments following an assault. This initial contact can provide an opening to refer the survivor for treatment or preventive intervention, either immediately or once the survivor indicates readiness. Research can be successfully implemented in the context of responsible outreach and clinical care, as it has been at the TRC.

Adoption of a Broad Range of Outcome Measures

The IOM Committee on Prevention of Mental Disorders proposed that the goal of preventive interventions should be to reduce significantly the proportion of individuals who develop the disorder in question (Mrazek & Haggerty, 1994). Given that incidence rates of depressive episodes in a given year are relatively low in the general population (6.6% in Kessler et al., 2003) and even among at-risk populations (e.g., 28.8% in Clarke et al., 2001), large sample sizes are generally required to attain adequate power in studies that assess rates of major depressive disorder as an outcome. Furthermore, analyses involving categorical outcomes (i.e., incidence) require more power to detect effects than do analyses with continuous outcomes. And, as Cuijpers (2003) notes, statistical power is also hampered by our limited knowledge concerning the causal pathways to mental disorders, as well as by the low specificity of identified risk factors (i.e., most individuals who possess risk factors do not develop the disorder in question).

Although we certainly advocate for continued research on prevention of major depressive episodes, we also welcome comments by a growing number of investigators regarding ways to expand the evaluation of successful prevention outcomes. As some investigators have noted, the reduction of depressive symptoms should be considered a critical outcome, perhaps equal in importance to the reduction of future depression incidence (Clarke, DeBar, Lynch, & Wisdom, 2003; Gillham, 2003). For instance, it has been found that adolescents with elevated depressive symptoms not only are at risk for future clinical disorder but also manifest similar levels of psychosocial dysfunction as those who meet criteria for major depression (Gotlib, Lewinsohn, & Seeley, 1995). As depressive symptomatology is assessed as a continuous variable, it poses fewer power constraints than does the analysis of incidence rates.

In addition, the enhancement of protective factors and assessment of positive outcomes has often been minimized or neglected in prevention research. Indeed, as advocates of positive psychology have argued, psychology has traditionally emphasized the study of pathology without paying adequate attention to the nature and role of human resiliency (Seligman & Csikszentmihalyi, 2000). Positive outcomes, such as subjective well-being (Park, 2003), increased motivation and energy (Kohn-Wood, 2003), and enhanced social skills (Parks & Herman, 2003) are vital to healthy functioning. As Park (2003) has noted, positive and negative affect are correlated but distinct, and positive outcomes may be relevant predictors of future functioning, including vulnerability to depressive relapse. In addition, as Clarke, Hawkins, Murphy, and Sheeber (1993) suggest, increased focus on positive capacities for coping may well enhance our ability to design effective prevention programs, as promotion of general healthy living skills may avert multiple mental health problems with a single intervention and may also be amenable to more widespread adoption in real-life settings.

Use of Multiple Prevention Levels

Indicated prevention studies have a substantial advantage over universal prevention studies with respect to statistical power (Cuijpers, 2003) because, by definition, incidence rates are higher among samples at indicated risk. (Selective prevention studies, although somewhat more inclusive, generally offer similar benefits.) As a result, these studies tend to require fewer participants and are often more easily conducted in clinical or medical settings. Findings from indicated prevention studies may help to convince governmental and other funding agencies of the importance and value of prevention programs. Thus, such studies may lay the groundwork for the kind of public support necessary for funding more inclusive prevention efforts.

We advocate continued use of indicated prevention research. However, given our stance that depression prevention should be a national public health priority (Le et al., 2003), indicated prevention programs alone are not likely to prove adequate. Parks and Herman (2003) argue that use of the full spectrum of preventive strategies is required to reduce population risk of depression. Universal and selective prevention programs have the advantage of potentially including a larger pool of individuals. Because our current methods of screening for depression risk are notably imperfect, casting a broader net may result in capturing individuals at risk who would not otherwise have been identified. For instance, Gillham (2003) reported that in her research, although children identified as high risk were three times more likely to score in the clinically depressed range, the *actual number* of children who scored in the clinical range was lower in the high-risk than low-risk group.

Not only are our measures of risk status imperfect, but in addition it is becoming increasingly clear that risk status is likely to be a complex and multiply determined construct. Thus, programs that target at-risk individuals based on single risk factors (e.g., elevated depressive symptoms) may not only fail to identify certain individuals who possess this risk factor but may also miss identifying individuals who are *not* at risk according to this criterion but may nevertheless be vulnerable due to other factors or circumstances (e.g., low SES, sexual abuse history). Thus, universal or selective programs that target various components of wellness enhancement may confer considerable benefits on more individuals than we would be able adequately to screen for indicated risk. Interestingly, a recent meta-analysis (Jané-Llopis et al., 2003) indicates that effect sizes have not differed significantly for universal, selective, and indicated preventive interventions with respect to reduction of depressive symptoms. A weighted mean effect size of 0.22 was obtained, indicating that each of the prevention levels is associated with an 11% improvement in depressive symptoms. Interventions that combine different levels of preventive intervention also appear promising (Jané-Llopis et al., 2003).

Different levels of prevention can be employed sequentially. Multitiered prevention programs in which the various levels of preventive intervention are offered on a continuum might offer universal preventive services first, followed by selective and then indicated services for individuals who fail to respond to initial intervention efforts. For instance, Parks and Herman (2003) cite the effective use of positive behavior supports in schoolwide initiatives to promote effective student behaviors. An entire school system (or network of school systems) implements a program of behavioral guidelines. Students who do not demonstrate positive outcomes receive more selective group interventions, and individualized care plans are offered to those who continue to have difficulty. This sort of strategy could be applied in schoolwide depression prevention programs for girls. Preadolescent girls could receive several sessions of cognitive-behavioral skills training designed to decrease depressive risk factors (e.g., ruminative coping) and promote resiliency factors (e.g., assertiveness). Students who manifested depressive symptoms even after the intervention could be assigned to receive a more intensive preventive intervention or referred for treatment, if needed.

In addition, efforts to ameliorate nationwide social, economic, and political norms that may currently contribute to women's risk for depression can be conceptualized as universal prevention strategies. For instance, media efforts to promote more realistic images of women and women's bodies, reductions in the wage gap between male and female workers, and improvements in parental leave policies may all have potential to positively impact risk factors for depression among women. When we begin to conceptualize the risk of depression as located not only within the individual (e.g., via biological and personality risk factors) but within the broader context of family, community, and society, it becomes clear that policies or social reform efforts that target harmful aspects of those broader contexts may have far-reaching impacts on the transmission of risk. Ultimately, multilevel prevention strategies may prove more powerful than interventions targeted only to individuals.

Adoption of Interdisciplinary Prevention Frameworks

All too often, biological psychiatry, psychology, epidemiology, sociology, and public health have pursued independent paths of theory and research. Prevention programs are likely to be far more effective at targeting complex and multiply-determined outcomes such as depression if they encompass methods and data from multiple disciplines. This is particularly true for community-based universal or selective prevention efforts in which macro-level, contextual factors are clearly implicated. Research that examines interactions among biological, psychological, and psychosocial factors (e.g., Kendler et al., 2002) is a promising step in this direction, but we still have a long way to go.

Moving beyond individual-level risk factors (e.g., cognitive distortions, low self-esteem) to macro-level risk factors (e.g., poverty, sexual and racial discrimination, income inequality) involves going beyond the traditional purview of psychological theory and research to incorporate perspectives from epidemiology, sociology, and public health. As some psychologists have noted (e.g., Brown, Abe-Kim, & Barrio, 2003), if we do not incorporate macro-level factors and interdisciplinary frameworks, we risk treating the symptoms without addressing the underlying causes. Women's lives merit consideration in context, taking into account the social and economic settings in which females live and in which they continue to be at a relative disadvantage to males with respect to power, social status, and resources. This sort of broader perspective has the potential to inform both theory and intervention development in very promising directions.

A growing number of researchers are advocating integrative theoretical perspectives with a multidisciplinary focus. For instance, in the field of social epidemiology, Krieger's ecosocial perspective (e.g., Krieger, 2001) weds biological and social analyses of population health to explain health disparities. This perspective employs multilevel, dynamic models that reflect the reciprocal influences among individual-level factors and macro-level, societal determinants of health (Krieger, 2001). For instance, Krieger proposes six interrelated pathways that create excess risk of hypertension among African Americans: economic and social deprivation, toxic substances and hazardous conditions, socially inflicted trauma, targeted marketing of commodities, inadequate health care, and resistance to racial oppression.

Developing this type of comprehensive model for negative mental health outcomes (e.g., depression) would expand our theoretical scope and our ability to explain mental health disparities across individuals of different genders, races, cultures, and socioeconomic strata. Such multilevel models would also point the way toward development of multilevel prevention strategies that target not only the individual (e.g., mood management skills) but also macro-level structures and processes, such as families, schools, and neighborhoods (e.g., modifying curricula, increasing community resources).

Interdisciplinary collaboration is important, not only from the perspective of model-building and intervention development but also with respect to dissemination and implementation of prevention programs. For instance, Clarke et al. (2003) argue persuasively that depression prevention researchers would do well to ally with other health advocates and policy makers in areas that are associated with depression, such as medical diseases (diabetes, cardiovascular disease) and adverse health behaviors (antismoking campaigns, violence prevention). Such collaborative partnerships can create additional resources and finances with which to fund prevention programs and raise public awareness of their importance.

Increased Attention to Race, Ethnicity, Culture, and Community

Roosa and his colleagues aptly note that a one-size-fits-all approach to prevention is unlikely to be effective (Roosa, Dumka, Gonzales, & Knight, 2002). Indeed, cultural and ethnic groups vary considerably in symptom presentation, attitudes toward mental health services, beliefs regarding health and disease, and risk and protective factors. Unfortunately, most psychological interventions to date have been designed and tested on European American populations (Bernal & Scharron-del-Rio, 2001). We must not allow prevention work to follow the same pattern.

Roosa and his colleagues suggest that research incorporating diverse populations should address (1) the values, beliefs, and goals particular to a group's ecological niche, (2) the way acculturation and enculturation processes shape group and individual-level risk and protective factors, (3) measurement equivalence of assessment tools used with diverse cultural groups, and (4) developing a process approach, in which trust is established within the community in question, community goals are acknowledged and incorporated into the research, and intervention programs are integrated into the community in such a way that they can be maintained over time, accruing long-term positive outcomes. Community research is not only a desirable method for studying cultural and ethnic factors in context, but it also has the potential to serve community needs and goals and, if successful, can result in the implementation of intervention programs that continue long after the research study has ended (see Ialongo, 2002).

In addition, because minorities are overrepresented among high-risk groups, we need more culturally competent researchers and culturally appropriate prevention programs (Roosa et al., 2002). The lack of people of color within academic research settings is a serious impediment to quality research on minority individuals and communities, and recruitment efforts have not yet succeeded in overcoming this problem. Eddy, Martinez, Morgan-Lopez, Smith, and Fisher (2002) advocate a community collaborative approach in which training and educational opportunities are offered to individuals within minority communities as a way of attracting minorities to the field of psychology and prevention research. With respect to research on depression prevention among women, the field would benefit from employing a community collaborative approach to involve women of diverse racial, ethnic, and cultural identities, socioeconomic backgrounds, and sexual orientations.

Nontraditional Delivery of Prevention Services

The recent strategic plan for mood disorders research crafted by the National Institute of Mental Health includes recommendations to develop and evaluate nontraditional mental health delivery approaches (Hollon et al., 2002). As our field becomes more committed to addressing the needs of underserved female populations and more adept at developing and delivering preventive interventions to these populations, we will need to be creative in moving beyond traditional service delivery methods. Many women do not have the means to attend weekly meetings at a mental health care clinic. Social stigma, lack of child care, unavailability of transportation, language barriers, and lack of knowledge about existing resources are all potential barriers to use of traditional services. Nontraditional service delivery methods have included interventions via the mail, use of paraprofessionals and figures within the community, use of self-help materials, and use of radio and television (e.g., Christensen, Miller, & Muñoz, 1978; Muñoz, VanOss Marin, Posner, & Perez-Stable, 1997). In addition, Muñoz and his colleagues are currently developing mood management tools for use in Internet-based smoking cessation trials (Lenert et al., 2003). Such interventions have the potential to be adapted for use with various problems and populations.

The Internet may become a valuable resource for reaching certain populations of women who may be reluctant to seek help in a mental health clinic. Currently, the Internet is used predominantly by individuals of higher socioeconomic status who have the resources to own or access computers. However, Muñoz and his colleagues are committed to devising ways of increasing access of low-income individuals to the Internet and providing them with training in its use. Thus, Web-based programs that instruct women in mood management and life coping skills may become a useful component of future prevention programs, either as an adjunct to interventions delivered in person or as a sole means of intervention.

SUMMARY AND RECOMMENDATIONS

In summary, we propose that depression prevention in women is feasible and should be accorded high priority by researchers and funding agencies. In addition, we have outlined a number of recommendations concerning how development and implementation of prevention efforts might proceed.

- Adolescence and pregnancy appear to represent critical developmental periods in which to prevent depression in women. A focus on these life stages may help reduce transmission of depression to the next generation.
- High-risk female subpopulations, such as women living in poverty and female survivors of trauma and abuse, merit increased attention in prevention research.
- Although reduction of depression incidence rates is extremely important, it is not the only valid measure of successful prevention. The use of multiple outcome measures, including depressive symptomatology and positive health outcomes, is recommended.
- Indicated prevention trials are necessary to advance the field, and we believe such trials should target the prevention of major depression in women. However, indicated trials are not sufficient. Universal and selective approaches involving schools and communities also merit increased attention and funding.
- Interdisciplinary collaboration is critical for developing accurate models of risk and effective prevention programs, as well as for implementing and disseminating such programs. Collaboration can facilitate the adoption of multilevel prevention strategies.
- Prevention research should include diverse populations (e.g., representative of various racial, ethnic, and cultural backgrounds, socioeconomic strata, and sexual orientations). Populations with documented increases in the prevalence of major depression from one generation to the next, such as Mexican immigrants, are particularly important groups to target.
- Prevention research should explore and utilize nontraditional service delivery methods to impact a broader range of individuals. These include radio, television, print media, and the Internet.

CONCLUSION

The depression prevention trials conducted by Clarke and his colleagues yielded a reduction in MDE incidence of approximately 44 to 68% (Clarke et al., 1995, 2001). This suggests that, if prevention programs of comparable efficacy were routinely available, we might be able to lower annual MDE incidence roughly *by half*. In addition to reducing the suffering of those afflicted, at least three other benefits would accrue. First, the prevention of first episodes of depression would prevent or delay the start of a chronic disease course, reducing potential chronic disability and health care costs. Second, we would reduce by some yet unknown proportion the number of deaths due to the major causes of preventable death attributable in part to depression, such as smoking, substance abuse, and deaths due to firearms, car accidents, and sexually transmitted disease. Finally, we would reduce the negative impact of depression on the children of women at risk, thereby potentially reducing transmission of depression to the next generation. The cumulative effect of these prevention outcomes on our society within two or three generations would be massive. It is time to launch a concerted effort to prevent depression, with a strong focus on women.

References

Abramson, L. Y., Alloy, L. B., & Metalsky, G. I. (1988). The cognitive diathesis-stress theories of depression. In D.C. McCann & N. Endler (Eds.), *Depression: New directions in theory, research, and practice*. Toronto: Wall & Thompson.

Austin, M. P. (2003). Targeted group antenatal prevention of postnatal depression: A review. *Acta Psychiatrica Scandinavica, 107*, 244–250.

Barnett, R. C., & Hyde, J. S. (2001). Women, men, work, and family: An expansionist theory. *American Psychologist, 56*, 781–796.

Bebbington, P. E., Dunn, G., Jenkins, R., Lewis, G., Brugha, T., Farrell, M., et al. (1998). The influence of age and sex on the prevalence of depressive conditions: Report from the National Survey of Psychiatric Morbidity. *Psychological Medicine, 28*, 9–19.

Belle, D., & Doucet, J. (2003). Poverty, inequality, and discrimination as sources of depression among U.S. women. *Psychology of Women Quarterly, 27*, 101–113.

Bernal, G., & Scharron-del-Rio, M. R. (2001). Are empirically supported treatments valid for ethnic minorities? Toward an alternative approach for treatment research. *Cultural Diversity and Ethnic Minority Psychology, 7*, 328–342.

Borowsky, S. J., Rubenstein, L. V., Meredith, L. S., Camp, P., Jackson-Triche, M., & Wells, K. B. (2000). Who is at risk of nondetection of mental health problems in primary care? *Journal of General Internal Medicine, 15*, 381–388.

Brown, C., Abe-Kim, J., & Barrio, C. (2003). Commentary: "Treatment is not enough: We must prevent depression in women," by H-H. Le, R. F. Muñoz, C. Ippen, & J. L. Stoddard (Eds.), *Prevention and Treatment, 6*, Article 18. Retrieved August 11, 2003, from http://journals.apa.org/prevention/volume6/pre0060018c.html.

Brugha, T. S., Wheatley, S., Taub, N. A., Culverwell, A., Friedman, T., Kirwan, P., et al. (2000). *Psychological Medicine, 30*, 1273–1281.

Burt, V. K., & Stein, K. (2002). Epidemiology of depression: Throughout the female life cycle. *Journal of Clinical Psychiatry, 63*, 9–15.

Cardemil, E. V., Reivich, K. J., & Seligman, M. E. P. (2002). The prevention of depressive symptoms in low-income minority middle school students. *Prevention and Treatment, 5*, Article 1. Retrieved August 11, 2003, from http://journals.apa.org/prevention/volume5/pre0050008a.html.

Caspi, A., Sugden, K., Moffitt, T. A., Taylor, A., Craig, I. W., Harrington, H., et al. (2003). Influence of life stress on depression: Moderation by a polymorphism in the 5-HTT gene. *Science, 301*, 386–389.

Christensen, A., Miller, W. R., & Muñoz, R. F. (1978). Paraprofessionals, partners, peers, paraphernalia, and print: Expanding mental health service delivery. *Professional Psychology*, 249–269.

Clarke, G. N., DeBar, L., Lynch, F., & Wisdom, J. (2003). Issues in the prevention of depression in women: A commentary on H-H. Le, R. F. Muñoz, C. Ippen, & J. L. Stoddard (Eds.), *Prevention and Treatment, 6*, Article 12. Retrieved August 11, 2003, from http://journals.apa.org/prevention/volume6/pre0060012c.html.

Clarke, G. N., Hawkins, W., Murphy, M., & Sheeber, L. B. (1993). School-based primary prevention of depressive symptomatology in adolescents: Findings from two studies. *Journal of Adolescent Research, 8*, 183–204.

Clarke, G. N., Hawkins, W., Murphy, M., Sheeber, L. B., Lewinsohn, P. M., & Seeley, J. R. (1995). Targeted prevention of unipolar depressive disorder in an at-risk sample of high school adolescents: A randomized trial of a group cognitive intervention. *Journal of the American Academy of Child and Adolescent Psychiatry, 34*, 312–321.

Clarke, G. N., Hornbrook, M., Lynch. F., Polen, M., Gale, J., Beardslee, W., et al. (2001). A randomized trial of a group cognitive intervention for preventing depression in adolescent offspring of depressed parents. *Archives of General Psychiatry, 58*, 1127–1134.

Cuijpers, P. (2003). Examining the effects of prevention programs on the incidence of new cases of mental disorders. *American Journal of Psychiatry, 160*, 1385–1391.

Cyranowski, J. M., Frank, E., Young, E., & Shear, M. K. (2000). Adolescent onset of the gender difference in lifetime rates of major depression. *Archives of General Psychiatry, 57*, 21–27.

Depression Guideline Panel. (1993). *Depression in primary care: Volume 2. Treatment of major depression* (Clinical Practice Guideline No. 5, AHCPR Publication No. 93-0552). Rockville, MD: U.S. Department of Health and Human Services, Public Health Service, Agency for Health Care Policy and Research.

Eddy, M. J., Martinez, C., Morgan-Lopez, A., Smith, P., & Fisher, P. A. (2002). Diversifying the ranks of prevention scientists through a community collaborative approach to education. *Prevention and Treatment, 5*, Article 3. Retrieved August 11, 2003, from http://journals.apa.org/prevention/volume5/pre0050003a.html.

Elliott, S. A., Leverton, T. J., Sanjack, M., Turner, H., Cowmeadow, P., Hopkins J., et al. (2000). Promoting mental health after childbirth: A controlled trial of primary prevention of postnatal depression. *British Journal of Clinical Psychology, 39*, 223–241.

Fergusson, D. M., Swain-Campbell, N. R., & Horwood, L. J. (2002). Does sexual violence contribute to elevated rates of anxiety and depression in females? *Psychological Medicine, 32*, 991–996.

Gillham, J. E. (2003). Targeted prevention is not enough. *Prevention and Treatment, 6*, Article 17. Retrieved August 11, 2003, from http://journals.apa.org/prevention/volume6/pre0060017c.html.

Gilman, S. E., Kawachi, I., Fitzmaurice, G. M., & Buka, S. L. (2002). Socioeconomic status in childhood and the lifetime risk of major depression. *International Journal of Epidemiology, 31*, 359–367.

Gore, S., Aseltine, R. H., & Colton, M. E. (1992). Social structure, life stress, and depressive symptoms in a high school-aged population. *Journal of Health and Social Behavior, 33*, 97–113.

Gotlib, I. H., Lewinsohn, P. M., & Seeley, J. R. (1995). Symptoms versus a diagnosis of depression: Differences in psychosocial functioning. *Journal of Consulting & Clinical Psychology, 63*, 90–100.

Hankin, B. L., & Abramson, L. Y. (2001). Development of gender differences in depression: An elaborated cognitive vulnerability-transactional stress theory. *Psychological Bulletin, 127*, 773–796.

Hankin, B. L., & Abramson, L. Y. (2002). Measuring cognitive vulnerability to depression in adolescence: Reliability, validity, and gender differences. *Journal of Clinical Child and Adolescent Psychology, 31*, 491–504.

Hays, R. D., Wells, K. B., Sherbourne, C. D., Rogers, W., & Spritzer, K. (1995). Functioning and well-being outcomes of patients with depression compared with chronic general medical illnesses. *Archives of General Psychiatry, 52*, 11–19.

Heilemann, M. V., Lee, K. A., & Kury, F. S. (2002). Strengths and vulnerabilities of women of Mexican descent in relation to depressive symptoms. *Nursing Research, 51*, 175–182.

Hollon, S. D., Muñoz, R. F., Barlow, D. H., Beardslee, W. R., Bell, C. C., Bernal, G., et al. (2002). Psychosocial intervention development for the prevention and treatment of depression: Promoting innovation and increasing access. *Biological Psychiatry, 52*, 610–630.

Howe, G. W. (2003). Next-generation trials for the prevention of depression: Lessons from longitudinal epidemiology and developmental psychopathology. *Prevention and Treatment, 6*, Article 14. Retrieved August 11, 2003, from http://journals.apa.org/prevention/volume6/pre0060014c.html.

Ialongo, N. (2002). Wedding the public health and clinical psychological perspectives as a prevention scientist. *Prevention & Treatment, 5*, Article 4. Retrieved August 11, 2003, from http://journals.apa.org/prevention/volume5/pre0050004a.html.

Ingram, R. E., Cruet, D., Johnson, B. R., & Wisnicki, K. S. (1988). Self-focused attention, gender, gender role, and vulnerability to negative affect. *Journal of Personality and Social Psychology, 55*, 967–978.

Ingram, R. E., Lumry, A., Cruet, D., & Sieber, W. (1987). Attentional processes in depressive disorders. *Cognitive Therapy and Research, 8*, 139–152.

Jané-Llopis, E., Hosman, C., Jenkins, R., & Anderson, P. (2003). Predictors of efficacy in depression prevention programmes. *British Journal of Psychiatry, 183*, 384–397.

Kahn, R. S., Wise, P. H., Kennedy, B. P., & Kawachi, I. (2000). State income inequality, household income, and maternal mental and physical health: Cross-sectional national survey. *British Medical Journal, 321,* 1311–1315.

Katz, J., Joiner, T. E., & Kwon, P. (2002). Membership in a devalued social group and emotional well-being: Developing a model of personal self-esteem, collective self-esteem, and group socialization. *Sex Roles, 47,* 419–431.

Kendler, K. S., Gardner, C. O., & Prescott, C. A. (2002). Toward a comprehensive developmental model for major depression in women. *American Journal of Psychiatry, 159,* 1133–1145.

Kendler, K. S., Thornton, L. M., & Prescott, C. A. (2001). Gender differences in the rates of exposure to stressful life events and sensitivity to their depressogenic effects. *American Journal of Psychiatry, 158,* 587–593.

Kessler, R. C. (2003). Epidemiology of women and depression. *Journal of Affective Disorders, 74,* 5–13.

Kessler, R. C., Berglund, P. A., Demler, O., Jin, R., Koretz, D., Merikangas, K. R., et al. (2003). The epidemiology of major depressive disorder: Results from the National Comorbidity Survey Replication (NCS-R). *Journal of the American Medical Association, 289,* 3095–3105.

Kessler, R. C., McGonagle, K. A., Nelson, C. B., Hughes, M., Swartz, M., & Blazer, D. G. (1994). Sex and depression in the National Comorbidity Survey. II: Cohort effects. *Journal of Affective Disorders, 30,* 15–26.

Keyes, C. L. M. (2004). The nexus of cardiovascular disease and depression revisited: The complete mental health perspective and the moderating role of age and gender. *Aging and Mental Health, 8,* 266–274.

Klonoff, E. A., Landrine, H., & Campbell, R. (2000). Sexist discrimination may account for well-known gender differences in psychiatric symptoms. *Psychology of Women Quarterly, 24,* 93–99.

Kohn-Wood, L. P. (2003). What is preventing prevention? Making the case for focusing on depression among women. *Prevention and Treatment, 6,* Article 13. Retrieved August 11, 2003, from http://journals.apa.org/prevention/volume6/pre0060013c.html.

Krieger, N. (2001). Theories for social epidemiology in the 21st century: An ecosocial perspective. *International Journal of Epidemiology, 30,* 668–677.

Le, H-H., Muñoz, R. F., Ghosh Ippen, C., & Stoddard, J. L. (2003). Treatment is not enough: We must prevent major depression in women. *Prevention and Treatment, 6,* Article 10. Retrieved August 11, 2003, from http://journals.apa.org/prevention/volume6/pre0060010a.html.

Lenert, L., Muñoz, R. F., Stoddard, J., Delucchi, K., Bansod, A., Skoczen, S., et al. (2003). Design and pilot evaluation of an Internet smoking cessation program. *Journal of the American Medical Informatics Association, 10,* 16–20.

Lorant, V., Deliège, D., Eaton, W., Robert, A., Philippot, P., & Ansseau, M. (2003). Socioeconomic inequalities in depression: A meta-analysis. *American Journal of Epidemiology, 157,* 98–112.

Maciejewski, P. K., Prigerson, H. G., & Mazure, C. M. (2001). Sex differences in event-related risk for major depression. *Psychological Medicine, 31,* 593–604.

Marks, M. N., Siddle, K., & Warwick, C. (2003). Can we prevent postnatal depression? A randomized controlled trial to assess the effect of continuity of midwifery care on rates of postnatal depression in high-risk women. *Journal of Maternal-Fetal and Neonatal Medicine, 13*, 119–127.

Mazure, C. M., Keita, G. P., & Blehar, M. C. (2002). *Summit on women and depression: Proceedings and recommendations.* Washington, DC: American Psychological Association. (Available online at www.apa.org/pi/wpo/women& depression.pdf.)

McGinnis, J. M., & Foege, W. H. (1993). Actual causes of death in the United States. *Journal of the American Medical Association, 270*, 2207–2212.

Mendelson, T., Le, H-N., Soto, J., Lieberman, A., & Muñoz, R. F. (2003, November). *Prevention of depression in low-income, pregnant women: An intervention development study.* Paper presented at the annual meeting of the Association for the Advancement of Behavior Therapy, Boston, MA.

Mirowsky, J., & Ross, C. E. (1989). *Social causes of psychological distress.* New York: Aldine de Gruyter.

Mrazek, P., & Haggerty, R. (1994). *Reducing risks for mental disorders: Frontiers for preventive intervention research.* Washington, DC: National Academy Press.

Muñoz, R. F. (2001). How shall we ensure that the prevention of onset of mental disorder becomes a national priority? *Prevention and Treatment, 4.* Retrieved August 11, 2003, from http://journals.apa.org/prevention/volume4/pre0040026c.html.

Muñoz, R. F., Ghosh-Ippen, C., Le, H-L., Lieberman, A. F., Diaz, M. A., La Plante, L. (2001a). The Mothers and Babies Course: A Reality Management Approach [Participant Manual]. Retrieved September 26, 2004, from http://www.medschool.ucsf.edu/latino/pdf/manuals.aspx.

Muñoz, R. F., Ghosh-Ippen, C., Le, H-L., Lieberman, A. F., Diaz, M. A., La Plante, L. (2001b). El Curso de "Mamás y Bebés: Como Construir una Realidad Saludable" [Cuaderno para Estudiantes]. Retrieved September 26, 2004, from http://www.medschool.ucsf.edu/latino/pdf/manuals.aspx.

Muñoz, R. F., Mrazek, P. J., & Haggerty, R. J. (1996). Institute of Medicine report on prevention of mental disorders – Summary and commentary. *American Psychologist, 51*, 1116–1122.

Muñoz, R. F., VanOss Marín, B. V., Posner, S. F., & & Pérez-Stable, E. J. (1997). Mood management mail intervention increases abstinence rates for Spanish-speaking Latino smokers. *American Journal of Community Psychology, 25*, 325–343.

Muñoz, R. F., & Ying, Y. (1993). *The prevention of depression: Research and practice.* Baltimore, MD: Johns Hopkins University Press.

Muñoz, R. F., Ying, Y. W., Bernal, G., Pérez-Stable, E. J., Sorensen, J. L., Hargreaves, W. A., et al. (1995). Prevention of depression with primary care patients: A randomized controlled trial. *American Journal of Community Psychology, 23*, 199–222.

Nolen-Hoeksema, S. (1990). *Sex differences in depression.* Stanford, CA: Stanford University Press.

Nolen-Hoeksema, S., Morrow, J., & Fredrickson, B. L. (1993). Response styles and the duration of episodes of depressed mood. *Journal of Abnormal Psychology, 102*, 20–28.

O'Hara, M. W. (1994). *Postpartum depression: Causes and consequences.* New York: Springer Verlag.

Park, N. (2003). Building wellness to prevent depression. *Prevention and Treatment, 6*, Article 16. Retrieved August 11, 2003, from http://journals.apa.org/prevention/volume6/pre0060016c.html.

Parks, A. C., & Herman, K. C. (2003). A sociocultural perspective on the primary prevention of depression. *Prevention and Treatment, 6*, Article 15. Retrieved August 11, 2003, from http://journals.apa.org/prevention/volume6/pre0060015c.html.

Petersen, R. W., & Quinlivan, J. A. (2002). Preventing anxiety and depression in gynaecological cancer: A randomized controlled trial. *British Journal of Gynaecology, 109*, 386–394.

Piccinelli, M., & Wilkinson, G. (2000). Gender differences in depression: A critical review. *British Journal of Psychiatry, 177*, 486–492.

Roosa, M. W., Dumka, L. E., Gonzales, N. A., & Knight, G. P. (2002). Cultural/ethnic issues and the prevention scientist in the 21st century. *Prevention & Treatment, 5*, Article 5. Retrieved August 11, 2003, from http://journals.apa.org/prevention/volume2/pre0020005a.html.

Seligman, M. E. P., & Csikszentmihalyi, M. (2000). Positive psychology: An introduction. *American Psychologist, 55*, 5–14.

Seligman, M. E. P., Schulman, P., DeRubeis, R. J., & Hollon, S. D. (1999). The prevention of depression and anxiety. *Prevention and Treatment, 2*, Article 8. Retrieved August 11, 2003, from http://journals.apa.org/prevention/volume2/pre0020008a.html.

Stamp, G. E., Williams, A. S., & Crowther, C. A. (1995). Evaluation of antenatal and postnatal support to overcome postnatal depression: A randomized, controlled trial. *Birth, 2*, 138–143.

Steiner, M., Dunn, E., & Born, L. (2003). Hormones and mood: From menarche to menopause and beyond. *Journal of Affective Disorders, 74*, 67–83.

Swim, J. K., Hyers, L. L., Cohen, L. L., & Ferguson, M. J. (2001). Everyday sexism: Evidence for its incidence, nature, and psychological impact from three daily diary studies. *Journal of Social Issues, 57*, 31–53.

Teti, D. M., & Gelfand, D. M. (1991). Behavioral competence among mothers of infants in the first year: The mediational role of maternal self-efficacy. *Child Development, 62*, 918–929.

U.S. Census Bureau. (2004, October). *DP-1. Profile of general demographic characteristics: 2000*. Retrieved October 4, 2004 from http://factfinder.census.gov/servlet/QTTable?_bm=y&-geo_id=01000US&-qr_name=DEC_2000_SF1_U_DP1&-ds_name=DEC_2000_SF1_U.

Vega, W. A., Kolody, B., Aguilar-Gaxiola, S., Alderete, E., Catalano, R., & Caraveo-Anduaga, J. (1998). Lifetime prevalence of DSM-III-R psychiatric disorders among urban and rural Mexican Americans in California. *Archives of General Psychiatry, 55*, 771–778.

Weiss, E. L., Longhurst, J. G., & Mazure, C. M. (1999). Childhood sexual abuse as a risk factor for depression in women: Psychosocial and neurobiological correlates. *American Journal of Psychiatry, 156*, 816–828.

Weissman, M. M., Bland, R. C., Canino, G. J., Faravelli, C., Greenwalk, S., Hwu, H-G., et al. (1996). Cross-national epidemiology of major depression and bipolar disorder. *Journal of the American Medical Association, 276*, 293–299.

Weissman, M. M., & Jensen, P. (2002). What research suggests for depressed women with children. *Journal of Clinical Psychiatry, 63*, 641–647.

Yu, D. L., & Seligman, M. E. P. (2002). Preventing depression in Chinese students. *Prevention and Treatment, 5*, Article 9. Retrieved August 11, 2003, from http://journals.apa.org/prevention/volume5/pre0050009a.html.

Zlotnick, C., Johnson, S. L., Miller, I. W., Pearlstein, T., & Howard, M. (2001). Postpartum depression in women receiving public assistance: Pilot study of an interpersonal-therapy-oriented group intervention. *American Journal of Psychiatry, 158*, 638–640.

19

Women and Depression

Research, Theory, and Social Policy

Jean A. Hamilton and Nancy Felipe Russo

The World Health Organization (WHO) has found gender differences in depression in all regions of the globe and has identified clinical depression as a leading cause of disease-related disability among the world's women (Demyttenaere et al., 2004). Depressive illness is known to be associated with (Hudson et al., 2003) and to affect the clinical course of certain medical illnesses (or risk factors for illness; Frasure-Smith, Lesperance, & Talajic, 1993; McEwen, 2003; McEwen & Lasley, 2002). Further, clinical depression has been found to have multiple social roots associated with inequality, powerlessness, and devaluation, including poverty and hunger, discrimination, poor working conditions, violence, and unwanted pregnancy (Desjarlais, Eisenberg, Good, & Kleinman, 1995; McGrath, Keita, Strickland, & Russo, 1990; Ustun & Sartorius, 1995; WHO, 2000). Given the personal, social, and economic impact of depression on women, their families, and society, policy issues take on new urgency.

The fact that depression is so strongly associated with powerlessness and devaluation makes a feminist analysis that values women and seeks to empower them particularly relevant (Worell, 2001). Feminists across the disciplines have taken the lead in fostering research and treatment approaches to women's mental health (Carmen, Russo, & Miller, 1981; Rieker & Carmen, 1984; Russo, 1984, 1995; White, Russo, & Travis, 2001a). In doing so they have begun to identify gender-related aspects of women's roles and life circumstances that contribute to women's risk for depression.

The 1990 American Psychological Association (APA) Presidential Task Force on Women and Depression (McGrath, et al., pp. 36–39) report summarized some of this work. That group highlighted the impact of poverty, discrimination, victimization, reproductive events, and interpersonal relationships at home and at work, as well as biological and psychological factors and styles of coping with stress. These themes were echoed at the more recent Depression Summit (Mazure, Keita, & Blehar, 2002).

There is no doubt that adversity and stressors in women's lives contribute to the excess of depression in women compared to men. The questions now revolve around identifying the mechanisms that govern the relationship between gender-related stressors and women's mental health outcomes.

As the research presented in this handbook has shown, there is consensus on the view that women have higher rates of depression than men and the difference cannot be accounted for by artifact or measurement error. The question is now, *Why* are rates of depression for women higher than those of men? Finding the answer is a necessary condition for developing effective social policies to address the gender gap in depression as a public health issue and ultimately will be key to effective treatment and prevention of depression in women. The chapters in this volume have considered a host of biological, psychological, and social factors that affect the development, diagnosis, treatment, and prevention of depression in women. As a body of work, they demonstrate that there is no easy (much less, a single) answer to why there is a gender gap in rates of depression, and scientific findings are incomplete. Nonetheless, it is clear that gender differences in the onset and course of depression cannot be entirely explained by genetic or hormonal systems.

We argue here that understanding the gender gap in depression begins with the recognition that gender is a complex multilevel cultural construct. Finding needed answers will require new theories that articulate how various aspects of gender mediate and moderate the interactive effects of social, psychological, and biological factors to determine the onset and course of depression. In this chapter, after presenting epidemiological findings of particular relevance, we discuss the theoretical and methodological implications of recognizing gender as a cultural construct, highlight needed areas for research, and consider policy implications of current knowledge, some of which is discussed in previous chapters of this volume. In the process we selectively update previous policy recommendations.

The epidemiological picture provided by Kessler (this volume), along with the social, political, and economic models (Marecek, Lennon, and Brayboy-Jackson & Williams, all in this volume) and public health models of risk (Sterk, Theall, & Elifson, this volume), support the need for a social and cultural perspective. These chapters illuminate deficiencies in conceptualization as well as methodology in previous research that need to be addressed for advancement to occur. Several findings have particular significance for our analysis:

- There is a gender difference in depressed mood that begins in childhood, but the gender gap in depression rates begins to emerge in early adolescence (ages 11–14). These gender differences set the stage for higher rates of depressive episodes for women compared to men.

- Prior psychiatric history mediates gender differences in depressive disorders, making understanding the depressogenic effects of the events of early childhood and adolescence a necessary condition for preventing women's depression over the life cycle.
- The extent to which biological versus social factors contribute to the emergence of the gender difference in depression in early adolescence depends on the social context; but what it is about the social context that specifically promotes or reduces the gender difference in depressive disorders is not well understood.
- Women's day-by-day patterns of being depressed suggest that women may be more likely to experience short-term depressive episodes than men, possibly as a response to stressors in their lives.
- Comorbidity research suggests that the pathways to depression differ for women and men in gender-stereotyped ways. Specifically, diagnoses of anxiety and depression are more likely to be found together for women whereas depression and substance abuse are more likely to be paired for men.
- Cross-nationally, there is a 15-fold (or higher) variation in rates of depression, suggesting that cultural differences may be important.
- Cross-nationally, rates of depressive symptoms (which can predict the clinical syndrome) vary with indicators (albeit crude) of gender inequality (Arrindell, Steptoe, & Wardle, 2003).
- Rates of depressive symptoms and the clinical syndrome vary within nations by regions (urban versus rural), between states within the United States that are ranked according to indicators of gender inequality (see especially, Chen, Subramanian, Acevedo-Garcia, & Kawachi, 2005), and by neighborhoods (Ross, 2000).

Experiencing a mix of stressors that combine danger and loss is strongly related to the development of comorbid anxiety and depression (Brown, Harris, & Eales, 1993), pointing to the need for analyses that consider gender-related stressors. Such analyses are needed to explain the links between women's social position and poor mental health outcomes and to design effective public policies for intervention. We argue that conceptualizing the complex interplay of the biological, psychological, social, and structural factors that result in women's high risk for depression begins at the cultural level.

GENDER AND CULTURE

The past two decades have witnessed a revolutionary change in interest and understanding of the relationship of culture to personality and the self (Lehman, Chiu, & Schaller, 2004; Triandis & Suh, 2002) and psychopathology (Kleinman, 1986, 1988; Kleinman & Good, 1985; Lopez & Guarnaccia,

2000). These productive new areas of research have resulted in theoretical insights as well as specific research findings. Researchers have begun to "unpack" culture (Lopez & Guarnaccia, 2000, p. 573) and have shown the importance of understanding just what it is about our social worlds that that affects risk for psychopathology. What is all too often neglected in a cultural analysis, however, is the recognition that gender is a social construct that is a product of culture. Gender needs to be unpacked as well as culture.

Gender is the social construct that defines the meaning of female and male in a particular culture. In Western society, a person's gender is typically assigned at birth based on biological sex, which may be defined anatomically or genetically, depending on the situation. Through the process of gender role socialization, gender shapes the way we construe our selves, and the way we process information about our world is filtered through that gendered self-construal (Cross & Madsen, 1997).

Gender shapes the relations between women and men and determines the gender-appropriateness of behavioral, psychological, and social characteristics of individuals over the life cycle. It has many interconnected elements – including gendered traits, emotions, values, expectations, norms, roles, environments, and institutions – that change and evolve within and across cultures and over time (Bourne & Russo, 1998). Gender is also a master (or a meta-) status in society and it typically accords women with less power, privilege, and resources than men in ways that have profound implication for women's mental health.

Gender does not exist in isolation from other dimensions of social difference. Researchers need to develop a "diversity mindfulness" that recognizes that the effects of gender may differ depending on its mix with other social identities (Russo & Vaz, 2001, p. 280). Age, ethnicity, race, sexual orientation, class, physical ability, and size are other social dimensions associated with stigmatized identities that may elicit prejudice and discrimination, confer differential access to power and privilege, and converge with gender to magnify or diminish risk for depression in women (see Braboy-Jackson, this volume, for discussion of race/ethnicity and cultural issues in depression).

Gender's rules (i.e., expected behaviors, rewards, and sanctions for violating those expectations) change over the life cycle. Sometimes there is abrupt change as a result of discrete life events such as losing one's virginity, getting married, or having one's first child. The fact that the gender difference in depression begins to emerge in preadolescence underscores the importance of understanding the impact of childhood experiences as contributors to risk and resilience over the life cycle for designing effective prevention-related social policies.

Gender organizes women's roles at home and work in ways that place extraordinary burdens on women while at the same time limiting their access to coping resources. Marital inequality undermines women's mental health and increases their risk for depression (Steil, 1997). The revolutionary changes in women's status and participation in the workforce occurring over the past 50 years are yet to be accompanied by a concomitant sharing of responsibilities in the family. As Janice Steil has observed, women continue to "do more of the work of the home and relationships and provide better emotional support for husbands than husbands provide for them" (Steil, 2001, p. 348). She describes how gendered inequalities in family structures limit the aspirations of girls and make wives financially vulnerable to their husbands. She also describes how such inequalities create gender differences in perceived entitlements, give different meanings to the resources women and men bring to their relationships, and legitimizes a gendered world of work that does not provide adequate support to working parents (Okin, 1989).

Lennon (this volume) provides a more in-depth discussion of the complexities of the relationship between work and depression, social policies affecting that relationship, and implication of findings for treatment, intervention, and public policy, including health, educational, and macro-level economic policies. Here we only add emphasis to the importance of increasing attention to women in traditionally female jobs having characteristics known to produce occupational stress, that is, those with high workload combined with role conflict and ambiguity, repetitive tasks, low control over work procedures and products, job insecurity, and poor workplace relationships (Swanson, Piotrkowski, Keita, & Becker, 1997). Public policy recommendations aimed at alleviating occupational stress are contained in the APA Women's Health Agenda (APA, 1996), some of which are highlighted at the end of this chapter.

Women's gender-related mental health risks resulting from poverty and from social and political inequality associated with women's roles and status interact with other gender-related personal and social risk factors for depression, including experience of childhood sexual abuse, rape, other forms of violence and victimization, prejudice, and discrimination. Recently, leaders in depression research have begun to develop more complex frameworks that apply a developmental perspective to understanding gender differences in depression over the life cycle and that recognize the impact and interaction of gendered roles and life events (e.g., Nolen-Hoeksema & Girgus, 1994).

Such developmental models need to incorporate the structural and symbolic aspects of gender in the analysis, however. The social systems that make up society are also cultural products that are gendered, that is, reflect cultural assumptions about the way women and men should function in society. The roles, expectations, and access to power and resources in these systems create a web of inequality for girls and women with a powerful mental health impact that requires prevention and intervention at the policy level.

Kessler (this volume) discusses the inconsistencies in and implications of the epidemiological picture for understanding emergence of the gender gap in depression in puberty. Evidence for the need for research at multiple levels, including the social system level, is seen in the impact of the structure of the school system on when gender differences in low self-esteem emerge (in the seventh grade in a system with a middle school compared to the ninth grade in a system with 4-year high school; Simmons & Blyth, 1987). Interventions at the structural level cannot be undertaken without reference to the public policies that create the structures of concern.

Culture engenders power and privilege and contributes to women's risk of depression in complex ways that have yet to be fully articulated (White, Russo, & Travis, 2001b). Mechanisms of power and privilege are found in the structures, policies, and procedures of systems and organizations and do not require the conscious will of an oppressor to have impact. These power dynamics can function in a culture as a "nonconscious ideology" (Bem & Bem, 1970, p. 84) that becomes incorporated into gendered customs, norms, and laws that even well-intentioned individuals may not challenge. Their impact is reflected in the relationship of societal inequality to women's mental health found in cross-cultural comparisons.

Gender and Depression: Inequality Matters

The United Nations collects and reports indicators of gender inequality (GI) cross-nationally (United Nations, 2004). As examples, GI indicators that are summarized by sex for many countries include average age of first marriage, rates of first- and second-level school participation and illiteracy, percentage representation in the labor force and in managerial positions, and rates of partner violence against women. Of these, we will consider violence.

Thinking about violence as a part of culture can help to put its relationship to women's mental health in a revealing new light. That is, some (and indeed, much) of the violence against women may have the function of structuring power relations between genders, thus operating as a culturally based mechanism for the control of women. Cross-cultural studies have documented the use of violation of gender role norms as a rationale for male violence and abuse. As Heise, Ellsberg, and Gottemoeller (1999) have documented, in many cultures, women are confined to traditional roles of housekeeping and child care. Meanwhile, male gender roles allow for flexibility (as long as the family is financially provided for) while entitling men to their wives' respect and obedience. Violations of male role entitlements become triggers for violence against women. Research has found that a wide range of women's gender role transgressions can trigger male violence, including wifely disobedience, talking back, not preparing food on schedule, failing to provide adequate care for home or children, raising questions about money, going somewhere without permission, refusing sex, or asking about girlfriends or showing suspicions of infidelity (Heise et al., 1999).

Cross-cultural evidence that violence functions as a form of social control for women is found in research documenting links between controlling behaviors and intimate partner violence in developing countries (Kishor & Johnson, 2004; United Nations Children's Fund, 2000). However, even in developed countries such as the United States, evidence suggests that abusive men connect masculinity with being able to control and dominate their partners (Goodrum, Umberson, & Anderson, 2001).

The conceptualization of gender as a dynamic cultural construct that operates at multiple levels has profound implications for research, clinical training, service delivery, prevention efforts, and public policy. Understanding gender's complex relationship to depression requires amending traditional biomedical models of mental health, crossing interdisciplinary boundaries, and asking questions at multiple levels and from multiple perspectives, including the individual, family, society, and culture. Unfortunately, one of the greatest barriers to advancing knowledge in this area is that even when new theoretical questions are envisioned the ability to answer them can be affected by the limitations of existing research methods.

Methodological Issues

Advances in theory and method go hand in hand. As Kessler (this volume), has pointed out, there are a number of methodological issues in previous depression research – including recall bias and the failure to distinguish between the initial onset and recurrence of depressive episodes – that undermine an understanding of gender effects on depression. Studying recurrence, in particular, requires developing and applying more sophisticated methods to examine interactive and reciprocal effects of events that occur in real-world contexts.

Both the WHO collaborative study (Ustun & Sartorius, 1995) and the National Comorbidity Study (Kessler et al., 1994) have examined both lifetime and 12-month comorbidity rates of depression with other disorders and found women to have higher rates of comorbidity compared to men. The gender difference in comorbidity patterns of depression suggests that the conceptualization of depressive and mood disorders as well as their measurement may need to be rethought.

It now appears, for example, that bipolar disorder may be underrecognized in women presenting with depression (Kack, 2004). Errors in diagnosis may lead to suboptimal treatment. Further, the clinical criteria for defining "remission" have required revision. Residual symptoms predict relapse. Despite the widely cited statistic that most patients achieve remission with treatment, this typically referred to only a 50% reduction in symptoms. It appears that women may be over-represented among those having residual symptoms (Karp et al., 2005).

We must also keep in mind that current quantitative measures of depression depend on self-report, interview, and diagnostic questions that are based on preexisting expert opinions about the nature and scope of the disorder. Such opinions reflect cultural assumptions about normality that may or may not reflect the realities of women's lives. As Stoppard and others have argued, there is a continuing need to develop innovative research methods that return attention to women's experiences and lived realities if knowledge is to progress (Gergen, 1988; Kimmel & Crawford, 1999; Russo & Dabul, 1994; Stoppard, 2000; White et al., 2001b).

Consequently, in addition to studying depression via prestructured, standardized measures that reflect current culturally bound (and perhaps gender-biased) conceptualizations of depression, women's experiences need to be explored by qualitative research methods that allow greater examination of women's sense of self and the symbolic and discursive conditions affecting women's experiences. Feminist critiques of science have led to a call for the use of a wider variety of research methods, including discourse analysis, ethnography, existential-phenomenological inquiry, focus groups, interviews, and narrative investigations (Kimmel & Crawford, 1999).

Researching gender's mental health impact requires respecting the Gestalt adage "the whole is more than the sum of its parts." Traditional reductionistic laboratory research methods that have served as the prototype for the scientific ideal are not equipped to answer many of the critically important questions about the relationship of gender to mental health. Reductionistic approaches cannot illuminate workings of gender any more than studying the properties of a hydrogen molecule can illuminate the functioning of ocean currents. In addition to the fact that the effects of any particular variable may be mediated or moderated by the presence of others, the meanings of events cannot be determined without reference to their context.

Feminists and others have long argued that failure of reductionistic experimental methods to account for meanings of behaviors and events that are derived from the social context has retarded the advancement of knowledge about women's lives and circumstances (Sherif, 1979; White et al., 2001b). The validity of their critique is confirmed by research on the dynamics of stress and coping that has documented the central role that cognitive appraisals play in determining mental health outcomes of stressful experiences (Lazarus & Folkman, 1984; Lazarus, 1993). Stress and coping models provide alternatives to traditional biomedical models and may be potentially useful in illuminating gender's relationship to depression, but the potential for these models cannot be realized without studying women's lives *in context* (or, as laboratory scientists might say, *in situ*) and identifying how various aspects of gender shape the appraisal process (Russo & Green, 1993).

Research on the impact of sexism has found that the negative impact of such events on mental health is not captured by traditional measures (e.g., life event scales) designed to assess proximal causes of distress, and the impact of such events is substantial (Krieger, Rowley, Herman, Avery, & Phillips, 1993; Landrine, Klonoff, Gibbs, Manning, & Lund, 1995). Experiencing sexist events (e.g., being treated unfairly by employer, called a sexist name like "bitch" or "cunt," being exposed to degrading sexual jokes) has been found to explain gender differences among college women and men in symptoms of depression, anxiety, and somatization (Klonoff, Landrine, & Campbell, 2000). Such findings suggest that qualitative and quantitative studies designed to (a) develop new measures of stressful life events over the life cycle as well as to (b) examine women's subjective experiences of sexism would contribute to explaining some of the gender gap in depression as well as other mood disorders.

In developing new measures it may be necessary to develop new conceptualizations and classifications of stressors depending on their meaning in the context of male–female power dynamics that affect appraisals of threat and coping resources. Distinguishing between social and other types of stress is an advance, but the implications of the fact that a social stressor has both immediate and long-term *threat* implications have yet to be fully recognized. Social stressors may be more likely to be associated with the potential for stigma and social exclusion for women, making social support a coping resource of direct relevance for dealing with the issues posed by the particular stressor, with resulting physiological benefits. Methods that assess number and *quality* of events are important. For example, a higher number of events associated with humiliation, entrapment, and bereavement have been found to contribute to increased risk for depression in women (Fullilove, 2002).

Consequently, comparing stress responses between men and women requires a complex gendered analysis. In particular, new theories of sex differences in stress responses, such as that posed by Taylor et al. (2000) do not sufficiently consider that the qualities of social stressors and of the women's social context may be determinants of what has been described as the tend-and-befriend response in women. Klein, Corwin, and Ceballos (this volume) discuss the biology of the stress response and discuss such theories in more detail.

Another advance in stress research that has implications for understanding the comorbidity of depression with other medical illnesses (Hudson et al., 2003) is McEwen's (2003; McEwen & Lasley, 2002) work on allostasis, which refers to the long-term psychobiological effects of adaptive coping to change and stress. We have long known that *chronic* stress is a risk factor for depression, but McEwen and others have demonstrated that adaptation to stress (even when coping responses are successful) has cumulative and, therefore, long-term effects.

As one example, the risk for thyroid disease is about twice as high in women compared to men; stress is known to adversely affect thyroid function (Cremaschi, Gorelik, Klecha, Lysionek, & Genaro, 2000; Matos-Santos et al., 2001). Further, borderline thyroid dysfunction (often termed subclinical hypothyroidism) is related to depression (Haggerty & Prange, 1995) and to nonresponsiveness to usual treatments (Hickie, Bennett, Mitchell, Wilhelm, & Orlay, 1996). Explanations such as allostasis are creative and interesting and merit further testing. Unfortunately, efforts to scrutinize such claims can be undermined by biological biases in the policies of funding agencies. Researchers may also be attracted to the alluring simplicity of biological and evolutionary explanations that are unable to be fully challenged by empirical tests because they rest on unspecified genetic mechanisms and unexamined assumptions about historical events (Funder, 2001).

In any case, the gender gap in depression will not be fully explained by solely focusing on stress caused by *proximal* life events. As described earlier, gender shapes the larger context as well. In identifying the contribution of adversity and stress in women's lives to their excess of depression, measures that typically assess proximal causes of distress in women's lives (e.g., life event scales), neglect the *distal conditions* that "govern the allocation of social and material resources in relation to gender" (Stoppard, 2000, p. 84).

Political and economic conditions affect the meaning of events such as single motherhood, divorce, and unemployment. When the gender-related *meaning* of experiences is addressed, it is critical to balance attention to cognitive paradigms with consideration of *structural conditions* and the *cultural discourse* that justifies and perpetuates the disadvantaged status of women in society. Without such a balance, the pervasive impact of stigma, abuse, and violence in women's lives can go unnoticed (Stoppard, 2000).

In conclusion, understanding the interactive and reciprocal contribution(s) of separable elements of gender to risk for depression requires innovations in theory and method at multiple levels that have implications for research, training, prevention, and public education policies discussed further. It is imperative that research questions raised by new theories drive the methods to answer them. As White et al. (2001b) have argued, "the days of distorting the question to fit the method should be put behind us" (p. 275). Application of a multiplicity of qualitative, quantitative, and field approaches is needed to answer questions about gender and to generate the complex understanding of how gender affects women's mental health over the life cycle. To be successful, however, the historical neglect and underestimation of the effects of women's unequal and devalued status in society and culture on research, treatment, prevention, and public policy efforts must be overcome.

It should also be remembered that innovation in methods to overcome reductionism and to study lives in context will not accomplish the goal of illuminating the impact of gender on the lives of women and men if those methods are limited or biased in other ways not considered here (DeVault,

1996). The complex and often subtle influence of gender bias on all stages of the research process has long been recognized (e.g., APA Task Force on Nonsexist Research, 1988), but the need to be vigilant in examining the effect of gender on all stages of the scientific enterprise – in methods old and new – continues.

SELECTED CRITICAL RESEARCH FINDINGS

Many aspects of the experiences of females that stand out as contributors to the gender gap in depression are covered in previous chapters in this volume. We have additional observations to make in four areas: (1) intimate violence, including childhood physical and sexual abuse, (2) unwanted pregnancy and its resolution, (3) sexualized objectification, and (4) stigma.

Gendered Intimate Violence

As discussed previously, gendered intimate violence is a pervasive threat to women's physical and mental health that takes on many forms, including childhood physical and sexual abuse, acquaintance rape, courtship violence and wife battering, marital rape, and stalking. The link between such violence and negative mental health consequences – including increased rates of depression and anxiety among women – is found in community as well as psychiatric samples (Anderson, Martin, Mullen, Romans, & Herbison, 1993; Bifulco, Brown, & Adler, 1991; Brown & Anderson, 1991; Finkelhor, Hotaling, Lewis, & Smith, 1990; Mullen, Romans-Clarkson, Walton, & Herbison, 1988; Pribor & Dinwiddie, 1992; Waller, 1994).

One of the most significant contributions of the women's movement has been to transform intimate violence against women from a private issue to a public concern (Koss et al., 1994). It is now recognized that such violence has widespread and long-lasting mental health effects, including clinical depression, anxiety, posttraumatic stress disorder, and substance use. It also interferes with women's ability to become educated and gain employment, undermining their access to important psychological, social, and economic coping resources (Koss, Bailey, Yuan, Herrera, & Lichter, 2003; see Penze, Heim, & Nemeroff, this volume, for a discussion of the relationship of trauma and depression).

Depression and anxiety increase with ongoing violence (Sutherland, Bybee, & Sullivan, 1998) and decrease as violence diminishes or stops (Campbell & Sullivan, 1994), suggesting that gendered violence could contribute causally to the gender gap in depression rates. Revictimization complicates the picture (Beitchman et al., 1992). Women who experienced sexual abuse in childhood show increased risk for experiencing rape and other forms of victimization as adults (Koss et al., 2003; Resnick, Acierno & Kilpatrick, 1997; Russell, 1986; Russo & Denious, 2001; Wyatt, Guthrie, & Notgrass, 1992).

Research on rape victims who have no histories of sexual abuse suggests that exposure to gendered violence may have mental health effects beyond those explained by sexual abuse history. A study that compared rape victims with no abuse histories to victims of a severe life-threatening but nonsexual event (e.g., car accident, violent robbery, physical assault) found rape victims to have high rates of a variety of symptoms, including depressed mood, distressing dreams, and difficulty in falling asleep. The authors concluded that the sexual nature (we would argue "gendered meaning") of the event made the difference between groups (Faravelli, Giugni, Salvatori, & Ricca, 2004).

Stalking has begun to be recognized as a significant social problem and questions about this phenomenon now appear on the National Violence Against Women Survey (Tjaden & Thoennes, 1998). Using a stalking definition that required high levels of fear, as much as 8% of women, compared with 2% of men, reported being stalked sometime in their life. Yet little is known about the mental health effects of this experience and the extent to which the quality of the experience is gendered.

As the Women's Health Research Agenda (APA, 1996) points out, the powerful role that sociocultural factors play in promoting and maintaining violence against women provides a theoretical rationale for policy development related to education and prevention. Such efforts need to be informed by research, including ethnographic research, which identifies gender-related factors that (1) predispose men to be violent against women (e.g., beliefs about a male entitlement that generate anger when confronted with a woman's perceived insubordination), (2) trigger violent behaviors, and (3) mitigate negative consequences to the perpetrator. In addition the combined and independent physical and mental health effects of various forms of violence on depression need to be assessed, with particular emphasis on the effects of stalking, which is a neglected area.

In addition to direct effects on mental health, exposure to violence lowers women's ability to control many areas of their lives that have implications for mental health, including the ability to pursue educational and income opportunities and maintain close, supportive connections with family and friends. Here we focus on one area: Unintended and unwanted pregnancy. Discussions of the relationship of reproductive events and depression, including postpartum depression, can be found in several other chapters in this volume, including Somerset, Newport, Ragan, and Stowe and Goodman and Tully.

Pregnancy: Unintended and Unwanted

In 1994, 49% of pregnancies in the United States were unintended and the majority of them unwanted; about 54% of those unintended pregnancies were terminated by abortion. The highest rates of such pregnancies were found in women who were between 18 and 24 years of age, poor, unmarried, Black, or Hispanic – groups already affected by social disadvantage and at higher risk for depression (Henshaw, 1998). Unwanted pregnancy has widespread mental health impact – on children unwanted during pregnancy (David, Dytrych, & Matejcek, 2003), on other family members (e.g., Barber, Axinn, & Thornton, 1999), and on society (Berk, Sorenson, Wiebe, & Upchurch, 2003). Here we focus on the relationship of unwanted pregnancy to the mental health of women.

Unintended and unwanted pregnancy is stressful life event, whether terminated in birth or abortion (Russo, 1992). Pregnancies that are wanted by the woman but unwanted by the partner may be particularly stressful. For example, a study of 124 cohabiting couples experiencing an unintended first pregnancy found that pregnancies associated with the highest risk for postpartum depressive symptoms were those considered intended by females and unintended by their partners (Leathers & Kelley, 2000).

With regard to risk for clinical depression among women with unwanted pregnancies, the well-designed research points to preexisting mental health as the most powerful predictor of postpregnancy mental health, however the pregnancy is resolved (Adler et al., 1992; Russo, 1992; Russo, David, Adler, & Major, 2005). This suggests shared risk factors for depression and unintended pregnancy need to be investigated. Chief among gender-related shared risk factors is exposure to intimate violence. There is a strong and complex relationship between intimate violence and unwanted pregnancy (see Russo & Denious, 1998, for a discussion). Such pregnancies are highly correlated with exposure to intimate violence, including childhood physical and sexual abuse, rape, and partner violence (Dietz et al., 2000; Russo & Denious, 2001; Wyatt et al., 1992).

Research findings based on interviews with mothers of newborns in 18 states found a direct association between exposure to violence and wantedness of pregnancy: 12% of new mothers with unwanted pregnancies (no births wanted at the time or in the future) reported having been "physically hurt" by their husband or partner during the 12 months before delivery, compared to 8% of mothers with mistimed pregnancies and 3.2% of new mothers with intended pregnancies. Highest rates of injury (16.6%) were found among unmarried new mothers with unwanted pregnancies (Gazmararian et al., 1995).

These findings probably *underestimate* the link between violence and unwanted pregnancy because the most unwanted of pregnancies are likely to end in abortion. Women who report abortions have higher rates of violence in their lives than other women. In one study based on a national telephone survey, 31.1% of women reporting an abortion also reported childhood physical or sexual abuse, compared to 13.6% of women not reporting an abortion (Russo & Denious, 2001). The differential is what is important, because this method likely underestimates the prevalence of both violence and abortion due to underreporting. The likelihood of a history of violence or abuse was found to be even higher in a sample of abortion patients, where reporting of abortion was not an issue: 39.5% (Glander, Moore, Michiellutte, & Parsons, 1998).

In addition to increased likelihood of experiencing intimate violence, women with unintended pregnancy may experience other co-occurring negative circumstances (e.g., partner leaving in response to the pregnancy, financial strain). Although not associated with clinically significant depression, there is also the *stress of being stigmatized* for terminating the pregnancy (Major & Gramzow, 1999), being exposed to clinic protesters, and experiencing harassment at clinic sites (Cozzarelli & Major, 1994, 1998).

Studying the relationship between unintended pregnancy and depression is complicated by a charged political context. Recent attacks on the integrity of science fueled by controversies over sexuality and abortion have involved attempts to manipulate scientific findings to serve a sociopolitical agenda in unprecedented ways (Mooney, 2004). Prevention programs that focus on promoting mental health by preventing unintended pregnancy would seem to be noncontroversial, given such programs should lead to a reduction in demand for abortion. However, opponents fear that such programs would encourage sexual behavior among unmarried persons. The willingness of conservative politicians to undermine the integrity of federal agencies established to serve the public's health is seen in the attacks that the Centers for Disease Control has experienced because it informed the public that there is no evidence that abortion causes breast cancer or that condoms, when used properly, can prevent a variety of sexually transmitted infections (Denious & Russo, 2005).

Despite a federal policy context that appears hostile to research that may lead to undesired findings, there are still many research questions about the shared precursors of depression and unwanted pregnancy that do not involve voluntary sex between unmarried persons or abortion issues. It is important that the researchers pursue such questions at the same time that efforts are undertaken to protect the integrity of the scientific process. In the meantime, educational efforts that will increase public understanding of the importance of maintaining the integrity of the science funded and disseminated by federal agencies provide an important foundation for ensuring a research climate in which controversial issues can be studied. Given the chilling climate for research on so many questions of importance for understanding depression in women, it is hoped that support for research on controversial questions may be found at the state level or from private foundations. Health professionals also have important roles to play in marshalling such support.

Sexualized Objectification

Feminist theorists (e.g. Beauvoir, 1978; Berger, 1972) have long argued that women often take an observer's perspective on their physical selves.

Researchers have been slow to examine the mental health consequences of such self-conscious body monitoring on mental health. The body dissatisfaction and shame that can result from objectified body consciousness in a culture in which women's bodies play a central role in defining their worth and in which one "can never be too rich or too thin" have profound mental health implications (Denious, Russo, & Rubin, 2005; Joiner, Schmidt, & Wonderlich, 1997; Key et al., 2002; Lin & Kulik, 2002).

The link between women's objectification experiences (OE) and depression was first documented in a series of studies led by Hamilton and colleagues, who found that OE accounted for 25% of the variance in depressive symptoms among a sample of college women (Burnett, Baylis, & Hamilton, 1994). That study also demonstrated that OE predicted the clinical syndrome of depression, as assessed by the Center for Epidemiologic Studies Depression Scale (CES-D) scores (at or above the usual cutoff of 16). In follow-up studies (Burnett, 1995; Burnett & Hamilton, 1996), they examined the relationship between the OE, daily hassles, coercive sexual experiences, and depressive symptoms in college women. OE significantly predicted psychological outcomes, including depressive symptoms and dysphoria, above and beyond those associated with the standard hassles scale even when effects of coercive sexual experiences were controlled.

Congruent with findings from Landrine et al. (1995) described earlier, a second study also found that the most important contributors to the effect of OE frequency were being called degrading, gender-stereotyped names and being the target of offensive (sexualized) gestures (Burnett, 1995). However, externalized self-perceptions (e.g., "I tend to judge myself by how I think other people see me") moderated this relationship. Taken as a whole, this research suggests that frequency of objectification experiences predicts depressive symptoms above and beyond its association with more severe sexual abuse, with this effect moderated by the tendency to rely on the opinions and evaluations of others (Burnett, 1995; Burnett & Hamilton, 1996; Fredrickson, Roberts, Wolf, & Hamilton, 1994; Hamilton, 1996).

There has been a great deal of subsequent theoretical (Fredrickson & Roberts, 1997), methodological (McKinley, 1995; McKinley & Hyde, 1996), and substantive work that has found objectification to be a powerful influence on women's thoughts, feelings, and behaviors in Western culture (Klonoff & Landrine, 1995; Klonoff et al., 2000; Landrine et al., 1995; Landrine & Klonoff, 1997; Travis & Meginnis-Payne, 2001). Janet Swim and her colleagues (2001) found that sexist incidents increased women's feelings of anger and depression and decreased their state of self-esteem. Tiggemann and Kuring (2004) also found that objectification increases women's depressive symptoms.

Denious, Russo, and Rubin (2005) suggest that the stigma and resulting shame that can result from objectification experiences may explain the link of objectification experiences with depression and eating disorders. This suggestion is congruent with the finding that externalized self-perceptions moderate the relationship of OE to depression. As they point out, shame may be triggered by the stigma of not meeting cultural appearance standards, but it is an emotion that implicates the entire self. Feeling shame as a result of failing to meet appearance standards is but one way that gendered stigmatization can affect women's mental health. Unfortunately, research has focused on the relationship of *body shame* to eating disorders, and little is known about its relationship to depression (Noll & Fredrickson, 1998).

In summary, further research on the dynamics of gender and its relationship to objectified body consciousness in a body-conscious culture is needed to inform treatment and prevention efforts. In particular, basic research is needed on the processes that underlie the dynamics of stigmatization, shame, and depression fueled by objectification processes.

Stigma

The Depression Summit recommendations called for research and public education efforts aimed at reducing stigma associated with mental illness, underscoring the point that such stigmas deter women from recognizing they are depressed and from seeking the mental health services they require (Mazure et al., 2000). The recommendations did not recognize that stigma affects women's depression in a variety of other ways, however.

Stigma is a cultural construct that is relationship and context specific. Attributes, experiences, and behaviors that are stigmatized vary with culture and over time. Stigmas label people as different and devalued by others, leading to low status and loss of power and discrimination (Link & Phelan, 2001). They may be visible or hidden, controllable or uncontrollable, and linked to an individual characteristic or group membership. Stigma poses a threat to one's personal and social identity and can affect mental health through several psychological and interpersonal mechanisms, including activation of stereotypes, expectancy effects, social exclusion, discrimination, and other forms of negative treatment.

Stigmatization may also undermine mental health by interfering with perceiving or obtaining social support from others (Crocker, Major, & Steele, 1998; Major & O'Brien, 2004). Disclosing stressful life events is an important part of the coping process (Tait & Silver, 1989), and for stigmas that can be concealed, the process of concealment may increase risk for physical and psychological problems (Pennebaker, 1995). Disclosure may result in social rejection and exclusion that have powerful negative effects as well, however, making management of stigma a highly stressful process.

As can be seen in the discussions on intimate violence, unwanted pregnancy, and sexual objectification, stigmatization can affect women's willingness to seek help for dealing with stigmatizing stressful events, undermine their mental health, and increase risk for depression. Depending on the norms of the community, women can be stigmatized for being victims of childhood sexual abuse, being raped, losing their virginity, staying single, having more than one sexual partner, having a child when not married, being childless, being battered by their husbands, having an abortion, placing a child for adoption, and working when her children are young, among other things. These are just a few examples of a variety of gender-related events associated with stigma that have implications for mental health. Basic research on both gender and stigma is needed to refine their conceptualization and measurement so that the relationships among gender, stigma, and depression can be adequately understood.

Full understanding of the gender gap in depression requires knowing (1) the extent to which gender shapes the events in women's lives that are stigmatized, and (2) how social inequality undermines women's ability to overcome stigmatization. Although the relationship of stigma to psychological and interpersonal relationships has burgeoned in recent years, inadequate attention has been paid to structural issues that affect access to power, status, and coping resources.

As Link and Phelan (2001) have emphasized, stigma depends on power – social, political, and economic – but power dynamics are all too often overlooked in analyses of stigma. As they point out, "there is a tendency to focus on the attributes associated with those [stigmatizing] conditions rather than on power differences between people who have them and people who do not" (Link & Phelan, p. 375). They argue that stigma is dependent on the power to label and negatively stereotype members of stigmatized groups, lower their status, and limit their access to major life domains such as education, jobs, housing, and health care.

How the power dynamics of gender relate to the processes of stigmatization has yet to be fully examined. The relationship of gender role transgressions to various forms of stigma and the mental health effects of stigma and its associated sanctions in particular need to be more fully understood. That knowledge can provide a foundation for building women's resistance to stigmatizing cultural messages. Research on the relationship of stigma to self-esteem suggests that the causal dynamics are complex. For example, there is a relationship between greater exposure to sexism to threat appraisals and reductions in self-esteem among pessimistic women that is not found among optimistic women (Kaiser, Major, & McCoy, 2004).

At the same time, research on how to foster social change that could reduce stigmatization as a force for the enforcement of gender role norms is needed. In particular there is a need to focus on change that will promote women's ability to pursue educational and income opportunities and develop the psychological and social resources needed to cope with stigma and its effects (e.g., resources such as self-esteem, self-efficacy, social support). As Link and Phelan (2001) point out, intervening in the stigma process either requires (1) producing fundamental changes in beliefs and attitudes or (2) changing the power relations that enable dominant groups to act on those stigmatizing beliefs and attitudes. The potential for success of either approach will depend on knowledge of the interrelationships among gender, stigma, and mental health.

THE NEED FOR INTEGRAL THEORIES

New biopsychosocial models that integrate physical and mental dimensions of health are needed to guide research in new directions. Integrative frameworks such as those found in the emerging interdisciplinary area of integral science offer new ways of thinking about how biological, psychological, social, and cultural forces interact to produce gender differences in depression. The aim of integral science is to generate a new understanding of the functioning of complex systems – human, biological, and social. This synthesis produces a vision of the human mind and body as physically integrated elements of an energy/information processing system that is assumed to follow universal patterns and principles, once they are understood (Goerner, 1999, 2004).

Hamilton and others (2004, pp. 295–296) have recently applied integral science to health-related studies, arguing that the resulting shift in medicine's worldview would integrate diverse efforts and improve health-related research and delivery of health care. Integral science is consistent with the position of feminists who have argued for the need to take a multilevel approach to research on mental heath. New integrative models that encompass physical, psychological, social, and cultural dimensions of health are needed to guide research in new directions. This approach could be especially fruitful in understanding gender's influence on depression through its relationship to fatigue.

Bioenergetics and Depression: A Woman's Work Is Never Done

The central question for this handbook concerns the origins of the gender gap in rates of depression. We believe that one of the key findings lies in the consistency of *fatigue* and *insomnia* as the two most pervasive symptoms of depression worldwide. For example, "loss of energy" and "insomnia" were present in most persons at all sites in a cross-national epidemiological study of major depression (Weissman et al., 1996). Insomnia can lead to drowsiness and to feeling unrested in the morning.

Tiredness (Popay, 1992) and fatigue are highly prevalent in women's lives overall (Stoppard, 2000, p. 107–109). In one community survey of women's primary health concerns, fatigue was ranked first by nearly 28% of women, and it appeared among the top 10 concerns of over 80% (Stewart, Abbey, Meana, & Boydell, 1998). Fatigue significantly predicts depressive symptom scores and reports of sleep dysfunction (Lavidor, Weller, & Babkoff, 2003). Fatigue is frequently nonresponsive to treatments for depression (Menza, Marin, & Opper, 2003) and possibly increases risk for relapse of depressive illness. In a recent clinical trial, both fatigue and sleep dysfunction remained elevated, compared to all other symptoms of depression, after 6 months of antidepressant treatment (Greco, Eckert, & Kroenke, 2004).

A focus on fatigue leads us to examine the biology of energy (Hamilton, Phillips, & Green, 2004). The study of bioenergetics is a growing field within the basic sciences and both primate and human behavioral ecology. Biologically, life cannot exist without energy. In fact, it has been said that "energy is the motor of life" (Heninger, 2001). Further, stress targets cellular energy carriers (Parsons, 2003).

In the basic health sciences – which include psychology and medicine, among many others – we can refer to energy at various levels. As examples, we can measure energy at the level of

- mitochondrial DNA and the biochemistry of ADP/ATP
- energy storage and release (as with glycogen and in fat cells)
- glucose and oxygen utilization and efficiency by the brain (as with various imaging techniques) or by muscles
- availability of vitamins and minerals that affect oxygen distribution and utilization (e.g., iron and haemoglobin; B12 and folate)
- basal metabolic rate (e.g., thyroid regulation) and body composition (Hamilton, 1995; Hamilton, Grant, & Jensvold, 1996)
- deployment of attention, which involves accessing and allocating a limited resource (Hamilton, Haier, & Buchsbaum, 1984; Kahneman, 1973).

The nervous system is highly energy-dependent. Energy substrates (primarily glucose) and oxygen are essential to brain–body functioning. Mitochondria are the energy-producing organelles of the cell, providing energy for all cellular functions.

We can think of energy regulation as having two main parts: *intake* (including both intake and creation of energy substrates, and storage); and *output* (including both accessing energy and governing its release). In living organisms, these two main aspects must be in balance for optimal functioning. An excess of output compared to input leads to energy depletion. We argue that energy depletion can give rise to fatigue. Fatigue and energy depletion are, in turn, depressogenic. The proposed theory is supported by existing data and can be even further operationalized, as briefly outlined below. (Some of the proposed measures may fall into more than one conceptual category in the model, depending on whether the measure is a proximal or distal contributor to the effect.) Ultimately, energy functioning needs to be "sustainable" over time.

Considering output first, energy might be drained by (a) excess energy expenditures and (b) the cumulative effects of coping with chronic stressors. As examples, excess energy expenditure among women might be assessed by measures including the following:

- excess total hours of labor (including excess work due to poverty) (Baker, Kiger, & Riley, 1996; Bird, 1999; Bird & Fremont, 1991; Cooper, 1980, cited in Frankenhaeser, Lundberg, & Chesney, 1991; Hochschild, 1997; Hochschild & Machung, 1989; United Nations, 2004)
- excess of calorie-burning physical labor (United Nations, 2004)
- excess emotional work (e.g., nurturing and caregiving are stereotypically female sorts of labor and can lead to high rates of burnout and depression)
- excess stress due to childbirth (which is an enormous physiological stressor and which can deplete iron stores) and excess work by new mothers postpartum (where much of the work could and should be done by others)

Second, energy is drained by the cumulative effects of coping with chronic stressors (McEwen, 2003; McEwen & Lasley, 2002). It is well known that we are better adapted for acute as opposed to chronic stress. Stress promotes adaptation (allostasis), but a perturbed diurnal rhythm or failed shutoff of mediators after stress leads, over time, to wear and tear on the body and brain (allosteric load). Allosteric load thus refers to the cost of adaptation. The effects of allosteric load can be seen in studies of posttraumatic stress disorder and mood disorders (McEwen, 2003). As examples, excess allosteric stress load among women might be assessed by measures including

- sleep deprivation or insufficient sleep (which can lead to elevations in glucose and cortisol)
- sedentary lifestyle (which impairs the muscles' ability to burn glucose)
- lack of *enjoyable* physical activity, which is promoted by stereotypically female hobbies and leisure activities (which are too often sedentary and typically occur in the home or other private, restricted places (Henderson, 1994)
- lack of health-promoting physical activity, which occurs in women partly due to exhaustion and perceived lack of time (Stewart et al., 1998) and the presence of children (Chen & Millar, 2001), especially given the gendered nature of parenting
- excess anxiety (e.g., related in part to fears of violence, which can be intensified by experiences of sexism and objectification or due to past violence and posttraumatic stress disorder), which can impair sleep

Further, Hamilton and Yeates (in press) have proposed that the following factors – known to be more prominent in women's lives compared to men's – may also be markers for allostatic load:

- violence, which creates anxiety and interferes with sleep
- objectification and stigmatization, both of which can create anxiety

- the stress of ambiguity (e.g., as women we continually ask ourselves was that a compliment or was that sexual harassment or bullying?), along with culturally sanctioned mixed messages (e.g., gendered double-binds and Catch-22 behavioral scripts), which can lead to what Nolen-Hoeksema (1987, 2003) has termed *rumination*, or thinking too much.

Shifting to a consideration of input, energy might be drained by (a) deficient energy intake or regulation, (b) lack of leisure and putative *restorative* health functions, (c) problems with allocation and efficiency (lack of flow experiences, energy conservation, lifestyle sustainability), and (d) presumed lack of access to (or creation of) positive sources of energy. As examples, deficient energy input might be assessed by measures including

- quality and amount of food intake (United Nations, 2004)
- blood levels of iron, hemoglobin, B12, and folate – even when values are arbitrarily (and erroneously) defined as normal (Hintikka, Tolmunen, Tanskanen, & Viinamaki, 2003; Verdon et al., 2003; Waalen, Felitti, & Beutler, 2002)
- blood indices of thyroid functioning (e.g., TSH, free T_4) – even when laboratory values are arbitrarily (and erroneously) defined as normal (Haggerty & Prange, 1995; Hickie et al., 1996).

We might also assess a relative lack of leisure and restorative health functions (i.e., some of which may help buffer the effects of allostatic stress load) by measures including

- gender gap in leisure time (Abrams, 1997; Bird, 1999; Bird & Fremont, 1991; Hochschild, 1997; Hochschild & Machung, 1989; Maushart, 2001; Stewart et al., 1998)
- lack of leisure time spent in physical-health-promoting activities (Chen & Millar, 2001; Henderson, 1994)
- lack of time for self-care (Henderson, 1996; Stewart et al., 1998)
- sleep duration and quality (as these are thought to be restorative processes)
- lack of time and space for *uninterrupted* flow experiences and other types of positive experiences that may be especially salient for women (Abrams, 1997; Csikszentmihalyi, 1990/1991; Hamilton, 1986a, 1986b, 1989, 1991, 1993; Hamilton et al., 1984; Hamilton, Rathunde, & Csikszentmihalyi, 1998; Lawlor et al., 2002; Maushart, 2001)
- gendered life events and conditions that may interfere with sexual pleasure (Kaschak & Tiefer, 2002; Laumann et al., 1999; Tiefer, Brick & Kaplan, 2003).

We now know that *positive indicators of mental health* are critical to understanding the global burden of depression. Keyes (2003) has documented, for example, that *languishing* (even in the absence of depression, per se) confers substantial morbidity; some of his findings are consistent with the allosteric theory of stress, especially in older women. His findings have important implications for diagnostic practices.

Hamilton and Yeates (in preparation) have theorized that positive experiences (e.g., as occurring with intrinsically motivating "flow" activities) may increase energy efficiency and access to energy. Hamilton and others (Hamilton et al., 1984; Hamilton et al., 1998) have documented neurophysiological correlates of flow that support the "efficiency theory of flow." Based on these and other studies, Hamilton (1991) proposed an "effort theory of depression."

It appears that women differ from men in usual types of flow activities, and Hamilton (1993) proposed that there are cultural and gender-based barriers to flow in women. As examples, measures related to efficiency and access to (or, moment to moment creation of) and allocation of energy might include

- blood glucose, oxygen, or other parameters suggesting efficiency of information processing (especially during flow experiences)
- gender-role-related, internalized prohibitions against flow experiences in women (Abrams, 1997; Hamilton, 1986a, 1986b; Henderson, 1996)
- languishing (although there do not appear to be sex-related differences in rates of this state)
- "obligatory" housework does not appear to confer health benefits usually associated with physical activity (Lawlor et al., 2002; Maushart, 2001)

In summary, Hamilton and Yeates (in preparation) have proposed a new, integral theory to explain the gender gap in rates of depression: the *energy depletion theory*. It is obvious that there are gender differences in many of the factors listed above. Moreover, impressive sex-related differences in related biological and clinical measures have been documented as well.[1]

---

[1] In general, rates of fatigue can be related to longer working hours, sleep deprivation, poor sleep quality (e.g., related to trauma; Krakow et al., 2000), inadequate nutrition, and a variety of physical illnesses. Sleep dysfunction is one of the earliest occurring (prodromal) symptoms of depression.

In a large, community-based sample, the rates of fatigue were consistently about two times higher in women compared to men, with prolonged fatigue occurring at a rate of 5–7.7% and chronic fatigue at 2.7–4.2% (Jason et al., 1999). In a population-based study of insufficient sleep in adults (sleep deprivation, defined as a difference of 1 hour in self-reports of sleep needed and sleep length), the prevalence was 1.4 times higher in women compared to men (23.9% in women and 16.2% in men; Hublin, Kaprio, Partinen, & Koskenvuo, 2001). The cost of sleep deprivation accumulates over time. Although short-term sleep deprivation effects may be milder in women than in men, women appear to need more sleep to recover (Corsi-Cabrera, Sanchez, Portilla, Villanueva, & Perez-Garci, 2003).

Taken together, these differences uniformly point in the predicted direction: an excess of energy depletion in women compared to men.[2] Further, the gender gap in such measures cross-culturally supports the hypothesized relationship between energy depletion and gender inequality (Hamilton & Yeates, 2004; Hamilton & Yeates, 2005b, c).

The present authors emphasize that the energy depletion theory of depression in women, although truly biopsychosocial and integrative, is inherently and explicitly both cultural and economic in nature. Thus, the theory leads away from merely targeting individual women as recipients of medicalized (e.g., pharmacological) interventions, but rather it points toward public-health-related interventions, such as changing the gendered nature of culture, particularly as the latter is manifested in inequitable economic arrangements and in social structures.

The Appreciation of Women's Experience vs. Only Giving the Diagnosis of Depression

As Hamilton (1995) has pointed out previously, a new look at diagnostic issues regarding women and depression is needed. Stoppard (2000) argues that trying to be a "good woman" can be exhausting and demoralizing. Consider, for example, the single mother, living in poverty, working outside the home, caring for small children in the home, and living in a culture

Depression is not related to fatigue that responds to rest and sleep (Lavidor et al., 2003). A community-based twin study found that fatigue was related to depression and to stressful life events; genetic effects were found to be especially important in women and environmental effects in men (Sullivan, Kovalenko, York, Prescott, & Kendler, 2003).

Fatigue, as felt in the body, is often (but not always) related to exercise. There is some evidence that women feel less fatigue with exertion compared to men because they better utilize oxidative pathways (D. Russo, 2002; Thomas, Granat, & Fernhall, 1997). An increase in nighttime overall oxygen use (VO2) is associated with stimulant-induced insomnia and fatigue. Using positron emission tomography to measure brain glucose utilization, there is evidence that women have higher resting brain metabolic rates compared to men (Andreason, Zametkin, Guo, Baldwin, & Cohen, 1994).

[2] A counterargument might be that women are *not* depleted because they have high rates of obesity in Western, industrialized nations. Obesity would seem to imply a relative excess of calories that are above those that are actually used; such calories are stored in fat. An answer to this objection lies in evolutionary biology: Women are designed to store fat in times of adversity (e.g., food scarcity). In theory, women's propensity toward fat storage might be so that a fetus or infant might make it through times of (temporary) scarcity through mobilization of the mother's femoral fat pads. But the precise trigger for heightened fat storage may include either stress or allosteric load, even if calories are plentiful. That is, women's lived experience (even in relatively egalitarian, modern society) may be so chronically stressful that our bodies go into the heightened fat storage mode just because of commonplace, gender-related adversity. Thus, there are likely to be excess energetic costs of living against the grain of male-dominant culture. Further, when obesity occurs, everyday activities require more energy. Many women experience stress-induced over-eating, which may be a side-effect of our excess "allosteric load."

of gender inequality. The lethargy and fatigue, apparent social withdrawal, lack of interest in (previous) usual pursuits, demoralization, anxiety, and rumination that such a woman experiences are symptoms that a clinician might diagnose as depression. Further, a clinician might rule out anemia and thyroid dysfunction in such a woman based on arbitrary (and inappropriate) laboratory norms. Or, we might be only partially understood, being labelled as excessively ruminating; whereas, in fact, we are predictably "spinning our wheels" energetically because of gender-binds. According to Stoppard (2000, p. 109), however, a better (overall) understanding might be that the prototypical depression in such a woman is her "response to insoluble dilemmas in her life." Thus, she implies that women are not just depressed because they "think too much," but rather because they face impossible demands that are culturally prescribed, with inadequate resources. Hamilton and Yeates (in preparation) agree with Stoppard but go one step further, arguing that women "do too much" in a social context that provides them with "too little" energetic resources in facing virtually insoluble challenges: Taken together, women's lives are too often energy depleting and thus depressogenic. Women both need and deserve more energetically sustainable lives.

PUBLIC POLICY: ISSUES AND OVERALL RECOMMENDATIONS

Social policies with implications for women's depression overlap with and are embedded in policies directed at improving women's health in general, and congruence in policy recommendations at multiple levels is needed for effective implementation. The Summit on Women's Depression (Mazure et al., 2002) developed cogent recommendations for public policy that encompassed areas of research and funding; prevention, treatment, and services research; public education; and prevention, treatment, and service delivery. These, as well as policy recommendations developed during the deliberations of the recent interdisciplinary women's health conferences (APA, 1996), provide the basis for the public policy development. A comprehensive approach is needed to develop a network of public policies designed to understand the effects of sex and gender on the etiology, diagnosis, treatment, and prevention of depression in women and men. Here we focus on what is needed to understand and ameliorate the gender gap in depression.

Research and Training

Research policies need to foster the development of innovative models and methods for understanding women's depression. We have identified a number of areas meriting priority in research. Current approaches to research on women's depression are lacking on at least three dimensions. First, the gendered nature of factors fostering risk and resilience for depression is not adequately acknowledged. Basic research is needed on how various aspects of gender shape women's biological, psychological, and social responses to stress (including allostatic load) and adversity over the life cycle. Second, a focus at the level of the individual fails to take into account the interaction of the person with the situation. This focus stems, in part, from the methodological limitations imposed by research that counts behaviors or symptoms without regard for their meaning or context. Third, research tends to focus on intrapersonal predictors (biological and psychological) and immediate situational cues. Situational, dyadic, social network, and sociocultural variables that interact with individual difference variables and affect the meaning of specific acts and events are neglected (White et al., 2001b).

New biopsychosocial models of research that view behavior as situated and contextualized can redress this situation, particularly interdisciplinary models that examine how gender-related cognitive and socioemotional factors are shaped by power differentials in relationships as well as other less personal social networks that sustain systems of inequality. In thinking about the relationship of gender to depression, it is important to keep in mind that gender is a dynamic construct that interacts with ethnicity and other social identities, including age, race and ethnicity, social class, and sexual orientation, among others, in shaping responses over the life course (Krieger et al., 1993; Russo & Vaz, 2001).

Women are more likely to experience a number of major chronic and painful diseases and other disabling conditions than men. This differential pattern of illness and disability contributes to the gender gap in rates of depression. Basic work on the development of biopsychosocial models that integrate physical and mental dimensions of health will be particularly helpful in understanding dynamics of depression in women and its relationship to disability (Cook, 2003; Karp et al., 2005).

Research and training funding priorities should include increasing knowledge about the multilevel effects of gender (and the new methods needed to study it) among current and future generations of mental health researchers. The pace of innovative research advances will be inextricably tied to training in new research methods as well as opportunities for interdisciplinary collaboration. Including women in research trials and conducting research on women's health has been an advance. Gendered multilevel theoretical analyses of the causes of mental disorder applied in research designed to generate a knowledge base that will inform treatment and prevention approaches are needed.

Diagnosis, Treatment, and Service Delivery

Mental disorder has often been defined in terms of person's ability to function "normally," irrespective of the toxicity of the environment (White et al., 2001b). More research is needed on the gendered assumptions about what is normal that may affect concepts of mental disorder, the construction of diagnostic categories, and the development of instruments that assess them (Caplan & Cosgrove, 2004). Research is needed on diagnostic errors that may occur more frequently in women than in men, perhaps leading to poor treatment outcomes or to residual symptoms. Substantial progress has been made in neurophysiological models of mental disorder, but as Widiger and Sankis (2000) have pointed out, the imbalance of attention "might be at the expense of equally valid psychological models" (p. 377). Advances in neuroscience are not substitutes for understanding the etiological significance of the meaning of events and the need for diagnosis and treatment to reflect the impact of that meaning. An interdisciplinary team of experts has called specifically for an increase in research in the *behavioral substrates* of mood and mood-regulation (Davidson et al., 2002).

The focus here is on the gender gap in depression, in which there are higher rates for women, but men have higher rates among substance use disorders, which are more congruent with their gender roles (Russo, 1985). The extent to which this differential reflects differential etiology or bias in diagnosis has yet to be fully understood. More research is needed on how gender inequality undermines the mental (and physical) health of both women and men (Krieger et al., 1993).

Gendered responses to treatment need to be more fully understood. Antidepressants, psychotherapy, or a combination of the two are typically used to treat depression. Recent research has found moderating effects of childhood trauma on the effects of such treatments for depression. Overall, the combination was found most effective, with antidepressant and psychotherapy treatments alone being equally effective. However, patients (both women and men) with higher exposure to childhood trauma, including physical and sexual abuse, responded better to psychotherapy alone than to antidepressant treatment, and the combination treatment was only slightly superior to psychotherapy alone.

There is increasing evidence that certain laboratory cutoff scores may be insensitive to health-related effects in women. There needs to be greater attention to the relationship between subclinical thyroid dysfunction and anemia, especially when assessing and treating women. Adjunctive treatments that are consistent with the bioenergetic theory of depression include thyroid, iron, or B12 and stimulants or stimulant-like drugs. For example, women benefit more than men from antidepressant treatment used in conjunction with thyroid (Cytomel, T3), even if measures of thyroid function are in the normal range (see Hamilton, 1995; Hamilton et al., 1996). It has been shown that iron supplementation is useful for fatigue in women even if the laboratory values are normal (Verdon et al., 2003). Responsivity to antidepressant therapy has also been correlated with levels of B12, even within the normal range (Hintikka et al., 2003).

Adult attention deficit disorder (AADD) is often comorbid with depression, yet AADD is often overlooked in girls and women (Solden, 1995). A useful adjunctive treatment for women with severe and recurrent depression, especially when there is a familial history of attention deficit hyperactivity disorder, has been methylphenidate (Hamilton, Phillips, & Sloan, 2003). There is reason to believe that there are sex-related differences in response to certain stimulants (Hamilton, Alagna, & Pinkel, 1984).

Keck (2004) suggested clues for detecting bipolar disorder in patients presenting with a depressive episode; many of these clues occur more frequently in women compared to men.

Another issue that may be critical to diagnosis is the morbidity associated with languishing, even in the absence of depression (Keyes, 2003). One important implication of Keye's work and that of others (Seligman, 2002) is that indicators of positive mental health and well-being must be incorporated into routine clinical assessments.

Although the authors suggest that psychotherapy may be essential in treating patients with chronic major depression who have histories of childhood trauma, they did not conduct a gendered analysis of such trauma and its impact of on responsiveness to treatment and remission. Even so, the findings are provocative and underscore the importance of taking exposure to adverse life events into account when evaluating gendered responses to treatment (see Penza, Heim, & Nemeroff, this volume, for more discussion of these issues).

The goal of therapy has been to alleviate symptoms to enable patients to function in their environments, however unhealthy those environments might be. In contrast, feminist approaches to therapy and other forms of treatment have gone beyond symptom reduction to focus on the development of strength, resiliency, and positive well-being (Hamilton & Yeates, in preparation; Keyes, 2003; Worell, 2001). Treatment is thus closely allied with prevention efforts. As Mendelson and Muñoz (this volume) point out, the problem of women's depression is of such alarming proportions that the likelihood of treatment services ever being adequate is low – strong and effective prevention efforts are needed.

Prevention

Research is needed to articulate the multilevel impact of gendered inequality and forms of social control (including violence and stigma) on women's depression. An understanding of the relationship of the social fabric to mental health puts a more complex perspective on women's depression. In that new, more complex vision, the roles of inequality, physical and sexual violence and victimization, exploitation, poverty, prejudice and discrimination, and stigmatization become salient. The impact of these and other sociocultural factors on psychological and social functioning (including physical and mental energy) become central in addressing treatment and prevention issues.

Mendelson and Muñoz (this volume) consider current research findings related to many of these areas. They identify a need for more multilevel dynamic theories to guide prevention research and program development and identify seven areas they view as important for depression prevention. Their cogent analysis articulates the value of multilevel interdisciplinary approaches as well as the need to expand the range of prevention outcomes incorporated into prevention efforts. We will not repeat their excellent policy recommendations, except to emphasize that successful implementation will rest on thoughtful analyses of how the social and political inequality of women (1) operates to increase risk for depression and (2) will affect the impact of prevention efforts in the *specific* contexts of the targeted populations of interest.

The emergence of the gender difference in depression during adolescence makes prevention efforts that target child and adolescent populations a high priority. As Lerner and Galombos (1998) point out, changing relations between the individuals and their contexts "constitute the basic process of development in adolescence . . . [and] underlie both positive and negative outcomes that occur" (p. 416). They emphasize that comprehensive intervention programs "can help at-risk youth integrate in a coherent way the biological, cognitive, emotional, and social changes they are experiencing, and to form a useful self-definition" (Lerner & Galombos, p. 439). They present a powerful argument for the need to consider the variety of individual and contextual factors that shape the adolescent's sense of self and relationships with others. Ironically, they do not consider gender as a contextual factor to be included in that consideration, illustrating the blind spot that is all too often found in the design of prevention efforts. This is despite substantial research demonstrating the impact of gender on friendships and romantic relationships in adolescence (Giordano, 2003).

Whether prevention programs will receive the gender analysis they require for success has yet to be seen. In 1978, the Subpanel on the Mental Health of Women of the President's Commission on Mental Health called for primary prevention efforts to counter the impact of gender inequality on women's mental health (Subpanel, 1978). Subsequently, prevention efforts continued to exclude consideration of women's issues despite a host of follow-up activity, including the development of a women's mental health research agenda for the National Institute of Mental Health (NIMH; Eichler & Parron, 1987; Russo, 1984, 1985).

In 1994, NIMH released a report that articulated a national research agenda for preventing mental disorders with two elements that undermined advancement in understanding and preventing depression in women (NIMH Prevention Research Steering Committee, 1994). First, in keeping with the biomedical biases of the National Institutes of Health (NIH), studies of general competence building and mental health promotion were rejected. Yet, as Reppucci and colleagues have observed (Reppucci, Woolard, & Fried, 1999), evidence suggests that such programs have great promise for preventing mental disorder. Second, prevention studies that advocated social and political change to achieve social equality for disadvantaged groups were rejected, which mandated that the basic roots of the gender gap in depression be ignored (Reppucci et al., 1999). The policy recommendations of the recent Summit on Women and Depression, which include substantial recommendations for addressing issues related to social inequalities experienced by women, raise the hope that times may change, but it is too soon to evaluate outcomes (Mazure et al., 2000).

Health scientists and professionals have special responsibilities for advocating for change in health policies in areas of their expertise. When prevention is the policy goal, however, societal policies that address social and political equality for women and minorities in society become essential components of prevention strategies at the policy level. As the report of the Depression Summit points out, relevant policies encompass paid family and medical leave, child care (including increased partner or community support postpartum), unemployment benefits, minimum wage level, universal health care, and flexible workplace policies. This does not mean that health scientists and professionals must become economists or sociologists. It does mean increasing interdisciplinary collaboration and incorporating multilevel approaches in designing prevention problems. It also means looking beyond the borders of our specific interests and joining with others to articulate the impact that broader social policies have on mental health to policy makers and the public.

Public Education

Public education efforts of scientific and professional societies need to look beyond diagnosis, treatment, and service delivery issues and recognize the need to promote public understanding of (a) the link between women's inequality and risk for depression and (b) societal changes that are required for effective prevention. As one example, there is a need for increased social and labor-related supports for new mothers.

There is also a need to educate the public about the reasons that it is important to preserve the integrity of the scientific process. Research dissemination needs to go beyond publication in scientific journals and be translated in forms that are useful to clinicians as well as to the community at large. The public information offices of WHO, the Office of Women's Health at NIH, and the public policy offices of professional and scientific associations provide models for public education efforts. If such efforts are to be maximally effective, however, the mental health impact of social inequality on women and men must be communicated to the public in ways that are understood and that will lead to positive social change and improved mental health for all.

CONCLUSION

Despite clear evidence that psychological, social, and economic factors play critical roles in the development and course of physical and mental disorders, translating research into changes in training, practice, community interventions, and public policy is slow. Understanding why requires an analysis of the way that public policies are developed, who the stakeholders are in the policy process, and recognition of the investment that major players, including pharmaceutical companies, have in obsolete biomedical models of health. Because gender is a dynamic cultural construct that reflects the interplay of biological, psychological, and social factors, the blinders of a monocular biological perspective will be particularly detrimental to the understanding of relationship of gender to mental health for both men and women. Indeed, until gender's widespread mental health impact is recognized and a true commitment to integrative biopsychosocial models of health is made, the knowledge needed to optimize the relationship between gender and health will continue to elude us.

References

Abrams, R. (1997). *The Playful Self*. London: Fourth Estate.

Adler, N. E., David, H. P., Major, B. N., Roth, S., Russo, N. F., & Wyatt, G. (1992). Psychological factors in abortion: A review. *American Psychologist, 47*, 1194–1204.

American Psychiatric Association. (1994). *Diagnostic and statistical manual of mental disorders* (4th ed.). Washington, DC. American Psychiatric Association.

American Psychological Association. (1996, February). *Research agenda for psychosocial and behavioral factors in women's health*. Washington, DC. American Psychiatric Association Retrieved from http://www.apa.org/pi/wpo/research.html.

Anderson, J., Martin, J., Mullen, P., Romans, S., & Herbison, P. (1993). Prevalence of childhood sexual abuse experiences in a community sample of women. *Journal of the American Academy of Child and Adolescent Psychiatry, 32*, 911–919.

Andreason, P. J., Zametkin, A. J., Guo, A. C., Baldwin, P., & Cohen, R. M. (1994). Gender-related differences in regional cerebral glucose metabolism in normal volunteers. *Psychiatry Research, 51*, 175–183.

Arrindell, W. A., Steptoe, A., & Wardle, J. (2003). Higher levels of state depression in masculine than feminine nations. *Behavior Research and Therapy, 41*, 809–817.

Baker, R., Kiger, G., & Riley, P. J. (1996). Time, dirt, and money: The effects of gender, gender ideology, and type of earner marriage on time, household-task, and economic satisfaction among couples with children. *Journal of Social Behavior and Personality, 11*, 161–177.

Bandura, A. (1977). Self-efficacy: Toward a unifying theory of behavioral change. *Psychological Review, 84*, 191–215.

Barber, J. S., Axinn, W. G., & Thornton, A. (1999). Unwanted childbearing, health, and mother-child relationships, *Journal of Health and Social Behavior, 40*, 231–257.

Bay-Cheng, L., Zucker, A. N., Stewart, A. J., & Pomerleau, C. S. (2002). Linking femininity, weight concern, and mental health among Latina, Black and White women. *Psychology of Women Quarterly*, *26*, 36–45.

Beauvoir, S. (1978). *The second sex*. Translated by H. M. Parshley. New York: Knopf.

Beitchman, J. H., Zucker, K. J., Hood, J. E., DaCosta, G. A., Akman, D., & Cassavia, E. (1992). A review of the long-term effects of child sexual abuse. *Child Abuse & Neglect*, *16*, 101–118.

Bem, S., & Bem, D. (1970). Training women to know her place: The Power of Non-conscious Ideology. In D. J. Bem (Ed.), *Beliefs, attitudes, and human affairs* (pp. 84–96). Belmont, CA: Wadsworth.

Berger, J. (1972). *Ways of seeing*. London: Penguin.

Berk, R. A., Sorenson, S. B., Wiebe, D. J., & Upchurch, D. M. (2003). The legalization of abortion and subsequent youth homicide: A time series analysis. *Analyses of Social Issues & Public Policy*, *3*, 45–64.

Bifulco, A., Brown, G. W., & Adler, Z. (1991). Early sexual abuse and clinical depression in adult life. *British Journal of Psychiatry*, *159*, 115–122.

Bird, C. E. (1999). Gender, household labor, and psychological distress: The impact of the amount and division of housework. *Journal of Health and Social Behavior*, *40*, 32–45.

Bird, C. E., & Fremont, A. M. (1991). Gender, time use, and health. *Journal of Health and Social Behavior*, *32*, 114–129.

Bourne, L. E., Jr., & Russo, N. F. (1998). *Psychology: Behavior in context*. New York: Norton.

Brown, G. R., & Anderson, B. (1991). Psychiatric morbidity in adult in-patients with childhood histories of sexual and physical abuse. *American Journal of Psychiatry*, *148*, 55–61.

Brown, G. W., Harris, T. O., & Eales, M. J. (1993). Aetiology of anxiety and depressive disorders in an inter-city population: 2. Comorbidity and adversity. *Psychological Medicine*, *23*, 155–165.

Brown, L. S. (2002). Feminist therapy and EMDR: A theory meets practice. In I. F. Shapiro (Ed.), *EMDR as an integrative psychotherapy approach: Experts of diverse orientations explore the paradigm prism* (pp. 263–287). Washington, DC: American Psychological Association.

Burnett, R. (1995). *Gendered objectification experiences: Construct validity, implications for depression and phenomenology*. Unpublished dissertation, Duke University, Durham, NC. (Summary available at www.JeanHamiltonMD.com.)

Burnett, R., Baylis, S., & Hamilton, J. A. (1994, May). Objectification experiences and relational orientation predict depression in young women. Presented (and published in Book of Abstracts) at the *APA/Psychological conference "Psychosocial and Behavioral Factors in Women's Health: Creating an Agenda for the 21st Century"*. American Psychological Association, Washington, DC. (Abstract and manuscript available at www.JeanHamiltonMD.com.)

Burnett, R., & Hamilton, J. (1996, September). [Objectification predicts depression.] *A women's health conference book of abstracts* (p. 103), American Psychological Association, Washington, DC. (Also available at www.JeanHamiltonMD.com.)

Campbell, J. C., & Sullivan, C. M. (1994). Relationship status of battered women over time. *Journal of Family Violence*, *9*, 99–111.

Caplan, P. J., & Cosgrove, L. (Eds.) (2004). Bias in psychiatric diagnosis. Livingston, NJ: Jason Aronson.

Carmen, E. M., Russo, N. F., & Miller, J. B. (1981). Inequality and mental health. *American Journal of Psychiatry, 138,* 1319–1339.

Chaisson, E. (1987). *The life era.* New York: Atlantic Monthly Press.

Chen, J., & Millar, W. J. (2001). Starting and sustaining physical activity. *Health Report, 12* (4), 33–43.

Chen, Y. Y., Subramanian, S. V., Acevedo-Garcia, D., & Kawachi, I. (2005). Women's status and depressive symptoms: A multilevel analysis. *Social Science and Medicine, 60,* 49–60.

Cook, J. (2003). Depression, disability, and rehabilitation services for women. *Psychology of Women Quarterly, 27,* 121–129.

Corsi-Cabrera, M., Sanchez, A., Portilla, Y., Villanueva, Y., & Perez-Garci, E. (2003). Effect of 38 h of total sleep deprivation on the waking EEG in women: Sex differences. *International Journal of Psychophysiology, 50,* 213–224.

Cozzarelli, C., & Major, B. (1994). The effects of anti-abortion demonstrators and pro-choice escorts on women's psychological responses to abortion. *Journal of Social and Clinical Psychology, 13,* 404–427.

Cozzarelli, C., & Major, B. (1998). The impact of antiabortion activities on women seeking abortions. In L. J. Beckman & S. M. Harvey (Eds.), *The new civil war: The psychology, culture, and politics of abortion* (pp. 28). Washington, DC: American Psychological Association.

Cremaschi, G. A., Gorelik, G., Klecha, A. J., Lysionek, A. E., & Genaro, A. M. (2000). Chronic stress influences the immune system through the thyroid axis. *Life Sciences, 67* (26), 3171–9.

Crocker, J. (2002). The costs of seeking self-esteem. *Journal of Social Issues, 58,* 597–615.

Crocker, J., Major, B., & Steele, C. (1998). In S. Fiske, D. Gilbert, & G. Lindzey (Eds.), *Handbook of social psychology* (Vol. 2, pp. 504–553). Boston: McGraw-Hill.

Cross, S. E., & Madsen, L. E. (1997). Models of the self: self-construals of and gender, *Psychological Bulletin, 122,* 5–37.

Csikszentmihalyi, M. (1990/1991). Flow. *In The psychology of optimal experience.* New York: Harper Collins.

David, H. P., Dytrych, Z., & Matejcek, M. (2003). Born unwanted: Observations from the prague study. *American Psychologist, 58,* 224–229.

Davidson, R. J., Lewis, D. A., Alloy, L. B., Amaral, D. G., Bush, G., Cohen, J. D., et al. (2002). Neural and behavioral substrates of mood and mood regulation. *Biological Psychiatry, 52,* 478–502.

Denious, J., & Russo, N. F. (2005). Controlling birth: Science, politics, and public policy. *Journal of Social Issues, 61,* 181–191.

Denious, J., Russo, N. F., & Rubin, L. (2005). The role of shame in socio- and sub-cultural influences on disordered eating. In M. Paludi (Ed.), *Praeger guide to the psychology of gender.* Westport, CT: Proeger.

Denmark, F. L., Russo, N. F., Frieze, I., & Sechzer, J. (1988). Guidelines for avoiding sexism in psychological research: A report of the Committee on Nonsexist Research. *American Psychologist, 43,* 582–585.

Demyttenaere, K., Bruffaerts, R., Posada-Villa, J., Gasquet, I., Kovess, V., Lepine, J. P., et al. (2004). Prevalence, severity, and unmet need for treatment of mental disorders in the World Health Organization World Mental Health Surveys. *Journal of the American Medical Association, 291,* 2581–2590.

Desjarlais, R., Eisenberg, L., Good, B., & Kleinman, A. (1995). *World mental health, problems and priorities in low income countries.* New York: Oxford University Press.

DeVault, M. L. (1996). Talking back to sociology: Distinctive contributions of feminist methodology. *Annual Review of Sociology, 22,* 29–50.

Dietz, P., Spitz, A. M., Anda, R. F., Williamson, D. G., McMahon, P. M., Santelli, J. S., et al. (2000). Unintended pregnancy among adult women exposed to abuse or household dysfunction during their childhood, *Journal of the American Medical Association, 282,* 1259–1364.

Eagly, A. H., & Wood, W. (1999). The origins of sex differences in human behavior: Evolved dispositions versus social roles. *American Psychologist, 54,* 408–423.

Eichler, A., & Parron, D. L. (1987). *Women's mental health: Agenda for research.* Rockville, MD: National Institute of Mental Health.

Faravelli, C., Giugni, A., Salvatori, S., & Ricca, V. (2004). Psychopathology after rape. *American Journal of Psychiatry, 161,* 1483–1485.

Finkelhor, D. (1987). The sexual abuse of children: Current research reviewed. *Psychiatric Annals, 17,* 233–241.

Finkelhor, D., Hotaling, D., Lewis, G., & Smith, I. A. (1990). Sexual abuse in a national survey of adult men and women: Prevalence, characteristics, and risk factors. *Child Abuse and Neglect, 14,* 19–28.

Folkman, S. (1984). Personal control and stress and coping processes: A theoretical analysis. *Journal of Personality and Social Psychology, 46,* 839–852.

Folkman, S., Lazarus, R. S., Dunkel-Schetter, C., DeLongis, A., & Gruen, R. (1986). Dynamics of a stressful encounter: Cognitive appraisal, coping, and encounter outcomes. *Journal of Personality and Social Psychology, 50,* 992–1003.

Forsythe, C. J., & Compas, B. E. (1987). Interaction of cognitive appraisals of stressful events and coping: Testing the goodness of fit hypothesis. *Cognitive Therapy and Research, 11,* 473–485.

Frank, E., Kupfer, D. J., Perel, J. M., Cornes, C., Jarrett, D. B., Mallinger, A. G., et al. (1990). Three-year outcomes for maintenance therapies in recurrent depression. *Archives of General Psychiatry, 47,* 1093–1099.

Frankenhaeser, M., Lundberg, U., & Chesney, M. (1991). *Women, work and health: Stress and opportunities.* New York: Plenum.

Frasure-Smith, N., Lesperance, F., & Talajic, M. (1993). Depression following myocardial infarction: Impact on 6-month survival. *Journal of the American Medical Association, 270,* 1819–1825.

Fredrickson, B., & Roberts, T-A. (1997). Objectification theory: Towards an understanding of women's lived experiences. *Psychology of Women Quarterly, 21,* 173–206.

Fredrickson, B., Roberts, T-A., Wolf, R., & Hamilton, J. (1994, May). *Social construction of the female body: Objectification theory as a model of lifespan gendered experience.* Unpublished manuscript, Duke University, Durham, NC.

Fullilove, M. (2002). Social and economic causes of depression. *Journal of Gender Specific Medicine, 5*(2), 38–41.

Funder, D. (2001). Personality. *Annual Review of Psychology, 52,* 197–221.

Gammell, D. J. (1996). *Women's experiences of depression: A qualitative analysis.* Unpublished paper, University of New Brunswick, Canada.

Gammell, D. J., & Stoppard, J. M. (1999). Women's experiences of treatment of depression: Medicalization or empowerment? *Canadian Psychology, 4*, 112–128.

Gazmararian, J. A., Adams, M. M., Saltzman, L. E., Johnson, C. H., Bruce, F. C., Marks, J. S., et al. (1995). The relationship between pregnancy intendedness and physical violence in mothers of newborns. *Obstetrics & Gynecology, 85*, 1031–1038.

Gergen, M. M. (1988). Toward a feminist metatheory and methodology in the social sciences. In M. Gergen (Ed.), *Feminist thought and the structure of knowledge* (pp. 87–104). New York: New York University Press.

Giordano, P. E. (2003). Relationships in adolescence. *Annual Review of Sociology, 29*, 257–81.

Glander, S. S., Moore, M. L., Michiellutte, R., & Parsons, L. H. (1998). The prevalence of domestic violence among women seeking abortion. *Obstetrics & Gynecology, 91*, 1002–1006.

Glied, S., & Kofman, S. (1995, March). *Women and mental health: Issues for health reform* [background paper]. New York: The Commonwealth Fund, Commission on Women's Health.

Goerner, S. J. (1999). *After the clockwork universe: The emerging science and culture of integral society*. Edinburgh, Scotland: Floris Books.

Goerner, S. J. (2004). How integral science changes our social stories by changing the picture. *World Futures: The Journal of General Evolution, 6*, 273–286.

Good, G. E., & Sherrod, N. B. (2001). The psychology of men and masculinity: Research status and future directions. In R. K. Unger (Ed.), *The handbook of the psychology of women and gender* (pp. 201–214). New York: Wiley.

Goodrum, S., Umberson, D., & Anderson, K. L. (2001). The batterer's view of the self and others in domestic violence. *Sociological Inquiry, 71*, 221–241.

Greco, T., Eckert, G., & Kroenke, K. (2004). The outcome of physical symptoms with treatment of depression. *Journal of General Internal Medicine, 19*, 813–818.

Haggerty, J. J., & Prange, A. J. (1995). Borderline hypothyroidism and depression. *Annual Review of Medicine, 46*, 37–46.

Hamilton, J. A. (1986a, March 4). Concentration is likened to euphoric states of mind. *New York Times*, Sect. C, p. 1.

Hamilton, J. A. (1986b, May 16). Intrinsic rewards. *Washington Post*, p. D5.

Hamilton, J. A. (1989, October 8). The power of concentration. *New York Times*, p. 6H26.

Hamilton, J. A. (1991, May). *Effort theory of depression*. Unpublished manuscript, University of Texas Southwestern Medical School, Dallas, TX. (Manuscript available at www.JeanHamiltonMD.com.)

Hamilton, J. A. (1993, Fall). In L. Waters & V. Robinson, The Gender/body/self workshop. *Women's Studies at Duke University*, pp. 1,2,3,6. University Archives, Perkins Library, Duke University, Durham, NC. (Also available at www.JeanHamiltonMD.com.)

Hamilton, J. A. (1995). Sex and gender as critical variables in psychotropic drug research. In B. Brown, P. Rieker, & C. Willie (Eds.), *Racism and sexism and mental health* (pp. 297–350), Pittsburgh: University of Pittsburgh Press.

Hamilton, J. A. (1996, September). Objectifying events as gendered, chronic hassles. *Women's Health Conference Book of Abstracts* (p. 103). Washington, DC: American Psychological Association. (Also available at www.JeanHamiltonMD.com.)

Hamilton, J. A. (2002, November). Root causes of depression: Epidemiology and risk factors. Retrieved from www.aahp.org. Washington, DC: American Association of Health Plans.

Hamilton, J. A., Alagna, S. W., & Pinkel, S. (1984). Gender differences in antidepressant and activating-drug effects on self-perceptions. *Journal of Affective Disorders, 7*, 235–243.

Hamilton, J. A., Grant, M., & Jensvold, M. F. (1996). Sex and treatment of depressions: When does it matter? In J. A. Hamilton, M. Jensvold, E. Rothblum, & E. Cole (Eds.), *Psychopharmacology of women: sex, gender and hormonal considerations* (pp. 241–260). Washington, DC: American Psychiatric Press.

Hamilton, J. A., Haier, R. J., & Buchsbaum, M. S. (1984). Intrinsic enjoyment and boredom coping scales: Validation with personality, evoked potential and attention measures. *Personality and Individual Differences, 5*, 183–193.

Hamilton, J. A., & Jensvold, M. (1995). Sex and gender as critical variables in feminist psychopharmacology research and pharmacotherapy. In J. A. Hamilton, M. Jensvold, E. Rothblum, & E. Cole (Eds.), *Psychopharmacology from a feminist perspective* (pp. 9–30). Binghamton, NY: Haworth Press.

Hamilton, J. A., Phillips, K. L., & Green, A. (2004). Integral medicine and health. *World Futures: The Journal of General Evolution, 6*, 295–302.

Hamilton, J. A., Phillips, K. L., & Sloan, L. (2003, September). *Integral science and theory applied to a woman's health case study of the affective spectrum: concept mapping and beyond.* Workshop presented at the American College of Women's Health Physician's Annual Meeting, Santa Fe Community and NIH Center for Excellence in Women's Health, Santa Fe, NM. (Also available at www.JeanHamiltonMD.com.)

Hamilton, J. A., Rathunde, K., & Csikszentmihalyi. (1998, April). *Theta activity is associated with intrinsic motivation.* Unpublished manuscript, presented, Department of Psychology, Duke University, Durham, NC. (Manuscript available at www.JeanHamiltonMD.com.)

Hamilton, J. A., & Yeates, M. (in preparation). *All work and no play makes Jane a dull girl. A new theory to explain the gender gap in depression.* (Abstract available at www.JeanHamiltonMD.com.)

Hamilton, J. A., & Yeates, M. (in preparation). Running on empty: Women, depression, and what to do about it.

Hamilton, J. A., & Yeates, M. (2005a). The bioenergetic depletion theory of women and depression. Manuscript submitted for publication.

Hamilton, J. A., & Yeates, M. (2005b, abstract). An integral theory of women's health: "Energy Depletion" (and Renewal), Part 1 of 2. *Journal of Women's Health, 14* (4):377.

Hamilton, J. A., & Yeates, M. (2005c, abstract). The "Energy Depletion" theory applied to depressive illness, Part 2 of 2. *Journal of Women's Health, 14* (4):377

Harkness, K. L., & Monroe, S. M. (2002). Childhood adversity and the endogenous vs. nonendogenous distinction in women with major depression. *American Journal of Psychiatry, 159*, 387–393.

Hebl, M. R., King, E. B., & Lin, J. (2004). The swimsuit becomes us all: Ethnicity, gender, and vulnerability to self-objectification. *Personality and Social Psychology Bulletin, 30*, 1322–1331.

Heise, L., Ellsberg, M., & Gottemoeller, M. (1999, December). Ending violence against women, *Population Reports,* Series L (11), pp. 1–45.

Henderson, K. A. (1994). Perspective on analyzing gender, women, and leisure. *Journal of Leisure Research, 26,* 119–137.

Henderson, K. A. (1996). One size doesn't fit all: The meaning of women's leisure. *Journal of Leisure Research, 28* (3), 139–154.

Heninger, K. (2001). The deprivation syndrome is the driving force of phylogeny, ontogeny, and oncogeny. *Reviews of Neuroscience, 1,* 217–287.

Henshaw, S. K. (1998). Unintended pregnancy in the United States. *Family Planning Perspectives, 30,* 24–29, 46.

Hickie, I, Bennett, B., Mitchell, P., Wilhelm, K., & Orlay, W. (1996). Clinical and subclinical hypothyroidism in patients with chronic and treatment-resistant depression. *Australia and New Zealand Journal of Psychiatry, 30,* 246–252.

Hintikka, J., Tolmunen, T., Tanskanen, A., & Viinamaki, H. (2003). High vitamin B12 level and good treatment outcome may be associated in depressive disorder. *British Medical College of Psychiatry, 3,* 17–22. Retrieved from http://www.biomedcentral.com/content/pdf/1471-244X-3-17.pdf.

Hobfoll, S. E., & Lieberman, J. R. (1987). Personality and social resources in immediate and continued stress resistance among women. *Journal of Personality and Social Psychology, 52,* 8–26.

Hochschild, A. (1997). *The time bind.* New York: OWL Book, Henry Holt & Co.

Hochschild, A., with A. Machung (1989). *The second shift.* New York: Avon Books.

Holahan, C. J., & Moos, R. H. (1986). Personality, coping, and family resources in stress resistance: A longitudinal analysis. *Journal of Personality and Social Psychology, 51,* 389–395.

Hublin, C., Kaprio, J., Partinen, M., & Koskenvuo, M. (2001). Insufficient sleep – a population based study in adults. *Sleep, 24,* 392–400.

Hudson, J. I., Mangweth, B., Pope, H. G. Jr, De Col, C., Hausmann, A., Gutweniger, S., et al. (2003). Family study of affective spectrum disorder. *Archives of General Psychiatry, 60,* 170–177.

Huselid, B. F., & Cooper, M. L. (1994). Gender roles as mediators of sex differences in expression of pathology. *Journal of Abnormal Psychology, 103,* 595–603.

Jason, L. A., Jordan, K. M., Richman, J. A., Rademaker, A. W., Huang, C., McCready, W., et al. (1999). A community-based study of prolonged fatigue and chronic fatigue. *Journal of Health Psychology, 4,* 9–26.

Johnson, J., Weissman, M. M., & Klerman, G. L. (1992). Service utilization and social morbidity associated with depressive symptoms in the community. *Journal of the American Medical Association, 267,* 1478–1483.

Joiner, T. E, Schmidt, N. B., & Wonderlich, S. A. (1997). Global self-esteem as contingent on body satisfaction among patients with bulimia nervosa: Lack of diagnostic specificity? *International Journal of Eating Disorders, 21,* 67–76.

Kahneman, D. (1973). *Attention and effort.* New York: Prentice-Hall.

Kaiser, C. R., Major, B., & McCoy, S. K. (2004). Expectations about the future and the emotional consequences of perceiving prejudice. *Personality and Social Psychology Bulletin, 27,* 254–263.

Karp, J. E., Scott, J., Houck, P. et al. (2005). Pain predicts longer time to remission during treatment of recurrent depression. *Journal of Clinical Psychiatry, 66* (5), 591–597.

Kaschak, E., & Tiefer, L. (Eds.) (2002). A new view of women's sexual problems, Haworth Press.

Katon, W., & Sullivan, M. D. (1990). Depression and chronic mental illness. *Journal of Clinical Psychiatry, 51*, 3–14.

Keck, P. E. (2004 Summer). Current and emerging strategies: Recognizing bipolar disorder in the depressed patient. www.measurecme.org (site visited December, 2004).

Kessler, R., & McLeod, J. (1985). Sex differences in vulnerability to undesirable life events. *American Sociological Review, 49*, 620–631.

Kessler, R. C., McGonagle, K. A., Zhao, S., Nelson, C. B., Hughes, M., Eshleman, S., et al. (1994). Lifetime and 12-month prevalence of DSM-III-R psychiatric disorders in the United States: Results from the National Comorbidity Survey. *Archives of General Psychiatry, 51*, 8–19.

Key, A., George, C. L., Beattie, D., Stammers, K., Lacey, H., & Waller, G. (2002). Body image treatment within an inpatient program for anorexia nervosa: The role of mirror exposure in the desensitization process. *International Journal of Eating Disorders, 31*, 185–91.

Keyes, C. L. (2003). Complete mental health: An agenda for the 21st century. In C. L. M. Keyes & J. Haidt (Eds.), *Flourishing: Positive psychology and the life well-lived* (pp. 293–312). Washington, DC: American Psychological Association.

Kimmel, E., & Crawford, M. (Eds.). (1999). *Innovations in feminist psychological research.* New York: Cambridge University Press.

Kishor, S., & Johnson, K. (2004). *Profiling domestic violence: A multi-country study.* Calverton, MD: ORC Macro.

Kleinman, A. (1986). *The social origins of distress and disease.* New Haven, CT: Yale University Press.

Kleinman, A. (1988). *Rethinking psychiatry: From cultural category to personal experience.* New York: Free Press.

Kleinman, A., & Good, B. (Eds.). (1985). *Culture and depression: Studies in the anthropology and cross-cultural psychiatry of affect and disorder.* Berkeley: University of California Press.

Klonoff, L., Landrine, H., & Campbell, R. (2000). Sexist discrimination may account for well-known symptoms. *Psychology of Women Quarterly, 24*, 93–99.

Klonoff, L., & Landrine, H. (1995). The Schedule of Sexist Events: A measure of lifetime and recent sexist discrimination in women's lives. *Psychology of Women Quarterly, 24*, 439–472.

Knudson-Martin, C., & Mahoney, A. R. (1996). Gender dilemmas and myth in the construction of marital bargains: Issues for marital therapy. *Family Process, 35*, 137–153.

Koss, M. P., Bailey, J., Yuan, N. P., Herrera, V. M., & Lichter, E. L. (2003). Depression and PTSD in the survivors of male violence: Research and training initiatives to facilitate recovery. *Psychology of Women Quarterly, 27*, 130–142.

Koss, M. P., Goodman, L. A., Browne, A., Fitzgerald, L., Keita, G. P., & Russo, N. F. (1994). *No safe haven: Male violence against women at home, at work, and in the community.* Washington, DC: American Psychological Association.

Krakow, B., Artar, A., Warner, T. D., Melendez, D., Johnston, L., Hollifield, M., et al. (2000). Sleep disorder, depression, and suicidality in female sexual assault survivors. *Crisis, 21* (4), 163–170.

Krieger, N., Rowley, D. L., Herman, A. A., Avery, B., & Phillips, M. T. (1993). Racism, sexism, and social class: Implications for studies of health, disease, and well-being. *American Journal of Preventive Medicine, 9* (suppl 2), 82–122.

Landrine, H., & Klonoff, E. A. (1997). *Discrimination against women: Prevalence, consequences, remedies.* Thousand Oaks, CA: Sage.

Landrine, H., Klonoff, E. A., Gibbs, J, Manning, V., & Lund, M. (1995). Physical and psychiatric correlates of gender discrimination: An application of the Schedule of Sexist Events. *Psychology of Women Quarterly, 19*, 473–492.

Laumann, E. O., Paik, A, & Rosen, R. C. (1999). Sexual Dysfunction in the United States: Prevalence and Predictors *Journal of the American Medical Association, 281*, 537–544.

Lavidor, M., Weller, A., & Babkoff, H. (2003). How sleep is related to fatigue. *British Journal Health Psychology, 8* (pt. 1), 95–105.

Lawlor, D. A., Taylor, M., Bedford, C., Ebrahim, S. (2002). Is housework good for health? *Journal of Epidemiology and Community Health, 56* (6), 473–478.

Lazarus, R., & Folkman, S. (1984). *Stress, appraisal, and coping.* New York: Springer.

Lazarus, R. S. (1993). Coping theory and research: Past, present, and future. *Psychosomatic Medicine, 55*, 234–247.

Leathers, S. J., & Kelley, M. A. (2000). Unintended pregnancy and depressive symptoms among first-time mothers and fathers. *American Journal of Orthopsychiatry, 70*, 523–531.

Lehman, D. R., Chiu, C., & Schaller, M. (2004). Psychology and culture. *Annual Review of Psychology, 55*, 689–714.

Lepore, S. J. (1992). Social conflict, social support, and psychological distress: Evidence of cross-domain buffering effects. *Journal of Personality and Social Psychology, 63*, 857–867.

Lerner, R. M., & Galambos, N. L. Adolescent development: Challenges and opportunities for research, programs, and policies. *Annual Review of Psychology, 49*, 413–46.

Lin, L. F., & Kulik, J. A. (2002). Social comparison and women's body satisfaction. *Basic and Applied Social Psychology, 24*, 115–123.

Link, B. G., & Phelan, J. C. (2001). Conceptualizing stigma. *Annual Review of Sociology, 27*, 363–85.

Lopez, S. R., & Guarnaccia, P. J. J. (2000). Cultural psychopathology: Uncovering the social world of mental illness. *Annual Review of Psychology, 51*, 571–98.

Major, B., Cozzarelli, C., Sciacchitano, A. M., Cooper, M. L., Testa, M., & Mueller, P. M. (1990). Perceived social support, self-efficacy, and adjustment to abortion. *Journal of Personality and Social Psychology, 59*, 452–463.

Major, B., & Gramzow, R. H. (1999). Abortion as stigma: Cognitive and emotional implications of concealment. *Journal of Personality and Social Psychology, 77*, 735–745.

Major, B., Mueller, P., & Hildebrandt, K. (1985). Attributions, expectations, and coping with abortion. *Journal of Personality and Social Psychology, 48*, 585–599.

Major, B., & O'Brien, L. T. (2004). The social psychology of stigma. *Annual Review of Psychology, 56*, 1–29.

Major, B., Richards, C., Cooper, M. L., Cozzarelli, C., &, Zubek, J. (1998). Personal resilience, cognitive appraisals, and coping: An integrative model of adjustment to abortion. *Journal of Personality and Social Psychology, 74*, 735–752.

Major, B., Zubek, J. M., Cooper, M. L., Cozzarelli, C., & Richards, C. (1997). Mixed messages: Implications of social conflict and social support within close relationships for adjustment to a stressful life event. *Journal of Personality and Social Psychology, 72,* 1349–1363.

Matos-Santos, A., Nobre, E. L., Costa, J. G., Noguera, P. J., Macedo, A., Galvao, A., et al. (2001). Relationship between the number and impact of stressful life events and the onset of Graves' disease and toxic nodular goiter. *Clinical Endocrinology, 55,* 15–9.

Maushart, S. (2001). *Wifework: What marriage really means for women.* New York: Bloomsbury.

Mazure, C., Keita, G. P., & Blehar, M. C. (2002). *Summit on women's depression: Proceedings and recommendations.* Washington, DC: American Psychological Association Women's Programs Office.

McEwen, B. S. (2003). Mood disorders and allosteric load. *Biological Psychiatry, 54,* 200–207.

McEwen, B. S., & Lasley, E. N. (2002). *The end of stress as we know it.* Washington, DC: Joseph Henry Press.

McGrath, E., Keita, G. P., Strickland, B. R., & Russo, N. F. (1990). *Women and depression: Risk factors and treatment issues.* Washington, DC: American Psychological Association.

McHugh, M. D., Koeske, R. D., & Frieze, I. H. (1986). Issues to consider in conducting nonsexist psychological research: A guide for researchers. *American Psychologist, 41,* 879–890.

McKinley, N. M. (1995). Objectified body consciousness scale. In J. K. Thompson (Eds.), *Exacting beauty: Theory, assessment, and treatment of body image disturbance* (pp. 233–234). Washington, DC: American Psychological Association.

McKinley, N. M., & Hyde, J. S. (1996). The objectified body consciousness scale: Development and validation. *Psychology of Women Quarterly, 20,* 181–215.

Menza, M., Marin, H., & Opper, R. S. (2003). Residual symptoms in depression: Can treatment be symptom-specific? *Journal of Clinical Psychiatry, 64,* 516–523.

Mooney, C. (2004, October). Research and destroy: How the religious right promotes its own "experts" to combat mainstream science. *Washington Monthly.* Retrieved from http://www.washingtonmonthly.com/features/2004/0410.mooney.html.

Mullen, P. E., Romans-Clarkson, S. E., Walton, V. A., & Herbison, P. (1988). Impact of sexual and physical abuse on women's mental health. *Lancet, 16,* 841–845.

National Center for Health Statistics, Centers for Disease Control and Prevention. (1994). *Health, United States 1995.* Hyattsville, MD: U.S. Public Health Service.

National Institute of Mental Health Prevention Research Steering Committee. (1994). *The prevention of mental disorders: A national research agenda.* Washington, DC: NIMH.

Nemeroff, C., Heim, C. M., Thase, M. E., Klein, D. N., Rush, J. A., Schatzberg, A. F., et al. (2003). Differential responses to psychotherapy versus pharmacotherapy in patients with chronic forms of major depression and childhood trauma. *Proceedings of the National Academy of Sciences, 100,* 14293–14296.

Nolen-Hoeksema, S. (1987). *Sex differences in depression.* Stanford, CA: Stanford University Press.

Nolen-Hoeksema, S. (2003). *Women who think too much: How to break free of overthinking and reclaim your life.* New York: Henry Holt.

Nolen-Hoeksems, S., & Girgus, J. S. (1994). The emergence of gender differences in depression during adolescence. *Psychological Bulletin, 115*, 424–443.

Noll, S. M. & Fredrickson, B. I. (1998). A mediational model linking self-objectification, body shame, and disordered eating. *Psychology of Women Quarterly, 22*, 623–636.

Okin, S. (1989). *Justice, gender, and the family.* New York: Basic Books.

Parsons, P. A. (2003). From the stress theory of aging to energetic and evolutionary expectations for longevity. *Biogerontology, 4* (2), 63–73.

Pennebaker, J. W. (1995). *Emotion, disclosure, and health.* Washington, DC: American Psychological Association.

Perlis, M. L., Giles, D. E., Buysse, D. J., Tu, X., & Kupfer, D. J. (1997). Self-reported sleep disturbance as a prodromal symptom in recurrent depression. *Journal of Affective Disorders, 42*, 209–212.

Piccinelli M., & Homen F. G. (1997). *Gender differences in the epidemiology of affective disorders and schizophrenia.* Geneva: World Health Organization.

Popay, J. (1992). "My health is all right, but I'm just tired all the time": Women's experience of ill health. In H. Roberts (Ed.), *Women's health matters* (pp. 99–120). London: Routledge.

Pribor, E. F., & Dinwiddie, S. H. (1992). Psychiatric correlates of incest in childhood. *American Journal of Psychiatry, 149*, 52–56.

Reppucci, N. D., Woolard, J. L., & Fried, C. S. (1999). Social, community, and preventive interventions. *Annual Review of Psychology, 50*, 387–418.

Resnick, H. S., Acierno, R., & Kilpatrick, D. G. (1997). Health impact of interpersonal violence. 2: Medical and mental health outcomes. *Behavioural Medicine, 23*, 65–78.

Rieker, P. P., & Carmen, E, (Eds.). (1984). *The gender gap in psychotherapy: Social realities and psychological processes.* New York: Plenum Press.

Ross, C. E. (2000). Neighborhood disadvantage and adult depression. *Journal of Health and Social Behavior, 41*, 177–187.

Russell, D. E. H. (1986). The secret trauma: Incest in the lives of girls and women. New York: Basic Books.

Russo, D. (2002, April). *Gender difference in fatigue explained by the use of the body's aerobic pathways.* Presented at the Annual Meeting of the American Physiological Society, New Orleans, LA.

Russo, N. F. (1984). Women in the mental health delivery system: Implications for policy, research, and practice. In L. Walker (Ed.), *Women and mental health policy. Sage yearbooks in Women's Policy Studies, 9*, 21–41.

Russo, N. F. (Ed.). (1985). A women's mental health agenda. Washington, DC: American Psychological Association.

Russo, N. F. (1992). Psychological aspects of unwanted pregnancy and its resolution. In J. D. Butler & D. F. Walbert (Eds.), *Abortion, medicine, and the law* (4th ed., pp. 593–626). New York: Facts on File.

Russo, N. F. (1995). Women's mental health: Research agenda for the twenty-first century. In B. Brown, B. Kramer, P. Rieker, & C. Willie (Eds.), *Mental health, racism, and sexism* (pp. 373–396). Pittsburgh: University of Pittsburgh Press.

Russo, N. F., & Dabul, A. (1994). Feminism and psychology: A dynamic interaction. In Trickett, E. J., Watts, R., & Birman, D. *Human diversity: Perspectives on people in context* (pp. 81–100). San Franciso: Jossey-Bass.

Russo, N. F., David, H. D., Adler, N. E., & Major, B. (2005). *Abortion and mental health: A scientific assessment of conflict in claims.* Unpublished manuscript.

Russo, N. F., & Denious, J. (1998). Understanding the relationship of violence against women to unwanted pregnancy and its resolution. In L. J. Beckman & S. M. Harvey (Eds.), *The new civil war: The psychology, culture, and politics of abortion* (pp. 211–234). Washington, DC: American Psychological Association.

Russo, N. F., & Denious, J. E. (2001). Violence in the lives of women having abortions: Implications for public policy and practice. *Professional Psychology: Research and Practice, 32,* 142–150.

Russo, N. F., & Green, B. L. (1993). Women and mental health. In F. L. Denmark & M. A. Paludi. (Eds.), *Psychology of women: A handbook of issues and theories* (pp. 379–436). Westport: Greenwood Press.

Russo, N. F., Horn, J., & Schwartz, R. (1992). U.S. abortion in context: Selected characteristics and motivations of women seeking abortion. *Journal of Social Issues, 48,* 182–201.

Russo, N. F., & Vaz, K. (2001). Addressing diversity in the Decade of Behavior: Focus on women of color. *Psychology of Women Quarterly, 25,* 280–294.

Scheier, M. F., & Carver, C. S. (1985). Optimism, coping and health: Assessment and implications of generalized outcomes expectancies. *Health Psychology, 4,* 219–247.

Scheier, M. F., & Carver, C. S. (1987). Dispositional optimism and physical well-being: The influence of generalized outcome expectancies on health. *Journal of Personality, 55,* 169–210.

Scheier, M. F., Weintraub, J. K., & Carver, C. S. (1986). Coping with stress: Divergent strategies of optimists and pessimists. *Journal of Personality and Social Psychology, 51,* 1257–1264.

Schreiber, R. (1996). (Re)defining my self: Women's process of recovery from depression. *Qualitative Health Research, 6,* 469–491.

Seligman, M. E. P. (2002). *Authentic happiness.* New York: Free Press.

Sherif, C. W. (1979). Bias in psychology. In J. A. Sherman & E. T. Beck (Eds.), *The prism of sex: Essays in the sociology of knowledge* (pp. 93–134). Madison: University of Wisconsin Press.

Silver, R. C., & Wortman, C. B. (1980). Coping with undesirable life events. In J. Garber & M. E. P. Seligman (Eds.), *Human helplessness* (pp. 279–340). San Diego, CA: Academic Press.

Simmons, R. G., & Blyth, D. A. (1987). *Moving into adolescence: The impact of pubertal change and school context.* New York: Aldine de Gruyter.

Smolak, L., & Munstertieger, B. F. (2002). The relationship of gender and voice to depression and eating disorders. *Psychology of Women Quarterly, 26,* 234–241.

Solden, S. (1995). *Women with attention deficit disorder.* Grass Valley, CA: Underwood Books.

Steil, J. (1997). *Marital equality: Its relationship for the well-being of husbands and wives.* Thousand Oaks, CA: Sage.

Steil, J. (2001). Family forms and member well-being: A research agenda for the Decade of Behavior. *Psychology of Women Quarterly, 25,* 344–363.

Stewart, D., Abbey, S., Meana, M., & Boydell, K. M. (1998). What makes women tired? A community sample. *Journal of Women's Health, 7,* 69–76.

Stice, E., Haywood, C., Cameron, R. P., Killen, J. D., & Taylor, C. B. (2000). Body-image and eating disturbance predict onset of depression among female adolescents: A longitudinal study. *Journal of Abnormal Psychology, 109,* 438–444.

Stoppard, J. M. (2000). *Understanding depression: Feminist social constructionist approaches.* London: Routledge.

Subpanel on the Mental Health of Women. (1978). *Women.* Task panel reports submitted to the President's Commission on Mental Health (Vol. 3, pp. 1022–1177). Washington, DC: Government Printing Office.

Sullivan, P. F., Kovalenko, P., York, T. P., Prescott, C. A., & Kendler, K. S. (2003). Fatigue in a community sample of twins. *Psychological Medicine, 33,* 263–81.

Sutherland, C., Bybee, D., & Sullivan, C. (1998). The long-term effects of battering on women's health. *Women's Health, 4,* 41–70.

Swanson, N., Piotrkowski, C. S., Keita, G. P., & Becker, A. B. (1997). Occupational stress and women's health. In S. Gallant, G. P. Keita, & Royak-Schaler (Eds.), *Psychosocial and behavioral factors in women's health: A handbook for medical educators, practitioners, and psychologists.* (pp. 147–159). Washington, DC: American Psychological Association.

Swim, J. K., & Cohen, L. L. (1997). Overt, covert, and subtle sexism. *Psychology of Women Quarterly, 21,* 103–118.

Swim, J. K., & Hyers, L. L. (1999). Excuse me – What did you say? Women's public and private responses to sexist remarks. *Journal of Experimental Social Psychology, 35,* 68–88.

Swim, J. K., Hyers, L. L., Cohen, L. L., & Ferguson, M. J. (2001). Everyday sexism: Evidence for its incidence, nature, and psychological impact from three daily diary studies. *Journal of Social Issues, 57,* 31–54.

Tait, R., & Silver, R. C. (1989). Coming to terms with major negative life events. In J. S. Uleman & J. A. Bargh (Eds.), *Unintended thought* (pp. 351–382). New York: Guilford Press.

Taylor, R. R., Jason, L. A., & Jahn, S. C. (2003). Chronic fatigue and sociodemographic characteristics as predictors of psychiatric disorders in a community-based sample. *Psychosomatic Medicine, 65,* 896–890.

Taylor, S. E., Klein, L. C., Lewis, B. P., Gruenewald, T. L., Gurung, R. A. R., & Updegraff, J. A. 2000). Female responses to stress: Tend-and-befriend, not fight-or-flight. *Psychological Review, 107,* 411–429.

Thomas, D. O., Granat, H., & Fernhall, B. (1997). Factors related to changes in running economy during a 5 km run differ among men and women. *Medicine and Science in Sports and Exercise, 29* (5, supplement abstract).

Tiefer, L., Brick, P., Kaplan, M. (2003). A new view of women's sexual problems, New View Campaign, LTiefer@mindspring.com.

Tiggemann, M., & Kuring, J. K. (2004). The role of body objectification in disordered eating and depressed mood. *British Journal of Clinical Psychology, 43,* 299–311.

Tjaden, P., & Thoennes, N. (1998, April). *Stalking in America: Findings from the National Violence Against Women Survey.* Research in brief (169592) Washington, DC: U.S. Department of Justice, National Institute of Justice.

Tolman, D. L., & Brown, L. M. (2001). Adolescent girls' voices: Resonating resistance in body and soul. In R. K. Unger (Ed.), *Handbook of the psychology of women and gender* (pp. 133–154). New York: Wiley.

Tomaka, J., Blascovich, J., Kelsey, R. M., & Leitten, C. L. (1993). Subjective, physiological, and behavioral effects of threat and challenge appraisal. *Journal of Personality and Social Psychology, 65*, 248–260.

Travis, C. B., & Meginnis-Payne, K. L. (2001). Beauty politics and patriarchy: The impact on women's lives. In J. Worrell (Ed.), *Encyclopedia of women and gender* (pp. 189–200), San Diego, CA: Academic Press.

Triandis, H. C., & Suh, E. M. (2002). Cultural influences on personality. *Annual Review of Psychology, 53*, 133–60.

United Nations. (2004). Promoting Gender Equality and Women's Empowerment. Retrieved from www.unfpa.org.

United Nations Children's Fund. (2000). Monitoring the situation of children and women. www.unicef.org.

Ustun, T. B., & Sartorius, N. (1995). *Mental illness in general health care: An international study*: London: Wiley for the World Health Organization.

Valentiner, D. P., Holahan, C. J., & Moos, R. H. (1994). Social support, appraisals of event controllability, and coping: An integrative model. *Journal of Personality and Social Psychology, 66*, 1094–1102.

Verdon, F., Burnand, B., Fallab Stubi, C-L., Bonard, C., Graff, M., Michaud, A., et al. (2003). Iron supplementation for unexplained fatigue in non-anemic women: Double blind randomized placebo controlled trial. *British Medical Journal, 326*, 1124–1128.

Waalen, J., Felitti, V., & Beutler, E. (2002). Haemoglobin and ferritin concentrations in men and women: Cross-sectional study. *British Medical Journal, 325*, 137.

Waller, G. (1994). Childhood sexual abuse and borderline personality disorder and eating disorders. *Child Abuse and Neglect, 18*, 97–101.

Weissman, M. M., Bland, R. C., Canino, G. J., Faravelli, C., Greenwald, S., Hwu, H. G., et al. (1996). Cross-national epidemiology of major depression and bipolar disorder. *Journal of the American Medical Association, 276*, 293–299.

Weissman, M., Myers, J., Thompson, W. D., & Belanger, A. (1986). Depressive symptoms as a risk factor for mortality and for major depression. In L. Erhlenmayer-Kimling & N. Miller (Eds.), *Life span research on the prediction of psychopathology* (pp. 251–260). Hillsdale, NJ: Lawrence Erlbaum.

Wells, J. D., Hobfoll, S. E., & Lavin, J. (1997). Resource loss, resource gain, and communal coping during pregnancy among women with multiple roles. *Psychology of Women Quarterly, 21*, 645–662.

White, J., Russo, N. F., & Travis, C. B. (Eds.). (2001a). Feminism and the Decade of Behavior. [Special Issue]. *Psychology of Women Quarterly, 25* (4).

White, J., Russo, N. F., & Travis, C. B. (2001b). Feminism and the Decade of Behavior: Overview. *Psychology of Women Quarterly, 25*, 267–280.

Wickramaratne, P. J., Warner, V., & Weissman, M. M. (2000). Selecting early onset MDD probands for genetic studies: results from a longitudinal high risk study. *American Journal of Medical Genetics, 96*, 93–101.

Widiger, T. A., & Sankis, L. M. (2000). Adult psychopathology: Issues and controversies. *Annual Review of Psychology, 51*, 377–404.

Worell, J. (2001). Feminist interventions: Accountability beyond symptom reduction. *Psychology of Women Quarterly, 25*, 335–343.

World Health Organization. (2000). *Women's mental health: An evidence based review*. Geneva, Switzerland.

Wyatt, G. E., Guthrie, D., & Notgrass, C. M. (1992). Differential effects of women's child sexual abuse and subsequent sexual revictimization. *Journal of Consulting & Clinical Psychology, 60*, 167–173.

Yonkers, K., & Hamilton, J. A. (1995). Psychotropic medications. In M. Weissman (Ed.), *Psychiatry update: Annual review* (Vol. 13, pp. 147–178, Section II: Women's Health Care). Washington, DC: American Psychiatric Press.

Zubenko, G., Maher, B., Hughes, H. B. 3rd, Zubenko, W. N., Stiffler, J. S., Kaplan, B. B., et al. (2003). Genome-wide linkage survey for genetic loci influence the development of depressive disorder in families with recurrent, early-onset major depression. *American Journal of Medical Genetics, 123B*, 1–18.

Author Index

Subject Index

NSFH. *See* National Survey of Families and
  Households
nursing homes, 130
nutrition, 140

obesity, 101, 109, 135, 148, 149, 501
objectification, 493, 498–499
obsessive-compulsive disorder (OCD), 51,
  328
OCD. *See* obsessive-compulsive disorder
off-time hypothesis, 110
openness, 181
optimism, 156
oral contraceptives, 46, 428
osteoporosis, 135, 436–437
overgeneralization, 156
oxytocin
  AVP and, 204
  citalopram and, 207–208
  fight-or-flight response, 204
  HPA-axis and, 203
  postpartum depression and, 208
  puberty and, 456
  social support and, 207, 210
  stress and, 205
  tend-and-befriend response and, 207

panic disorder (PD), 68, 366, 433
paraventricular nuclei (PVN), 204
parenting, 115, 243, 365. *See also* family;
  motherhood 115, 243, 364
Parkinsonism, 133–134
Partial Trial Therapy, 432
passivity, 332
pathoplastic effect, 184, 185–186
PD. *See* panic disorder
PDS. *See* Pubertal Development Scale
Pediatric Research in Office Settings
  (PROS), 99, 102
peers, 115
Personal Responsibility and Work
  Opportunity Reconciliation Act
  (PRWORA), 316
personality, 28, 176
  dependency and, 181 *See* dependent
    persons
  disorders of, 182, 189–190
  elderly and, 136
  gender differences, 176
  instruments for, 185–186
  optimism and, 155
  pathoplastic effect, 184

pessimism and, 155
self-concept and, 153 *See* self-concept
sociotropy and, 152
traits, 63, 176, 183, 189, 298
treatment and, 188
personalization, 156
personological theory, 297, 298–299
pessimism, 155, 156, 158
PET. *See* position emission tomography
pharmacotherapy, 189–190, 207, 424. *See also*
  antidepressants 189–190, 207, 427
phencyclidine (PCP), 391
phenytonin, 422
phobias, 433
physical activities, 140
physical disability, 151
pituitary gland, 201, 203, 204
PMDD. *See* premenstrual dysphoric
  disorder
PME. *See* premenstrual exacerbation
PMS. *See* premenstrual syndrome
Positive Behavior Supports, 467
positive indicators, 500, 505
positron emission tomography (PET), 371
post-colonial theorists, 302
post-traumatic stress disorder (PTSD),
  291–292, 360, 361, 369
postpartum depression (PPD), 42, 243–244,
  436
  anemia and, 209
  biology of, 75
  bipolar disorder and, 76
  cognitive behavioral therapy, 75–76
  heritability of, 252
  HPA-axis and, 208
  major depression and, 24
  mood disorders and, 73
  oxytocin and, 208
  parasuicide and, 265
  PPS and, 42
  prevention of, 459, 461
  risk factors, 73, 74
  thyroid dysfunction and, 209
  treatment of, 75
poverty, 318, 320–321, 455, 483
power relations, 331, 479, 483, 495, 503
PPD. *See* postpartum depression
prefrontal cortex, 48
pregnancy, 24, 243–244, 428, 462, 463
  abortion, 491, 492
  antidepressants, 254, 436
  childbearing, 62